DIGITAL VIDEO: AN INTRODUCTION TO MPEG-2

Barry G. Haskell,
Atul Puri, and
Arun N. Netravali

Join Us on the Internet

WWW: http://www.thomson.com
EMAIL: findit@kiosk.thomson.com

thomson.com is the on-line portal for the products, services and
resources available from International Thomson Publishing (ITP).
This Internet kiosk gives users immediate access to more than 34 ITP
publishers and over 20,000 products. Through *thomson.com* Internet
users can search catalogs, examine subject-specific resource centers
and subscribe to electronic discussion lists. You can purchase ITP
products from your local bookseller, or directly through *thomson.com*.

Visit Chapman & Hall's Internet Resource Center for information on our new publications,
links to useful sites on the World Wide Web and an opportunity to join our e-mail
mailing list. Point your browser to: **http://www.chaphall.com/chaphall.html** or
http://www.chaphall.com/chaphall/electeng.html for Electrical Engineering

---

A service of

Digital
Multimedia
Standards
Series

DIGITAL VIDEO: AN INTRODUCTION TO MPEG-2

Barry G. Haskell
Head, Image Processing Research Department, AT&T Labs

Atul Puri
Principal Member of Technical Staff,
Image Processing Research Department, AT&T Labs

Arun N. Netravali
Vice President of Research, Bell Labs,
Lucent Technologies

CHAPMAN & HALL

 INTERNATIONAL THOMSON PUBLISHING

New York · Albany · Bonn · Boston · Cincinnati · Detroit · London · Madrid · Melbourne
Mexico City · Pacific Grove · Paris · San Francisco · Singapore · Tokyo · Toronto · Washington

Cover design: Curtis Tow Graphics

Printed in the United States of America

Chapman & Hall
115 Fifth Avenue
New York, NY 10003

Chapman & Hall
2-6 Boundary Row
London SE1 8HN
England

Thomas Nelson Australia
102 Dodds Street
South Melbourne, 3205
Victoria, Australia

Chapman & Hall GmbH
Postfach 100 263
D-69442 Weinheim
Germany

International Thomson Editores
Campos Eliseos 385, Piso 7
Col. Polanco
11560 Mexico D.F
Mexico

International Thomson Publishing–Japan
Hirakawacho-cho Kyowa Building, 3F
1-2-1 Hirakawacho-cho
Chiyoda-ku, 102 Tokyo
Japan

International Thomson Publishing Asia
221 Henderson Road #05-10
Henderson Building
Singapore 0315

1 2 3 4 5 6 7 8 9 10 XXX 01 00 99 98 97

Library of Congress Cataloging-in-Publication Data

Haskell, Barry G.
    Digital video : an introduction to MPEG-2 / Barry G. Haskell, Atul
Puri, and Arun N. Netravali.
        p.   cm.
    Includes bibliographical references and index.
    ISBN 0-412-08411-2
    1. Digital video.   2. Video compression -- Standards.   3. Coding
theory.   I. Puri, Atul.   II. Netravali, Arun N.   III. Title.
TK6680.5.H37   1996
621.388'33--dc20   96-14018
                                                    CIP

British Library Cataloguing in Publication Data available

"Digital Video: An Introduction to MPEG-2" is intended to present technically accurate and authoritative information from highly regarded sources. The publisher, editors, authors, advisors, and contributors have made every reasonable effort to ensure the accuracy of the information, but cannot assume responsibility for the accuracy of all information, or for the consequences of its use.

To order this or any other Chapman & Hall book, please contact **International Thomson Publishing, 7625 Empire Drive, Florence, KY 41042.** Phone: (606) 525-6600 or 1-800-842-3636. Fax: (606) 525-7778. e-mail: order@chaphall.com.

For a complete listing of Chapman & Hall titles, send your request to **Chapman & Hall, Dept. BC, 115 Fifth Avenue, New York, NY 10003.**

To our families
Ann, Paul, and Andrew
Diane
Chitra, Ilka, and Ravi

CONTENTS

PREFACE

On November 11, 1994 the ISO* Moving Picture Experts Group, known as MPEG, approved the Audio/Video digital compression standard known as MPEG-2. Two months later on January 27, 1995, the ITU-T† Rapporteurs Group on ATM Video approved the H.310 series of standards, which incorporates MPEG-2 into a coherent framework for two-way digital Audio/Video communications. It is probably not an understatement to say that these standards represent one of the most versatile, and perhaps most complicated, communication standards ever written.

In this book we first explore the history, fundamentals, and details of MPEG-2. The treatment includes most of the specifics needed to implement an MPEG-2 *Decoder*, including the syntax and semantics of the coded bitstreams. However, MPEG-2 *Encoders* are not specified by the standard. Indeed the optimization of encoder parameters is an art that many vendors keep as a closely held secret. Thus, in this book we can only outline the fundamentals of encoder design and algorithm optimization.

We then launch into MPEG-2 applications with chapters on digital video delivery media, interactive television, High-Definition Television (HDTV), three dimensional stereoscopic television and VLSI implementation architectures. We finish up with a progress report on MPEG-4, followed by an Appendix containing a resource guide of MPEG Web sites.

Although the basic framework has been established, MPEG-2 continues to evolve and to be applied in a wide variety of applications ranging from video conferencing to High-Definition Television. This only serves to reinforce the old adage that the more useful a standard is, the more likely it is to be changed.

---

*International Organization for Standards.
†International Telecommunications Union—Telecommunications Sector.

ACKNOWLEDGMENTS

It is with great pleasure and gratitude that we acknowledge the advice, support, and criticism that we received from many quarters.

We are very thankful that a number of experts in the field found the time to read and critique early portions of the manuscript. They include R. Aravind, Ted Darcie, Nahrain Gehani, Paul Haskell, Clive Holborow, James Johnston, Richard Kollarits, Ajay Luthra, Joan Mitchell, Banu Ozden, Avi Silberschatz, Guy Story, and Edward Szurkowski.

We thank Dr. Horng-Dar Lin for writing Chapter 16, Robert Schmidt for producing performance data, Eric Petajan for help in Chapter 14, and Tristan Savatier, Andria Wong and Joseph Worth for help in Annex A.

For help in preparing the manuscript, we are extremely indebted to Suzanne Fernandez, Patricia DiCuollo, MaryAnn Cottone, and Cecilia Scheckler.

Finally, a heartfelt thanks to our colleagues in MPEG, without whom this prodigious effort could not have occurred. Special appreciation goes to the MPEG Convenor, Dr. Leonardo Chiariglione, whose tireless stimulation, encouragement, and yes sometimes resounding vociferation was an inspiration to us all.

1

Introduction to Digital Multimedia, Compression, and MPEG-2

1.1 THE DIGITAL REVOLUTION

It is practically a cliché these days to claim that all electronic communications is engaged in a digital revolution. Some communications media such as the telegraph and telegrams were always digital. However, almost all other electronic communications started and flourished as analog forms.

In an analog communication system, a transmitter sends voltage variations, which a receiver then uses to control transducers such as loudspeakers, fax printers, and TV cathode ray tubes (CRTs), as shown in Figure 1.1. With digital, on the other hand, the transmitter first converts the controlling voltage into a sequence of 0s and 1s (called *bits*), which are then transmitted. The receiver then reconverts or *decodes* the bits back into a replica of the original voltage variations.

The main advantage of digital representation of information is the robustness of the bitstream. It can be stored and recovered, transmitted and received, processed and manipulated, all virtually without error. The only requirement is that a zero bit be distinguishable from a one bit, a task that is quite easy in all modern signal handling systems. Over the years, long and medium distance telephone transmissions have all become digital. Fax transmission changed to digital in the 1970s and 80s. Digitized entertainment audio on compact disks (CDs) has completely displaced vinyl records and almost replaced cassette tape. Digital video satellite is available in several countries, and digital video disks have just arrived in the market. Many other video delivery media are also about to become digital, including over-the-air, coaxial cable, video tape and even ordinary telephone wires. The pace of this conversion is impressive, even by modern standards of technological innovation.

Almost all sensory signals are analog at the point of origination or at the point of perception. Audio and video signals are no exception. As an example, television, which was invented and standardized over fifty years ago, has remained mostly analog from camera to display, and a plethora of products and services have been built around these analog standards, even though dramatic changes have occurred in technology. However, inexpensive integrated circuits, high-speed communication networks, rapid access dense storage media, and computing architectures that can easily handle video-rate data are now rapidly making the analog standard obsolete.

Digital audio and video signals integrated with computers, telecommunication networks, and consumer products are poised to fuel the information revolution. At the heart of this revolution is the digital compression of audio and video signals.

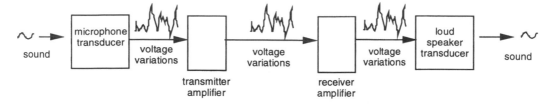

Analog Transmission System

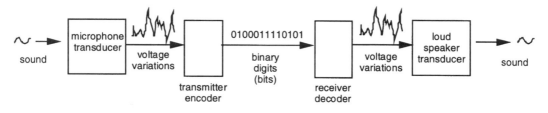

Digital Transmission System

Fig. 1.1   Analog communication systems send voltage variations that track in analogous fashion the waveforms produced by transducers such as microphones, fax scanners, and so forth. Digital systems first convert or *encode* the waveforms into a string of 0s and 1s called *bits*. The bits are then sent to a receiver, which *decodes* the bits back into an approximation of the original waveform.

This chapter summarizes the benefits of digitization as well as some of the requirements for the compression of the multimedia signals.

1.1.1   Being Digital

Even though most sensory signals are analog, the first step in processing, storage, or communication is usually to digitize the signals into a string of bits. Analog-to-digital converters that digitize such signals with required accuracy and speed have become inexpensive over the years. The cost or quality of digitization therefore is no longer an issue. However, simple digitization usually results in a *bandwidth expansion*, in the sense that transmitting or storage of these bits often takes up more bandwidth or storage space than the original analog signal. In spite of this, digitization is becoming universal because of the relative ease of handling the digital signal versus the analog. In particular, signal processing for enhancement, removal of artifacts, transformation, compression, and so forth is much easier to do in the digital domain using a growing family of specialized integrated circuits. One example of this is the conversion from one video standard to another (e.g., NTSC to PAL*). Sophisticated adaptive algorithms required for good picture quality in standards conversion can be implemented only in the digital domain. Another example is the editing of digitized signals. Edits that require transformation (e.g., rotation, dilation of pictures or time-warp for audio) are significantly more difficult in the analog domain.

The density of today's digital storage media continues to increase, making storage of digital multimedia signals practical. With digital storage,

---

*NTSC and PAL are analog composite color TV standards. Japan and North America use NTSC. Most of Europe, Australia etc. use PAL.

the quality of the retrieved signal does not degrade in an unpredictable manner with multiple reads as it often does with analog storage. Also, with today's database and user interface technology, a rich set of interactions is possible only with stored digital signals. For example, with medical images the quality can be improved by noise reduction and contrast enhancement algorithms. Through the use of pseudo color, the visibility of minute variations and patterns can be greatly increased.

Mapping the stored signal to displays with different resolutions in space (number of lines per screen and number of samples per line) and time (frame rates) can be done easily in the digital domain. A familiar example of this is the conversion of film,* which is almost always at a different resolution and frame rate than the television signal.

Digital signals are also consistent with the evolving telecommunications infrastructure. Digital transmission allows much better control of the quality of the transmitted signal. In broadcast television, for example, if the signal were digital, the reproduced picture in the home could be identical to the picture in the studio, unlike the present situation where the studio pictures look far better than pictures at home. Finally, analog systems dictate that the entire television system from camera to display operate at a common clock with a standardized display. In the digital domain considerable flexibility exists by which the transmitter and the receiver can negotiate the parameters for scanning, resolution, and so forth, and thus create the best picture consistent with the capability of each sensor and display.

1.2   THE NEED FOR COMPRESSION

With the advent of fax came the desire for faster transmission of documents. With a limited bitrate available on the PSTN,[†] the only means available to increase transmission speed was to reduce the average number of bits per page, that is, *compress*

the digital data. With the advent of video conferencing, the need for digital compression[1] was even more crucial. For video, simple sampling and binary coding of the camera voltage variations produces millions of bits per second, so much so that without digital compression, any video conferencing service would be prohibitively expensive.

For entertainment TV, the shortage of over-the-air bandwidth has led to coaxial cable (CATV[‡]) connections directly to individual homes. But even today, there is much more programming available via satellite than can be carried on most analog cable systems. Thus, compression of entertainment TV would allow for more programming to be carried over-the-air and through CATV transmission.

For storage of digital video on small CD sized disks, compression is absolutely necessary. There is currently no other way of obtaining the quality demanded by the entertainment industry while at the same time storing feature length films, which may last up to two hours or longer.

Future low bitrate applications will also require compression before service becomes viable. For example, wireless cellular video telephony must operate at bitrates of a few dozen kilobits per second, which can only be achieved through large compression of the data. Likewise, video retrieval on the InterNet or World Wide Web is only feasible with compressed video data.

1.2.1   What Is Compression?

Most sensory signals contain a substantial amount of redundant or superfluous information. For example, a television camera that captures 30 frames per second from a stationary scene produces very similar frames, one after the other. Compression attempts to remove the superfluous information so that a single frame can be represented by a smaller amount of finite data, or in the case of audio or time varying images, by a lower data rate.

Digitized audio and video signals contain a significant amount of *statistical redundancy*. That is,

---

*Most commercial movie film operates at 24 frames per second, whereas U.S. Broadcast TV operates at approximately 30 frames per second.

[†]Public Switched Telephone Network.

[‡]Community Antenna Television.

samples are similar to each other so that one sample can be predicted fairly accurately from another. By removing the predictable or similarity component from a stream of samples, the data rate can be reduced. Such statistical redundancy can be removed without destroying any information whatsoever. That is, the original uncompressed data can be recovered exactly by various inverse operations. Unfortunately, the techniques for accomplishing this depend on the probabilistic characterization of the signal. Although many excellent probabilistic models of audio and video signals have been proposed, serious limitations continue to exist because of the nonstationarity of the signal statistics. In addition, statistics may vary widely from application to application. A football game captured by a video camera shows very different correlations of samples compared to the head and shoulders view of people engaged in video telephony.

The second type of superfluous data is the information that a human audio–visual system can neither hear nor see. We call this *perceptual redundancy*. If the primary receiver of the multimedia signal is a human (rather than a machine as in the case of some pattern recognition applications), then transmission or storage of the information that humans cannot perceive is wasteful. Unlike statistical redundancy, the removal of information based on the limitations of the human perception is irreversible. The original data cannot be recovered following such a removal. Unfortunately, human perception is very complex, varies from person to person, and depends on the context and the application. Therefore, the art and science of compression still has many frontiers to conquer even though substantial progress has been made in the last two decades.

1.2.2 Advantages of Compression

The biggest advantage of compression is in data rate reduction. Data rate reduction reduces transmission costs, and where a fixed transmission capacity is available, results in a better quality of multimedia presentation. As an example, a single 6-

MHz analog* cable TV channel can carry between four and ten digitized, compressed, audio–visual programs, thereby increasing the overall capacity (in terms of the number of programs carried) of an existing cable television plant. Alternatively, a single 6-MHz broadcast television channel can carry a digitized, compressed High-Definition Television (HDTV) signal to give a significantly better audio and picture quality without additional bandwidth. We shall discuss in detail the applications of compressed digital TV in Chapters 12 through 15.

Data rate reduction also has a significant impact on reducing the storage requirements for a multimedia database. A CD-ROM will soon be able to carry a full length feature movie compressed to about 4 Mbits/s. Thus, compression not only reduces the storage requirement, but also makes stored multimedia programs portable in inexpensive packages. In addition, the reduction of data rate allows processing of video-rate data without choking various resources (e.g., the main bus) of either a personal computer or a workstation.

Another advantage of digital representation/compression is for packet communication. Much of the data communication in the computer world is by self-addressed packets. Packetization of digitized audio–video and the reduction of packet rate due to compression are important in sharing a transmission channel with other signals as well as maintaining consistency with telecom/computing infrastructure. The desire to share transmission and switching has created a new evolving standard, called Asynchronous Transfer Mode (ATM), which uses packets of small size, called *cells*. Packetization delay, which could otherwise hinder interactive multimedia, becomes less of an issue when packets are small. High compression and large packets make interactive communication difficult, particularly for voice.

1.3 CONSIDERATIONS FOR COMPRESSION

The algorithms used in a compression system depend on the available bandwidth or storage

---

*NTSC uses 6-MHz analog bandwidth. PAL uses more.

capacity, the features required by the application, and the affordability of the hardware required for implementation of the compression algorithm (encoder as well as decoder). This section describes some of the issues in designing the compression system.

1.3.1 Quality

The quality of presentation that can be derived by decoding the compressed multimedia signal is the most important consideration in the choice of the compression algorithm. The goal is to provide near-transparent coding for the class of multimedia signals that are typically used in a particular service.

The two most important aspects of video quality are *spatial* resolution and *temporal* resolution. Spatial resolution describes the clarity or lack of blurring in the displayed image, while temporal resolution describes the smoothness of motion.

Motion video, like film, consists of a certain number of frames per second to adequately represent motion in the scene. The first step in digitizing video is to partition each frame into a large number of *picture elements*, or *pels** for short. The larger the number of pels, the higher the spatial resolution. Similarly, the more frames per second, the higher the temporal resolution.

Pels are usually arranged in rows and columns, as shown in Fig. 1.2. The next step is to measure the brightness of the red, green, and blue color components within each pel. These three color brightnesses are then represented by three binary numbers.

1.3.2 Uncompressed versus Compressed Bitrates

For entertainment television in North America and Japan, the NTSC[†] color video image has 29.97 frames per second; approximately 180 visi-

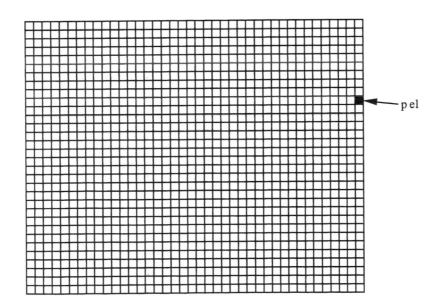

Video Frame

Fig. 1.2 Frames to be digitized are first partitioned into a large number of *picture elements*, or *pels* for short, which are typically arranged in rows and columns. A row of pels is called a *scan line*.

---

*Some authors prefer the slightly longer term *pixels*.
†National Television Systems Committee. It standardized composite color TV in 1953.

Table 1.1   Raw and compressed bitrates for film, NTSC, PAL, HDTV, videophone, stereo audio, and five-channel audio. Compressed bitrates are typically lower for slow moving drama than for fast live-action sports.

	Video Resolution (pels × lines × frames/s)	Uncompressed Bitrate (RGB)	Compressed Bitrate
Film (USA and Japan)	(480 × 480 × 24Hz)	133 Mbits/s	3 to 6 Mbits/s
NTSC video	(480 × 480 × 29.97Hz)	168 Mbits/s	4 to 8 Mbits/s
PAL video	(576 × 576 × 25Hz)	199 Mbits/s	4 to 9 Mbits/s
HDTV video	(1920 × 1080 × 30Hz)	1493 Mbits/s	18 to 30 Mbits/s
HDTV video	(1280 × 720 × 60Hz)	1327 Mbits/s	18 to 30 Mbits/s
ISDN videophone (CIF)	(352 × 288 × 29.97Hz)	73 Mbits/s	64 to 1920 kbits/s
PSTN videophone (QCIF)	(176 × 144 × 29.97Hz)	18 Mbits/s	10 to 30 kbits/s
Two-channel stereo audio		1.4 Mbits/s	128 to 384 kbits/s
Five-channel stereo audio		3.5 Mbits/s	384 to 968 kbits/s

ble scan lines per frame; and requires approximately 480 pels per scan line in the red, green, and blue color components. If each color component is coded using 8 bits (24 bits/pel total), the bitrate produced is ≈168 Megabits per second (Mbits/s). Table 1.1 shows the raw uncompressed bitrates for film, several video formats including PAL,* HDTV,† and videophone, as well as a few audio formats.

We see from Table 1.1 that uncompressed bitrates are all very high and are not economical in many applications. However, using the MPEG compression algorithms these bitrates can be reduced considerably. We see that MPEG is capable of compressing NTSC or PAL video into 3 to 6 Mbits/s with a quality comparable to analog CATV and far superior to VHS videotape.

1.3.3   Bitrates Available on Various Media

A number of communication channels are available for multimedia transmission and storage. PSTN modems can operate up to about 30 kilobits/second (kbits/s), which is rather low for generic video, but is acceptable for personal videophone. ISDN‡ ranges from 64 kbits/s to 2 Mbits/s, which is often acceptable for moderate quality video.

Table 1.2   Various media and the bitrates they support.

Medium	Bitrate
PSTN modems	Up to 30 kbits/s
ISDN	64 to 1920 kbits/s
LAN	10 to 100 Mbits/s
ATM	135 Mbits/s or more
CD-ROM (normal speed)	1.4 Mbits/s
Digital video disk	9 to 10 Mbits/s
Over-the-air video	18 to 20 Mbits/s
CATV	20 to 40 Mbits/s

LANs§ range from 10 Mbits/s to 100 Mbits/s or more, and ATM** can be many hundreds of Mbits/s. First generation CD-ROMs†† can handle 1.4 Mbits/s at normal speed and often have double or triple speed options. Second generation digital video disks operate at 9 to 10 Mbits/s depending on format. Over-the-air or CATV channels can carry 20 Mbits/s to 40 Mbits/s, depending on distance and interference from other transmitters. These options are summarized in Table 1.2.

1.4   COMPRESSION TECHNOLOGY

In picture coding, for example, compression techniques have been classified according to the groups

---

*Phase Alternate Line, used in most of Europe and elsewhere.
†High-Definition Television. Several formats are being used.
‡Integrated Services Digital Network. User bitrates are multiples of 64 kbits/s.
§Local Area Networks such as Ethernet, FDDI, etc.
**Asynchronous Transfer Mode. ATM networks are packetized.
††Compact Disk–Read Only Memory.

Table 1.3 Picture coding primitive elements. Pulse code modulation (PCM) is simple binary coding of pels. Differential pulse code modulation (DPCM) sends differences of pels.

Coded Element	Encoding	Example
Picture element	PCM, DPCM	Wirephoto
Block of elements	Spatial transform	JPEG, subband
Translated blocks	Motion compensation	MPEG-1, -2
Regions	Object-based coders	MPEG-4 (1998)
Named content	3-D scene elements	(Research only)

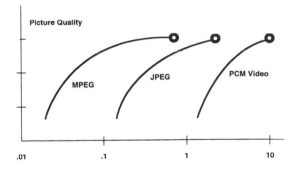

Fig. 1.3 Approximate measure of picture quality versus bitrate in bits/pel of several video compression algorithms.

of pels on which they operate. Table 1.3 shows a simplified view of this. Encoding using a few pels is considerably simpler than encoding that requires a complex spatial–temporal processing of groups of pels. However, performance is lower.

Over the years, to gain compression efficiency (that is, picture quality for a given bitrate), the compression algorithms have become more and more complex. Figure 1.3 shows a rough qualitative comparison of the relative picture quality at various compression ratios for three different coding methods.

It is important to note that the less complex the algorithm, the more rapid the degradation of picture quality with only small changes in compression ratio. In general, entertainment applications using broadcast, cable, or satellite TV have tended to require a much higher quality of compressed signal than video telephony, video groupware, and video games. Since most entertainment signals have a significantly larger amount of motion than video telephone signals, a smaller compression ratio results for the high quality that is required. In any case, both the compression algorithm and the compression ratio need to be tailored for the picture quality that is required for the application at hand.

1.4.1 Robustness

As the redundancy from the multimedia signal is removed by compression, each compressed bit becomes more important in the sense that it affects a larger number of samples of the audio or picture signal. Therefore, an error either in transmission or storage of the compressed bit can have deleterious effects for either a large region of the picture or over an extended period of time.

For noisy digital transmission channels, video compression algorithms that sacrifice efficiency to allow for graceful degradation of the images in the presence of channel errors are better candidates. Some of these are created by merging source and channel coding to optimize the end-to-end service quality. A good example of this is portable multimedia over a low-bandwidth wireless channel. Here, the requirements on compression efficiency are severe owing to the lack of available bandwidth. Yet a compression algorithm that is overly sensitive to channel errors would be an improper choice. Of course, error correction is usually added to an encoded signal along with a variety of error concealment techniques, which are usually successful in reducing the effects of random isolated errors. Thus, the proper choice of the compression algorithm depends on the transmission error environment in which the application resides.

1.4.2 Interactivity

Both consumer entertainment and business multimedia applications are characterized by picture scanning and browsing. In the home, viewers switch to the channels of their choice. In the business environment, people get to the information of their choice by random access using, for example, on-screen menus. In the television of the future, a much richer interaction based on content rather than channel switching or accessing a different picture will be required.

Many multimedia offerings and locally produced video programs often depend on the concatenation of video streams from a variety of sources, sometimes in real time. Commercials are routinely inserted into nationwide broadcasts by network affiliates and cable headends. Thus, the compression algorithm must support a continuous and seamless assembly of these streams for distribution and rapid switching of images at the point of final decoding. It is also desirable that these simple edits as well as richer interactions occur on compressed data rather than reconstructed sequences.

In general, a higher degree of interactivity requires a compression algorithm that operates on a smaller group of pels. MPEG, which operates on spatio-temporal groups of pels, is more difficult to interact with than JPEG,[2] which operates only on spatial groups of pels. As an example, it is much easier to fast forward a compressed JPEG bitstream than a compressed MPEG bitstream. In a cable/broadcast environment or in an application requiring browsing through a compressed multimedia database, a viewer may change from program to program with no opportunity for the encoder to adapt itself. It is important that the buildup of resolution following a program change take place quite rapidly so that the viewer can make a decision to either stay on the program or change to another depending on the content the viewer wishes to watch.

1.4.3 Compression and Packetization Delay

Advances in compression have come predominantly through better analysis of the signal in question. For picture coding this is exemplified in Table 1.3. As models have progressed from element-based to picture blocks to interframe regions, efficiency has grown rapidly. Correspondingly, the complexity of the analysis phase of encoding has also grown, resulting in the encoding delay becoming an important parameter. A compression algorithm that looks at a large number of samples and performs very complex operations usually has a larger encoding delay.

For many applications, such encoding delay at the source is tolerable, but for some it is not. Broadcast television, even in real time, can often admit a delay in the order of seconds. However,

teleconferencing or multimedia groupware can tolerate a far smaller delay. In addition to the encoding delay, modern data communications that packetize the information also introduce delay. The more efficient the compression algorithm, the larger is the delay that is introduced by packetization, since the same size packet carries information about many more samples of the multimedia signal. Thus, the compression algorithm must also take into consideration the delay that is tolerable for the application.

1.4.4 Symmetry

A cable, satellite, or broadcast environment has only a few transmitters that compress, but a large number of receivers that have to decompress. Similarly, multimedia databases that store compressed information usually compress it only once. However, the retrieval of this information may happen thousands of times by different viewers. Therefore, the overall economics of many multimedia applications is dictated to a large extent by the cost of decompression. The choice of the compression algorithm ought to make the decompression extremely simple by transferring much of the cost to the transmitter, thereby creating an *asymmetrical* algorithm. The analysis phase of a compression algorithm, which routinely includes motion analysis (done only at the encoder), naturally makes the encoder more expensive. In a number of situations, the cost of the encoder is also important (e.g., camcorder, videotelephone). Therefore, a modular design of the encoder that is able to trade off performance with complexity, but that creates data decodable by a simple decompressor, may be the appropriate solution.

1.4.5 Multiple Encoding

In a number of instances, the original multimedia signal may have to be compressed in stages or may have to be compressed and decompressed several times. In most television studios, for example, it is necessary to store the compressed data and then decompress it for editing as required. Such an edited signal is then compressed and stored again. Any multiple coding–decoding cycle of the signal is bound to reduce the quality of the multimedia signal, since artifacts are introduced every time the signal is coded. If the application requires such multiple codings, then a higher quality compression is required, at least in the several initial stages.

1.4.6 Scalability

A compressed multimedia signal can be thought of as an alternative representation of the original uncompressed signal. From this alternative representation, it is desirable to create presentations at different resolutions (in space, time, amplitude, etc.) consistent with the limitations of the equipment used in a particular application. For example, if a high-definition television signal compressed to 24 Mbits/s can be simply processed to produce a lower resolution and lower bitrate signal (e.g., NTSC at 6 Mbits/s), the compression is generally considered to be *scalable*. Of course, the scalability can be achieved in a brute force manner by decompressing, reducing the resolution, and compressing again. However, this sequence of operations introduces delay and complexity, and results in a loss of quality. A common compressed representation from which a variety of low-resolution or higher resolution presentations can be easily derived is desirable. Such scalability of the compressed signal puts a constraint on the compression efficiency in the sense that algorithms with the highest compression efficiency usually are not very scalable.

1.5 VIDEOPHONE AND COMPACT DISK STANDARDS— H.320 AND MPEG-1

Digital compression standards for video conferencing were developed in the 1980s by the CCITT, which is now known as the ITU-T. Specifically, the ISDN video conferencing standards are known collectively as H.320, or sometimes P*64 to indicate that it operates at multiples of 64 kbits/s. The video coding portion of the standard is called H.261 and codes pictures at a *Common Intermediate Format* (CIF) of 352 pels by 288 lines. A lower resolution of 176 pels by 144 lines, called QCIF, is available for interoperating with PSTN videophones.

In the late 1980s, a need arose to place motion video and its associated audio onto first generation CD-ROMs at 1.4 Mbits/s. For this purpose, in the late 1980s and early 1990s, the ISO MPEG committee developed digital compression standards[4] for both video and two-channel stereo audio. The standard[3,4,8] is known colloquially as MPEG-1 and officially as ISO 11172. The bitrate of 1.4 Mbits/s available on first generation CD-ROMs is not high enough to allow for full-resolution TV. Thus, MPEG-1 was optimized for the reduced CIF resolution of H.320 video conferencing.

1.6 THE DIGITAL ENTERTAINMENT TV STANDARD—MPEG-2

Following MPEG-1, the need arose to compress entertainment TV for such transmission media as satellite, cassette tape, over-the-air, and CATV. Thus, to have available digital compression methods[4-6,8,10] for full-resolution Standard Definition TV (SDTV) pictures such as shown in Fig. 1.4a or High Definition TV (HDTV) pictures such as shown in Fig. 1.4b, ISO developed a second standard known colloquially as MPEG-2 and officially as ISO 13818. Since the resolution of entertainment TV is approximately four times that of videophone, the bitrate chosen for optimizing MPEG-2 was 4 Mbits/s.

1.6.1 Other Functionalities of MPEG-2

In addition to simply compressing audio and video with high quality and low bitrate, a number of other capabilities[10] are needed for full multimedia service:

Random Access: For stored multimedia or Interactive Television (ITV) applications such as shown in Fig. 1.5 we often need quick access to the interior of the audiovisual data without having to start decoding from the very beginning or waiting a long time for the decoder to start up.

(a)

(b)

Fig. 1.4   a) Example of a full resolution Standard Definition TV (SDTV) picture. b) Example of a High Definition TV (HDTV) picture. Courtesy of Richard Kollarits.

Trick Modes: Users of stored multimedia want all of the features they now have on VCRs. These include fast-play forward and reverse, slow-play forward and reverse, and freeze-frame.

Multicast to Many Terminal Types: Some applications may need to simultaneously broadcast to many different terminal types over a variety of communication channels. Preparing and storing a different bitstream for each of them is impractical. Thus, a single bit stream is desired that is *scalable* to the various requirements of the channels and terminals.

Multiple Audio and Video: Multiple audios are needed, for example, if many languages are to be provided simultaneously. Multiple videos may also be required if closeups and wide area views are to be sent simultaneously in the same program.

Compatible Stereoscopic 3D: Compatible 3D for pictures such as shown in Fig. 1.6, is needed so that viewers with 3D displays can see 3D program-

Fig. 1.5   Example of an Interactive Television (ITV) picture.

Fig. 1.6   Example of left and right eye views of a stereoscopic 3D picture. Courtesy of C.C.E.T.T., Rennes, France.

ming,[9] while those with normal TVs can see the usual monocular display.

These features of MPEG-2 are summarized in Table 1.4. How all these features can be combined into a single compression and multiplexing standard for utilization in a variety of applications is the subject of the following chapters.

Table 1.4   Features offered by MPEG-2.

MPEG2 Features
Random access
Trick modes
Multicast to many terminal types
Multiple audio and video
Compatible stereoscopic 3D

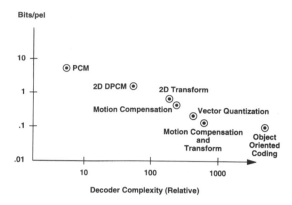

Fig. 1.7   Bits/pel versus complexity of video decoding for several video compression algorithms.

1.7   COMPLEXITY/COST

Since cost is directly linked to complexity, this aspect of a compression algorithm is the most critical for the asymmetrical situations described previously. The decoder cost is most critical. Figure 1.7 represents an approximate tradeoff between the compression efficiency and the complexity under the condition that picture quality is held constant at an 8-bit PCM level. The compression efficiency is in terms of compressed bits per Nyquist sample. Therefore, pictures with different resolution and bandwidth can be compared simply by proper multiplication to get the relevant bitrates. The complexity allocated to each codec should not be taken too literally. Rather, it is an approximate estimate relative to the cost of a PCM codec, which is given a value of 5.

The relation of cost to complexity is controlled by an evolving technology, and codecs* with high complexity are quickly becoming inexpensive through the use of application-specific video DSPs and submicron device technology. In fact, very soon fast microprocessors will be able to decompress the video signal entirely in software. Figure 1.8 shows the computational requirements in millions of instructions per second (MIPS) for encoding and decoding of video at different resolutions. It is clear that in the near future a standard resolution (roughly 500 line by 500 pel TV signal) will be

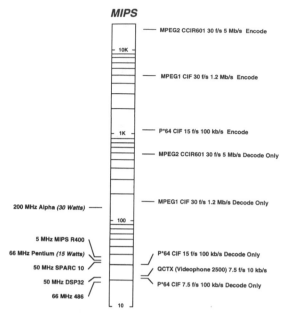

Fig. 1.8   Computational requirements in millions of instructions per second (MIPs) for video encoding and decoding at different image resolutions.

decoded entirely in software for even the MPEG compression algorithm.

On the other hand, the promise of object-based compression systems, in which we recognize objects and code them as a whole instead of coding each pel, will increase the complexity by several orders of magnitude, thereby requiring specialized hardware support. Similarly, higher resolution images (e.g., HDTV[7]) will raise the processing speed requirement even higher. Some object-oriented approaches are only marginally harder to decode, but may require far more analysis at the compressor. Others envisage wholly new architectures for picture reconstruction, thereby requiring a far more complex decoder. The real challenge is to combine the different compression techniques to engineer a cost-effective solution for the given application.

In this chapter, we have examined the advantages of digital representation of the multimedia signal that usually originates as analog signals. Digital representation is consistent with the evolving

---

*Coder-decoder.

computing/telecommunications infrastructure and to a large extent gives us media independence for storage, processing, and transmission. The bandwidth expansion resulting from digitization can be more than compensated for by efficient compression that removes both the statistical and the perceptual redundancy present in the multimedia signal. Digital compression has many advantages, most of which come as a result of reduced data rates. However, when digital compression is bundled as a part of an application, many constraints need to be attended to.

These constraints impact directly on the choice of the compression algorithm. Many of these constraints are qualitatively examined in this chapter. A more detailed analysis of these will be made in the subsequent chapters.

REFERENCES

1. A. N. NETRAVALI and B. G. HASKELL, *Digital Pictures—Representation, Compression and Standards*, 2nd edit., Plenum Press, New York, 1995.

2. W. B. PENNEBAKER and J. L. MITCHELL, *JPEG—Still Image Compression Standard*, Van Nostrand Reinhold, New York, 1993.

3. J. L. MITCHELL et al, MPEG-1, Chapman & Hall, New York, 1997.

4. L. CHIARIGLIONE, "The Development of an Integrated Audiovisual Coding Standard: MPEG," *Proceedings of the IEEE*, Vol. 83, No. 2, (February 1995).

5. D. J. LE GALL, "MPEG Video: The Second Phase of Work," International Symposium: Society for Information Display, pp. 113–116 (May 1992).

6. A. N. NETRAVALI and A. LIPPMAN, "Digital Television: A perpective," *Proceedings of the IEEE*, Vol. 83, No. 6 (June 1995).

7. R. HOPKINS, "Progress on HDTV Broadcasting Standards in the United States," *Signal Processing: Image Commun.*, Vol. 5, pp. 355–378 (1993).

8. A. PURI, "MPEG Video Coding Standards," Invited Tutorial: International Society for Circuits and Systems (April 1995).

9. A. KOPERNIK, R. SAND, and B. CHOQUET, "The Future of Three-Dimensional TV," International Workshop on HDTV'92, Signal Processing of HDTV, IV pp. 17–29 (1993).

10. S. OKUBO, "Requirements for High Quality Video Coding Standards," *Signal Processing: Image Commun.*, Vol. 4, No. 2, pp 141–151 (April 1992).

2

Anatomy of MPEG-2

After many fits and starts, the era of Digital Television[27] is finally here. Although a variety of audio and video coding standards predate MPEG standardization activity, the MPEG-1 and the MPEG-2 standards are truly unique integrated audio-visual standards.[25,28]

Neither the MPEG-1 nor the MPEG-2 standards prescribe which encoding methods to use, the encoding process, or details of encoders. These standards only specify formats for representing data input to the decoder, and a set of rules for interpreting these data. These formats for representing data are referred to as *syntax* and can be used to construct various kinds of valid data streams referred to as *bitstreams.* The rules for interpreting the data are called *decoding semantics;* An ordered set of decoding semantics is referred to as a *decoding process.* Thus, we can say that MPEG standards specify a decoding process; However, this is still different than specifying a decoder implementation.

Given audio or video data to be compressed, an encoder must follow an ordered set of steps called the *encoding process.* This encoding process, however, is not standardized and typically varies, since encoders of different complexities may be used in different applications. Also, since the encoder is not standardized, continuing improvements in quality are still possible because of encoder opti-

mizations even after the standard is complete. The only constraint is that the output of the encoding process result in a syntactically correct bitstream that can be interpreted by following the decoding semantics by a standards compliant decoder.

2.1 MPEG-1

Although we plan to discuss the anatomy of MPEG-2, we first need to understand the anatomy of MPEG-1. This will allow us to make comparisons between the standards to clearly show similarities and differences, and wherever appropriate, compatibilities between the two standards.

The MPEG-1 standard is formally referred to as ISO 11172 and consists of the following parts:

- 11172-1: Systems
- 11172-2: Video
- 11172-3: Audio
- 11172-4: Conformance
- 11172–5: Software

Initially, the MPEG-1 standard was intended for video coding at bitrates of about 1.2 Mbit/s with stereo audio coding at bitrates of around 250 kbits/s.

2.1.1 MPEG-1 Systems

The MPEG-1 Systems[1] part specifies a system coding layer for combining coded audio and video data and to provide the capability for combining with it user-defined private data streams as well as streams that may be specified at a later time. To be more specific,[12] the MPEG-1 Systems standard defines a packet structure for multiplexing coded audio and video data into one stream and keeping it synchronized. It thus supports multiplexing of multiple coded audio and video streams where each stream is referred to as an *elementary stream*. The systems syntax includes data fields to assist in parsing the multiplexed stream after random access and to allow synchronization of elementary streams, management of decoder buffers containing coded data, and identification of timing information of coded program. The MPEG-1 Systems thus specifies syntax to allow generation of systems bitstreams and semantics for decoding these bit streams.

Since the basic concepts of timing employed in MPEG-1 Systems are also common to MPEG-2 Systems, we briefly introduce the terminology employed. The Systems Time Clock (STC) is the reference time base, which operates at 90 kHz, and may or may not be phase locked to individual audio or video sample clocks. It produces 33-bit time representations that are incremented at 90 kHz. In MPEG Systems, the mechanism for generating the timing information from decoded data is provided by the *Systems Clock Reference* (SCR) fields that indicate the current STC time and appear intermittently in the bitstream spaced no further than 700 ms apart. The presentation playback or display synchronization information is provided by *Presentation Time Stamps* (PTSs), that represent the intended time of presentation of decoded video pictures or audio frames. The audio or video PTS are sampled to an accuracy of 33 bits.

To ensure guaranteed decoder buffer behavior, MPEG Systems employs a Systems Target Decoder (STD) and Decoding Time Stamp (DTS). The DTS differs from PTS only in the case of video pictures that require additional reordering delay during the decoding process.

2.1.2 MPEG-1 Video

The MPEG-1 Video[2] standard was originally aimed at coding of video of SIF resolution (352 × 240 at 30 noninterlaced frames/s or 352 × 288 at 25 noninterlaced frames/s) at bitrates of about 1.2 Mbit/s. In reality it also allows much larger picture sizes and correspondingly higher bitrates. Besides the issue of bitrates, the other significant issue is that MPEG includes functions to support interactivity such as fast forward (FF), fast reverse (FR), and random access into the stored bitstream. These functions exist since MPEG-1 was originally aimed at digital storage media such as video compact discs (CDs).

The MPEG-1 Video standard basically specifies the video bitstream syntax and the corresponding video decoding process. The MPEG-1 syntax supports encoding methods[13,14,15] that exploit both the spatial redundancies and temporal redundancies. Spatial redundancies are exploited by using block based Discrete Cosine Transform (DCT) coding of 8 × 8 pel blocks followed by quantization, zigzag scan, and variable length coding of runs of zero quantized indices and amplitudes of these indices. Moreover, quantization matrix allowing perceptually weighted quantization of DCT coefficients can be used to discard perceptually irrelevant information, thus increasing the coding efficiency further. Temporal redundancies are exploited by using motion compensated prediction, which results in a significant reduction of interframe prediction error.

The MPEG-1 Video syntax supports three types of coded pictures, Intra (I) pictures, coded separately by themselves, Predictive (P) pictures, coded with respect to immediately previous I- or P-pictures, and Bidirectionally Predictive (B) pictures coded with respect to the immediate previous I- or P-picture, as well as the immediate next I- or P-picture. In terms of coding order, P-pictures are causal, whereas B-pictures are noncausal and use two surrounding causally coded pictures for prediction. In terms of compression efficiency, I-pictures are least efficient, P-pictures are somewhat better, and B-pictures are the most efficient. In typical MPEG-1 encoding[13,15] an input video sequence is divided into units of group-of-pictures (GOPs), where each GOP consists of an arrangement of one I-picture, P-pictures, and B-pictures. A GOP serves as a basic access unit, with the I-picture serving as the entry point to facilitate random access. Each picture is divided further into one or more *slices* that offer a mechanism for resynchronization and thus limit the propagation of errors. Each slice is composed of a number of macroblocks; each macroblock is basically a 16 × 16 block of luminance pels (or alternatively, four 8 × 8 blocks) with corresponding chrominance blocks. MPEG-1 Video encoding is performed on a macroblock basis. In P-pictures each macroblock can have one motion vector, whereas in B-pictures each macroblock can have as many as two motion vectors.

2.1.3 MPEG-1 Audio

The MPEG-1 Audio[3] standard basically specifies the audio bitstream syntax and the corresponding audio decoding process. MPEG-1 Audio is a generic standard and does not make any assumptions about the nature of the audio source, unlike some vocal-tract-model coders that work well for speech only. The MPEG syntax permits exploitation of perceptual limitations of the human auditory system while encoding, and thus much of the compression comes from removal of perceptually irrelevant parts of the audio signal. MPEG-1 Audio coding[16,23] allows a diverse assortment of compression modes and in addition supports features such as random access, audio fast forwarding, and audio fast reverse.

The MPEG-1 Audio standard consists of three Layers—I, II, and III. These Layers also represent increasing complexity, delay, and coding efficiency. In terms of coding methods, these Layers are somewhat related, since a higher Layer includes the building blocks used for a lower Layer. The sampling rates supported by MPEG-1 Audio are 32, 44.2, and 48 kHz. Several fixed bitrates in the range of 32 to 224 kbits/s per channel can be used. In addition, Layer III also supports variable bitrate coding. Good quality is possible with Layer I above 128 kbits/s, with Layer II around 128 kbit/s, and with Layer III at 64 kbit/s. MPEG-1 Audio has four modes: mono, stereo, dual with two separate

channels, and joint stereo. The optional joint stereo mode exploits interchannel redundancies.

A polyphase filter bank is common to all Layers of MPEG-1 Audio coding. This filter bank subdivides the sudio signal into 32 equal-width frequency subbands. The filters are relatively simple and provide good time-resolution with a reasonable frequency-resolution. To achieve these characteristics, some exceptions had to be made. First, the equal widths of subbands do not precisely reflect the human auditory system's frequency-dependent behavior. Second, the filter bank and its inverse are not completely lossless transformations. Third, adjacent filter bands have significant frequency overlap. However, these exceptions do not impose any noticeable limitations, and quite good audio quality is possible. The Layer I algorithm codes audio in frames of 384 samples by grouping 12 samples from each of the 32 subbands. The Layer II algorithm is a straightforward enhancement of Layer I; it codes audio data in larger groups (1 152 samples per audio channel) and imposes some restrictions on possible bit allocations for values from the middle and higher subbands. The Layer II coder gets better quality by redistributing bits to better represent quantized subband values. The Layer III algorithm is more sophisticated and uses audio spectral perceptual entropy coding and optimal coding in the frequency domain. Although based on the same filter bank as used in Layer I and Layer II it compensates for some deficiencies by processing the filter outputs with a modified DCT.

2.2 MPEG-2

After this brief overview of the parts of the MPEG-1 standard, we are now well prepared to delve into a discussion of the various parts of the MPEG-2 standard. This standard is formally referred to as ISO 13818 and consists of the following parts:

- 113818-1: Systems
- 113818-2: Video
- 113818-3: Audio
- 113818-4: Conformance
- 113818-5: Software

- 113818-6: Digital Storage Media—Command and Control (DSM-CC)
- 113818-7: Non Backward Compatible (NBC) Audio
- 113818-8: 10-Bit Video (This work item has been dropped!)
- 113818-9: Real Time Interface
- 113818-10: Digital Storage Media—Command and Control (DSM-CC) Conformance

We now present a brief overview of the aforementioned parts of the MPEG-2 standard.

2.2.1 MPEG-2 Systems

Since the MPEG-1 standard was intended for audio–visual coding for Digital Storage Media (DSM) applications and since DSMs typically have very low or negligible transmission bit error rates, the MPEG-1 Systems part was not designed to be highly robust to bit errors. Also, the MPEG-1 Systems was intended for software oriented processing, and thus large variable length packets were preferred to minimize software overhead.

The MPEG-2 standard on the other hand is more generic and thus intended for a variety of audio–visual coding applications. The MPEG-2 Systems[4] was mandated to improve error resilience plus the ability to carry multiple programs simultaneously without requiring them to have a common time base. Additionally, it was required that MPEG-2 Systems should support ATM networks. Furthermore, MPEG-2 Systems was required to solve the problems addressed by MPEG-1 Systems and be somewhat compatible with it.

The MPEG-2 Systems[17,18] defines two types of streams: the Program Stream and the Transport Stream. The Program Stream is similar to the MPEG-1 Systems stream, but uses a modified syntax and new functions to support advanced functionalities. Further, it provides compatibility with MPEG-1 Systems streams. The requirements of MPEG-2 Program Stream decoders are similar to those of MPEG-1 System Stream decoders. Furthermore, Program Stream decoders are expected to be forward compatible with MPEG-1 System Stream decoders, i.e., capable of decoding MPEG-1 System Streams. Like MPEG-1 Systems

decoders, Program streams decoders typically employ long and variable-length packets. Such packets are well suited for software based processing and error-free environments, such as when the compressed data are stored on a disk. The packet sizes are usually in range of 1 to 2 kbytes chosen to match disk sector sizes (typically 2 kbytes); however, packet sizes as large as 64 kbytes are also supported. The Program Stream includes features not supported by MPEG-1 Systems such as scrambling of data; assignment of different priorities to packets; information to assist alignment of elementary stream packets; indication of copyright; indication of fast forward, fast reverse, and other trick modes for storage devices; an optional field for network performance testing; and optional numbering of sequence of packets.

The second type of stream supported by MPEG-2 Systems is the Transport Stream, which differs significantly from MPEG-1 Systems as well as the Program Stream. The Transport Stream offers robustness necessary for noisy channels as well as the ability to include multiple programs in a single stream. The Transport Stream uses fixed length packets of size 188 bytes, with a new header syntax. It is therefore more suited for hardware processing and for error correction schemes. Thus the Transport Stream stream is well suited for delivering compressed video and audio over error-prone channels such as coaxial cable television networks and satellite transponders. Furthermore, multiple programs with independent time bases can be multiplexed in one Transport Stream. In fact the Transport Stream is designed to support many functions such as asynchronous multiplexing of programs, fast access to desired program for channel hopping, multiplex of programs with clocks unrelated to transport clock, and correct synchronization of elementary streams for playback. It also allows control of decoder buffers during startup and playback for both constant bitrates and variable bitrate programs.

A basic data structure that is common to the organization of both the Program Stream and Transport Stream data is called the Packetized Elementary Stream (PES) packet PES packets are generated by packetizing the continuous streams of compressed data generated by video and audio

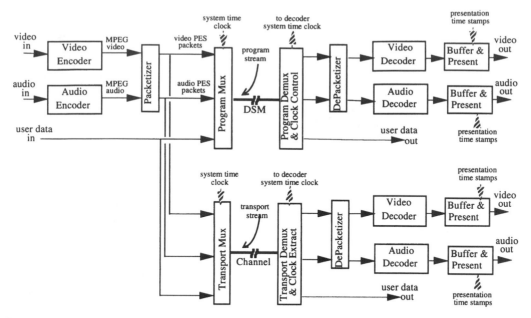

Fig. 2.1 MPEG-2 Systems Multiplex showing Program and Transport Streams.

(i.e., elementary stream) encoders. A Program Stream is generated simply by stringing together PES packets from the various encoders with other packets containing necessary data to generate a single bitstream. A Transport Stream consists of packets of fixed length consisting of 4 bytes of header followed by 184 bytes of data, where data is obtained by chopping up data in PES packets.

In Fig. 2.1 we illustrate both types of MPEG-2 Systems, the ones using Program Stream multiplex and the ones using Transport Stream multiplex. An MPEG-2 System is also capable of combining multiple sources of user data along with the MPEG encoded audio and video. The audio and video streams are packetized to form audio and video PES packets that are then sent to either a Program Multiplexer or a Transport Multiplexer, resulting in a Program Stream or Transport Stream, respectively. As mentioned earlier, Program Streams are intended for error-free environments such as DSMs, whereas Transport Streams are intended for noisier environments such as terrestrial broadcast channels.

Transport Streams are decoded by a Transport Demultiplexer (which includes a clock extraction mechanism), unpacketized by a DePacketizer, and

sent for audio and video decoding to Audio and Video Decoders. The decoded signals are sent to respective Buffer and Presentation units that output them to a display device and speaker at the appropriate times. Similarly, if Program Streams are employed, they are decoded by a Program Stream Demultiplexer, DePacketizer, and sent for decoding to Audio and Video Decoders. The decoded signals are sent to respective Buffer and Present and await presentation. Also, as with MPEG-1 Systems, the information about systems timing is carried by Clock Reference fields in the bitstream that are used to synchronize the decoder Systems Time Clock (STC). The presentation of decoded output is controlled by Presentation Time Stamps (PTSs) that are also carried by the bitstream.

2.2.2 MPEG-2 Video

This part of MPEG-2, part 2, addresses coding of video and is referred to as MPEG-2 Video. Originally the MPEG-2 Video[5] standard was primarily intended for coding of interlaced video of standard TV resolution with good quality in the bitrate range of 4 to 9 Mbit/s. However, the scope of MPEG-2 Video was considerably revised[20] to include video of higher resolutions such as HDTV at higher bitrates as well as hierarchical video coding for a range of applications. Furthermore, it also supports coding of interlaced video of a variety of resolutions.

Actually, MPEG-2 Video does not standardize video encoding; only the video bitstream syntax and decoding semantics are standardized. The standardization process for MPEG-2 Video consisted of a competitive phase followed by a collaborative phase. First an initial set of requirements for MPEG-2 video was defined and used as the basis for a "Call for Proposals" which invited candidate coding schemes for standardization. These schemes underwent formal subjective testing and analysis to determine if they met the listed requirements. During the collaborative phase, the best elements of the top performing schemes in the competitive phase were merged to form an initial *Test Model*. A Test Model contains a description of an encoder, bitstream syntax, and decoding semantics. The encoder description is not to be standardized, but only needed for experimental evaluation of proposed techniques.

Following this, a set of *Core Experiments* were defined that underwent several iterations until convergence was achieved. During the time that various iterations on the Test Model[19] were being performed, the requirements for MPEG-2 Video also underwent further iterations.

In Fig. 2.2 we show a simplified codec[19,21] for MPEG-2 Nonscalable Video Coding. The MPEG-2 Video Encoder shown for illustration purposes consists of an Inter frame/field DCT Encoder, a Frame/field Motion Estimator and Compensator, and a Variable Length Encoder (VLE). Earlier we had mentioned that MPEG-2 Video is optimized for coding of interlaced video, which is why both the DCT coding and the motion estimation and compensation employed by the Video Encoder are frame/field adaptive. The Frame/field DCT Encoder exploits spatial redundancies, and the Frame/field Motion Compensator exploits temporal redundancies in interlaced video signal. The coded video bitstream is sent to a systems multiplexer, Sys Mux, which outputs either a Transport or a Program Stream.

The MPEG-2 Decoder in this codec consists of a Variable Length Decoder (VLD), Inter frame/field DCT Decoder, and the Frame/field Motion Compensator. Sys Demux performs the complementary function of Sys Mux and presents the video bitstream to VLD for decoding of motion vectors and DCT coefficients. The Frame/field Motion Compensator uses a motion vector decoded by VLD to generate motion compensated prediction that is added back to a decoded prediction error signal to generate decoded video out. This type of coding produces video bitstreams called nonscalable, since normally the full spatial and temporal resolution coded is the one that is expected to be decoded. Actually, this is not quite true, since, if B-pictures are used in encoding, because they do not feedback into the interframe coding loop, it is always possible to decode some of them, thus achieving temporal resolution scaling even for nonscalable bitstreams. However, by nonscalable cod-

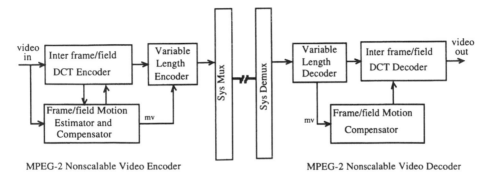

Fig. 2.2 A generalized codec for MPEG-2 Nonscalable Video Coding.

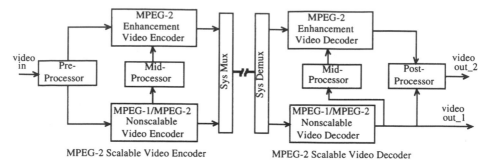

Fig. 2.3 A generalized codec for MPEG-2 Scalable Video Coding.

ing we usually mean that we have not gone out of our way to facilitate scalability.

As you may have guessed by this time, the MPEG-2 Video standard specifies the bitstream syntax and decoding process not only for single layer (MPEG-1 like) video coding but also for scalable video coding. Scalability is the property that allows decoders of various complexities to be able to decode video of resolution/quality commensurate with their complexity from the same bitstream. We will discuss a lot more about MPEG-2 scalable video coding in Chapter 9, but for now we illustrate the general principle by introducing the generalized scalable codec structure shown in Fig. 2.3. In MPEG-2 Video there are many different types of scalability and thus it is difficult to represent all of them with a single generalized codec. Our generalized codec structure basically implies spatial and temporal resolution scalability.

Input video goes through a Pre-Processor that produces two video signals, one of which (the *Base Layer*) is input to an MPEG-1 or MPEG-2 Nonscalable Video Encoder and the other (the *Enhancement Layer*) input to an MPEG-2 Enhancement Video Encoder. Some processing of decoded video from MPEG-1 or MPEG-2 Nonscalable Video Encoder may be needed in the Mid-Processor depending on specific scalability. The two bitstreams, one from each encoder, are multiplexed in Sys Mux (along with coded audio and user data). Thus it becomes possible for two types of decoders to be able to decode a video signal of quality commensurate with their complexity, from the same encoded bitstream. For example, if an MPEG-1 or MPEG-2 Nonscalable Video Decoder is employed, a basic

video signal can be decoded. If in addition, an MPEG-2 Enhancement Video Decoder is employed, an enhanced video signal can be decoded. The two decoded signals may undergo further processing in a Post-Processor.

Since the MPEG-2 Video standard, to be truly generic, had to include a bitstream and semantics description for a variety of coding methods and tools, the implementation of all its features in every decoder was considered too complex and thus the standard was partitioned into Profiles. In MPEG-2 Video, applications with somewhat related requirements are addressed via Profiles; a Profile typically contains a collection of coding techniques (or tools) designed to address a set of such applications. We will discuss a lot more about Requirements and Profiles in Chapter 11.

The MPEG-2 Video standard is a syntactic superset of the MPEG-1 Video standard and is thus able to meet the requirement of forward compatibility, meaning that an MPEG-2 video decoder should be capable of decoding MPEG-1 video bitstreams. The requirement of backward compatibility, meaning that subsets of MPEG-2 bitstreams should be decodable by existing MPEG-1 decoders, is achieved via the use of scalability and supported in specific profiles. To confirm that the MPEG-2 Video standard met its quality objective for various Profiles and for individual applications within a profile, a series of verification tests have been conducted in association with other international standardization bodies and a number of organizations. Thus far, these tests confirm that MPEG-2 Video does indeed meet or exceed the performance bounds of its target applications. MPEG-2 Video quality has also been judged sufficient for HDTV.[34] Recently, newer applications of MPEG-2 Video have also emerged, necessitating combinations of coding tools requiring new Profiles.

At the time of writing of this chapter in early 1996, two new amendments in MPEG-2 video, each adding a new profile, are in progress. One of the two amendments is nearly complete, and the second one is expected to be completed by the end of 1996. Although MPEG-2 video already includes various coding techniques (decoding techniques, really), in reality, these techniques must be included in one of the agreed Profiles also (these profiles can even be defined after completion of the standard). Next, we briefly introduce what application areas are being targeted by these two amendments.

The first amendment to MPEG-2 Video involves a Profile that includes tools for coding of a higher resolution chrominance format called the 4:2:2 format. We will discuss more about this format in Chapter 5; for now, we need to understand why it was relevant for MPEG-2 video to address applications related to this format. When work on MPEG-2 Video was close to completion, it was felt that MPEG-2 video was capable of delivering fairly high quality, and thus it could be used in professional video applications. In sorting out needs of such applications, besides higher bitrates, it was found that some professional applications require higher than normal chrominance spatial (or spatio-temporal) resolution, such as that provided by the 4:2:2 format while another set of applications required higher amplitude resolution for luminance and chrominance signal. Although coding of the 4:2:2 format video is supported by existing tools in the MPEG-2 Video standard, it was necessary to verify that its performance was equal to or better than that of other coding schemes while fulfilling additional requirements such as coding quality after multiple codings and decodings.

The second amendment[32] to MPEG-2 Video involves a Profile that includes tools for coding of a number of simultaneous video signals generated in imaging a scene, each representing a slightly different viewpoint; imaging of such scenes is said to constitute multiviewpoint video. An example of multiviewpoint video is stereoscopic video which uses two slightly different views of a scene. Another example is generalized 3D video where a large number of views of a scene may be used. We will discuss more about this profile in Chapter 5 and a lot more about 3D/stereoscopic video in Chapter 15; for now, as in the case of 4:2:2 coding we need to understand why it was relevant to address multiviewpoint coding applications. As mentioned earlier, MPEG-2 Video already includes several Scalability techniques that allow layered coding which exploits correlations between layers. As the result of a recent rise in applications in video games, movies, education, and entertainment, efficient coding of multiviewpoint video such as stereoscopic video is becoming increasingly important. Not surprisingly, this involves exploiting correlations between different views, and since scalability techniques of MPEG-2 Video already included such techniques, it was considered relatively straightforward to enable use of such techniques via definition of a Profile (which is currently in progress).

2.2.3 MPEG-2 Audio

Digital multichannel audio systems employ a combination of p front and q back channels; for example, three front channels—left, right, center—and two back channels—surround left and surround right—to create surreal experiences. In addition, multichannel systems can also be used to provide multilingual programs, audio augmentation for visually impaired individuals, enhanced audio for hearing impaired individuals, and so forth.

The MPEG-2 Audio[6] standard consists of two parts, one that allows coding of multichannel audio signals in a forward and backward compatible manner with MPEG-1 and one that does not. Part 3 of the MPEG-2 standard is forward and backward compatible with MPEG-1. Here forward compatibility means that the MPEG-2 multichannel audio decoder can decode MPEG-1 mono or stereo audio signals. Backward compatibility (BC) means that an MPEG-1 stereo decoder can reproduce a meaningful downmix of the original five channels from the MPEG-2 audio bitstream.

While forward compatibility is not as hard to achieve, backward compatibility is a much more

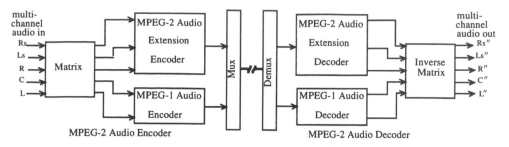

Fig. 2.4 A generalized codec for MPEG-2 Backward Compatible (BC) Multichannel Audio Coding.

difficult and requires some compromise in coding efficiency. However, backward compatibility does offer the possibility of migration to multichannel audio without making existing MPEG-1 stereo decoders obsolete.

Initially, the primary requirements identified for MPEG-2 Audio were multichannel coding and compatibility with MPEG-1 Audio. Candidate coding schemes were invited via a "Call for Proposals." In response, the various proposed solutions were found to be very close to each other, and the collaborative phase was begun without the need for competitive tests.

A generalized codec[24] structure illustrating MPEG-2 Multichannel Audio coding is shown in Fig. 2.4. A Multichannel Audio consisting of five signals, left (L), center (C), right (R), left surround (Ls), and right surround (Rs) is shown undergoing conversion by use of a Matrix operation resulting in five converted signals, two of which are encoded by an MPEG-1 Audio Encoder to provide compatibility with the MPEG-1 standard. The remaining three signals are encoded by an MPEG-2 Audio Extension Encoder. The resulting bitstreams from the two encoders are multiplexed in Mux for storage or transmission.

Since it is possible to have coded MPEG-2 Audio without coded MPEG Video, a generalized multiplexer Mux is shown. However, in an MPEG Audio–Visual System, the specific Mux and Demux used would be MPEG-2 Sys Mux and MPEG-2 Sys Demux. At the decoder, an MPEG-1 Audio Decoder decodes the bitstream input to it by Sys Demux and produces two decoded audio signals; the other three audio signals are decoded by an MPEG-2 Audio Extension Decoder. The

decoded audio signals are reconverted back to original domain by using an Inverse Matrix and represent approximated values indicated by L'', C'', R'', Ls'', Rs''.

To verify the syntax two type of tests were conducted: first, bitstreams produced by software encoding were decoded in non-real-time by software decoders, and second, bitstreams produced by software encoders were decoded in real-time. Tests were also conducted comparing the performance of MPEG-2 Audio coders that maintain compatibility with MPEG-1 to that which are not backward compatible. It has been found that for the same bitrate, compatibility does impose a quality loss which may be difficult to justify in some applications. Hence, it has been found necessary to also include a nonbackward compatible coding solution, which is referred to as MPEG-2 Part 7-NBC Audio; further elaboration regarding this work, which is currently in progress, is provided in our discussion on part 7.

Both the MPEG-1 and MPEG-2 Audio coding is discussed in much more detail in Chapter 4.

2.2.4 MPEG-2 Conformance

This part of MPEG-2, part 4, specifies Conformance[7] to the MPEG-2 standard; it is also called Compliance. It specifies how tests can be designed to verify whether bitstreams and decoders meet the requirements as specified in parts 1, 2, and 3. These tests can be used for a variety of purposes, for example, manufacturers of encoders and their customers can ensure whether an encoder produces valid bitstreams, manufacturers of decoders and their customers can verify if decoders meet requirements, and applications users can use tests to verify whether characteristics of a bitstream meet the application requirements.

Since MPEG-2, like MPEG-1, does not specify an encoder, and for that matter not even a specific decoder implementation but only bitstream syntax and decoding semantics, it is essential to know numerical bounds that various interim computations in the decoding process must satisfy so that it can be assured that the same bitstream decoded on different decoders will provide nearly identical results. If certain numerical bounds in interim processing calculations are not met, decoding on different decoders may produce not only slightly different results, but worse, certain errors in calculations may accumulate over time, eventually resulting in significant degradation in quality. This part of the standard therefore specifies bounds to be met in various steps of video and audio decoding so that errors do not accumulate. Being able to identify a range of numerical bounds on every calculation that goes on in an audio and video decoding process in real hardware is a challenging task since it implicitly requires a knowledge of types of errors and causes of such errors in terms of hardware operations that occur. Furthermore, it is difficult to be ever sure that all potential restrictions on error accumulations have been identified. Nevertheless, MPEG-2 technology has undertaken a significant effort to devise conformance bounds and to design a suite of test bitstreams that have been generated with great care to illustrate potential difficulties in conformance of a decoder. These bitstreams are referred to as "evil bitstreams," and as its name suggests are designed to test conformance of decoders to the MPEG-2 standard almost to the breaking point.

Since part 4 is intended to provide conformance for all the three primary parts of MPEG-2—Systems, Video, and Audio—it provides bitstream characteristics, bitstream tests, decoder characteristics, and decoder tests for these three parts.

In Fig. 2.5 we show an example of a setup for testing video decoder compliance. It is assumed that a Reference Decoder exists that satisfies the criterion for conformance to the MPEG-2 standard. Although ideally conformance testing should be possible on every type of decoder, MPEG-2 specification concerns only the decoding process from a bitstream of a digital output. For example, in testing MPEG-2 Video conformance for decoders supporting MPEG-1 like chrominance resolution (4:2:0), it is expected that the digital output from the Decoder under Test be 4:2:0 for comparison with the Reference Decoder providing a 4:2:0 digital output. In fact the proposed procedure can also be used for decoder tests when the digital output is 4:2:2 (as may be the case when higher chroma resolution is obtained by upsampling 4:2:0 for display); however, the results on the 4:2:2 signal need to be carefully interpreted (assuming that 4:2:0 resolution is coded in the bitstream) since chroma conversion from 4:2:0 to 4:2:2 is outside of the scope of the standard.

The Statistical Analysis operation generates a report on the difference file for each conformance bitstream input. The report consists of first, a histogram of pel differences per field between Reference Decoder and Decoder under Test and second, the number and addresses of erroneous macroblocks and the erroneous component in each erroneous macroblock. Furthermore, if more than 90% of macroblocks are declared erroneous in a picture for a specific bitstream, a complete break-

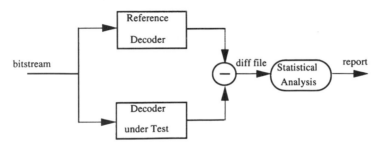

Fig. 2.5   Setup for testing Decoder Compliance.

down is said to have occurred. Other types of breakdown are also identified such as a short breakdown (the decoder does not decode the end of picture and resynchronizes on the next picture) and long breakdown (the decoder loses the information of a picture and resynchronizes on the next I-picture, next GOP, or next sequence startcode). In case of probable breakdown on a conformance bitstream a new stream is created by concatenation of an easy bitstream followed by the conformance bitstream followed by an easy bitstream. A visual test is then conducted to determine if concatenated parts are visually different and if the decoder restarts after a breakdown.

Besides the tests for video decoders, a number of tests are also specified for audio decoders. A suite of audio test sequences has also been standardized and covers various layer decoders. Testing can be done as for video by comparing the output of a Decoder under Test with the output of a Reference Decoder. To be called an MPEG-2 Audio decoder, the decoder shall provide an output such that the root-mean-square level of difference signal between the output of the Decoder under Test and the supplied Reference Decoder output is less than a specified amount for a supplied sine sweep signal. In addition, the difference signal magnitude must have less than a maximum absolute value. Further, to be called a limited accuracy MPEG-2 Audio decoder, a less strict limit is placed on the root-mean-square level of difference signal.

This part is expected to achieve the final status of International Standard by the end of the first quarter of 1996.

2.2.5 Software

This part of MPEG-2, part 5, is referred to as Software Simulation.[8] As the name suggests, it consists of software simulating Systems, Video, and Audio, the three principal parts of the MPEG-2 standard.

The MPEG-2 Systems Software consists of software for encoding and decoding of both the Program and the Transport Streams.

The MPEG-2 Video Software mainly consists of software for encoding and decoding of nonscalable video. In addition, it also includes some software for scalable encoding and decoding.

The MPEG-2 Audio Software consists of software for encoding and decoding. Two different psychoaccoustical models are included in the encoder.

MPEG-2 Software has in early 1996 achieved the final status of International Standard. The source code included in this part of MPEG-2 is copyrighted by the ISO.

2.2.6 DSM-CC

Coded MPEG-2 bitstreams may typically be stored in a variety of digital storage media such as CD-ROM, magnetic tape and disks, optical disks, magneto-optical disks, and others. This presents a problem for users wishing to access coded MPEG-2 data stored on a Digital Storage Media (DSM), since each DSM usually has its own control command language, requiring the user to know many such languages. Moreover, the DSM may be either local to the user or at a remote location. When remote, a common mechanism to access various DSM over a network is needed; otherwise the type of DSM being accessed has to be conveyed to the user. This, however, may not even be known in many cases. A solution to these two problems is offered by MPEG-2 DSM-CC.[1,9,10]

MPEG-2 DSM-CC is a set of generic control commands independent of the type of DSM. The DSM-CC is intended for MPEG applications that need to access a variety of DSM without the need to know details of each DSM. Thus, to allow a set of basic functions specific to MPEG bitstreams on a DSM, control commands are defined as a specific application protocol. The resulting control commands do not depend on the type of the DSM or whether a DSM is local or remote, the network transmission protocol, or the operating system with which it is interfacing. The control command functions can be performed on MPEG-1 systems bitstreams, MPEG-2 Program streams, or MPEG-2 Transport Streams. Some example functions are connection, playback, storage, editing, remultiplexing etc. A basic set of control commands is also included as an informative (nonmandatory) annex in MPEG-2 Systems, part 1.

Depending on the requirements of an application and the complexity that can be handled, DSM control commands can be divided into two categories. In the first category are a set of very basic operations such as selection of stream, play, and store commands. The stream selection command is used to request a specific bitstream and a specific operation mode on that bitstream. The play command is used to request playing of a selected bitstream and involves specification of speed and direction of play to accommodate various trick modes such as fast forward, fast reverse, pause, resume, step through, or stop. The store command is used to request recording of a bitstream on DSM.

The second category consists of a set of more advanced operations such as multiuser mode, session reservation, server capability information, directory information, bitstream editing, and others. With the multiuser mode command, more than one user is allowed to access the same server within a session. With the session reservation command, a user can request a server for a session at a later time. The server capability information command allows the user to be notified of the capabilities of the server; example capabilities include playback, fast forward, fast reverse, slow motion, storage, demultiplex, remultiplex, and others. The directory information command allows user access to information about directory structure and specific attributes of a bitstream such as bitstream type, IDs, sizes, bitrate, entry points for random access, program descriptors, and others. Typically, all this information may not be available through an Application Program Interface (API). Bitstream editing functions allow creation of new bitstreams by insertion of a portion of one bitstream into another bitstream, deletion of a portion of a bitstream, and correctly concatenating portions of various bitstreams.

In Fig. 2.6 we show a simplified relationship of DSM-CC with MHEG (Multimedia and Hypermedia Experts Group standard) Scripting Language, and DSM-CC. The MHEG standard is basically an interchange format for multimedia objects between applications. MHEG specifies a class set that can be used to specify objects containing monomedia information, relationships between objects, dynamic behavior between objects, and

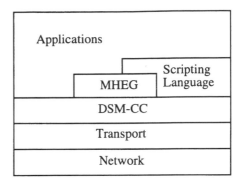

Fig. 2.6 DSM-CC centric View of MHEG, Scripting Language, and Network.

information to optimize real-time handling of objects. MHEG's classes include the content class, composite class, link class, action class, script class, descriptor class, container class, and result class. Furthermore, the MHEG standard does not define an API for handling of objects on its classes nor does it define methods on its classes. Finally, one of the important things to note is that MHEG explicitly supports scripting languages through the use of a Script Class; however, it does not specify a scripting language of its own.

Returning to our discussion on DSM-CC and its relationship to MHEG, applications may access DSM-CC either directly or through an MHEG layer. Moreover, scripting languages may be supported through an MHEG layer on top of DSM-CC. DSM-CC protocols form a layer higher than Transport Protocols. Examples of Transport Protocols are TCP, UDP, and MPEG-2 Transport Stream/Program Streams.

DSM-CC provides access for general applications, MHEG applications, and scripting languages to primitives for establishing or deleting network connections using User-Network (U–N) Primitives and communication between a Client and a Server across a network using User–User (U–U) Primitives. U–U operations may use a Remote Procedure Call (RPC) protocol. Both U–U and U–N operations may employ a message passing scheme that involves a sequence of bit-pattern exchanges.

In Fig. 2.7 we show such a stack of DSM protocols consisting of primitives allowing communications between User–User and User–Network.

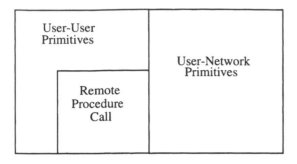

Fig. 2.7 DSM-CC protocol stack.

Next in Fig. 2.8 we clarify DSM-CC User–Network and User–User interaction. A Client can connect to a Server either directly via a Network or through a Resource Manager located within the Network.

A Client setup typically is expected to include a Session Gateway, which is a User–Network interface point, and a library of DSM-CC routines, which is a User–User interface point. A Server setup typically consists of a Session Gateway, which is a User–Network interface point, and a Service Gateway, which is a User–User interface point. Depending on the requirements of an application both a User–User connection and a User–Network connection can be established.

DSM-CC may be carried as a stream within an MPEG-1 Systems Stream, an MPEG-2 Transport Stream, or an MPEG-2 Program Stream. Alternatively, DSM-CC can also be carried over other transport networks such as TCP or UDP Internet Protocols.

DSM-CC is expected to achieve the final status of International Standard by mid 1996.

2.2.7 MPEG-2 NBC Audio

Earlier, while discussing part 3 of MPEG-2, we mentioned that MPEG-2 audio comes in two parts, one that is backward compatible with MPEG-1 (part 3), and one that is not. As you may have guessed, this part, part 7, is not backward compatible with MPEG-1 and is thus referred to as Non Backward Compatible (NBC) Audio. Thus this part of MPEG-2 is intended for applications that do not require compatibility with MPEG-1 stereo and thus do not need to compromise on coding efficiency at all.

MPEG-2 NBC Audio is expected to support the sampling rates, audio bandwidth, and channel configurations of MPEG-2 Audio, but will operate at bitrates from 32 kbit/s to the bitrate required for high-quality audio. This work item was started late, around the time when work for part 3 was in an advanced stage, and so this work item is currently still in progress.[26,30] It is expected to achieve a draft standard status by the end of 1996.

The results from NBC Audio tests[33] indicate that notably better quality is obtained by nonbackward compatible coding as compared to backward compatible coding. About 10 organizations participated in NBC Audio competitive tests. A collaborative phase is currently in progress, including the definition of an initial Reference Model that is the framework for conducting core experiments. It is

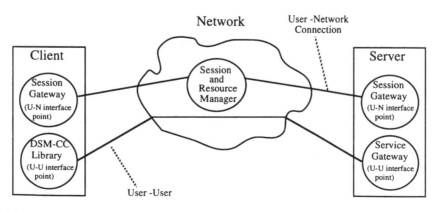

Fig. 2.8 DSM-CC User–Network and User–User interaction.

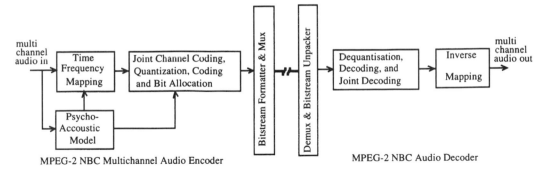

Fig. 2.9   A generalized Codec for MPEG-2 Non Backward Compatible (NBC) Multichannel Audio Coding.

expected to undergo several iterations before finally achieving convergence.

In Fig. 2.9 we show a simplified Reference Model configuration of the NBC Audio coder under consideration. We now briefly introduce the structure employed without going into specific details.

Multichannel audio undergoes conversion of domain in Time-Frequency Mapping, whose output is subject to various operations currently being optimized via experiments such as Joint Channel Coding, Quantization, Coding, and Bit Allocation. A psychoacoustical model is employed at the encoder. A Bitstream Formatter generates the bitstream for storage or transmission. At the decoder, an inverse operation is performed by a Bitstream Unpacker, which is followed by Dequantization, Decoding, and Joint Decoding. Finally, an Inverse Mapping is performed to go back from frequency domain to time domain representation, resulting in the reconstructed multichannel audio output.

NBC Audio is expected to achieve the mature stage of Committee Draft by mid 1996 and the final status of International Standard by the end of the first quarter of 1997.

2.2.8   10-Bit Video (Cancelled!)

Earlier, during the discussion of part 2, we had mentioned that when MPEG-2 video work was close to completion, it was envisaged that the MPEG-2 video would be useful in professional applications because of its ability to achieve high quality. The two different types of requirements for these applications were found to be higher spatial resolution for chrominance (4:2:2 format) and higher amplitude resolution for luminance and chrominance, called the 10-bit video. In 10-bit video, as the name suggests, each video sample is digitized to 10-bit accuracy rather than the normal 8-bit accuracy. Specific applications of 10-bit video are production of high-quality video and motion picture material, archiving, specialized visual effects, and efficient transmission and distribution. Although there was sufficient initial interest[29] in this part of MPEG-2, the work on this part had to be abandoned. This was partly due to insufficient availability of systems for capture, storage, processing, and display of 10-bit video. Another reason had to do with limitations in resources of organizations participating in MPEG-2 owing to various continuing work items in parallel. Moreover, some organizations felt that at least for the near term, applications of 10-bit video were too specialized and lacked a broad base in consumer markets.

This work item has now officially been removed from the MPEG-2 program of work.

2.2.9   MPEG-2 RTI

This part of MPEG-2, part 9,[11] is related to MPEG-2 Systems, part 1, and is primarily intended for applications where the MPEG-2 System may be used for real-time interchange of data such as in telecommunications applications. It specifies a Real-Time Interface (RTI) between Channel Adapters and MPEG-2 Systems Transport Stream decoders. The RTI makes it possible to specify network interconnects and to verify the performance of various products that include MPEG decoders.

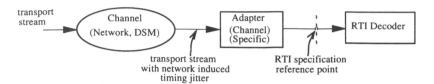

Fig. 2.10 Configuration for illustrating Real-Time Interface (RTI).

Such interconnects may be specified for networks such as LAN, ATM, and others by other organizations and standards bodies.

Thus with the specification of an RTI, it will be possible to build network adaptation layers that guarantee required timing performance and also guarantee that decoders will have correct behavior in terms of buffers and timing recovery. The RTI is an optional supplement to the System's Transport Stream System Target Decoder (STD) model. In Fig. 2.10 we show an example configuration for illustrating the real-time interface specification.

In this example configuration, a transport stream generated by an MPEG encoder passes over a channel such as a network and through a channel adapter on its way to an RTI decoder. The transport stream in going through the network is assumed to suffer from some timing jitter. Then the reference point at which the RTI specification really applies is between the channel adapter and the RTI decoder. The exact value of acceptable jitter depends on the application.

It is expected that RTI will achieve the final stage of International Standard by the end of the first quarter of 1996.

2.2.10 MPEG-2 DSM-CC Conformance

This part of MPEG-2, part 10, is a relatively new part and is related to part 6, called DSM-CC. Part 10 of MPEG-2 was originally aimed to be an exchange format for multimedia scripts; it was then referred to as Reference Scripting format (RSF) or more specifically the DSM-RSF, DSM-RSF was expected to be closely linked to the Multimedia and Hypermedia Experts Group (MHEG) standard which mainly deals with presentation aspects of coded audio-visual data. The MHEG standard, although it does not explicitly include a

scripting language, does implicitly support such language.

With DSM-CC, it was envisaged[31] to be possible to code a script controlling presentation aspect of the audio-visual data along with the coded representation (MPEG bitstreams) of the audio-visual data. DSM-RSF was thus expected to play the role of a specialized multimedia scripting language. Although many such languages exist, it was felt that DSM-RSF was necessary to take advantage of special features of MPEG Audio and Video coding as well as that of DSM-CC.

The need for this part of the MPEG-2 standard arose during the DSM-CC work and thus this part of the standard was started late. During a recent (March '96) MPEG meeting, this part was renamed from DSM-RSF to DSM-CC conformance. Thus the original goal of this part has been revised and thus this part will mainly provide a means to verify if a specific DSM-CC implementation is conformant to part 6 MPEG-2 or not. Part 10 is expected to achieve the stage of Committee Draft by July 1996 and the final status of International Standard by March 1997.

2.3 SUMMARY

In this chapter we have presented a brief overview of the various parts of the MPEG-2 standard and shown the relationship of parts of MPEG-2 to parts of MPEG-1 where appropriate. As the reader may have guessed, from among the various parts of the MPEG-2 standard, this book primarily focuses on MPEG-2 Video. In the following chapters we will certainly be discussing coding algorithms, syntax, semantics, profiles, and implementation aspects of MPEG-2 Video. However, since we believe that any discussion about an MPEG standard would be incomplete without a working knowledge of MPEG Audio and MPEG Systems, we include one chapter on MPEG-2 Audio and another on MPEG-2 Systems. We will then introduce example applications of MPEG-2 and discuss characteristics of transmission channels and storage media that may be used to transmit or store MPEG compressed audio and video.

We now summarize a few key points discussed in this chapter.

- Systems, Video, and Audio are three primary parts of the MPEG-1 and the MPEG-2 standards. Conformance and Software are two other parts of the MPEG-1 and MPEG-2 standards. The MPEG-2 standard has four other parts: Digital Storage Media-Control and Comand (DSM-CC), Non Backward Compatible Audio (NBC Audio), Real-Time Interface (RTI), and DSM-CC Conformance. Another part of MPEG-2 called 10-bit video has been dropped.

- MPEG-2 Systems comes in two forms. The first one, called the Program Stream, is MPEG 1 like and intended for error-free media. The second one, called the Transport Stream, is intended for noisier environments such as terrestrial and satellite channels.

- MPEG-2 Video is a syntactic superset of MPEG-1 Video. It is based on motion compensated DCT coding, B-pictures, and a Group-of-Pictures structure as in MPEG-1 Video. Furthermore, MPEG-2 Video supports tools for efficient coding of interlace. MPEG-2 Video also allows scalable video coding. Compatibility is a specific form of scalability that can allow interoperability with MPEG-1 Video. From an applications standpoint, MPEG-2 video is organized as profiles, where each profile contains a subset of all the coding tools of MPEG-2 Video.

- MPEG-2 Audio comes as two parts. The first is Backward Compatible (BC) with MPEG-1, and the second is Non Backward Compatible (NBC) with MPEG-1. MPEG-2 BC Audio allows coding of multichannel (such as five-channel) audio in a backward compatible manner so that from an MPEG-2 bitstream an MPEG-1 stereo decoder can reproduce a meaningful downmix of five channels to deliver correct sounding stereo. On the other hand, MPEG-2 NBC Audio achieves notably higher coding performance as it does not impose the restriction of compatibility with MPEG-1; this part of MPEG-2 was started late and is currently in progress.

- MPEG-2 Conformance specifies tests to verify compliance of bitstreams and decoders claiming to meet the requirements specified for MPEG-2 Systems, Video, and Audio. MPEG-2 Software, also referred to as Technical Report, is a C-programming implementation of the MPEG-2 Systems, Video, and Audio standards. The implementation includes a limited encoder and a full decoder for all three parts. This software is copyrighted by ISO.

- MPEG-2 DSM-CC specifies protocols to manage interaction of users with stored MPEG-1 and MPEG-2 bitstreams. A very basic type of interaction consists of actions of DSM in response to commands by a user; Annex A in MPEG-2 Systems supports such single user, single DSM applications. The DSM-CC, however, includes more extensive protocols to extend the basic idea of interaction to heterogeneous networked environments and interactive-video applications. Another part of MPEG-2 that is related to DSM is currently ongoing and is referred to as MPEG-2 DSM-CC Conformance. This part was originally intended to specify formats for Reference Scripts and was expected to be related to Multimedia and Hypermedia Experts Group Standard (MHEG) which deals with presentation aspects. However, it is now expected to deal mainly with compliance aspects of MPEG-2 DSM-CC.

- MPEG-2 RTI imposes some restrictions on MPEG-2 Systems and in particular Transport Streams. It specifies timing of the real-time delivery of bytes of the Transport Stream as a Real-Time Interface. This part does not override any of the requirements of MPEG-2 Systems, and compliance to this part of the standard is optional. One of the important parameters for RTI specification is the network induced timing jitter, which may depend on the application.

REFERENCES

1. "Coding of Moving Pictures and Associated Audio for Digital Storage Media at up to about 1.5

Mbit/s," *ISO/IEC 11172–1:Systems* (November 1991).

2. "Coding of Moving Pictures and Associated Audio for Digital Storage Media at up to about 1.5 Mbit/s," *ISO/IEC 11172–2: Video* (November 1991).

3. "Coding of Moving Pictures and Associated Audio for Digital Storage Media at up to about 1.5 Mbit/s," *ISO/IEC 11172–3: Audio* (November 1991).

4. "Generic Coding of Moving Pictures and Associated Audio Information: Systems," ISO/IEC 13818–1 : Draft International Standard (November 1994).

5. "Generic Coding of Moving Pictures and Associated Audio Information: Video," ISO/IEC 13818–2 : Draft International Standard (November 1994).

6. "Generic Coding of Moving Pictures and Associated Audio Information: Audio," ISO/IEC 13818–3 : International Standard (November 1994).

7. CONFORMANCE EDITING COMMITTEE, "Generic Coding of Moving Pictures and Associated Audio: ISO/IEC 13818–4: Conformance," *ISO/IEC/ JTC1/SC29/WG11 N0742* (July 1994).

8. "Generic Coding of Moving Pictures and Associated Audio Information: Software," ISO/IEC 13818–5: Draft Technical Report, March 1995.

9. S. J. HUANG, "MPEG Digital Storage Media (DSM) Control Command," *Signal Proc. Image Commun.*, pp. 521–524, (February 1995).

10. M. S. GOLDMAN, D. F. Hooper, and C. Adams, "ISO/IEC 13818–6 MPEG-2 Digital Storage Media Command & Control (DSM-CC) Tutorial," Tokyo (July 1995).

11. SYSTEMS EDITING COMMITTEE, "Generic Coding of Moving Pictures and Associated Audio: Real-Time Interface Specification," *ISO/IEC/JTC1/SC29/ WG11 N0809* (November 1994).

12. A. G. MACINNIS, "The MPEG Systems Coding Specification," *Signal Proc. Image Commun.* pp. 153–159, (April 1992).

13. VIDEO SIMULATION MODEL EDITING COMMITTEE, "MPEG Video Simulation Model 3 (SM3)" (July 1990).

14. D. J. LE GALL, "MPEG: A Video Compression Standard for Multimedia Applications," *Commun. ACM* 34:47–58 (April 1991).

15. A. PURI, "Video Coding Using the MPEG-1 Compression Standard," *Society for Information Display: International Symposium*, pp. 123–126, Boston (May 1992).

16. D. PAN, "An Overview of the MPEG/Audio Compression Algorithm," *Proc. SPIE Digital Video Compression on Personal Computers: Algorithms and Technologies*," 2187:260–273, San Jose (February 1994).

17. A. G. MACINNIS, "MPEG-2 Systems," *Proc. SPIE Digital Video Compression on Personal Computers: Algorithms and Technologies*, 2187:274–278, San Jose (February 1994).

18. C. E. HOLBOROW, "MPEG-2 Systems: A Standard Packet Multiplex Format for Cable Digital Services," *Proc. Society of Cable TV Engineers–Conference on Emerging Technologies*, Phoenix (January 1994).

19. TEST MODEL EDITING COMMITTEE, "Test Model 5," *ISO/IEC JTC1/S29/WG11/N0400* (April 1993).

20. D. J. LE GALL, "MPEG Video: The Second Phase of Work," *Society for Information Display: International Symposium*, pp. 113–116, Boston (May 1992).

21. A. PURI, "Video Coding Using the MPEG-2 Compression Standard," *Proc. SPIE Visual Communication and Image Processing*, Boston (November 1993).

22. M. M. STOJANCIC and C. NGAI, "Architecture and VLSI Implementation of the MPEG-2:MP@ML Video Decoding Process," *SMPTE* J. (February 1995).

23. D. PAN, "A Tutorial on MPEG/Audio Compression," *IEEE Multimedia*, pp. 60–74 (Summer 1995).

24. P. NOLL, "Digital Audio Coding for Visual Communications," *Proceedings of the IEEE*, 83 (6) (June 1995).

25. L. CHIARIGLIONE, "Development of Multi-Industry Information Technology Standards: The MPEG Case," *Proceedings of the International Workshop on HDTV'93*, pp. 55–64 (October 1993).

26. ADHOC GROUP ON NBC SYNTAX DEVELOPMENT, "Report of Adhoc Group on NBC Syntax Development," *ISO/IEC/JTC1/SC29/WG11 MPEG96/ 0626* (January 1996).

27. A. N. NETRAVALI and A. LIPPMAN, "Digital Television: A Perspective," *Proceedings of the IEEE*, 83 (6) (June 1995).

28. L. CHIARIGLIONE, "The Development of an Integrated Audiovisual Coding Standard: MPEG," *Proceedings of the IEEE*, 83, (2) (February 1995).

29. A. T. ERDEM and M. I. SEZAN, "Compression of 10-bit Video Using Tools of MPEG-2," *Signal Proc. Image Commun.* pp. 27–56, (March 1995).

30. L. CHIARIGLIONE, "Report of the 30th Meeting," *ISO/IEC/JTC1/SC29/WG11   N0890* (March 1995).

31. C. READER, "MPEG Update, Where It's at, Where It's Going," *19th International Broadcaster's Symposium,* pp. 117–127 (June 1995).

32. VIDEO SUBGROUP, "Proposed Draft Amendment No. 3 to 13818–2 (Multi-view Profile)," *ISO/IEC JTC1/SC29/WG11 Doc. N1088* (November 1995).

33. AUDIO SUBGROUP, "NBC Time/Frequency Module Subjective Tests: Overall Results," *ISO/IEC/JTC1/SC29/WG11 N0973,* Tokyo (July 1995).

34. United States Advanced Television Systems Committee, "Digital Television Standard for HDTV Transmission," ATSC Standard (April 1995).

3

MPEG-2 Systems

Part 1 of ISO 13818 is called *Systems*. As we shall see, this term has many meanings and applies to many different contexts. In fact, Systems is the framework that allows MPEG-2 to be used in an extraordinary number of applications. For example, audio may be sent in several languages. Video may be sent at several resolutions and bitrates. Multiple views or 3D stereoscopic video may be sent that is completely compatible with ordinary MPEG-2 video.

3.1 MULTIPLEXING

The main function of MPEG-2 Systems is to provide a means of combining, or *multiplexing*, several types of multimedia information into one stream that can be either transmitted on a single communication channel or stored in one file of a DSM.* Since MPEG data are always byte aligned, the bitstreams are actually bytestreams, and we refer to them simply as *streams*.

There are two methods in wide use for multiplexing data from several sources into a single stream. One is called Time Division Multiplexing (TDM), which basically assigns periodic time slots to each of the elementary streams of audio, video,

data, etc. H.320 uses TDM in its multiplexing standard, which is called H.221.

MPEG-1 Systems[1] and MPEG-2 Systems use the other common method of multiplexing, called *packet multiplexing*. With packet multiplexing, data packets from the several elementary streams of audio, video, data, etc. are interleaved† one after the other into a single *MPEG-2 stream*, as shown in Fig. 3.1. Elementary streams and MPEG-2 streams can be sent either at constant bitrate (CBR) or at variable bitrates (VBR) simply by varying the lengths or the frequency of the packets appropriately. In particular, elementary streams can be sent as VBR, even though the multiplexed MPEG-2 stream itself is transmitted as CBR. This enables some elementary streams that temporarily do not require very many bits to give up their proportionate channel capacity in favor of other elementary streams that temporarily require more than their share of the overall channel capacity. This functionality is called *statistical multiplexing*.

3.2 SYNCHRONIZATION

With multimedia information, there is a strong requirement to maintain proper synchronization

---

*Digital Storage Medium.

†Packet interleaving may or may not be periodic.

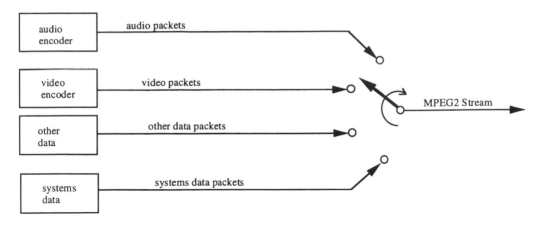

Fig. 3.1 Packet Multiplexing. At the multiplexer, packets from several elementary streams of audio, video, data, and so forth are multiplexed into a single MPEG-2 stream.

between the elementary streams when they are decoded and passed to their various output displays or transducer devices.

With TDM, the delay from encoder to decoder is usually fixed. However, in packetized systems the delay may vary owing both to the variation in packet lengths and frequencies during multiplexing. Moreover, if audio and video are prepared and edited separately prior to multiplexing, some means for restoring proper synchronization must be provided. For both MPEG-1 and MPEG-2, synchronization is achieved through the use of *Time Stamps* and *Clock References*. Time Stamps are 33 bit data fields indicating the appropriate time, according to a *Systems Time Clock* (STC) that a particular *Presentation Unit* (video pictures, audio frames, etc.) should be decoded by the decoder and presented to the output device. Clock References are 42-bit data fields indicating to the demultiplexer what the STC time should be when each clock reference is received.

There are two types of time stamps. A *Decoding Time Stamp* (DTS) indicates when the associated Presentation Unit* is to be decoded. A *Presentation Time Stamp* (PTS) indicates when the decoded Presentation Unit is to be passed to the output device for display. If the decoding and presentation time

are the same[†] for a particular Presentation Unit, then either no time stamp is sent or only the PTS is sent. If no time stamp is sent, then the presentation (and decoding) time is extrapolated based on the presentation time for the previous Presentation Unit and the known sampling rate for that source.

Synchronization and decoding requires buffers at the decoder to temporarily store the compressed data until the correct time of decoding, as shown in Fig. 3.2. When the decoder STC advances to the time specified by the DTS, the corresponding Presentation Unit is removed from the buffer and passed to the decoder. If the PTS is later than the DTS, the decoded Presentation Unit waits inside the decoder until the STC advances to the PTS time.[‡] Otherwise, the decoded Presentation Unit is presented to the display immediately. It is the responsibility of encoders and multiplexers to ensure that data are not lost due to buffer overflow at the receiver.

3.3 ELEMENTARY STREAMS

Elementary streams consist of compressed data from a single source (e.g., audio, video, data, etc.)

---

*MPEG defines *Access Units* for DTS timing. Here we use the term Presentation Unit to apply to both Presentation and Access Units since the difference is minor.

[†]PTS and DTS normally differ only in the case of video.

[‡]Practical MPEG receivers may add a delay to the DTS and PTS values for implementation purposes, in which case buffer sizes must be increased.

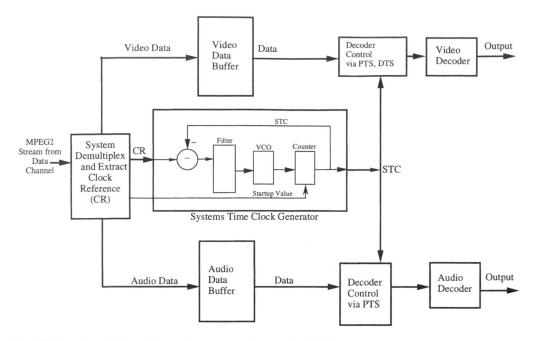

Fig. 3.2   MPEG receiver. A demultiplexer switch separates the incoming MPEG-2 stream into its component elementary streams. Compressed elementary stream Presentation Units are stored in buffers, where they wait until the proper time for decoding, as specified by a decoding time stamp (DTS) and the System Time Clock (STC). The STC generator is a simple phase locked loop (PLL) that tracks incoming *Clock References* (CRs). PLLs and CRs will be described later.

plus ancillary data needed for synchronization, identification, and characterization of the source information. The elementary streams themselves are first packetized into either constant-length or variable-length packets to form a *Packetized Elementary Stream* (PES). Each PES packet* consists of a header followed by stream data called the *payload*.

The PES header begins with one of 66 possible PES Start-Codes. MPEG-2 also has many other types of Start-Codes that are used for delineation of various other types of data. All MPEG-2 Start-Codes begin with a **start_code_prefix** (Psc), which is a string of 23 or more[†] binary 0s, followed by a binary 1, followed by an 8-bit *Start-Code ID* that is usually specified in hexadecimal.[‡] Furthermore, the audio and video compressed data are designed to rarely if ever contain 23 consecutive

zeros, thus easing random access to Start-Codes as well as speeding recovery after the occurrence of transmission errors.

In PES Start-Codes, the 8-bit Start-Code ID is called the **stream_id.** It ranges in value from 0xBD to 0xFE and serves to label the stream, as well as to specify the type of the stream, as shown in Table 3.1. Different stream_ids may be used to distinguish between separate streams of the same type.

Following the stream_id in the PES header are **PES_packet_length,** flags and indicators for various features and, finally, optional data whose presence is determined by the stream type as well as the values of the flags and indicators. The syntax of the PES header is shown in Fig. 3.3.

The semantics of the PES header variables[§] are as follows:

---

*MPEG also defines other *specialized* streams for various purposes, as we shall see later. They are also packetized, but they are not usually called PESs.

†More that 23 zeros may be used in video streams, if desired, for byte alignment or padding.

‡Hexadecimal numbers are denoted by the prefix "0x." For example, 0xBC = binary "1011 1100." Numbers without prefixes or quotes are decimal.

§Flag and indicator variables are 1 bit, unless otherwise indicated.

start_code_prefix 23 or more binary 0s, followed by a 1. This is the same for all MPEG Start-Codes.

stream_id (8 bits) Hexadecimal values ranging from 0xBD to 0xFE. Defines ID type according to Table 3.1. May also be used as an ID number for different streams of the same type.

PES_packet_length (16 bits) The number of bytes that follow. For Transport Stream packets carrying video, a value of 0 indicates an unspecified length.

padding_bytes Fixed 8-bit values equal to 0xFF, which are discarded by the decoder.

marker_bits Fixed valued bits usually equal to all binary 1s.

Table 3.1 MPEG-2 Systems Start-Code IDs and stream_ids. Stream_ids range from 0xBD to 0xFE*. Program Streams (PS) will be described later. Entitlement Control Messages (ECM) and Entitlement Management Messages (EMM) are part of the Conditional Access, that is, scrambling control.

Start-Code ID's and stream_id's (hexadecimal)	Data or Stream Type
B9	PS End Code
BA	PS Pack Header
BB	PS System Header
BC	PS Program Map Table (PS-PMT)
BD	private_stream_1
BE	padding_stream
BF	private_stream_2
C0 to DF	MPEG1 or MPEG2 audio
E0 to EF	MPEG1 or MPEG2 video
F0	ECM private
F1	EMM private
F2	DSM Command & Control
F4	ITU-T type A (defined by ITU-T)
F5	ITU-T type B
F6	ITU-T type C
F7	ITU-T type D
F8	ITU-T type E
F9	ancillary_stream
FA to FE	reserved streams
FF	PS Directory

*MPEG's definition of elementary stream is somewhat looser than ours. MPEG includes values 0xBC and 0xFF as stream_ids, whereas we prefer to call them Start-Code IDs, since they never occur in PES packets.

PES_scrambling_control 2 bits, indicates the scrambling mode, if any.

PES_priority 1 indicates the priority of this PES packet. 1 indicates high priority.

data_alignment_indicator 1 indicates the payload begins with a video Start-Code or audio syncword.

copyright 1 indicates the PES packet payload is copyrighted.

original_or_copy 1 indicates the PES packet payload is an original.

PTS_DTS_flags 2 bit field. Values of 2 and 3 indicate a PTS is present. 3 indicates a DTS is present. 0 means neither are present.

ESCR_flag 1 indicates a clock reference (called ESCR) is present in the PES packet header.

ES_rate_flag 1 indicates a bitrate value is present in the PES packet header.

DSM_trick_mode_flag 1 indicates the presence of an 8-bit trick mode field.

additional_copy_info_flag 1 indicates the presence of copyright information.

PES_CRC_flag 1 indicates a CRC[†] is present in the PES packet.

PES_extension_flag 1 indicates an extension of the PES packet header.

PES_header_data_length (8-bits) Specifies the total number of bytes following in the optional fields plus stuffing data, but not counting PES_packet_data. The presence or absence of optional fields is indicated by the above indicators.

PTS (presentation time stamp) Presentation times for presentation units.

DTS (decoding time stamp) Decoding times for presentation units. PTS and DTS can differ only for video.

ESCR Elementary stream clock reference for PES Streams (see below).

ES_rate (elementary stream rate) The bitrate of a PES Stream. Normally, ESCR and ES_rate are used only if a PES is transmitted all by itself, in which case it is called a PES Stream.

trick_mode_control Indicates which trick mode (fast forward, rewind, pause, etc.) is applied to the video.

[†]Cyclic Redundancy Check, a parity check for error detection purposes.

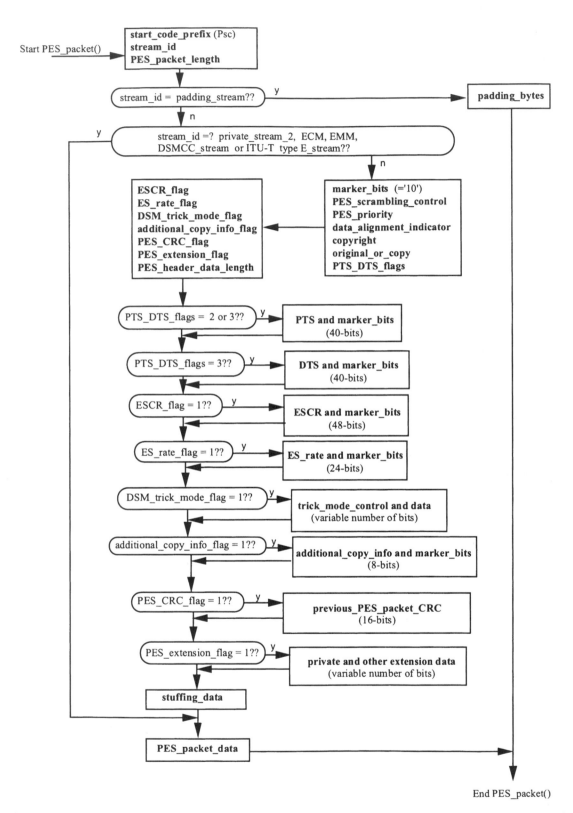

Fig. 3.3 Syntax of the PES-Packet header for several stream types. A bold face variable name means simply read in the value of the variable.

additional_copy_info Contains private data for copyright information.

previous_PES_packet_CRC (16 bits) Contains the CRC of the previous PES packet, exclusive of the PES packet header.

extension_data May include buffer sizes, sequence counters, bitrates, etc.

stuffing data Fixed 8-bit values equal to 0xFF that may be inserted by the encoder. It is discarded by the decoder. No more than 32 stuffing bytes are allowed in one PES header.

PES_packet_data Contiguous bytes of data from the elementary stream.

3.4 PROGRAMS AND NONELEMENTARY STREAMS

PES packets from various elementary streams are combined to form a *Program*. A program has its own STC, and all elementary streams in that program are synchronized using PTSs and DTSs referenced to that STC.

MPEG has defined three nonelementary streams for transmitting or storing programs. The first is called a *Program Stream* and is aimed at applications having negligible transmission errors. The second is called a *Transport Stream* and is aimed at applications having nonnegligible transmission errors. The third has the strange appellation *PES Stream** and is useful for certain operations in specific implementations; however, it is not recommended for transmission or storage because of error resilience and random access limitations.

3.4.1 Program Streams

A Program Stream carries one *Program* and is made up of *Packs* of multiplexed data. A Pack consists of a pack header followed by a variable number of multiplexed PES packets from the various elementary streams plus other descriptive data to be described below. In a PS, each elementary stream must have its own unique stream_id.

A Pack Header, shown in Fig. 3.4, consists of a *Pack Start-Code*, followed by a 42-bit *Systems Clock Reference* (SCR), which is used to reset the decoder STC, followed by the assumed PS bitrate (called *program_mux_rate*), followed finally by optional stuffing bytes.

Program Streams (PSs) are suitable for Digital Storage Media (DSM) or transmission networks that have either negligible or correctable transmission errors.

It is the responsibility of the multiplexer to guarantee that Program Streams are decodable by all standard MPEG decoders. For this to be possible, MPEG specifies certain rules for multiplexers in the form of a *Program Stream–System Target Decoder* (PS-STD). The PS-STD is a hypothetical program decoder, shown in Fig. 3.5, that is synchronized exactly with the encoder. It consists of a demultiplexer, system control, STC, and elementary stream buffers and decoders.

Bytes from the multiplexed Program Stream are fed to the PS-STD at the rate specified in the Pack header.† The demultiplexer switch then routes the PES packets to the appropriate decoder buffers or Systems control as indicated by the Start-Code IDs and stream_ids.

The STC is a 42-bit counter incrementing at a rate of 27MHz.‡ Each received SCR specifies the desired value of the STC at the exact arrival time of that SCR at the PS-STD. However, in real systems the SCR may have a small error. Thus, most practical program receivers will use a Phase Locked Loop (PLL), as shown in Fig. 3.2, to control their STCs. PLLs will be covered later in this chapter.

Time Stamps are used for PS-STD control. A coded Presentation Unit is removed from its buffer and passed to its decoder at the time specified by its DTS, which is either included in the stream or obtained from the PTS. The decoded Presentation Unit is then passed to the output device at the time specified by its PTS, which is either included in the stream or extrapolated from a past PTS.

The multiplexer must construct the Program Stream so that PS-STD buffers never overflow.

---

*Some members of MPEG deny the existence of PES Streams.

†There may be gaps between packs, during which no data is transferred.

‡A tolerance of ±810Hz is allowed in the STC.

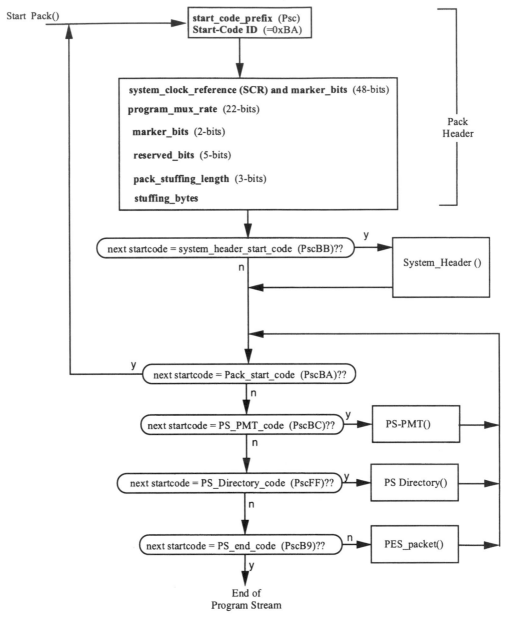

Fig. 3.4 Syntax of a Pack. A Program Stream (PS) is made up of Packs of multiplexed PES packets from the various elementary streams plus other descriptive data to be described later. Each pack begins with a pack header. The PS ends with Start-Code ID 0xB9.

Also, buffers are allowed to underflow only in certain prescribed situations. The most straightforward way to guarantee this is for the encoder/multiplexer to keep track of the timing and buffer fullness in each of the elementary stream decoders, as will be described later.

In addition to PES packets, PS Packs also contain other descriptive data called *PS Program Specific Information* (PS-PSI), which defines the program and its constituent parts. One specialized type of PS-PSI is called the *System Header,* which if present must directly follow the pack header. In the PS Sys-

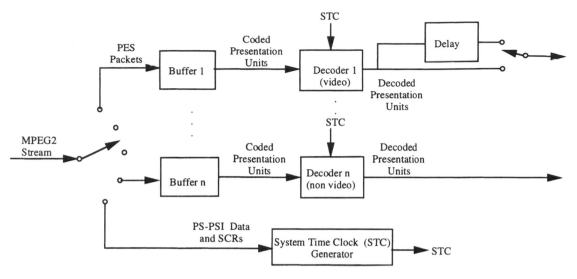

Fig. 3.5 Program Stream–System Target Decoder (PS-STD).

tem Header certain specifications are provided such as bitrate bound, buffer sizes for the various elementary streams, and so forth. The syntax of the PS System Header is shown in Fig. 3.6. Note that it is different than the PES format of Fig. 3.3.

The semantics of the PS System Header are as follows:

header_length (16 bits) The length in bytes of the following data.

rate_bound (22 bits) Upper bound on the program_mux_rate field.

audio_bound (6 bits) Upper bound on the number of active audio streams.

fixed_flag 1 indicates fixed bitrate for the MPEG-2 stream.

CSPS_flag 1 indicates constrained parameters (a set of limitations on bitrate and picture size).

system_audio_lock_flag Indicates the audio sampling is locked to the STC.

system_video_lock_flag Indicates the video sampling is locked to the STC.

video_bound (5 bits) Upper bound on the number of video streams.

packet_rate_restriction_flag Indicates packet rate is restricted.

stream_id Elementary stream ID for the following data.

P-STD_buffer_bound (14 bits) Upper bound on the PS-STD buffer size specified in the PES packet

extension_data. A practical decoder buffer size should exceed that specified by *P-STD_buffer_bound*.

Another specialized type of PS-PSI is called the *PS Program Map Table** (PS-PMT), shown in Fig. 3.7, which contains information about the program and its constituent elementary streams. The semantics for the PS-PMT are as follows:

program_map_table_length (16 bits) The number of bytes that follow.

current_next_indicator 1 indicates this PS-PMT is currently valid. (More correctly, it will be valid as soon as the final byte is received.) 0 indicates that it is not yet valid, but is about to be valid and is being sent for preparatory purposes only.

program_map_table_version (5 bits) Version number, which increments with each change.

program_stream_info_length (16 bits) Number of bytes in the program descriptors (to be described later).

program_descriptors( ) Descriptive information for this program (to be discussed later).

---

*MPEG uses the term "Program Stream Map." We use "PS Program Map Table" for clarity later in the chapter.

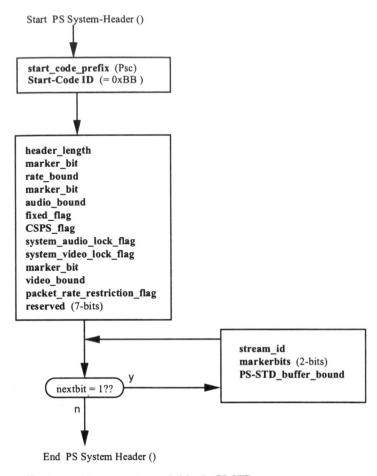

Fig. 3.6   The PS System Header provides parameters needed for the PS-STD.

elementary_stream_map_length (16 bits)   Total number of bytes in all of the elementary stream (ES) data that follows, not including CRC.

stream_type (8 bits)   Type of the elementary stream according to Table 3.2.

elementary stream_id (8 bits)   Each PS Elementary Stream requires a unique stream_id (see Table 3.1).

elementary_stream_info_length (16 bits)   Number of bytes in the ES descriptors for this ES.

ES_descriptors( )   Descriptive information for this ES (to be discussed later).

CRC_32 (32 bits)   CRC for the entire program stream map.

Another specialized type of PS-PSI is called the *Program Stream Directory*, which contains locational information for points of random access within a stored stream. Usually, it lists selected video I-pictures, their locations and sizes, plus time stamps. However, it can also be used for audio.

3.4.2 Transport Streams

Transport Streams (TSs) are defined for transmission networks that may suffer from occasional transmission errors. They are similar to Program Streams in that they consist of multiplexed data from Packetized Elementary Streams (PESs) plus other descriptive data. However, one more layer of packetization is added in the case of TS. In general, relatively long, variable-length PES packets are further packetized into shorter TS packets of length 188 bytes. A constant packet length makes error recovery easier and faster. Another distinc-

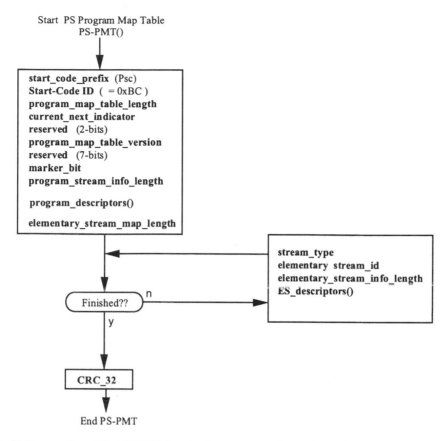

Start PS Program Map Table
PS-PMT()

start_code_prefix (Psc)
Start-Code ID (= 0xBC)
program_map_table_length
current_next_indicator
reserved (2-bits)
program_map_table_version
reserved (7-bits)
marker_bit
program_stream_info_length

program_descriptors()

elementary_stream_map_length

stream_type
elementary stream_id
elementary_stream_info_length
ES_descriptors()

Finished??   n   y

CRC_32

End PS-PMT

Fig. 3.7   The PS Program Map Table (PS-PMT) describes the program and its constituent elementary streams.

Table 3.2   Elementary stream types.

stream_type	Description
0x00	Reserved
0x01	MPEG1 Video
0x02	MPEG2 video
0x03	MPEG1 Audio
0x04	MPEG2 Audio
0x05	private_sections (see Fig. 3.11)
0x06	PES packets containing a private stream (see Fig. 3.3)
0x07	MHEG
0x08	DSM CC
0x09	ITU-T Rec. H.222.1
0x0A	ISO/IEC 13818–6 type A
0x0B	ISO/IEC 13818–6 type B
0x0C	ISO/IEC 13818–6 type C
0x0D	ISO/IEC 13818–6 type D
0x0E	auxiliary
0x0F–0x7F	Reserved
0x80–0xFF	User Private

tion between TS and PS is that TS can carry several programs, each with its own STC. Program Specific Information (TS-PSI) is also carried differently in TSs than in PSs.

Each TS packet consists of a TS Header, followed optionally by ancillary data called the *Adaptation Field,* followed typically by some or all of the data from one PES packet.* The TS Header consists of a sync byte, flags and indicators, Packet Identifier (PID), plus other information for error detection, timing, etc., as shown in Fig. 3.8. PIDs are used to distinguish between elementary streams. Thus, in a Transport Stream, elementary streams need not have unique **stream_id** values.

The sematics for the TS packet are as follows:

sync_byte   (8 bits) A fixed value 0x47.

transport_error_indicator   1 indicates an uncorrectable bit error exists in the current TS packet.

*Data from more than one PES packet is not allowed in a TS packet. Thus, a new PES packet requires a new TS packet.

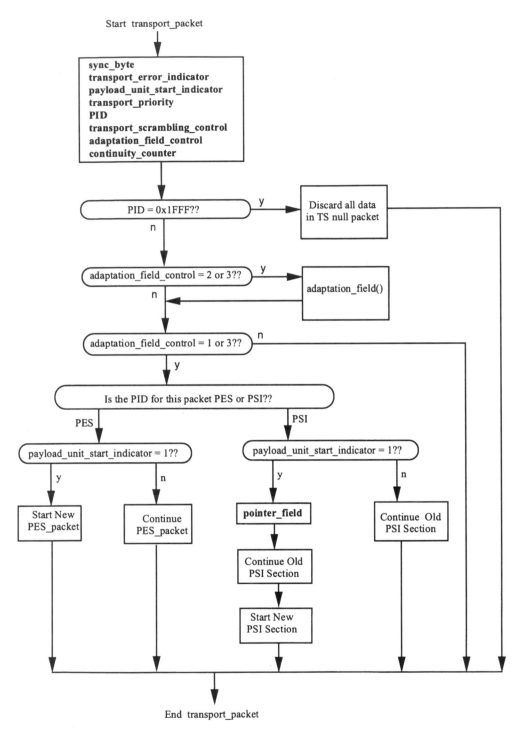

Fig. 3.8 Transport Stream (TS) Packet Syntax.

payload_unit_start_indicator Indicates the presence of a new PES packet or a new TS-PSI Section.

transport_priority 1 Indicates a higher priority than other packets.

PID 13-bit packet IDs. Values 0 and 1 are preassigned, while values 2 to 15 are reserved. Values 0x0010 to 0x1FFE may be assigned by the Program Specific Information (PSI), which will be described later. Value 0x1FFF is for null packets.

transport_scrambling_control (2 bits) Indicates the scrambling mode of the packet payload.

adaptation_field_control (2 bits) Indicates the presence of an adaptation field or payload.

continuity_counter (4 bits) One continuity_counter per PID. It increments with each nonrepeated Transport Stream packet having the corresponding PID.

pointer_field (8 bits) Used only in TS-PSI packets that contain a new TS-PSI Section (to be described later). Indicates the number of bytes until the start of the new TS-PSI Section.

The Adaptation Field contains flags and indicators, STC timing information in the form of a *Program Clock Reference* (PCR), which is used in exactly the same way as the SCR of the Program Stream, plus other data as shown in Fig. 3.9.

Semantics of the adaptation field are as follows:

adaptation_field_length (8 bits) The number of bytes following.

discontinuity_indicator 1 indicates a discontinuity in the clock reference or the continuity counter or both.

random_access_indicator 1 indicates the next PES packet is the start of a video sequence or an audio frame.

elementary_stream_priority_indicator 1 indicates a higher priority.

PCR_flag 1 indicates the presence of a PCR.

OPCR_flag 1 indicates the presence of an original PCR.

splicing_point_flag Indicates that a splice_countdown is present.

transport_private_data_flag Indicates the presence of private_data bytes.

adaptation_field_extension_flag Indicates the presence of an adaptation field extension.

program_clock_reference (**PCR**) defined above.

original_program_clock_reference (**OPCR**) This PCR may be used when extracting an original single program from a multiprogram TS (use with caution).

splice_countdown (8 bits) Specifies the remaining number of Transport Stream packets, of the same PID, until a splicing point is reached. Splice points mark the end of an audio frame or a video picture.

transport_private_data_length (8 bits) The number of private_data bytes immediately following.

private_data_bytes Private data.

adaptation_field_extension_length (8 bits) The number of bytes of the extended adaptation field.

stuffing_bytes Fixed 8-bit values equal to 0xFF that may be inserted by the encoder. They are discarded by the decoder.

Within a TS, PIDs are used to distinguish between different streams and different PSI. Different streams may belong to different programs or the same program. In any event, they all have different PIDs.

The program descriptions plus the assignments of PESs and PIDs to programs are contained in specialized TS streams called *TS-Program Specific Information* (TS-PSI). The specialized TS-PSI streams are shown in Table 3.3.

For convenience as well as length limitations, some of the TS-PSI may be sent in *Sections*. If a TS packet contains the start of any TS-PSI Section, then the TS packet data starts with an 8-bit **pointer_field** that tells how many bytes are contained in a partial section at the beginning of the TS packet data. If there is no such partial section, then **pointer_field** = 0.

If the TS-PSI is spread over many TS packets and is sent periodically to assist random tuning to a broadcast TS, then the pointer_field enables rapid access to the very next TS-PSI Section. Again for convenience, after the end of any TS-PSI Section, stuffing bytes = 0xFF may be inserted until the end of the TS packet.

The first TS-PSI stream is called the *Program Association Table* (TS-PAT). The TS-PAT, shown in Fig. 3.10, is always carried in TS packets having PID = 0, which are typically the first TS packets read by a receiver. For each program in the TS, the TS-PAT

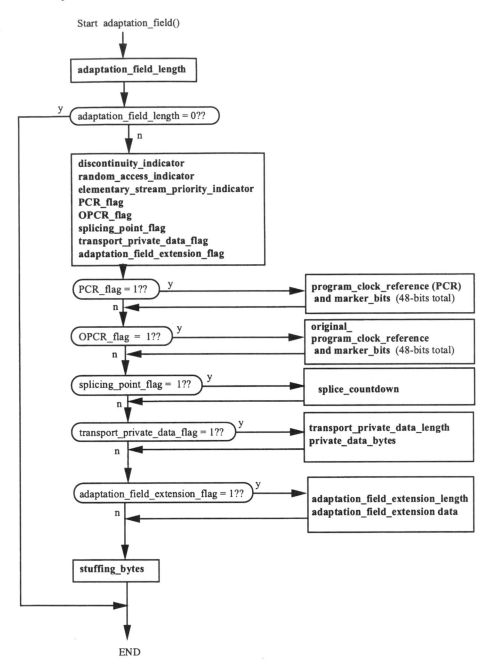

Fig. 3.9   Transport Stream Adaptation Field.

defines a unique nonzero **program_number** and an associated **TS_PMT_PID*** to be used later by packets that carry the Program Map Table (TS-PMT) for that program. Several (or in fact all) programs may use the same TS_PMT_PID for carrying their TS_PMTs, or each program could use a differ-

*This as well as other user defined PIDs are restricted to the range 0x0010 to 0x1FFE.

Table 3.3   Transport Stream–Program Specific Information (TS-PSI) is carried in specialized TS-PSI streams. Private data may be carried in Private_sections.

TS-PSI Type	PID Value (13-bits)	table_id (8-bits)	Function
Program Association Table (TS-PAT)	0x0000	0x00	Assigns Program Numbers and Program Map Table PID's
Network Information Table (TS-NIT)	Assigned in TS-PAT	0x40 to 0xFE	Specifies physical network parameters
Program Map Table (TS-PMT)	Assigned in TS-PAT	0x02	Specifies PID values for program components
Conditional Access Table (TS-CAT)	0x0001	0x01	Carries information and data for scrambling
Private_section	Assigned in TS-PAT	0x40 to 0xFE	Defined by application
Defined by DSM-CC	Assigned in DSM-CC	0x38 to 0x3F	Defined by DSM-CC
TS Description	0x0003	0x03	Optional TS Description (under study)

ent TS_PMT_PID. Programs may be added or deleted at any time* during the TS by retransmitting the entire TS-PAT with the version number increased[†] by 1. If there are many programs, the TS-PAT may be sent in several sections, with each section typically defining one or more programs.

The semantics of the TS-PAT are as follows:

table_id (8 bits)   Set to 0x00

section_syntax_indicator   Set to 1.

section_length (12 bits)   The number of bytes (≤1021) below in this section.

transport_stream_id (16 bits)   ID defined by the user.

version_number (5 bits)   Version number of the whole Program Association Table, that is, every section of a given TS-PAT has the same version_number. A new TS-PAT with an incremented version_number becomes *valid* when the last section is received.

current_next_indicator   1 indicates that this Program Association Table is currently valid. (More correctly, it will be valid as soon as the final section is received.) 0 indicates that it is not yet valid, but is about to be valid and is being sent for preparatory purposes only.

section_number (8 bits)   The number of this section. It is incremented by 1 for each additional section.

last_section_number (8 bits)   The number of the final section of the complete Program Association Table.

program_number (16 bits)   The program to which the following TS_PMT_PID is applicable. Program_number 0x0000 refers to the network PID.

TS_PMT_PID (13 bits)   The PID of packets that carry the Program Map Table (TS-PMT) for the above program_number. This PID may be used to carry TS-PMT's for several (or all) programs. It may also carry private data with the format of Fig. 3.11.

network_PID (13 bits)   The PID of the Network Information Table.

CRC_32 (32 bits)   CRC for this TS-PAT section.

stuffing_bytes   Fixed 8-bit values equal to 0xFF that may be inserted by the encoder. They are discarded by the decoder.

The second TS-PSI stream, which is optional, is called the *Network Information Table* (TS-NIT). Its TS packets use the **network_PID** defined in the TS-PAT. The Network Information Table carries network-specific data describing characteristics and features of the transmission network carrying the TS. Such data are very dependent on network implementation and are therefore considered *Private*. The TS-NIT is carried using the *Private_section* data structure shown in Fig. 3.11.

Other private data may also be carried with the Private_section structure by indicating a

---

*A programs **TS_PMT_PID** is not allowed to change during the program.

[†]After reaching 31, version numbers *wrap around*, that is, start again with 0.

Start TS Program Association Table (TS-PAT) Section

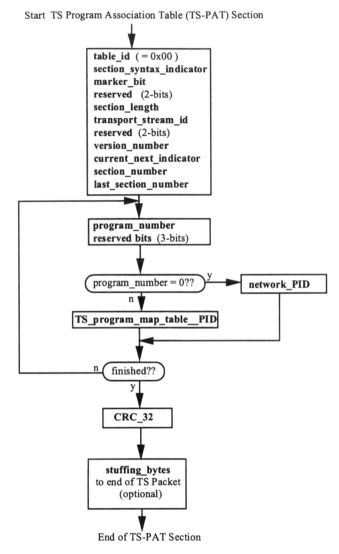

Fig. 3.10 One section of the TS Program Association Table (TS-PAT). The entire TS-PAT consists of one or more such sections. PID = 0x00 is always assigned to the TS-PAT.

stream_type = 0x05 (see Table 3.2) or by private definition of table_id values and then using a PID assigned for TS-PSI Sections.

The semantics of the TS Private_section are as follows:

table_id   Values from 0x40 to 0xFE. Chosen by the user.

section_syntax_indicator   1 indicates that this private section follows the generic section syntax.

private_indicator   User-definable flag.

private_section_length (12 bits)   The number of remaining bytes in this section (≤4093).

private_data_byte   User-definable.

table_id_extension (16 bits)   Defined by the user.

version_number (5 bits)   Version number only for this section and table_id. A new version with an incremented version_number becomes *valid* when the last byte of the section is received.

current_next_indicator   1 indicates that this section and table_id is currently valid. (More correctly, it will be valid as soon as the final byte is received.)

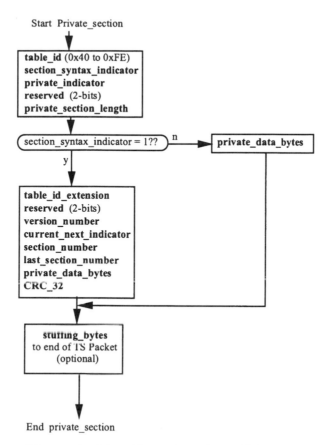

Start Private_section

table_id (0x40 to 0xFE)
section_syntax_indicator
private_indicator
reserved (2-bits)
private_section_length

section_syntax_indicator = 1?? —n→ **private_data_bytes**

y

table_id_extension
reserved (2-bits)
version_number
current_next_indicator
section_number
last_section_number
private_data_bytes
CRC_32

stuffing_bytes
to end of TS Packet
(optional)

End private_section

Fig. 3.11 A Transport Stream Private_section. It is used for carrying the Network Information Table (TS-NIT). It may also be used to carry other implementation data, network parameters or information outside the MPEG standard.

0 indicates that it is not yet valid, but is about to be valid and is being sent for preparatory purposes only.

section_number Same as TS-PAT.

last_section_number Same as TS-PAT.

CRC_32 Same as TS-PAT.

stuffing_bytes Fixed 8-bit values equal to 0xFF that may be inserted by the encoder. They are discarded by the decoder.

The third TS-PSI stream is comprised of *Program MAP Tables* (TS-PMT). Each TS-PMT, shown in Fig. 3.12, carries the defining information for one program as indicated by its *program_number*. This includes assignment of unique PID values* called **elementary_PIDs** to each packetized elementary stream (PES) or other type of stream in

the program along with the **stream_type** of each stream. It also carries descriptive and relational information about the program and the elementary streams therein. This information is carried in data structures called *Descriptors*. More will be said about Descriptors later.

The semantics of the TS-PMT are as follows:

table_id (8 bits) Set to 0x02

section_syntax_indicator Set to 1.

section_length (12 bits) The number of bytes (≤1021) following in this section.

program_number (16 bits) Same as TS-PAT.

version_number (5 bits) Version number only for this TS-PMT Section and for this program. A new version with an incremented version_number

---

*User-defined PIDs are restricted to the range 0x0010 to 0x1FFE.

Start TS Program Map Table (TS-PMT)

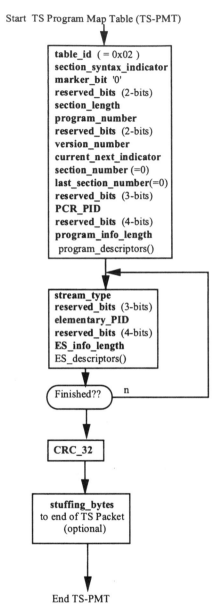

Fig. 3.12 Program Map Table (TS-PMT). Each program has a unique program_number and a corresponding TS-PMT that defines the constituent parts of the program. TS-PMTs are carried in TS packets having TS_PMT_PID identifiers defined in the TS-PAT.

becomes *valid* when the last byte of the new TS-PMT Section is received.

current_next_indicator 1 indicates that this section is currently valid. (More correctly, it will be valid as soon as the final byte is received.) 0 indicates

that it is not yet valid, but is about to be valid and is being sent for preparatory purposes only.

section_number (8 bits) Set to 0x00. The entire TS-PMT for a program must fit in one PSI Section.

last_section_number (8 bits) Set to 0x00.

PCR_PID (13 bits)   The PID that contains the PCR for this program.

program_info_length (12 bits)   The number of bytes of the program_descriptors( ) immediately following.

program_descriptors( )   Descriptive information for this program (to be discussed later).

stream_type (8 bits)   Elementary stream (ES) type as specified in Table 3.2.

elementary_PID (13 bits)   The PID that will carry this ES.

ES_info_length (12 bits)   The number of bytes in the ES_descriptors( ) for this ES.

ES_descriptors( )   Descriptive information for this ES (to be discussed later).

CRC_32   Same as TS-PAT.

stuffing_bytes   Same as TS-PAT.

TS-PMTs for programs that use the same TS_PMT_PID value may be concatenated* and sent in TS packets using that TS_PMT_PID. Also, private data may be sent in any TS_PMT_PID packets using the Private_section data structure of Fig. 3.14. Any program definition may be changed at any time during the TS by simply retransmitting its changed TS-PMT with its version number incremented by 1.

The fourth TS-PSI stream is called the *Conditional Access Table* (TS-CAT). The TS-CAT, shown in Fig. 3.13, is always carried in TS packets having PID = 1. The TS-CAT, which may be sent in Sections, carries entitlement and management information to limit access to authorized recipients. This data may be changed at any time during the TS by retransmitting the entire TS-CAT with the version number increased by 1.

The semantics of the TS-CAT are as follows:

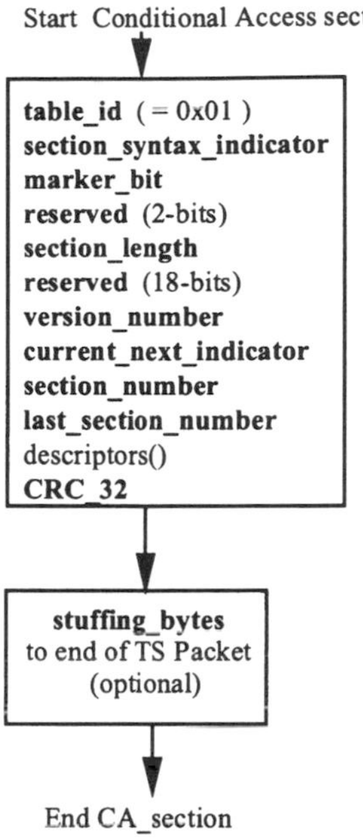

table_id (8 bits)   Set to 0x01.

section_syntax_indicator   Set to 1.

section_length (12 bits)   The number of bytes (≤1021) of this section that follow.

version_number (5 bits)   Version number of the whole Conditional Access Table, that is, every section of a given TS-CAT has the same version_number. A

Fig. 3.13   The Transport Stream Conditional Access Table (TS-CAT) is carried in TS packets having PID = 1. It provides information on scrambling of the multimedia data.

new TS-CAT with an incremented version_number becomes *valid* when the last section is received.

current_next_indicator   1 indicates that this Conditional Access Table is currently valid. (More correctly, it will be valid as soon as the final section is received.) 0 indicates that it is not yet valid, but is about to be valid and is being sent for preparatory purposes only.

section_number   Same as TS-PAT.

last_section_number   Same as TS-PAT.

CA_descriptors   To be discussed later.

CRC_32   Same as TS-PAT.

stuffing_bytes   Same as TS-PAT.

As with the Program Stream, a Transport Stream–System Target Decoder (TS-STD) is defined to enable multiplexers to create data streams that are decodable by all receivers. Also as in the the PS-STD, the TS-STD inputs data at a piecewise constant rate[†] between PCRs. However, the TS-STD attempts to take into account several

---

*TS-PMTs are considered unnumbered PSI Sections.

[†]However, unlike the PS, this rate is unspecified in the TS itself.

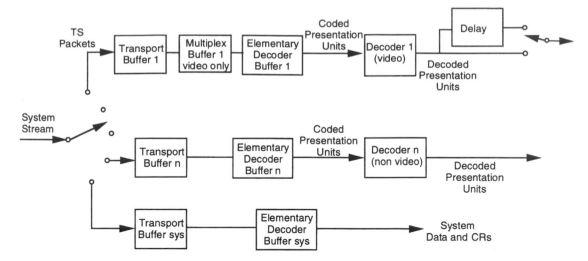

Fig. 3.14 Transport Stream–System Target Decoder (TS-STD). Three buffer types are defined: Transport Buffers (TB), Multiplex Buffers (MB), and Elementary Stream Buffers (EB). Only video uses MBs.

transmission impairments ignored by the PS-STD, as shown by Fig. 3.14.

The main difference is the definition of three types of buffers. The first type is the *Transport Buffer* (TB) of size 512 bytes. It inputs data at the full rate of the TS, but outputs data (if present) at a rate Rx that depends on data type and specification.

The second buffer type, used only for video, is the *Multiplexing Buffer* (MB), which is provided to alleviate the effects of TS packet multiplexing. Its output rate is specified in one of two ways. With the *Leak Method* the MB outputs data (if present and if EB is not full) at a rate Rbx that depends on the data type and specification. With the *Vbv_delay Method* the output rate is piecewise constant for each picture and is specified by the encoder using parameters* in the video bitstream.

The third buffer is the normal elementary stream decoder buffer (EB), whose size is fixed for audio and systems and specified in the video stream for video (see vbv_buffer_size). Several TS-STD transfer rates and buffer sizes of interest are shown for Main Profile Video, Audio, and Systems in Tables 3.4 and 3.5.

3.4.3 Descriptors

As we have seen, the PS and TS Program Map Tables carry descriptive and relational information about the program and the PESs therein. This information is carried in data structures called *Descriptors.*

For example, some programs may consist of multiple audios, in which several languages may be present in a movie. Also, several videos may sometimes be desired, such as several views of a sports contest.

Program_descriptors apply to programs, and ES_descriptors apply to elementary streams. Initially, a total of 256 possible descriptors are provided for, as shown in Table 3.6a. A few descriptors of interest are shown in Tables 3.6b to 3.6g.

In certain implementations the ITU-T timing descriptor allows for much more rapid receiver clock acquisition than can be achieved by using Clock References alone.

The semantics of the ITU-T timing descriptor are as follows:

SC_PESPktR (24 bits) If ≠ 0xFFFFFF, this parameter indicates that PES packets are generated by

---

*In real decoders some parameters may arrive too late to emulate the ideal TS-STD.

Table 3.4 Transport Stream–System Target Decoder (TS-STD) transfer rates for Main Profile Video,* Audio, and Systems streams.

Elementary Stream	TB Rate (Rx)	MB Rate (Rbx)
Low Level Video	4.8 Mbits/s	4 Mbits/s
Main Level Video	18 Mbits/s	15 Mbits/s
High 1440 Level Video	72 Mbits/s	MIN[1.05xbit_rate, 60Mbs]
High Level Video	96 Mbits/s	MIN[1.05xbit_rate, 80Mbs]
Audio	2 Mbits/s	n.a.
Systems data	1 Mbits/s	n.a.

*The parameters bit_rate and vbv_buffer_size are given in the video stream.

Table 3.5 Transport Stream–System Target Decoder (TS-STD) buffer sizes for Main Profile Video, Audio and Systems streams.

Elementary Stream	TB Size	MB Size	EB Size
Low Level Video	4096 bits	496 469 bits − EB size	vbv_buffer_size
Main Level Video	4096 bits	1 915 008 bits − EB size	vbv_buffer_size
High 1440 Level Video	4096 bits	320 000 bits	vbv_buffer_size
High Level Video	4096 bits	426 667 bits	vbv_buffer_size
Audio	4096 bits	n.a.	28 672 bits
Systems data	4096 bits	n.a.	12 288 bits

Table 3.6a A total of 256 descriptors are provided. The ITU-T has defined additional descriptors using descriptor_tags 64 to 69.

descriptor_tag (8-bits)	Descriptor Function
0 and 1	Reserved
2	video_descriptor
3	audio_descriptor
4	hierarchy_descriptor
5	registration_descriptor
6	data_alignment_descriptor
7	target_background_grid_descriptor
8	video_window_descriptor
9	CA_descriptor
10	ISO_639_language_descriptor
11	system_clock_descriptor
12	multiplex_buffer_utilization_descriptor
13	copyright_descriptor
14	maximum bitrate descriptor
15	private data indicator descriptor
16	smoothing buffer descriptor
17	STD_descriptor
18	IBP Picture descriptor
19 to 22	DSM − CC
23 to 63	Reserved
64 to 255	User Private

Table 3.6b Video descriptor syntax and semantics.

video_descriptor( )

Syntax	Semantics
descriptor_tag (8-bits)	= 2
descriptor_length (8-bits)	number of bytes following
multiple_frame_rate_flag	0 indicates a single frame-rate
frame_rate_code (4-bits)	same as video stream
MPEG_1_only_flag	1 indicates only MPEG1
constrained_parameter_flag	1 indicates constrained parameters
still_picture_flag	1 indicates only still pictures
MPEG_1_only_flag = 1?? if yes, quit	
profile_and_level (8-bits)	same as video stream
chroma_format (2-bits)	same as video stream
frame_rate_extension_flag	1 indicates frame-rate extensions
reserved bits	

Table 3.6c Audio descriptor syntax and semantics.

audio_descriptor( ) Syntax	Semantics
descriptor_tag (8-bits)	= 3
descriptor_length (8-bits)	number of bytes following
free_format_flag	same as audio stream
ID	same as audio stream
layer (2-bits)	same as audio stream
variable_rate_audio_indicator	1 indicates bit-rate may change
reserved (3-bits)	

Table 3.6d Hierarchy descriptor syntax and semantics.

hierarchy_descriptor( ) Syntax	Semantics
descriptor_tag (8-bits)	=4
descriptor_length (8-bits)	number of bytes following
reserved (4-bits)	
hierarchy_type (4-bits)	1 to 4 indicates video scalabilities
	5 indicates audio external stream
	6 indicates private stream
	15 indicates base layer
reserved (2-bits)	
hierarchy_layer_index (6-bits)	unique id for this layer
reserved (2-bits)	
hierarchy_embedded_layer_index (6-bits)	id for the next lower layer
reserved (2-bits)	
hierarchy_channel (6-bits)	transmission channel (optional)

Table 3.6e Conditional Access descriptor syntax and semantics.

CA_descriptor( ) Syntax	Semantics
descriptor_tag (8-bits)	= 9
descriptor_length (8-bits)	number of bytes following
CA_system_ID (16-bits)	type of CA system
reserved (3-bits)	
CA_PID (13-bits)	PID to which CA applies
private_data_bytes	

Table 3.6f IPB descriptor syntax and semantics.

IPB_Picture_descriptor( ) Syntax	Semantics
descriptor_tag (8-bits)	= 18
descriptor_length (8-bits)	number of bytes following
closed_gop_flag	1 indicates GOP before each I frame, and no prediction outside the GOP.
identical_gop_flag	1 indicates GOPs 2,3,4 . . . are all the same size
max_gop-length (14-bits)	maximum number of pictures in a GOP

Table 3.6g ITU-T Timing descriptor syntax.

ITU-T_timing_descriptor( ) Syntax
descriptor_tag (8-bits) (=69)
descriptor_length (8-bits)
SC_PESPktR
SC_TESPktR
SC_TSPktR
SC_byte_rate
vbv_delay_flag
reserved (33-bits)

mentary stream are generated by the encoder at a constant rate. SC_TESPktR is the ratio of 27MHz to the TES packet rate.

SC_TSPktR (24 bits)   If ≠ 0xFFFFFF, this parameter indicates a constant Transport Stream packet rate. SC_TSPktR is the ratio of 27MHz to the TS packet rate.

SC_byterate (30 bits) If ≠ 0x3FFFFFFF, this parameter indicates that PES bytes are generated by the encoder at a constant rate. SC_byterate is the ratio of 27MHz to (byte_rate/50), that is, SC_byterate = 1 350 000 000/byte_rate

vbv_delay_flag   "1" indicates that the video parameter vbv_delay may be used for timing recovery.

3.5 PRACTICAL SYSTEMS DECODERS

Practical Systems decoders must take into account several types of transmission and multiplexing impairments that are unknown at the time of encoding of the elementary streams. Aside from

the encoder at a constant rate. SC_PESPktR is the ratio of 27MHz to the PES packet rate.

SC_TESPktR (24 bits)   If ≠ 0xFFFFFF, this parameter indicates that TS packets for the specified ele-

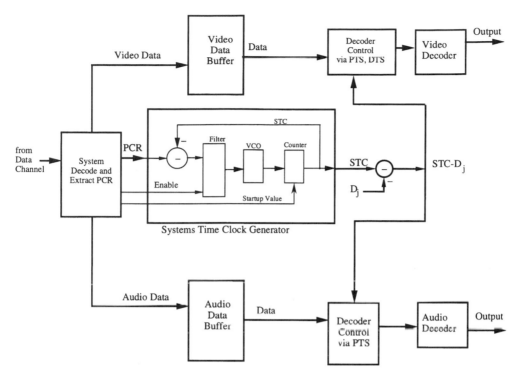

Fig. 3.15 Timing recovery in an MPEG-2 Systems Decoder.

transmission errors and packet losses, which the PES decoders must deal with, the most debilitating transmission effect for System decoders is probably due to transmission delay variation, also called jitter. This causes the received SCRs and PCRs to be in error by an amount dependent on the delay variation.

To smooth the effects of the PCR errors, practical decoders commonly utilize Phase Locked Loops (PLL) to recover stable STC timing. A typical timing recovery system is shown in Fig. 3.15. Here, the differences between the decoder STCs and the received PCRs are averaged (filtered) and the result is used to either speed up or slow down a Voltage Controlled Oscillator (VCO) that clocks the STC counter.

This design can also deal with small as well as large transmission delay jitter by adapting to whatever jitter is present in the received data stream. In this system, the received jitter is estimated as the peak positive value of the difference (STC −

PCR). Extra delay is then introduced to accommodate for the jitter by setting Dj equal to the peak jitter value. Increased buffer sizes are also needed in the presence of jitter.

3.6 PRACTICAL SYSTEMS ENCODERS

Practical Systems multiplexers must ensure that conforming MPEG-2 decoders are able to decode the multiplexed data streams. One mechanism for achieving this is shown in Fig. 3.16. Here the Systems Multiplex Controller keeps track of the fullness of the decoder and encoder buffers[2] and controls the elementary stream encoders to ensure that buffers do not overflow or unexpectedly underflow. In some implementations the decoder feeds back an estimate of the jitter that is encountered during transmission. The encoder can then adapt its

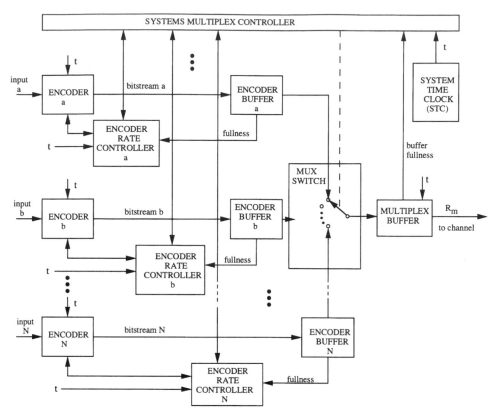

Fig. 3.16   Example of a Systems multiplexer.

buffer control to minimize the occurrence of over-flow or underflow due to jitter in transmission.

program to develop a strategy for appropriate display.

3.7 CONCLUSION

In this chapter we have seen that a variety of mechanisms are provided in MPEG-2 for multi-plexing Elementary Streams from one or more programs. Through the Program Map Tables, Program Descriptors, and Elementary Stream Descriptors, a decoder can discern the interrela-tionships between the elementary treams of a

REFERENCES

1. A. G. MacInnis, "The MPEG Systems Coding Specification," *Signal Processing: Image Commun.* pp. 153–159 (1992).

2. B. G. Haskell and A. R. Reibman, "Multiplexing of Variable Rate Encoded Streams," *IEEE Trans. Circuits and Systems for Video Technology,* 4(4): 417–424 (August 1994).

4

Audio

4.1 INTRODUCTION

Although this book deals mainly with video issues, we present this chapter to acquaint readers with the fascinating technology of audio coding. The methods employed here exploit the properties of human hearing to a far greater extent than video coding researchers have been able to utilize the characteristics of human vision.

The ITU has a plethora of speech compression and coding standards,[1] as shown in Table 4.1, reflecting its various needs for efficient speech transmission over a variety of telecommunication networks. For example, ITU-T G.711 is simple 8-bit PCM, 64 kbits/s coding of telephone quality audio with a nonlinearity introduced prior to A/D and following D/A. ITU-T G.722 utilizes DPCM coding of monaural AM radio quality audio to

compress the data again into 64 kbits/s. ITU-T G.728 employs even more compression technology, specifically Linear Predictive Coding and Vector Quantization, to squeeze telephone speech into 16 kbits/s. And the now finished G.723 is able to obtain reasonable quality speech at rates as low as 5.3 kbits/s.

ISO has relatively few audio compression standards, compared with ITU-T. ISO MPEG-1 audio,[2] officially known as ISO 11172-3, is basically for two-channel stereo. ISO MPEG-2 audio,[3] officially known as ISO 13818-3, extends this capability to five-channel surround sound, with the option of a sixth low-frequency channel. MPEG-2 audio comes in two flavors, one that is backward compatible* with MPEG-1, and one that is not.

Wideband audio has a considerably larger bandwidth than telephone speech. This causes the

Table 4.1 Commonly used ITU-T Monaural Speech Coding Standards.

ITU-T Designation	Bandwidth (Hz)	Sampling Rate (kHz)	Bit Rate (kbits/s)
G.711	200 to 3200	8	64
G.722	50 to 7000	16	64
G.721	200 to 3200	8	32
G.728	200 to 3200	8	16
G.723	200 to 3200	8	5.3 and 6.3

---

*Backward compatible means MPEG-1 can play MPEG-2-BC audio streams. This feature necessarily requires a tradeoff between MPEG-1 quality and MPEG-2 quality.

Table 4.2   Audio bitrates and total bitrates for Compact Disk (CD), Digital Audio Tape (DAT), Digital Compact Cassette[13] (DCC), and MiniDisk[14] (MD).

Digital Storage Medium	Audio Bandwidth (kHz)	Sampling Rate (kHz)	Stereo Audio Bitrate (kbits/s)	Coding
CD	20	44.1	1411.2	PCM
DAT	16	32.0	1024.0	PCM
ProDAT	16	44.1	1411.2	PCM
DAT	16	48.0	1536.0	PCM
DCC	20	44.1	384	PASC
MD	22	44.1	292	ATRAC

uncompressed data rates to be also much larger, as shown in Table 4.2. Uncompressed audio typically comprises 16 bits per sample and two channels of stereo for a total bitrate of 32 times the sampling rate. However, optical and magnetic storage devices are fairly error prone and require 100% to 200% of extra bits for error correcting redundancy. Thus, the total bitrate is two to three times the audio bitrate.

4.2   AUDIO COMPRESSION FUNDAMENTALS

Most wideband audio encoding, and MPEG in particular, heavily exploit the subjective *masking*

effects of the human ear to make any coding noise as inaudible as possible. More specifically, if a tone of a certain frequency and amplitude is present, then other tones or noise of similar frequency, but of much lower amplitude, cannot be heard by the human ear. Thus, the louder tone *masks* the softer tone, and there is no need to transmit the softer tone.

The maximum nonperceptible amplitude level of the softer tone is called the *Masking Threshold*. It depends on the frequency, amplitude, and time history of the louder (masker) tone, and is usually plotted as a function of the softer (maskee) noise frequency, as shown in Fig. 4.1.

In addition to the frequency masking described above, the human ear is also affected by *Temporal*

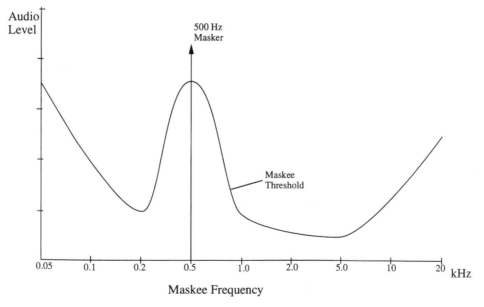

Fig. 4.1   With Frequency Masking, a louder tone masks softer tones of nearby frequency. The *Masking Threshold* is plotted as a function of the softer (Maskee) noise frequency. If the maskee amplitude is less than the Masking Threshold, the maskee is inaudible.

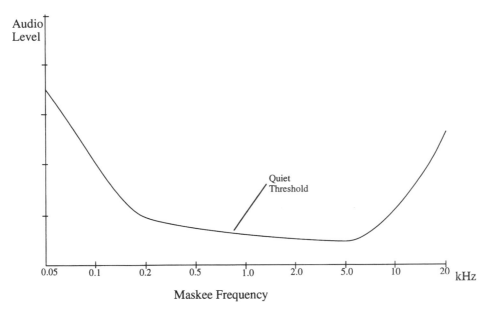

Fig. 4.2 The *Absolute Threshold*. A single tone or noise by itself is inaudable if it is below the Absolute Threshold.

Masking. That is, a loud tone of finite duration will mask a softer tone that follows it or very closely preceeds it in time.* Thus in Fig. 4.1, if the masker tone suddenly disappears, the masking Threshold does not simultaneously disappear. Instead, it dies away over a period of up to 200 ms, depending on masker amplitude and frequency.[†]

If the masker tone decreases in amplitude, then generally the Masking Threshold also decreases. However, there is a lower limit below which the Masking Threshold will not fall no matter how small the masker. This limit, shown in Fig. 4.2, is called the *Absolute Threshold*. A single tone or noise even with no masker is inaudible if it is below the Absolute Threshold.

With normal audio signals, a multitude of maskers are usually present at a variety of frequencies. In this case, by taking into account the Frequency Masking, Temporal Masking, and Absolute Threshold for all the maskers, the overall net *Masking Threshold* can be derived. Thus, the Masking Threshold is a time varying function of frequency that indicates the maximum inaudible noise at each frequency at a given time, as illustrated in Fig. 4.3.

The Masking Threshold is used by the audio encoder to determine the maximum allowable quantization noise at each frequency to minimize noise perceptibility.

Another technology used by wideband audio coders is called *Subband Coding*.[6] With this approach, shown in Fig. 4.4, the incoming digital PCM signal is first decomposed into N digital bandpass signals by a parallel bank of N digital bandpass *Analysis* filters. Each bandpass signal is then subsampled (or *decimated*) by a factor of N so that the total number of samples in all the subbands is the same as the PCM input. After decimation, each of the bandpass signals becomes, in fact, a low-frequency baseband signal that can be coded and compressed using any baseband compression algorithm.

Reconstruction of the original PCM signal is accomplished by first inserting $N - 1$ zero samples between each sample of the decimated subband signals, which increases the sampling rate of each bandpass signal to its original value. The *upsampled* bandpass signals are then fed to their respective bandpass *Synthesis* filters and finally summed to obtain the full bandwidth PCM output signal.

---

*Maskee tones that follow the masker are masked significantly more than maskee tones that precede the masker.
[†]Most of the decay is finished by about 30 ms.

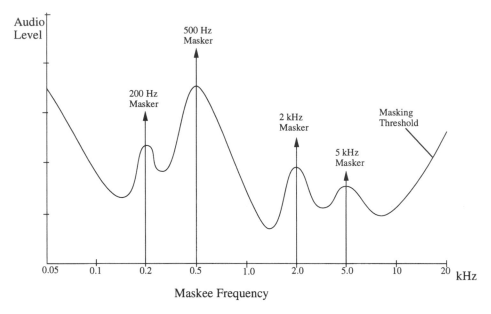

Fig. 4.3 With normal audio signals containing many maskers at a variety of frequencies, the overall net Masking Threshold is calculated from the Frequency Masking, Temporal Masking and Absolute Threshold for all the maskers. The Masking Threshold is a time varying function of frequency that indicates the maximum inaudible noise at each frequency.

Since practical filters can only approximate the ideal bandpass characteristic, each subsampled bandpass signal will contain some alias components that would normally prevent a distortion free reconstruction of the original PCM signal. It turns out, theoretically at least, that by using specially designed bandpass filters, the aliasing in the bandpass signals can be made to cancel when they are summed at the final output. One such set of filters is called *Quadrature Mirror Filters* (QMFs). A particularly efficient QMF filter bank implementation is possible through the use of *Polyphase Networks*.[7] Another such set of filters are called Lapped Orthogonal Transforms (LOTs) or Extended Lapped Transforms (ELTs).[8] These are basically block transforms with overlapping blocks.

Even with QMFs and ELTs, certain approximations and compromises are usually necessary, with the result that performance of the bandpass filters is never completely ideal. Also, with lossy compression coding, the bandpass signals themselves are rarely recovered with full accuracy at the decoder, which further contributes to non-ideal operation.

4.3 MPEG-1

In the first phase of MPEG, a compression standard for two-channel stereo audio was developed that gives near Compact Disk (CD) quality at a bitrate of 256 kbits/s. MPEG-1 is officially known as ISO/IEC 11172–3.

In principle, the standard does not specify the encoder, thus allowing for an evolution as technology progresses and costs decline. MPEG-1 only specifies the decoder, and even here the specification ignores post processing and error recovery.

The algorithm has three possible *Layers** of operation, namely I, II, or III. In principle, the encoder chooses the Layer depending on how much quality is needed or equivalently how much compression is desired. However, in some applica-

---

*Although MPEG uses the term Layers, they are in no way hierarchical. Audio Layers I, II, and III are three different coding algorithms.

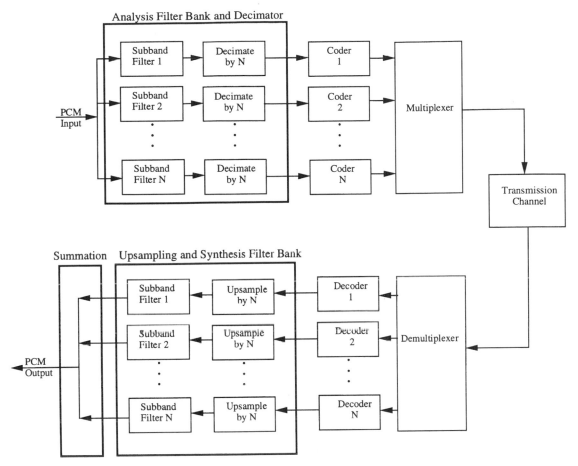

Fig. 4.4   Subband Audio Coder (single channel). The incoming signal is fed to a parallel bank of N digital bandpass *Analysis* filters to produce N digital bandpass signals. These are then decimated by a factor of N, coded and transmitted. Each coded bandpass signal thus consists of one sample for every N input PCM samples. At the decoder the bandpass signals are upsampled by a factor of N by the insertion of zeros, then fed to a bank of digital bandpass *Synthesis* filters and finally summed to produce the final PCM output.

tions, decoders only implement the first one or two layers, which restricts the encoder choices.

MPEG-1 can code a single channel, a stereo pair, or two independent (perhaps multilingual) channels. With a stereo pair, the two audio channels can be coded either separately or jointly. In the latter case, MPEG-1 provides rudimentary methods for exploiting the redundancies between the left and right channels for a modest bitrate reduction.[9]

MPEG-1 audio allows for initial sampling rates of 32, 44.1, and 48 kHz. A variety of fixed bitrates can be specified, ranging from 32 kbits/s up to 448 kbits/s depending on the Layer, or alternatively a

fixed unspecified bitrate is also possible. Pseudo-variable-bitrate operation is possible in Layer III.

An MPEG-1 audio encoder and decoder example for Layers I and II is shown in Fig. 4.5. A 16-bit PCM input signal is first fed to a polyphase, approximately ELT, analysis filter bank consisting of 32 equally spaced bandpass filters. Thus, each band has a width of $1/64$ of the PCM sampling rate.

The bandpass analysis filters are of order 511 and have fairly steep rolloff, providing more than 96 dB of alias suppression. As mentioned previously, ELTs with no distortion would ordinarily cancel the alias components at the decoder. However, with coding and compression, the subbands suffer

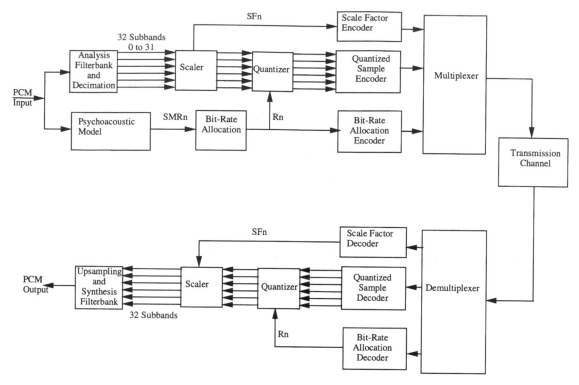

Fig. 4.5   MPEG-1 audio encoder and decoder for Layers I and II (single channel). The standard does not specify the encoder in order to allow for evolution over time. Thus, this is only an example.

from a certain level of quantization noise, and this can produce unwanted alias components if they are not attenuated as much as possible in the analysis filter bank. Following analysis, each of the 32 bandpass outputs is subsampled (decimated) by a factor of 32.

For each subband n ($n = 0$ to 31), the decimated samples are assembled into blocks of 12 consecutive samples, giving 32 parallel blocks Bn that correspond to $32 \times 12 = 384$ input samples.

For each block Bn, a *Scale_Factor* SFn is then chosen from a table of 63 allowable values, such that when the samples of the block are divided (i.e., normalized) by that Scale_Factor, they all have absolute value less than 1. This is typically done by first finding the sample with the largest absolute value, and then choosing the next higher value in the table for SFn.

4.3.1   Psychoacoustic Model

In parallel with these operations, the Masking Threshold is computed for each group of 384 input samples, that is, each group of 32 blocks Bn. Computation of the Masking Threshold, which is referred to as the *Psychoacoustic Model*, is an encoder-only operation and is not specified by the standard. However, MPEG does provide two examples of Psychoacoustic Models, the first of which we will now describe.* It consists of nine steps, as shown in Fig. 4.6:

1. Calculation of the Frequency Spectrum via a windowed FFT.† This generally uses a 512-point (Layer I) or 1024-point (Layer II) FFT with a raised cosine Hann window. The resulting spectral resolution is much finer than the subbands, giving 8 points per subband for Layer I and 16 points per

---

*Psychoacoustic Model 1 is the simpler of the two. Psychoacoustic Model 2 performs better for Layers II and III.
†Fast Fourier Transform.

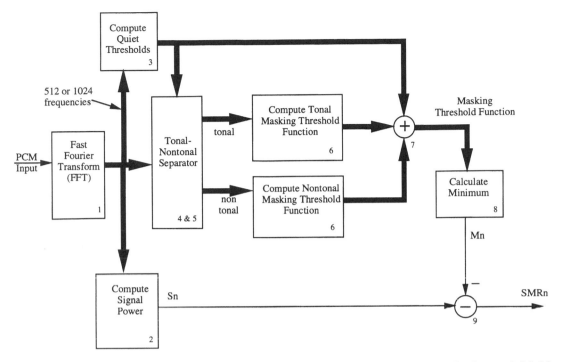

Fig. 4.6 Psychoacoustic Model suitable for MPEG-1 Layers I and II. The standard does not specify the Psychoacoustic Model. Thus, this is only an example.

subband for Layer II. This increased resolution is needed to accurately estimate masking, especially at the lower frequencies.

2. Computation of *Sound Pressure Level* (Sn) within each block Bn. Either the maximum FFT value in the block subband is used or a normalized value of Scale_Factor SFn is used, whichever is larger. For pure tones within the block subband the FFT value will be larger, whereas for noiselike sounds the normalized SFn may be larger.

3. Determination of the Absolute Threshold for each FFT component. These values are simply taken from empirically derived tables. For higher quality coding, that is, bitrates ≥96 kbits/s, the value is reduced by 12 dB.

4. Detecting the tonal (more sinusoidlike) and nontonal (more noiselike) components. A tone is basically an FFT component that is a local maximum surrounded by at most one or two other significant components within a predefined bandwidth of nearby frequencies.* The predefined bandwidths

are smaller than a subband for low-frequency tones and larger than a subband for high-frequency tones. Tonal components are identified, their frequencies noted, and their signal powers measured.

Each tonal component and its nearby frequencies in the predefined bandwidths are then removed from the set of FFT components to obtain the nontonal or noiselike components. These are then characterized by measuring the remaining signal power within each band of a contiguous set of *Critical Bands,* which are specified in empirically derived tables. Nontonal components are then assigned a representative frequency within each Critical Band.

5. Detecting only the relevant tonal and nontonal maskers. First, tonal or nontonal components less than the Absolute Threshold are discarded. Next, if two tonal components are closer in frequency than a specified amount, then the smaller component is discarded.

6. Computing the thresholds for each masker. For each relevant tonal masker, a masking threshold

---

*This is a signal processing definition of tones. It is not necessarily a good definition of tones as perceived by the human ear.

function is computed that is an empirically derived nonlinear function of masker signal power and masker frequency. For each relevant nontonal masker, a masking threshold function is also computed that is a different nonlinear function of masker power and frequency. A given amount of tonal masker power generally provides less masking than the same amount of nontonal masker power.

7. The Masking Threshold at each maskee frequency is then computed as a sum in the signal power domain* of the Absolute Threshold, the tonal masker thresholds, and the nontonal masker thresholds. Typically, Masking Thresholds are computed separately for each channel of a stereo pair. However, in some applications it may be possible to exploit further masking of one stereo channel by the other.

8. Calculating the minimum masking threshold value Mn in each block Bn. The minimum value of the Masking Threshold is computed for frequencies within the block Bn subband. Mn is then set to this value.

9. Computing the signal-to-minimum_mask ratio in each block.

$$SMRn = Sn - Mn \text{ (in dB)}.$$

Following the calculation of the SMR values, we then carry out bit assignments, quantization, and coding for each block Bn. In the case of stereo, this is done simultaneously for both channels, which allows for adaptive bit assignment between channels. Subband samples of each block Bn are then quantized by a uniform midstep quantizer[†] having the assigned number of levels. However, bit assignment and choice of the number of quantization levels in each block is done differently in Layers I and II, as we shall see later.

4.3.2 MPEG-1 Syntax and Semantics

Data from all blocks from both channels, including bit assignments Rn, Scale_Factors SFn,

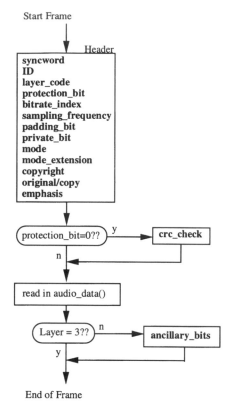

Fig. 4.7 Syntax of an MPEG-1 Audio Frame and Header. Two channels (one if single_channel) are carried in one Frame. A boldface variable by itself means simply read in the value of the variable.

coded samples, and so forth are then assembled into a *Frame* of data,[‡] preceded by a 32-bit Frame Header. The syntax of the MPEG-1 Frame and Header is shown in Fig. 4.7.

The semantics of the MPEG-1 Frame and Header are as follows:

syncword   12 bits, all 1 s.
ID   1 bit, indicates the algorithm. **ID** = 1 for MPEG-1 audio. **ID** = 0 for low rate extension of MPEG-1 as defined in MPEG-2.
layer_code   2 bits, indicates the Layer. **layer_code** = 4 − Layer, where Layer = 1, 2, or 3.
protection_bit   1 bit. 0 indicates **crc_check** is present. 1 indicates it is not.

---

*Individual masker and Absolute Thresholds are usually specified in dB = 10 log(power).

[†]A uniform midstep quantizer has an odd number N of levels and contains the zero level. The set of levels is $\{i \times Q$, for $i = 0, \pm1, \pm2, \pm3, \ldots \pm (N - 1)/2\}$, where Q is the step size. If samples to be quantized have a range of -1 to $+1$, then $Q = 2/N$.

[‡]Not to be confused with a video frame.

Table 4.3 Modes of audio for MPEG-1.

mode	channel assignment
00	stereo
01	joint_stereo (intensity_stereo for Layers I and II, ms_stereo for Layer III)
10	dual_channel
11	single_channel monophonic

bit-rate_index 4 bits, indicates the bitrate. 0 indicates a nonstandard bitrate. 14 standard values depend on Layer and ID.

sampling_frequency 2 bits, indicates the sampling frequency, which depends on ID.

padding_bit 1 bit; 1 indicates the frame contains additional bytes to control the mean bitrate.

private_bit bit for private use. This bit will not be used in the future by ISO.

mode 2 bits, according to Table 4.3. In stereo modes, channel 0 is the left channel.

mode_extension 2 bits, used in joint_stereo mode. In Layers I and II they indicate which subbands are in intensity_stereo and which are coded in normal stereo. In Layer III they indicate which type of joint stereo coding method is applied.

copyright 1 bit, 1 indicates copyright protected.

original/copy 1 bit; 1 indicates an original.

emphasis 2 bits, indicates the type of deemphasis.

crc_check 16-bit Cyclic Redundancy Code computed for Header. Used for optional error detection.

audio_data( ) Contains coded data for audio samples.

ancillary_data Contains ancillary data, if any.

4.3.3 Layer I Coding

Layer I is capable of compressing stereo audio to approximately 384 kbits/s with near CD quality. It utilizes 15 possible quantizers, with a new quantizer being selected for each block. Layer I assigns R_n bits per sample to be assigned to each block B_n, where either $R_n = 0$ or $R_n = 2$ to 15. If $R_n > 0$ bits per sample are assigned to a block, then the samples of that block (normalized by SF_n) are quantized by a uniform midstep quantizer having $2^{R_n} - 1$ levels and step size $Q_n = 2/(2^{R_n} - 1)$. If zero bits are assigned, all samples are set to zero.

For the optimization of subjective quality versus bitrate, we first note that, to a very good approximation, the ratio of signal power to quantization noise (SNR) depends only on the number of quantization levels.* In fact, MPEG provides tables of assumed SNR versus number of quantization levels for audio waveforms. For zero bits per sample (all samples set to the zero level), SNR = 0 dB, whereas for 15 bits per sample (32,767 levels), SNR = 92 dB.

For optimum quality within the constraints of available bitrate, we would like the ratio of noise power to masking threshold (NMR_n) to be fairly uniform over the subbands. If the bitrate is high enough to allow for $NMR_n < 0$ dB for all blocks, then the quantization noise will be imperceptible. However, this cannot always be achieved, especially for low bitrates.

The quantization bit assignment R_n for each block may be accomplished through an iterative procedure that starts out with zero bits per block, that is, $R_n = 0$ and $SNR_n = 0$. For each block B_n, we compute NMR_n from

$$NMR_n = SMR_n - SNR_n \text{ dB}$$

We then look for the block having maximum NMR_n, and increase its bit assignment R_n. If the old R_n was zero, then 6 bits for the Scale_Factor SF_n must also be take into account. We then recompute NMR_n and repeat the process iteratively until all bits available for coding are used up. The normalized samples of all blocks assigned $R_n > 0$ are then quantized and coded using a fixed length code[†] of R_n bits per sample.

The Layer I decoder is much simpler than the encoder, since it does not need the Psychoacoustic Mode or the bit assignment decision process. The decoder basically processes the received stream according to the syntax below, and then reproduces subband samples and PCM audio samples as described previously.

---

*For $R_n > 2$, SNR $\approx R_n \times 6$ dB is not a bad approximation. For $R_n = 2$, SNR ≈ 7 dB.

[†]Binary values 0 to $2^{R_n} - 2$ are used. The value $2^{R_n} - 1$ is reserved for synchronization.

4.3.4   Intensity_Stereo Mode

With the Stereo mode, the left and right channels are normally coded separately. However, in case there are not enough bits to achieve high quality in certain Frames, the Intensity_Stereo mode may be used. This mode exploits a characteristic of human stereo perception in which the stereo effect of middle and upper frequencies depends not so much on the different *content* of the left and right channels, but instead on the differing *amplitudes* of the two channels. Thus, to save bits in Intensity_Stereo mode, the middle and upper subbands of the left and right channel are added together, and only the resulting summed samples are quantized, coded, and sent as a single channel. However, to maintain the stereo effect, Scale_Factors are sent for both channels so that amplitudes may be controlled independently during playback.* The subband number sb_bound at which Intensity_Stereo coding begins is determined by the Frame Header **mode_extension** variable.

The syntax of the MPEG-1 Layer I audio_data( ) is shown in Fig. 4.8.

The semantics of the MPEG-1 Layer I audio_data( ) are as follows:

nch   Specifies the number of channels. nch = 1 for monophonic mode, and 2 for all other modes.

ch   Channel number, which ranges from 0 to nch-1.

sb_bound   Starting   subband   number   for intensity_stereo.

allocation[ch][sb] (4 bits)   Indicates the number of bits used to quantize and code the samples in subband sb of channel ch. For subbands in intensity_stereo mode the bitstream contains only one allocation per subband.

scalefactor[ch][sb] (6 bits)   Indicates the Scale_Factor of subband sb of channel ch. Actual values are found from tables.

sample[ch][sb][s] Coded data for samples in subband sb of channel ch.

4.3.5   Layer II Coding

Layer II is capable of compressing audio to 256 kbits/s with near CD quality. It utilizes all the compression methods of Layer I, but further reduces overhead redundancy by exploiting similarities in Scale_Factors of successive blocks in a subband, as well as by reducing the number of quantizer choices for higher frequency subbands. Layer II is somewhat more complex and has more delay than Layer I, however.

Unlike Layer I, which codes 32 blocks at a time (one block from each analysis filter), Layer II codes 96 blocks at a time (three blocks from each analysis filter). The first step is to check the three consecutive Scale_Factors that occur in the three blocks (b = 0, 1, 2) of each subband. If all three Scale_Factors are nearly identical, then only one value is sent. If two out of the three Scale_Factors are similar in value, then two Scale_Factor values are sent. Also, if large temporal masking is present, then either one or two Scale_Factor values may be sent. Otherwise, all three Scale_Factors must be sent. Scale_Factor selection side-information (**scfsi**) must also be sent to tell the decoder which Scale_Factors were sent.

The next redundancy to be reduced lies in the number of quantization levels that are allowed in the middle and upper subbands. Since the sample values in these bands are almost always small, coarser quantization rarely has an audible effect, and a saving in bit allocation side information is possible if only a few choices of quantization are allowed. In case a high level tone occurs, 16-bit quantization is always an option (7 bits for the lower bitrates). The bit allocation and quantization apply to all three blocks (36 samples) of each subband.

The next compression feature of Layer II is to allow two new quantizations, namely 5-level and 9-level. In addition, for these new quantizations plus the former 3-level quantization, *sample grouped coding* is used. With sample grouping, three consecutive quantized samples in a block form one *triplet* and are coded together using one code word. For 3-, 5-, and 9-level quantization, a triplet[†] is coded using a 5-, 7-, or 10-bit code word, respectively.

The quantization bit assignment for Layer II may be derived iteratively as with Layer I. However, for Layer II we also need to take into account

---

*The rate at which Scale_Factors are sent (once per block) is not high enough to reproduce excellent quality stereo. Thus, intensity_stereo is useful only for medium quality audio, for example, at bitrates well below 384 kbits/s.

[†]MPEG calls this triplet a granule. However, granule has another meaning in Layer III. Thus, we avoid its usage here.

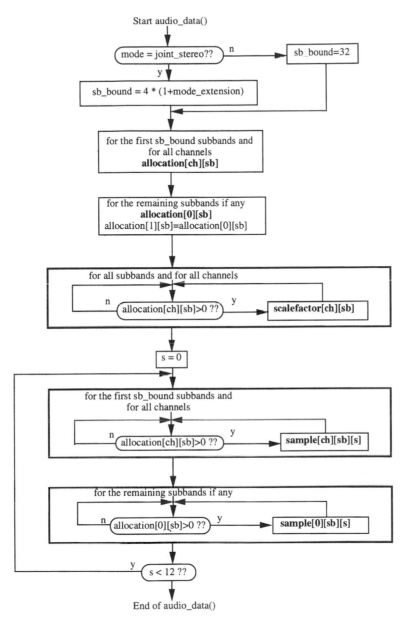

Fig. 4.8 Syntax for MPEG-1 audio_data( ) for Layer I. Each Frame represents 384 samples of input PCM audio.

the variable number of bits due to Scale_Factor coding, bit allocation, side information, and sample grouping.

The Layer II decoder is much simpler than the encoder. The decoder basically processes the received stream according to the syntax below, and then reproduces subband samples and PCM audio samples as described previously.

The syntax of the MPEG-1 Layer II audio_data( ) is shown in Fig. 4.9.

The semantics of the MPEG-1 Layer II audio_data( ) are as follows:

nch Specifies the number of channels. nch = 1 for monophonic mode, and 2 for all other modes.

ch Channel number, which ranges from 0 to nch-1.

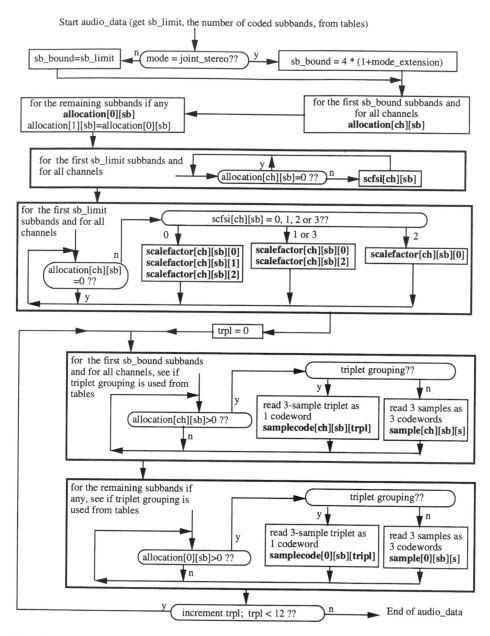

Fig. 4.9   Syntax for MPEG-1 audio_data( ) for Layer II. Each Frame represents 1152 samples of input PCM audio. Boldface means read in the data from the stream.

sb_limit   Number of subbands to be coded. Depends on bit and sampling rates.

sb_bound   Starting subband number for intensity_stereo.

trpl   3-sample triplet number.

allocation[ch][sb] 2 to 4 bits (from tables)   Specifies the quantizers used in subband sb in channel ch, whether three consecutive samples have been

grouped, and the number of bits per sample. Actual values are taken from tables.

scfsi[ch][sb] (2 bits)   Specifies the number of Scale_Factors for subband sb in channel ch and in which blocks they are valid.

scalefactor[ch][sb][b] (6 bits)   Indicates the Scale_Factor for block b of subband sb of channel ch. Actual values are found from tables.

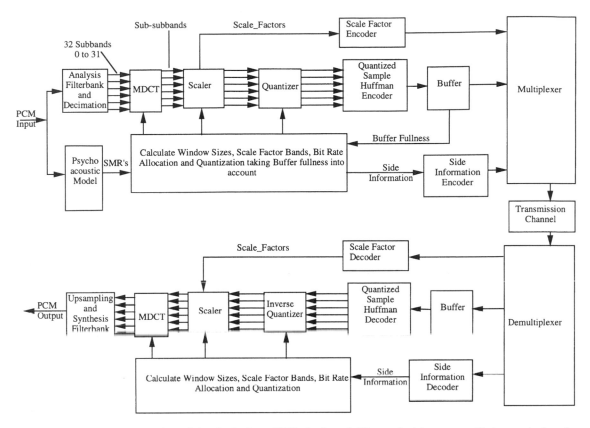

Fig. 4.10 MPEG-1 audio encoder and decoder for Layer III (single channel). The standard does not specify the encoder in order to allow for evolution over time. Thus, this is only an example.

samplecode[ch][sb][trpl] Coded data for three consecutive samples in triplet trpl in subband sb of channel ch.

sample[ch][sb][s] Coded data for sample s in subband sb of channel ch.

4.3.6 Layer III

Layer III is considerably more complex and utilizes many more features than Layers I and II. It is capable of compressing stereo audio to 128 kbits/s with adequate entertainment quality, but usually not CD quality. An example of an MPEG-1 Layer III coder and decoder is shown in Fig. 4.10.

The first improvement over Layer II is to increase the frequency resolution. This is achieved

by further processing each subband analysis filter output using a Modified Discrete Cosine Transform (MDCT). We call the MDCT outputs sub-subbands. The increased sub-subband resolution allows for better perceptual coding, especially at lower frequencies where the Masking Threshold may have considerable variation within a subband. It also allows for some cancellation of aliasing caused by the polyphase analysis subband filters within a subband.

The MDCTs can be switched to generate for each subband either 6 sub-subbands (called *short-window* MDCTs) or 18 sub-subbands* (called *long-window*[†] MDCTs). Long-windows give better frequency resolution, while short-windows give better time resolution. Also, during large transients, the

---

*The MDCT is a 50% overlapped transform. Thus, it is actually a 12-point or a 36-point transform, respectively.

[†]In MPEG Layer III, the terms *window* and *block* are used interchangeably. To avoid confusion with earlier terminology, we use only the term *window*.

lower two subband MDCTs can be switched to long-windows, while the upper 30 subband MDCTs are switched to short-windows. This assists dynamic adaptation of the coding while maintaining the higher frequency resolution for the low frequencies.

Adaptation to the transition of an MDCT between long- and short-windows is also possible. This is implemented by two special long-window MDCTs, one for the long-to-short transition and one for the short-to-long transition.

After formation of the sub-subbands, the samples are then assembled in frequency order into *Granules*. A Granule consists of one sample from each long-window sub-subband and three successive time samples from each short-window sub-subband, for a total of 576 samples. Because of the frequency ordering, the quantized sample amplitudes on the average decrease from beginning to end of the Granule. Two Granules make up each audio Frame, which corresponds to 1152 input PCM samples.

There are not enough bits to allow for a separate Scale_Factor for each sub-subband. Thus, the samples of each Granule are adaptively grouped into *Scale_Factor_Bands*, and one Scale_Factor is computed for all the samples in the associated Scale_Factor_Band. If for a particular Scale_Factor_Band, the Scale_Factors in both Granules of a Frame are identical, then the second Scale_Factor need not be transmitted. In this case, the corresponding Scale_Factor of the first Granule is used for both Granules. A further partitioning of scale factor bands into four *Classes* facilitates the transmission of this latter side information.

The Granule samples are then nonuniformly quantized. This is implemented by raising the absolute value of the samples to the power 0.75 before passing to a uniform quantizer. The quantized samples are then coded using variable-length codewords. For this purpose the samples are partitioned into five variable sized *Regions*, each Region using its own set of Huffman code tables especially designed for the statistics expected in that Region. The first three Regions are called *Big Values*

Regions. Within these Regions the quantized samples are coded in pairs, that is, two samples per Huffman codeword. The fourth Region of quantized samples (called the *Small Values** Region) consists entirely of zeros and ±1s. Samples in this Region are coded four at a time, that is, four samples per codeword. The fifth Region, corresponding to the highest frequencies, consists entirely of zeros and is not transmitted.

The Huffman coding process may generate a variable number of bits per Frame, thus requiring a buffer[†] to connect to a constant rate digital channel. In this case, feedback from the buffer is used to control the quantization to avoid emptying or overflowing the buffer. However, a more important use of the buffer concerns large temporal changes in the audio signal. In case of a sudden increase in audio amplitude, the buffer allows for a higher quality of coding just *before* the transient when noise is more audible, at the expense of lower quality coding *during* the transient when noise is less audible. This is called *Pre-echo* correction and results in a significant improvement in overall coding quality.

To improve error resilience, the Frame Header and some of the side information of Layer III are not necessarily placed at the beginning of the variable length audio data of each Frame. Instead the Header and partial side information may occur later. Specifically, they are placed exactly at the locations they would appear if the Frames were all of constant length. For this to work correctly, a pointer called **main_data_begin** is included after each Frame Header to indicate where the audio main_data for the corresponding Frame actually begins.

The Layer III decoder is much simpler than the encoder. In fact, the Layer III decoder is only slightly more complex that a Layer II decoder. The main increase in complexity is due to the overlapped MDCTs. Because of the overlap, samples from two successive inverse transforms must be added together to produce samples for a subband.

The decoder basically processes the received stream according to the syntax below, and then

---

*MPEG calls this the count1 region. Don't ask.

[†]Alternatively, a rate controller may be used to achieve a constant number of bits per frame.

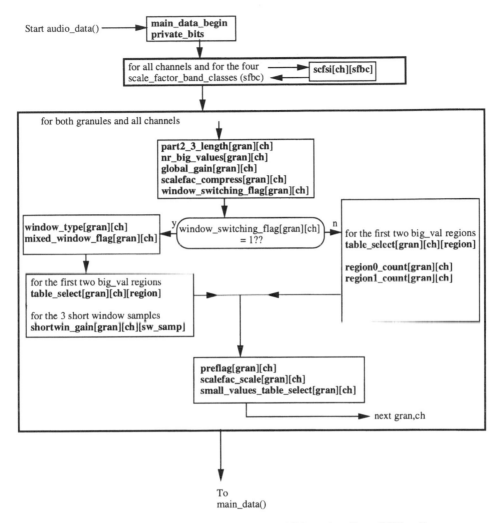

Fig. 4.11a Syntax for MPEG-1 audio_data( ). Each Frame represents 1152 samples of input PCM audio.

reproduces subband samples and PCM audio samples as described previously.

4.3.7 MS_Stereo mode

In addition to Intensity_Stereo Mode, Layer III supports an additional stereo mode called Middle-Side (MS) stereo. With this technique, the frequency ranges that would normally be coded as left and right are instead coded as Middle (left + right) and Side (left − right). This method exploits the fact that the Side channel can be coded with fewer bits, thus reducing the overall bitrate. If MS_Stereo and Intensity_Stereo are both used, then MS_Stereo applies only to the lower bands where Intensity_Stereo is not used.

The syntax of the MPEG-1 Layer III audio_data( ) is shown in Fig. 4.11.

The semantics of the MPEG-1 Layer III audio_data( ) are as follows:

nch   Specifies the number of channels. nch = 1 for monophonic mode, and 2 for all other modes.

ch   Channel number, which ranges from 0 to nch-1.

sfb   Scale_Factor_Band. Ranges from 0 to 20.

sfbc   Scale Factor Band Class. The first 6 scale factor bands are in class 0. Classes 1, 2, 3 contain five scale factor bands each.

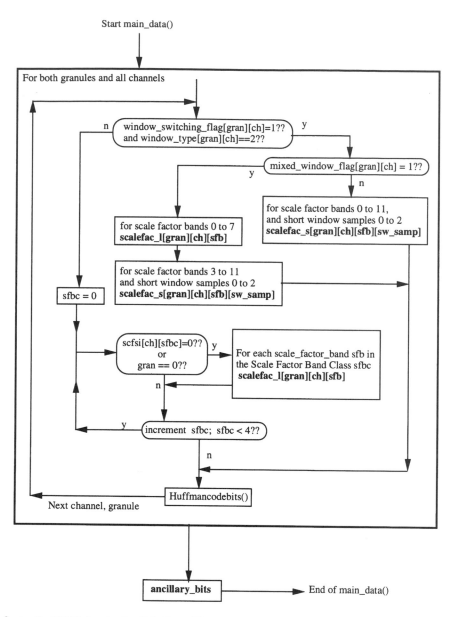

Start main_data()

For both granules and all channels

window_switching_flag[gran][ch]=1?? and window_type[gran][ch]==2??

mixed_window_flag[gran][ch] = 1??

for scale factor bands 0 to 11, and short window samples 0 to 2 **scalefac_s[gran][ch][sfb][sw_samp]**

for scale factor bands 0 to 7 **scalefac_l[gran][ch][sfb]**

for scale factor bands 3 to 11 and short window samples 0 to 2 **scalefac_s[gran][ch][sfb][sw_samp]**

sfbc = 0

scfsi[ch][sfbc]=0?? or gran == 0??

For each scale_factor_band sfb in the Scale Factor Band Class sfbc **scalefac_l[gran][ch][sfb]**

increment sfbc; sfbc < 4??

Huffmancodebits()

Next channel, granule

ancillary_bits　　　End of main_data()

Fig. 4.11b Syntax for MPEG-1 main_data() for Layer III.

gran　Granule number 0 or 1.

main_data_begin (9 bits) Specifies the location of the audio main_data of a frame as a negative offset in bytes from the first byte of the audio sync word. 0 indicates the main data starts after the Header and side-information.

private_bits　3 or 5 bits depending on mode, for private use.

scfsi[ch][sfbc] (1 bit) Scalefactor selection information, as in Layers I and II. But here it is for classes of Scale_Factor_Bands, instead of subbands.

part2_3_length[gran][ch] (12 bits) Specifies the number of main_data bits used for Scale_Factors and Huffman code data.

nr_big_values[gran][ch] (9 bits) The number of sample pairs in the lowest frequency region (big values region).

nr_small_values The number of sample quadruples in the mid frequency region (Small Values region). It is equal to the remaining samples in the granule, and is found from **part2_3_length** and **nr_big_values.**

global_gain[gran][ch] (8 bits) Specifies quantizer step size for the granule and channel.

scalefac_compress[gran][ch] (4 bits) Specifies the number of bits used for Scale_Factors. Actual values are taken from a Table.

window_switching_flag[gran][ch] (1 bit) Specifies a window other than the normal long window.

table_select[gran][ch][region] (5 bits) Specifies Huffman code tables for the first two big values regions.

region0_count (1 bits) Specifies number of samples in subregion 0 of the big_values region. Depends on **window_type** and **mixed_window_flag.**

region1_count (3 bits) Specifies number of samples in subregion 1 of the big_values region. Depends on **window_type** and **mixed_window_flag.**

region2_count The remaining samples of the big_values region. It is found from **nr_big_values, region0_count,** and **region1count.**

window_type[gran][ch] (2 bits) Specifies window type for the granule.

mixed_window_flag[gran][ch] (1 bit) Specifies that the two lowest subbands have a long window type, which may be different than the higher 30 subbands.

shortwin_gain[gran][ch][sw_samp] (3 bits) Specifies a gain offset from the global_gain for short window samples (sw_samp = 0 to 2) of short windows.

preflag[gran][ch] (1 bit) Specifies additional high frequency amplification.

scalefac_scale[gran][ch] (1 bit) Specifies quantization of Scale_Factors.

small_values_table_select[gran][ch] (1 bit) Specifies Huffman code tables for the Small Values region.

scalefac_l[gran][ch][sfb] (0 to 4 bits) Scale_Factors for long windows.

scalefac_s[gran][ch][sfb][sw_samp] (0 to 4 bits) Scale_Factors for each sample (sw_samp = 0 to 2) of a short window.

huffmancodebits( ) Huffman encoded data

4.4 MPEG-2 AUDIO

For two-channel stereo, the differences between MPEG-1 and MPEG-2 are minor. The initial PCM sampling rate choices are extended downward to include 16, 22.05, and 24 kHz. Also, the preassigned bitrates are extended to as low as 8 kbits/s. The lower sampling and bitrates are signalled by setting **ID** = 0 in the Header.

For the lower rates, MPEG-2 provides better quantization tables. Also, in Layer III, the coding of Scale_factors and intensity_mode stereo are improved.

4.4.1 MPEG-2 Backward Compatible (BC)

The most substantial change in MPEG-2 is the definition of a compression standard for five-channel surround-sound that is backward compatible (BC) with MPEG-1 audio. Thus, MPEG-1 audio players can receive and decode two channels of MPEG-2-BC audio, namely the left (L) and right (R) front speaker signals. However, MPEG-2-BC allows for up to four additional channels, namely front center (C), side/rear left surround (LS), side/rear right surround (RS), plus an optional low-frequency enhancement (LFE) channel.* MPEG-2-BC achieves Backward Compatibility by coding the L and R channels as MPEG-1, while the additional channels are coded as ancillary data in the MPEG-1 audio stream.

The L, C, R, LS, RS arrangement is often referred to as $^3/_2$ stereo, that is, 3 front and 2 rear channels. If LFE is included, it is called 5.1 channel stereo. The compressed data for these channels is called the *Multichannel* or mc coded data.

Other possible channel arrangements are $^3/_1$, $^3/_0$, $^2/_2$, and $^2/_1$, in addition to $^2/_0$ (MPEG-1 stereo) and

---

*The LFE frequencies typically range from 15 Hz to 120 Hz. One or more loudspeakers are positioned for LFE, which usually reproduces loud sound effects.

$^1/_0$ (MPEG-1 monaural). For the cases having three or less channels, it is possible to define a *Second Stereo Program*, perhaps for bilingual, consisting of a second left-right pair (L2, R2) that is carried on two of the unused channels. For example, $^3/_0$ + $^2/_0$ and $^2/_0$ + $^2/_0$ designate dual (perhaps bilingual) stereo programming on one MPEG-2-BC audio stream.

In the following, we will concentrate mainly on Layers I and II. Many of the concepts also carry over to Layer III, where they are coded more efficiently.

4.4.2   Subband Groups

Some of the adaptive and dynamic features of MPEG-2 are specified for Subband Groups instead of individual subbands or sub-subbands. Subband Groups for Layers I and II are defined in Table 4.4. Layer III has a similar definition, except that in Layer III they are called Scalefactorband_Groups.

4.4.2.1   *Matrixing for Backward Compatibility and Downward Compatibility**

In many cases, it is advantageous to transmit linear combinations of the five input audio channels instead of the input signals themselves. This is called *Matrixing*,[10] as shown in the matrix equation

Table 4.4   Subband Groups for MPEG-2 Layers I and II.

Subband Group for Layers I and II	Subbands in Subband Group
0	0
1	1
2	2
3	3
4	4
5	5
6	6
7	7
8	8 to 9
9	10 to 11
10	12 to 15
11	16 to 31

$$[\text{Lo Ro T2 T3 T4}] = [\text{L C R LS RS}] \times \mathbf{M}$$

where **M** is a 5×5 numerical matrix. The matrix converts the *Audio Multichannels* (L C R LS RS) into *Transmission Channels* (Lo Ro T2 T3 T4). Lo and Ro[†] are sent as normal MPEG-1 stereo channels, while T2, T3, T4, and LFE,[‡] if present, are sent as ancilliary data in the MPEG-1 stream. Matrixing can produce Lo and Ro that are better renditions of two-channel stereo, and therefore give better quality with MPEG-1 decoders. Also, transmission channels T2, T3, and T4 are often easier to compress by exploiting the subjective redundancies and correlations with Lo and Ro. However, better MPEG-1 quality can sometimes mean worse MPEG-2-BC quality when five-channel audio must be compressed.

For downward compatibility or other reasons, we may wish to code fewer than five transmission channels. For example, $^2/_2$ stereo could be produced by the matrixing:

$$[\text{Lo Ro T2 T3}] = [\text{L C R LS RS}] \times \mathbf{M}$$

where M is a 5×4 numerical matrix.

4.4.3   Dynamic Transmission Channel Switching

The matrix may be defined for all subbands, or individually for each subband group. The matrix definitions may also be changed dynamically for each new Frame.

4.4.4   Dynamic Crosstalk

Dynamic Crosstalk is the MPEG-2 implementation of intensity_stereo. It allows for dynamic deletion of sample bits in specified subband groups of specified Transmission Channels. As in MPEG-1, the missing samples are copied from another specified Transmission Channel. However, scale Factors are transmitted for the missing samples to

---

*Downward Compatibility is for decoders that cannot handle the full $^3/_2$ audio channel.
†MPEG sometimes uses T0 and T1 instead of Lo and Ro. However, the meaning is the same.
‡The optional LFE channel is never matrixed with other channels.

allow for independent amplitude control, as in MPEG-1 intensity_stereo.

4.4.5 Adaptive Multichannel Prediction

In Layers I and II, further compression of channels T2, T3, and T4 is allowed using predictive coding. The first eight subband groups may be coded predictively using up to three samples from T0 plus up to three samples from T1. Up to six predictors may be specified, depending on channel assignment and dynamic crosstalk modes. Layer III allows predictive coding of more channels using more samples.

4.4.6 Phantom Coding of the Center Channel

This refers to a compression method for the center channel, wherein the subbands above subband 11 are not transmitted. These are set to zero at the decoder, and the listener perceives high frequencies only from the other channels. In this case, the encoder may choose to matrix some of the center channel into the side channels prior to coding.

4.4.7 Multilingual Channels

MPEG-2-BC allows for up to seven so called *Multilingual* or ml channels. These are monophonic and would typically contain speech in different languages, commentary for the visually impaired, clean speech for the hearing impaired, educational side comments, and so forth. A multilingual channel might be self contained, or it might contain speech or singing that is to be added to the multichannel program signal. For example, the multichannel signal might be in $^2/_2$ format, while one of the multilingual channels is passed to the center channel speaker.

A maximum of seven multilingual channels can be carried on a single MPEG-2-BC stream. In addition, MPEG Systems allows up to 32 audio streams per program. Thus, in principle up to 224

multilingual channels could be carried in an MPEG-2 program.

4.4.8 Syntax and Semantics of MPEG-2-BC Audio

Space does not permit the inclusion of all details of the MPEG-2-BC audio stream. Thus, we will only describe the main structure and arrangement of data for the three Layers. As stated above, the Lo and Ro channels are sent as normal MPEG-1 data. The additional channels plus descriptive data, which is called the **mcml_extension( )**, replaces the ancillary data in the MPEG-1 stream. If there is insufficient capacity in the MPEG-1 stream for all of the **mcml_extension( )**, then the remainder may be carried in an *external stream** having a simple header and packet structure defined by MPEG-2. The external stream is treated by MPEG Systems as a separate audio stream, and normally requires the hierarchy descriptor in the Systems PSI.

In Layer I, three MPEG-1 Frames plus a packet from the external stream, if present, are required to carry the *mcml_extension( )*. In Layers II and III, only one MPEG-1 Frame is needed, plus a packet from the external stream, if present. The syntax of the MPEG-2-BC **mcml_extension** ( ) for Layers I, II and III is shown in Fig. 4.12.

The semantics of the MPEG-2-BC **mcml_extension( )** for Layers I, II and III are as follows:

mpeg-1_ancillary_data( ) Data for MPEG-1 decoders, which may be defined in the future.

ext_stream_present (1 bit) 1 indicates that an external stream exists, which contains the remainder of the data in case it does not all fit in one MPEG-1 compatible stream. This bit is redundant if Systems PSI is present.

n_ad_bytes 8 bits, indicating how many bytes are in the MPEG-1 compatible ancillary data in the external stream.

center 2 bits, indicating if a center channel exists and if phantom center coding is used.

center_limited[mch][sb] 1 indicates center channel subbands 12 and above are not sent.

---

*MPEG calls this the *extension* stream, which we believe is confusing.

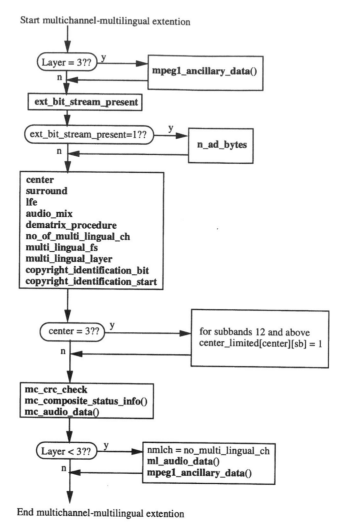

Start multichannel-multilingual extention

End multichannel-multilingual extention

Fig. 4.12 Syntax of the MPEG-2-BC **mcml_extension( ).** Two channels Lo and Ro (one if single_channel) are carried in the MPEG-1 portion of the Frame. The additional channels plus descriptive data is called the **mcml_extension( )** and replaces the ancillary data in the MPEG-1 stream. An external stream may be required to carry excess data. A boldface variable by itself means simply read in the value of the variable

surround 2 bits, indicating the presence of surround channels or the second stereo program.

lfe (1 bit) 1 indicates a low frequency enhancement channel is present.

audio_mix (1 bit) Indicates whether mixing for a large listening room, like a theatre, or for a small listening room. 1 indicates a small listening room.

dematrix_procedure 2 bits, indicating the dematrix procedure to be applied in the decoder.

no_of_multi_lingual_ch 3 bits, indicating the number of multilingual or commentary channels.

multi-lingual_fs (1 bit) 1 indicates the sampling frequency of the multilingual channels is half that of the main audio channels. 0 indicates both sampling frequencies are the same.

multi_lingual_layer (1 bit) Indicates whether Layer II multilingual or Layer III multilingual is used.

copyright_identification_start (1 bit) 1 indicates that a 72-bit copyright identification field starts in this frame to identify copyrighted material.

mc_crc_check (16 bits) Check word for error detection.

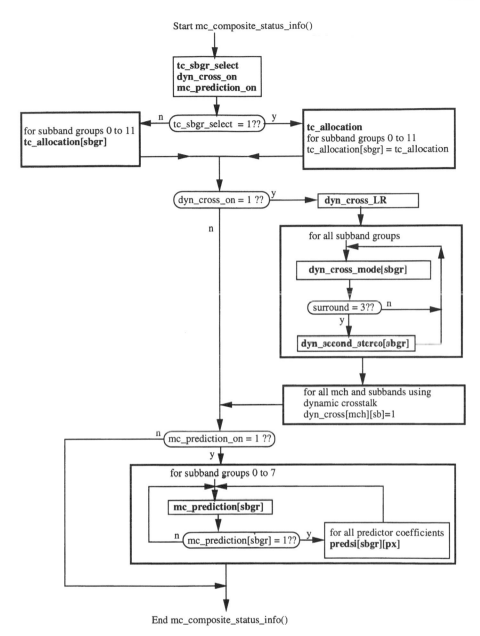

Fig. 4.13 Syntax for MPEG-2-BC **mc_composite_status_info**( ) for Layers I and II. Each Frame represents 1152 samples of input PCM audio.

mc_composite_status_info ( ) Specifies information on the multichannel coding mode.

mc_audio_data( ) Coded audio data for multichannel audio.

ml_audio_data( ) Coded audio data for multilingual audio.

The syntax of the MPEG-2-BC Multichannel Layers I and II **mc_composite_status_info**( ) is shown in Fig. 4.13.

The semantics of the MPEG-2-BC Multichannel Layers I and II **mc_composite_status_info**( ) are as follows:

tc   transmission channel.

sbgr   subband group number.

dyn_cross_on   1 bit, indicating whether dynamic crosstalk is used.

mc_prediction_on   1 bit, indicating whether mc_prediction is used.

tc_sbgr_select   1 bit, to indicate if the tc_allocation below is valid for all subbands or for individual subband groups.

tc_allocation, tc_allocation[sbgr] (1 bit)   Contains information on the multichannel allocation of the additional channels that are not in the MPEG-1 portion of the stream.

dyn_cross_LR   1 bit, indicating whether dynamic crosstalk copying shall be from Lo or Ro.

dyn_cross_mode[sbgr] (1 to 4 bits)   Indicates between which transmission channels dynamic crosstalk is active.

dyn_cross[mch][sb]   1 indicates no samples sent for multichannel mch and subband sb.

dyn_second_stereo[sbgr] (1 bit)   Indicates whether dynamic crosstalk is used in the second stereo program.

mc_prediction[sbgr] (1 bit)   Indicates whether multichannel prediction is used in subband group sbgr.

predsi[sbgr][px]   2 bits, predictor select information. Indicates whether the predictor indexed by px in subband group sbgr is used, and if so, how many predictor coefficients are transferred.

The syntax of the MPEG-2-BC Multichannel Layers I and II **mc_audio_data**( ) is shown in Fig. 4.14.

The semantics of the MPEG-2-BC Multichannel Layers I and II **mc_audio_data**( ) are as follows:

mch   Multichannel (mc) number.

trpl   Three-sample triplet number.

lfe_allocation (4 bits)   Specifies the quantiser used for the low-frequency enhancement channel.

allocation[mch][sb] (2 to 4 bits)   Specifies the quantiser used for the samples in subband sb of the multichannel mch.

scfsi[mch][sb] (2 bits)   Scale_Factor select information. Indicates the number of Scale_Factors transferred for subband sb of the multichannel mch. As in MPEG-1 Layer II, the frame is divided into three equal blocks (b) of 12 samples each.

delay_comp[sbgr][px] (3 bits)   Specifies a shift of subband samples for delay compensation in subband group sbgr and predictor index px.

pred_coef[sbgr][px][pci] (8 bits)   Coefficient of predictor with up to second order in subband group sbgr and predictor index px.

Lf_Scale_Factor (6 bits)   Scale_Factor for the low-frequency enhancement channel.

Scale_Factor[mch][sb][b] (6 bits)   Scale_Factor for block b of subband sb of multichannel mch.

Lf_sample[trpl] (2 to 15 bits)   Coded single sample of the low frequency enhancement channel. There are three normal samples for each LFE sample.

samplecode[mch][sb][trpl] (5 to 10 bits)   Coded three consecutive samples in triplet trpl of subband sb of multichannel extension channel mch.

sample[mch][sb][s] (2 to 16 bits)   Coded sample s of subband sb of multichannel extension channel mch.

The syntax of the MPEG-2-BC Multichannel Layers I and II **ml_audio_data**( ) is shown in Fig. 4.15.

The semantics of the MPEG-2-BC Multichannel Layers I and II **ml_audio_data** are as follows:

ntrip   Number of sample triplets. It is 12 or 6 depending on whether sampling is at full rate (multi_lingual_fs = 0) or half rate (multi_lingual_fs = 1), respectively.

mlch   Multilingual (ml) channel number.

allocation[mlch][sb] (2 to 4 bits)   Specifies the quantiser used for subband sb of multilingual channel mlch. The number of levels is taken from tables, depending on sampling and bitrates.

scfsi[mlch][sb] (2 bits)   Specifies the number of scalefactors for subband sb of multilingual channel mlch. As in MPEG-1 Layer II, the frame is divided into three equal blocks (b) of 12 or 6 samples each depending on whether sampling is at full or half rate, respectively. One, two, or three scalefactors may be sent.

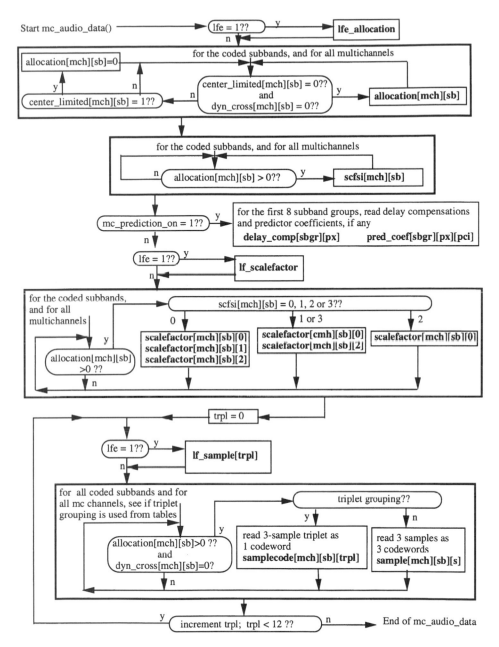

Fig. 4.14 Syntax for MPEG-2-BC **mc_audio_data( )** for Layers I and II. Each Frame represents 1152 samples of input PCM audio.

scalefactor[mlch][sb][b] (6 bits) Scale factor index for block b of subband sb of multilingual channel mlch. The actual scale factor is taken from tables.

samplecode[mlch][sb][trpl] (5 to 10 bits) Code word for three samples in triplet trpl of subband sb of multilingual channel mlch.

sample[mlch][sb][s] (2 to 16 bits) Coded representation of the sample s of subband sb of multilingual extension channel mlch.

This ends the description of MPEG-2 Layers I and II.

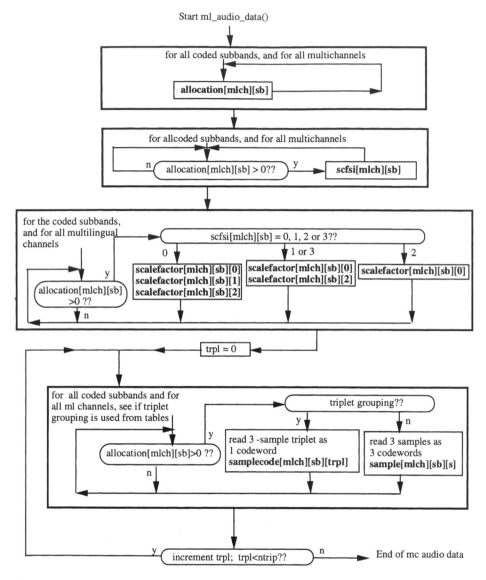

Fig. 4.15   Syntax for MPEG-2-BC **ml_audio_data**( ) for Layers I and II. Each Frame represents 1152 samples of input PCM audio.

The syntax and semantics of Layer III is more complex, reflecting its greater coding efficiency. Space does not permit its description here.

4.4.9   MPEG-2 Non Backward Compatible (NBC)

Providing backward compatibility with MPEG-1 necessarily places a constraint on the quality that can be achieved in the compression of five-channel audio. Moreover, the quality may not be easily improved by increasing the bitrate. For this reason, there is currently underway an effort to standardize a Non Backward Compatible audio compression algorithm that can achieve state-of-the-art audio quality with five channels. We designate it as MPEG-2-NBC audio.

Two algorithms have already been found to have better quality than MPEG-2-BC. They are the AC-3 system[11] from Dolby and the PAC algorithm[12] from AT&T.

REFERENCES

1. P. NOLL, "Wideband Speech and Audio Coding," IEEE Commun. pp. 34–44. (November 1993). See also P. NOLL, "Digital Audio Coding for Visual Communications," Proc. IEEE, pp. 925–943 (June 1995).

2. ISO/IEC 11172–3, "Coding of Moving Pictures and Associated Audio for Digital Storage Media at up to about 1.5 Mbits/s-Part 3: Audio. ISO/IEC JTC 1/SC 29. See also D. PAN, "A Tutorial on MPEG/Audio Compression," IEEE Multimedia, pp. 60–74 (Summer 1995).

3. ISO/IEC 13818–3, "Generic Coding of Moving Pictures and Associated Audio Information—Part 3: Audio." ISO/IEC JTC 1/SC 29.

4. A. HOOGENDORN, "Digital Compact Cassette," Proc. IEEE 82:554–563 (April 1994).

5. T. YOSHIDA, "The Rewritable MiniDisc System," Proc. IEEE 82:1492–1500 (October 1994).

6. N. S. JAYANT and P. NOLL, Digital Coding of Waveforms, Prentice Hall, Englewood Cliffs, NJ (1984).

7. P. P. VAIDYANATHAN, Multirate Systems and Filter Banks, Prentice Hall, Englewood Cliffs, NJ (1993). See also, "Polyphase Quadrature Filters—A New Subband Coding Technique," Proc. Int'l Conf. IEEE ASSP, IEEE Press, New York, pp. 1280–1283 (1984).

8. H. S. MALVAR, Signal Processing with Lapped Transforms, Artech House, Norwood, MA (1992).

9. J. D. JOHNSTON and A. J. FERREIRA, "Sum-Difference Stereo Transform Coding," Proc. IEEE ICASSP '92, II:569–572.

10. "Multichannel Stereophonic Audio System With and Without Accompanying Picture," ITU-R Recommendation 775, (November 1992).

11. "Digital Audio Compression (AC-3) Standard," published by the United States Advanced Television Systems Committee, April 12, 1995. Also, C. TODD et al., "AC-3: Flexible Perceptual Coding for Audio Transmission and Storage," AES 96th Convention (February 1994).

12. J. JOHNSTON et al., "The AT&T Perceptual Audio Coder (PAC)," Proceedings of the American Acoustic Society Conference, New York (October 1995).

13. A. HOOGENDORN, "Digital Compact Cassette," Proc. IEEE 82:554–563 (April 1994).

14. T. YOSHIDA, "The Rewritable MiniDisc System," Proc. IEEE 82:1492–1500 (October 1994).

5

Video Basics

Before we actually discuss video compression, and in particular, MPEG video compression, we need to understand the various aspects of the video signal itself and several related issues. Simply speaking, to understand how to compress video, we must have some understanding of how a video signal is generated, how the human visual system processes visual information, how a video signal is represented for storage[8] or transmission,[6] and how it can be converted between the various representations. We will certainly provide answers to these questions and much more. Before you begin wondering what this has to do with MPEG, remember that MPEG video compression is performed on color video signals represented in digital format. As explained in earlier chapters, digital format allows case of manipulation as well as robust storage[8,9] and transmission. Also, depending on the needs of an application and the bandwidth available, one of many video representations may be used in MPEG compression, which may be different from the representations made available by the camera used for imaging the scene. Furthermore, the video compressed by MPEG is eventually decoded for display, and if the decoded video uses a different representation than the display can handle, then again conversion is needed.

First we discuss the process of video imaging with emphasis on video scanning methods used for TV and film; image aspect ratio (IAR), which con-trols display shape; and gamma, which controls the relationship between voltage and intensity. Second, we discuss the human visual system, properties of colored light such as hue and saturation, the coordinate system used for representation of color video signals, and the composite and component systems used in various TV systems. Third, since MPEG compression uses digital video signals, we briefly discuss the generation of digital video from analog video signals. Fourth, we discuss specific digital component formats that MPEG coding uses. Fifth and sixth, we describe the various conversions[10,11] between the video formats used by MPEG; typically these conversions are needed in preprocessing prior to MPEG encoding or postprocessing after MPEG decoding. The last section discusses a unique form of conversion called mixing of video, which has potential for use in MPEG-1 and MPEG-2 applications involving overlaid text and graphics, as well as future MPEG based coding.

5.1 VIDEO IMAGING

To begin with, imaging is a process by which a representation of an actual scene is generated. Imaging can be of various types, such as normal photography, X-rays, electronic documents, electronic still pictures, motion pictures, and TV. We are obviously interested in the kind of imaging that is time varying as it can adequately represent the motion in a scene. Video can be thought of as comprised of a sequence of still pictures of a scene taken at various subsequent intervals in time. Each still picture represents the distribution of light energy and wavelengths over a finite size area and is expected to be seen by a human viewer. Imaging devices such as cameras allow the viewer to look at the scene through a finite size rectangular window. For every point on this rectangular window, the light energy has a value that a viewer perceives which is called its intensity. This perceived intensity of light is a weighted sum of energies at different wavelengths it contains.

The lens of a camera focuses the image of a scene onto a photosensitive surface of the imager, which converts optical signals into electrical signals. There are basically two types of video imagers: (1) tube-based imagers such as vidicons, plumbicons, or orthicons, and (2) the more recent, solid-state sensors such as charge-coupled devices (CCD). The photosensitive surface of the tube imager is typically scanned with an electron beam or other electronic methods, and the two-dimensional optical signal is converted into an electrical signal representing variations of brightness as variations in voltage. With CCDs the signal is read out directly.

5.1.1 Raster Scan

Scanning is a form of sampling of a continuously varying two-dimensional signal. Raster scanning[1-3] is the most commonly used spatial sampling representation. It converts a two-dimensional image intensity into a one-dimensional waveform. The optical image of a scene on the camera imager is scanned one line at a time from left to right and from top to bottom. The imager basically converts the brightness variations along each line to an electrical signal. After a line has been scanned from left to right, there is a retracing period and scanning of next line again starts from left to right. This is shown in Fig. 5.1.

In a television camera, typically an electron beam scans across a photosensitive target upon which light from a scene is focused. The scanning spot itself is responsible for local spatial averaging of the image since the measured intensity at a time

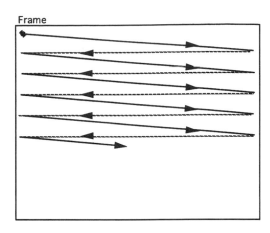

Fig. 5.1 Raster scan of an image

is actually a weighted average over the area of the spot. With the use of electron beam scanning, however, it is very hard to control the scanning spot filter function, with the result that the image is sometimes not sufficiently prefiltered. Optical scanning devices that use a beam of light are significantly better in this regard. Prefiltering can also be done in most cameras by a slight defocusing of the optical lens. With raster scanned display devices, filtering in the vertical direction is done by the viewer's eye or by adjustment of the scanning beam or optical components.

The smallest detail that can be reproduced in a picture is about the size of a picture element, typically referred to as a *pixel* (in computer terminology) or a *pel* (in video terminology). A sampled TV raster, for example, can be described as a grid of pels as shown in Fig. 1.2. The detail that can be represented in the vertical direction is limited by the number of scan lines. Thus, some of the detail in vertical resolution is lost as the result of raster scanning fall between the scan lines and cannot be represented accurately. Measurements indicate that only about 70% of the vertical detail is actually represented by scan lines. Similarly, some of the detail in the horizontal direction is lost owing to sampling of each scan line to form pels.

5.1.2 Interlace Raster Scan

The choice of number of scan lines involves a tradeoff among contradictory requirements of bandwidth, flicker, and resolution. Interlaced[1-3,5-6] scanning tries to achieve these tradeoffs by using frames that are composed of two *fields* sampled at different times, with lines of the two fields inter-leaved, such that two consecutive lines of a frame belong to alternate fields. This represents a verti-cal–temporal tradeoff in spatial and temporal res-olution. For instance, this allows slow-moving objects to be perceived with higher vertical detail while fast-moving objects are perceived at a higher temporal rate, albeit at close to half vertical resolu-tion. Thus the characteristics of the eye are well matched because, with slower motion, the human visual system is able to perceive the spatial details, whereas with faster motion spatial details are not as easily perceived. This interlace raster scan is shown in Fig. 5.2.

From a television transmission and display aspect, the interlace raster scan represents a rea-sonable tradeoff in the bandwidth necessary and the amount of flicker generated. It is worth noting that the tradeoffs that may have been satisfactory for television may not be satisfactory for computer displays, owing to the closeness of the screen to the viewer and the type of material typically displayed, that is, text and graphics. Thus, if interlaced TV displays were to be used with computers, the result would be an annoying large area flicker, interline flicker, line crawling, and other problems. To avoid these problems, computer displays use noninter-laced (also called progressive or sequential) displays with refresh rates of higher than 60 frames/s, typi-cally 72 frames/s.

Table 5.1 Equivalent Line Number of Film and TV Image Formats

Image Formats	Equivalent Line Number Range
NTSC TV	175 to 310
16 mm Film	200 to 300
35 mm Film	350 to 450
HDTV	300 to 515

5.1.3 Film versus Television

Originally, television was designed to capture the same picture definition as that of 16-mm film. Actually, the resolution of 16-mm film is signifi-cantly higher than that of normal TV, but the per-ceived sharpness is about the same owing to a bet-ter temporal response of normal TV. To compare the performance of quite different display systems, the concept of equivalent line number is useful. In Table 5.1 we show the equivalent line numbers for film and television.

Based on what we have said so far, it may appear that a television system can achieve the per-formance of film. In practice, however, there are too many ways in which the performance of a TV system can become degraded. In this context, it is worth noting that *frequency response*, which is the variation in peak-to-peak amplitude of a pattern of alternate dark and bright lines of varying widths, can be used to establish the performance of an imaging system. In reality, TV and film images never really appear the same owing to the differ-ence in shape of the overall frequency response. For instance, the frequency response of TV drops much faster, whereas the frequency response of film decays more gradually.

The response of the human eye over time is generally continuous,[1-3] that is, nerve impulses travel from the eye to the brain at the rate of about 1000 times per second. Fortunately, televi-sion does not need to provide new images of a scene at this rate as it can exploit the property of persistence of vision to retain the sensation of an image for a time interval even though the actual image may have been removed. Both TV and motion pictures employ the same principle—a scene is portrayed by showing a sequence of slightly different images (frames) taken over time to replicate the motion in a scene. For motion in a

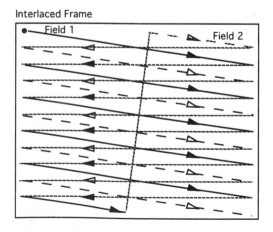

Fig. 5.2 Interlaced raster scan of an image

scene to appear realistic, the rate of imaging of frames should be high enough to capture the motion without jerkiness, and the rate of display should be high enough so that the captured motion can be replicated smoothly. The display rate must also be high enough so that persistence of vision works properly. One of the nuances of human vision is that if the images being presented to it are bright, the persistence of vision is actually shorter, thus requiring a higher repetition rate for display.

In the early days of motion pictures,[1-3] it was found that for smooth motion rendition, a minimum frame rate of 15 per second (15 frames/s) was necessary. Thus old movie equipment, including home movie systems, use 16 frames/s. However, because of a difficulty in representing action in cowboy movies of those days, the motion picture industry eventually settled on the rate of 24 frames/s, which is used to this day. Note that although 24 frames/s movies use 50% more film than 16 frames/s movies, the former was adopted as a standard as it was considered worthwhile. As movie theater projection systems became more powerful, the nuance of eye mentioned earlier, which requires a higher frame rate, began to come into play, resulting in the appearance of flicker. A solution that did not require a change in the camera frame rates of movies was to use a 2 blade rotating shutter in the projector, synchronized to the film advance, which could briefly cut the light and result in double the displayed frame rate to 48 frames/s. Today, even a 3 blade shutter to effectively generate a display rate of 72 frames/s is sometimes used since projection systems have gotten even more brighter.

The frame rate used for monochrome TV in Europe was in fact chosen to be slightly higher than that of film at 25 frames/s, in accordance with electrical power there, which uses 50 cycles/s (Hz) rate. Television in Europe uses systems such as PAL and SECAM, each employing 25 frames/s. It is common practice in Europe to run 24 frames/s movies a little faster at 25 frames/s. The PAL and SECAM based TV systems are currently in widespread use in many other parts of the world as well.

The frame rate of monochrome TV in the United States was chosen as 30 frames/s, primarily to avoid interference to circuits used for scanning and signal processing by 60-Hz electrical power. In 1953, color was added to monochrome by migration to an NTSC system in which the 30 frames/s rate of monochrome TV was changed by 0.1 percent to 29.97 frames/s owing to the requirement of precise separation of the video signal carrier from the audio signal carrier. Incidentally Japan and many other countries also employ the NTSC TV system.

For reasons other than those originally planned, the frame rate chosen in the NTSC TV system turned out to be a good choice since there is a basic difference in the way TV is viewed as compared to film. While movies are generally viewed in dark surroundings that reduce the human eye's sensitivity to flicker and the smoothness of motion, TV is viewed under condition of surround lighting, in which case the human visual system is more sensitive to flicker and the smoothness of motion. Thus, TV requires a higher frame rate than movies to prevent the appearance of flicker and motion-related artifacts.

One very important point is that unlike TV, which uses interlace, film frames are sequential (also called noninterlaced or progressive).

5.1.4 Image Aspect Ratio (IAR)

The image aspect ratio is generally defined as the ratio of picture width to height* and impacts the overall appearance of the displayed image. For example, the IAR of normal TV images is 4/3, meaning that the image width is 1.33 times the image height. This value was adopted for TV, as this format was already used and found acceptable in the film industry at that time, prior to 1953. However, since then the film industry has migrated to wide-screen formats with IARs of 1.85 or higher. Since subjective tests on viewers show a significant preference for a wider format than that used for normal TV, HDTV uses an IAR of 1.78, which is quite close to that of the wide-screen film format. Currently, the cinemascope film format provides one of the highest IARs of 2.35. Table 5.2

---

*Some authors, including MPEG, use height/width.

Table 5.2 Image Aspect Ratio (IAR) of Film and TV Image Formats

Image Formats	Aspect Ratio
NTSC, PAL and SECAM TV	1.33
16 mm and 35 mm Film	1.33
HDTV	1.78
Widescreen Film	1.85
70 mm Film	2.10
Cinemascope Film	2.35

compares aspect ratios of a variety of formats used in film and TV.

Sometimes there may be a mismatch such that an image that was intended for display at one IAR may have to be shown on a display of a different IAR. Figure 5.3 shows images of a 4:3 IAR on a 16:9 display and vice versa.

In the discussion of aspect ratio the concept of pel aspect ratio (PAR) is often quite useful. Although pels have no intrinsic shape of their own,

under some restrictions, they can be thought of as contiguous rectangles as shown in Fig. 1.2. MPEG-1 video allows specification of PAR through the parameter *pel_aspect_ratio*, a 4-bit code representing several discrete values of PAR.

MPEG-2 video interprets the same 4-bit code as *aspect_ratio_information*, and uses it to either specify that pels in a reconstructed frame are square (PAR = 1) or it specifies one of three values of IAR (see Sec. 10.2.2). When IAR is specified, there are two cases, one in which the reconstructed image is to be shown on the entire display, and another in which, depending on the display size and coded image size, either a portion of the reconstructed image is displayed or a full image is displayed on a portion of the screen.

In the first case, PAR can be calculated as follows:

$$PAR = IAR \times vertical\_size/horizontal\_size$$

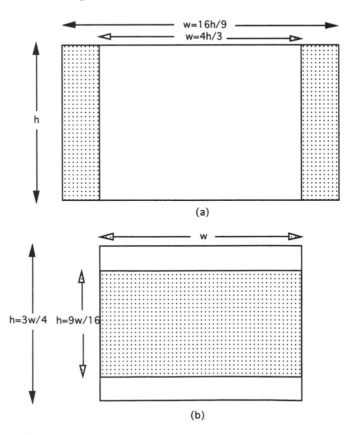

Fig. 5.3 Image aspect ratio (IAR) =w/h.

where horizontal_size is the width of the image in terms of pels and the vertical_size is the height of the image in lines. As an example consider a 4:3 IAR display, and suppose the image width is 720 pels and the image height is 486 lines. Then PAR can be calculated as follows.

$$PAR = {}^4/_3 \times {}^{486}/_{720} = 0.9$$

In the second case, PAR is calculated as follows:

$$PAR = IAR \times display\_vertical\_size/ \\ display\_horizontal\_size$$

This second case is signalled via *sequence_display_extension*( ) which carries information regarding *display_horizontal_size* and *display_vertical_size* which determine the size of the display rectangle. When this display rectangle is smaller than the image size it implies that only a portion of image is to be displayed, and conversely when the display rectangle is larger than the image size it implies that the image is displayed only on a portion of the display screen.

It is worth noting that computer displays use square pels (PAR = 1.0) and if material intended for TV is shown on a computer display, correction of the pel aspect ratio is necessary; otherwise, circles on a TV screen appear as ellipses on a computer display.

5.1.5 Gamma

Many TV cameras and all CRT-based displays have a nonlinear relationship between signal voltage and light intensity. The light intensity input to the camera or output by the display is proportional to the voltage raised to the power gamma and is given by the following generalized relationship:

$$B = cv^{\gamma} + b$$

where B is the light intensity, v is voltage, c is a gain factor, b is either cutoff (camera) or black level (cathode ray tube, CRT) light intensity, and γ typically takes values in the range of 1 to 2.5, although for some picture displays it can be as high as 3.0.

Typically, camera picture tubes or CCD sensors have a gamma of about 1 except for a vidicon type picture tube, which has a gamma of 1.7. The CRT displays use kinescopes, which typically have a gamma of 2.2 to 2.5.

To avoid gamma correction circuitry inside millions of TV receivers, gamma correction is done prior to transmission. For example, assuming the gamma of the camera to be 1 and that of the display to be 2.2, then the camera voltage is raised to a power of $^1/_{2.2} = 0.45$.

The result of gamma correction is that a gamma-corrected voltage is generated by the camera that can be transmitted and applied directly to the CRT. As a side benefit, the gamma-corrected voltage is also less susceptible to transmission noise.

In normal TV transmission, the signal from the camera is gamma corrected, modulated, and transmitted to the receiver as a radiofrequency (RF) signal, which is received by the receiver's antenna and demodulated and applied to the display. The main components of such a TV transmission system are shown in Fig. 5.4.

In our discussion thus far, we have considered only monochrome TV and have deferred a discussion about color TV so that we could introduce

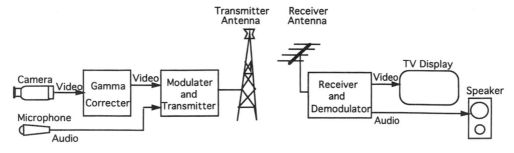

Fig. 5.4 Monochrome television analog transmission system

some aspects of human vision and how they are exploited in the design of a color TV system.

5.2   THE HUMAN VISUAL SYSTEM AND COLOR TV

With the background on video imaging developed thus far, we are now well prepared to delve into the intricacies of the human visual system, the perception of color, the representation of color, and further to specific color TV systems.

5.2.1   The Human Visual System

The human eye is a remarkably intricate system; its anatomy is shown in Fig. 5.5.

Incoming light bouncing off different objects is refracted by the cornea toward the pupil. The pupil is the opening of the iris through which all light enters the eye. The light again is refracted by the lens onto the back of the eyeball, forming an image on the retina. The retina consists of receptors sensitive to light called photoreceptors connected by nerve cells. To reach these photoreceptors, light must first pass through nerve cells. The photoreceptors contain chemical pigments that can absorb light and initiate a neural response. There are two types of photoreceptors, rods and cones. Rods are responsible for low light vision, while cones are responsible for details and color under normal light conditions, as in daylight. Both rods and cones enable vision when the amount of light is between the two extremes. Light absorbed by photoreceptors initiates a chemical reaction that bleaches the pigment, which reduces the sensitivity to light in proportion to amount of pigment bleached. In general, the amount of bleached pigment rises or falls depending on the amount of light. The visual information from the retina is passed via optic nerves to the brain. Beyond the retina, processing of visual information takes place en route to the brain in areas called the lateral geniculate and the visual cortex. Normal human vision is binocular, composed of a left-eye and a right-eye images. The left-eye image is processed by the right half of the brain, and the right-eye image by the left half of brain. We further discuss

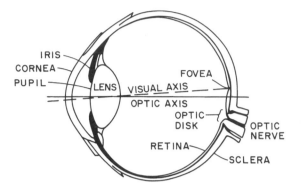

Fig. 5.5   Anatomy of the human eye

the intricacies of the human visual system, and in particular as it refers to binocular vision, in Chapter 15.

Light of various wavelengths produces a sensation called *color*. Different distributions generally result in the perception of different colors. A color is thus visible energy of a given intensity and of a given composition of wavelengths. The normal human visual system can identify differences in sensation caused by different colors and thus colors can be classified. In our earlier discussion of the human visual system we mentioned that the retina of the human eye contains photoreceptors called cones that are responsible for color vision. The human retina contains a color-sensitive area that consists of three sets of cones that are individually sensitive to red, green, and blue light. Thus the sensation of color experienced directly depends on the relative amounts of red, green, and blue light; for instance, if only blue and red cones are excited, the sensation is magenta. When different combinations of red, blue, and green cones are excited, different colors are sensed; a combination of approximately 30% red, 60% green, and 10% blue results in white light.

5.2.2   Hue and Saturation

The sensation known as color produced by visible light is also called *hue*. Visible light is electromagnetic radiation with a spectrum of wavelengths; a hue is determined by its dominant wavelength. Besides hue, *saturation* is another property of interest and represents the degree of purity of a color. For instance, a pure spectral color having a single wavelength has a saturation of 100%, while the saturation of white or gray light is zero. The three basic properties of light—*brightness*, *hue*, and *saturation*—can be completely defined by its spectral distribution, a plot of amplitude versus wavelength. Based on our earlier discussion about the sensation of color caused by the response of three types of color-sensitive cones in the retina of human eye, it can be stated that the perceived hue and saturation of two colors is exactly the same if they invoke the same response from the color-sensitive cones in the retina. The trichromatic theory of color vision implies that hue and saturation of nearly any color can be duplicated by an appropriate combination of the three *primary* colors.

5.2.3 Color Primaries

One of the requirements in the choice of primary colors is that they are independent and chosen such as to represent the widest range of hues and saturation by appropriate combinations of the three primaries. There are basically two kinds of color primaries, one composed of subtractive primaries and the other composed of additive primaries. Subtractive primaries are used in printing, photography, and painting, whereas additive primaries are used in color television.

Although a given color can be matched by an additive mixture of three primaries, when different observers match the same color, slight differences in the amount of primary colors needed to match the color arise. In 1931, the CIE defined a standard observer and three standard primaries by averaging the color matching data of a large number of observers with normal vision. The CIE chromaticity diagrams enclose a horseshoe-shaped region in which each point represents a unique color and is identified by its x and y coordinates. Three such points identify primary colors, and the triangle formed by these points represents the gamut of colors that can be represented by the display using these primaries. For example, the FCC has defined three such points that represent the color primaries for the NTSC system. Likewise three such points are defined for the PAL system.

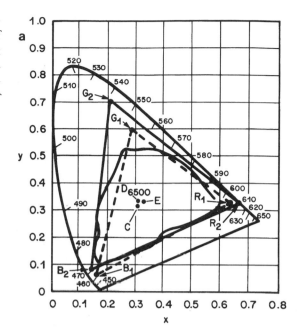

Fig. 5.6 Color space primaries for NTSC and PAL

The color space primaries for the NTSC and the PAL systems are shown in Fig. 5.6.

As mentioned earlier, in the United States in 1953 color was added to the monochrome TV system by migration to the NTSC system. This system has been adopted widely in North and South America, the Caribbean, and in parts of Asia. On the other hand the PAL and SECAM color TV systems originated in Europe and are in use in the remaining parts of the world. Although all three TV systems have basic differences in picture resolution and frame rates, they have considerable similarities as they all belong to the class of *composite systems*. A second class of TV systems is called *component systems* and thus far is used in most digital applications.

5.2.4 Composite Systems: NTSC, PAL, and SECAM

In composite systems, the luminance and chrominance signals are multiplexed within the same channel or on the same carrier signal. Composite formats[1,4,5,7] such as NTSC, PAL, or SECAM formats were designed to be compatible with the monochrome format. That is, monochrome receivers could receive color signal transmission but would display it as a compatible monochrome signal. Likewise, monochrome transmission could be received by color receivers.

A camera imaging a scene generates for each pel the three color weights RGB, which may be further processed for transmission or storage. At the receiver, the three components are sent to the display, which regenerates the contents of the scene at each pel from the three color components. For transmission or storage between the camera and the display a luminance signal Y representing brightness and two chrominance signals representing color are used. The need for such a transmission system arose with NTSC, where compatibility with monochrome receivers required a black-and-white signal, which is now referred to as the Y signal. It is well known that the sensitivity of the human eye is highest to green light, followed by that of red, and the least to blue light. The NTSC system exploited this fact by assigning a lower bandwidth to the chrominance signals as compared to the luminance, Y, signal. This made it possible to save bandwidth without losing color quality. The PAL and SECAM systems also employ reduced chrominance bandwidths.

5.2.4.1 The PAL System

The YUV color space of PAL is employed in one form or another in all three color TV systems. The basic YUV color space can be generated from gamma-corrected RGB (referred to in equations as $R'G'B'$) components as follows:

$$Y = 0.299R' + 0.587G' + 0.114B'$$

$$U = -0.147R' - 0.289G' + 0.436B' = 0.492(B' - Y)$$

$$V = 0.615R' - 0.515G' - 0.100B' = 0.877(R' - Y)$$

The inverse operation, that is, generation of gamma-corrected RGB from YUV components, is accomplished by the following:

$$R' = 1.0Y + 1.140V$$

$$G' = 1.0Y - 0.394U - 0.580V$$

$$B' = 1.0Y - 2.030U$$

Although our primary interest is in the component color space of PAL, a brief discussion about how composite encoding is performed may be useful.

The Y, U, and V signals in PAL are multiplexed in a total bandwidth of either 5 or 5.5 MHz. With PAL, both U and V chrominance signals are transmitted with bandwidth of 1.5 MHz. A color subcarrier is modulated with U and V via Quadrature Amplitude Modulation (QAM) and the composite signal is limited to the allowed frequency band, which ends up truncating part of the QAM signal. The color subcarrier for PAL is located at 4.43 MHz. PAL transmits the V chrominance component as $+V$ and $-V$ on alternate lines. The demodulation of the QAM chrominance signal is similar to that of NTSC, which we will cover a bit more in detail. Here it suffices to say that the recovery of the PAL chrominance signal at the receiver includes averaging of successive demodulated scan lines to derive the U and V signals.

5.2.4.2 The NTSC System

The NTSC color space of YIQ can be generated from the gamma-corrected RGB components or from YUV components as follows:

$$Y = 0.299R' + 0.587G' + 0.114B'$$

$$I = 0.596R' - 0.274G' - 0.322B'$$
$$= -(\sin 33°)\,U + (\cos 33°)\,V$$

$$Q = 0.211R' - 0.523G' + 0.311B'$$
$$= (\cos 33°)\,U + (\sin 33°)\,V$$

The inverse operation, that is, generation of gamma-corrected RGB components from the YIQ composite color space, can be accomplished as follows:

$$R' = 1.0Y + 0.956I + 0.621Q$$

$$G' = 1.0Y - 0.272I - 0.649Q$$

$$B' = 1.0Y - 1.106I + 1.703Q$$

In NTSC, the Y, I, and Q signals are all multiplexed into a 4.2 MHz bandwidth. Although the Y component itself takes a 4.2 MHz bandwidth, multiplexing all three components into the same

4.2 MHz becomes possible by interleaving luminance and chrominance frequencies, without too much interference with one another. This is done by defining a color subcarrier at approximately 3.58 MHz. The two chrominance signals I and Q are QAM modulated onto this carrier. The envelope of this QAM signal is approximately the saturation of the color, and the phase is approximately the hue. The luminance and modulated chrominance signals are then added to form the composite signal. The process of demodulation first involves comb filtering (horizontal and vertical filtering) of the composite signal to separate the luminance and a combined chrominance signal followed by further demodulation to separate the I and Q components.

5.2.4.3 The SECAM System

The SECAM composite color space is $YDrDb$ and can be generated from gamma-corrected $R'G'B'$ components:

$$Y = 0.299R' + 0.587G' + 0.114B'$$

$$Db = -0.450R' - 0.883G' + 1.333B' = 3.059U$$

$$Dr = -1.333R' + 1.116G' - 0.217B' = -2.169V$$

The inverse operation, that is, generation of gamma-corrected RGB from $YDrDb$, can be accomplished as follows:

$$R' = 1.0Y - 0.526Dr$$

$$G' = 1.0Y - 0.129Db + 0.268Dr$$

$$B' = 1.0Y + 0.665Db$$

In the SECAM system, the video bandwidth is 6 MHz. The chrominance signals are transmitted on alternate lines by use of Frequency Modulation (FM) of the two color carriers which are located at 4.25 MHz and 4.41 MHz. Owing to the higher allocated bandwidth of 6 MHz, the demodulation of chrominance signals is much simpler than with QAM used in PAL and NTSC. Further, because of FM, gain distortion of the signal is somewhat less.

5.2.4.4 Comparison of NTSC, PAL, and SECAM

In Table 5.3 we provide a summary of the TV parameters used in the composite systems. Composite color TV systems resulted from the necessity of fitting extra color information in the same bandwidth as that used by the monochrome TV system and maintaining compatibility with monochrome TV receivers. Although the tradeoffs made were quite reasonable, understandably, some artifacts resulted. One of the disadvantages of the composite system is that for some pictures there is crosstalk between luminance and chrominance signals that gives rise to cross-color and cross-luminance artifacts. Cross-color artifacts result when the luminance signal contains frequencies near the subcarrier frequency resulting in spurious color, whereas cross-luminance artifacts are caused by a change of the luminance phase from field to field near edges of objects in a scene due to imperfect cancellation of the subcarrier.

5.2.5 The Component System

In a component TV system, the luminance and chrominance signals are kept separate, such as on separate channels or at different times. Thus the component TV system is intended to prevent the crosstalk that causes cross-luminance and cross-

Table 5.3 Summary of parameters used in worldwide Color TV Standards

Parameter	NTSC	PAL	SECAM
Field Rate (Hz)	59.94	50	50
Lines per Frame	525	625	625
Gamma	2.2	2.8	2.8
Audio Carrier (MHz)	4.5	QAM	FM
Color Subcarrier (MHz)	3.57	4.43	4.25 (+U), 4.4 (−V)
Color Modulation Method	QAM	QAM	FM
Luminance Bandwidth (MHz)	4.2	5.0, 5.5	6.0
Chrominance Bandwidth (MHz)	1.3 (I), 0.6 (Q)	1.3 (U), 1.3 (V)	>1.0 (U), > 1.0 (V)

chrominance artifacts in the composite systems. The component system is preferable in all video applications that are without the constraints of broadcasting, such as studio, digital storage media, and so forth.

Although any of the previously discussed sets of component signals[1,4,5,7] can be used, of particular significance is the CCIR-601* digital component video format. The color space of this format is *YCrCb* and is obtained by scaling and offsetting the *YUV* color space. The conversion from gamma-corrected *RGB* components represented as 8-bits (0 to 255 range) to *YCrCb* is specified as follows:

$$Y = 0.257R' + 0.504G' + 0.098B' + 16$$

$$Cr = 0.439R' - 0.368G' - 0.071B' + 128$$

$$Cb = -0.148R' - 0.291G' + 0.439B' + 128$$

In these equations, Y is allowed to take values in the 16 to 235 range whereas Cr and Cb can take values in the range of 16 to 240 centered at a value of 128, which indicates zero chrominance.

The inverse operation, that is, generation of gamma-corrected *RGB* from *YCrCb* components, can be accomplished as follows:

$$R' = 1.164(Y - 16) + 1.596(Cr - 128)$$

$$G' = 1.164(Y - 16) - 0.813(Cr - 128)$$
$$- 0.392(Cb - 128)$$

$$B' = 1.164(Y - 16) + 2.017(Cb - 128)$$

In converting *YCrCb* to gamma-corrected *RGB*, the *RGB* values must be limited to lie in the 0 to 255 range as occasional excursions beyond nominal values of *YCrCb* may result in values outside of the 0 to 255 range.

The sampling rates for the luminance component Y and the chrominance components are 13.5 MHz and 6.75 MHz, respectively. The number of active pels per line is 720, the number of active lines for the NTSC version (with 30 frames/s) is 486 lines and for the PAL version (with 25 frames/s) is 576 lines. Using 8-bit per sample representation, the digital bandwidth of the uncompressed CCIR-601 signal can be calculated as 216 Mbit/s.

5.2.6 Color TV

We are now ready to discuss the color TV system[1-7] excluding details of transmission for simplicity. Fig. 5.7 shows both a composite color TV system and a component color TV system. The color space used in each case is referred to by the generalized notation of *YC1C2* and could be one of the various color spaces discussed.

As shown in Fig. 5.7, the RGB analog signal from the color camera undergoes gamma correction, resulting in a gamma corrected version

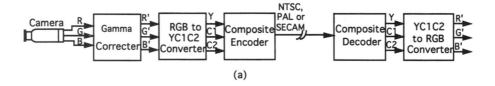

(a)

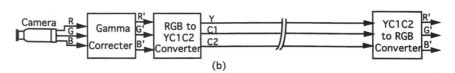

(b)

Fig. 5.7 Composite and component analog TV systems

---

*CCIR is now ITU-R.

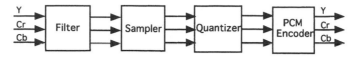

Fig. 5.8 Conversion of component analog TV signals to digital TV signals.

$R'G'B'$, which further undergoes color space conversion from $R'G'B'$ to $YC1C2$ depending on what $C1$ and $C2$ are (e.g., I and Q, U and V or something else) using the equations discussed. If the color TV system is composite, the three signals, Y, $C1$, and $C2$, undergo modulation encoding to generate a composite signal that is transmitted and at receiver undergoes the inverse operation from a single composite signal to three component signals and through $YC1C2$ color conversion back to $R'G'B'$ which is applied to display. On the other hand, in a component color TV system, the three signals are directly sent to the receiver without any composite encoding and are converted to $R'G'B'$ for display.

It is worth noting that we have assumed that the camera generates RGB and that the display uses the RGB signal. However, most home TV receivers directly take a composite signal, and any color conversions may be internal. Likewise, in nonbroadcast applications with inexpensive cameras, only a composite signal may be available at the camera output.

5.3 FROM ANALOG TO DIGITAL VIDEO

In the previous section we mentioned CCIR-601 component color signals in digital format and the digital bandwidth required. While it was not necessary to delve into how a digital representation can be obtained, now is probably a good time to do so. Furthermore, we will oversimplify the facts about digital conversion of analog video signal rather than get into detailed mathematical analysis.

The process of digitization of video consists of operations[1–7] of Prefiltering, Sampling, Quantization, and Encoding. To make our discussion more concrete, we will assume that we are dealing with three signals that form a CCIR-601 video format,

that is, Y, Cr, and Cb, and that these signals undergo the steps we just listed. The sequence of operations is illustrated in Fig. 5.8.

5.3.1 Filtering

The filtering operation employed here is also referred to as prefiltering, as this operation is performed prior to sampling. Prefiltering can reduce the unwanted excessive frequencies in the signal as well as noise in the signal. The simplest filtering operation involves simply averaging of the image intensity within a small area around the point of interest and replacing the intensity of the original point by the computed averaged intensity. Prefiltering can sometimes be accomplished by controlling the size of the scanning spot in the imaging system.*

In dealing with video signals, the filtering applied on the luminance signal may be different than that applied on chrominance signals owing to different bandwidths required.

5.3.2 Sampling

Next, the filtered signal is sampled at a chosen number of sampling points. The sampling operation can be thought of as a switch that closes at uniform duration and senses the value of the amplitude of the input signal, after which it opens again awaiting closure at the next appropriate time. The minimum rate at which an analog signal must be sampled is called the Nyquist rate and corresponds to twice that of the highest frequency in the signal. For NTSC system this rate is $2 \times 4.2 = 8.4$ MHz and for PAL this rate is $2 \times 5 = 10$ MHz. It is normal practice to sample at a rate higher than this for ease of signal recovery. The CCIR-601 signal employs 13.5 MHz for luminance and half of that rate for chrominance signals. This rate is an inte-

---

*For CCD cameras, electrical filters are used.

gral multiple of both NTSC and PAL line rates but is not an integral multiple of either NTSC or PAL color subcarrier frequency.

5.3.3 Quantization

The sampled signal, which is now simply a sequence of pel values, is quantized next. The quantization operation here means that when the input pel value is in a certain range of values, we assign and output a single fixed value representing that range. Thus a discretized representation of the pel values can be obtained. The process of quantization results in loss of information since the quantized output pel value is less accurate than the input value. The difference between the value of the input pel and its quantized representation is called quantization error. The choice of the number of levels of quantization involves a tradeoff of accuracy of representation and the visibility of noise in the signal due to quantization errors. The process of quantization is depicted in Fig. 5.9.

5.3.4 PCM Coding

The last step in analog to digital conversion is encoding of quantized values. Here, by encoding we simply mean obtaining a digital representation. The type of encoding that we are interested in is called PCM, Pulse Code Modulation. Often video pels are represented by 8-bit PCM codewords, which means that with this type of coding, it is possible to assign to the pel one of the $2^8 = 256$ possible values in the range of 0 to 255. For example, if the quantized pel amplitude is 68, the corresponding 8-bit PCM codeword is the sequence of bits 01000100 which obviously stands for $1 \times 2^6 + 1 \times 2^2 = 68$.

5.3.5 Digital Component Video

In this section we briefly introduce the various digital component video formats[10,11] typically used in MPEG video coding. MPEG video coding makes some assumptions about the color spaces as well as format of luminance and chrominance component signals, although it is quite flexible about the size of pictures themselves. The video resolutions and formats typically used at normal TV or lower resolution are all derived from the CCIR-601 format. We will discuss actual derivation of various formats in the next section; for now, we focus only on what each of the formats looks like.

5.3.6 The 4:2:0 Format

By the 4:2:0 video format we do not imply a specific picture resolution, only the relative rela-

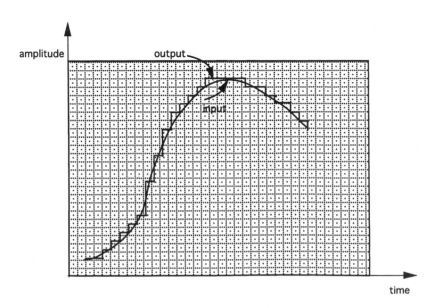

Fig. 5.9   Quantization of analog input to discrete output.

tionship between the luminance and chrominance components. In the MPEG-2 4:2:0 video format, in each frame of video, the number of samples of each chrominance component, *Cr* or *Cb,* is one-half of the number of samples of luminance, both horizontally and vertically. Alternatively, for every four samples of luminance forming a 2×2 array, two samples of chrominance, one a *Cr* sample and the other a *Cb* sample, exist. The exact location of chrominance samples with respect to luminance samples in a frame of 4:2:0 video is shown in Fig. 5.10.

MPEG-1 also uses the 4:2:0 video format, although there is a slight difference in location of the chrominance samples. In MPEG-1 the 4:2:0 video format has chrominance samples located midway between all four samples of luminance. During MPEG-2 video development, this slight change in location of MPEG-2 4:2:0 chrominance from the location of MPEG-1 4:2:0 chrominance was needed to maintain a consistent definition of 4:2:0, 4:2:2, and 4:4:4 formats employed by MPEG-2. As a matter of interest, the 4:2:0 format is sometimes referred to in the outside literature as the 4:1:1 format, since there is one *Cb* and one *Cr* sample per four luminance samples. However, this use should be avoided to reduce confusion owing

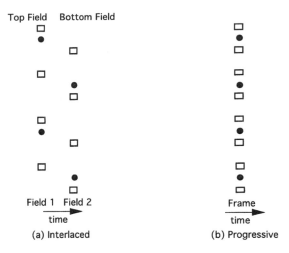

Fig. 5.11　Interlaced and Progressive representation of a 4:2:0 frame.

to the existence of an older 4:1:1 format that uses 4:1 reduction of chrominance resolution in the horizontal direction only, but full resolution in the vertical direction.

We now illustrate the vertical and temporal positions of samples in 4:2:0 by showing what amounts to the side view of a 4:2:0 frame, in Fig. 5.11. It is worth mentioning that a 4:2:0 frame may be interlaced or noninterlaced. The difference was pointed out earlier during our discussion on raster scanning — an interlaced frame is composed of lines of samples belonging to two different fields separated in time but taken together to form a frame, whereas in a noninterlaced or progressive frame there is no such separation in time between consecutive lines of samples.

Because of the special significance of 4:2:0 formats to MPEG-1 and MPEG-2 video coding, we now provide examples of specific image frame sizes used. The *Source Input Format* (SIF) is noninterlaced and uses image size of either 352×240 with 30 frames/s* or 352×288 with 25 frames/s and are correspondingly referred to as SIF-525 or SIF-625, since the 240-line format is derived from the CCIR-601 video format of 525 lines whereas the 288-line format is derived from the CCIR-601 video format of 625 lines. The two versions of CCIR-601 correspond to the two TV systems

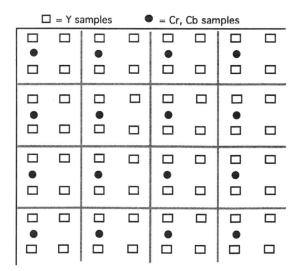

Fig. 5.10　Luminance and chrominance samples in a 4:2:0 video frame

---

*More precisely, 29.97 frames/s.

NTSC and PAL. The SIF format is typically used in MPEG-1 video coding. An interlaced version of SIF is called SIF-I and uses the same image size as SIF, that is either 352×240 with 30 frames/s or 352×288 with 25 frames/s and is used in layered MPEG-2 video coding (Chapter 9), although it is not as popular.

An interlaced video format that is somewhat more popular in MPEG-2 video coding is Half Horizontal Resolution (HHR) and uses an image size of either 352×480 with 30 frames/s or 352×288 with 25 frames/s. The interlaced format most often used in MPEG-2 video coding is 4:2:0 and uses an image size of either 720×480 with 30 frames/s or 720×576 with 25 frames/s. A word of caution about the image sizes mentioned in the various formats is in order, since MPEG video coding when using SIF, SIF-I, or HHI actually uses only 352 pels per line instead of 360 pels. This is so because the number of significant pels horizontally have to be a multiple of 16; this has to do with block sizes used in coding that we will discuss later. Of course, the chrominance resolution is half both horizontally and vertically in the case of each image resolution and format described. The various example image formats just discussed are shown in Fig. 5.12.

5.3.7 The 4:2:2 Format

In the 4:2:2 video format, in each frame of video, the number of samples per line of each chrominance component, *Cr* or *Cb*, is one-half of the number of samples per line of luminance. The chrominance resolution is full vertically, that is, it is same as that of the luminance resolution vertically. Alternatively, for every two samples of luminance forming a 2×1 array, two samples of chrominance, one a *Cr* sample and the other a *Cb* sample, exist. The exact location of chrominance samples with respect to luminance samples in a frame of 4:2:2 video is shown in Fig. 5.13.

MPEG-1 video coding allows only use of the 4:2:0 format; however, in specialized applications where higher chrominance resolution is required, MPEG-2 video coding also allows use of the 4:2:2 format. We now illustrate the vertical and temporal positions of samples in 4:2:2 by showing what amounts to the side view of a 4:2:2 frame, in Fig. 5.14. Although in principle a 4:2:2 frame may be interlaced or noninterlaced similar to a 4:2:0 frame, only an interlaced format is usually expected since the purpose of having full chrominance resolution vertically was mainly to eliminate color degradations associated with interlaced 4:2:0 chrominance.

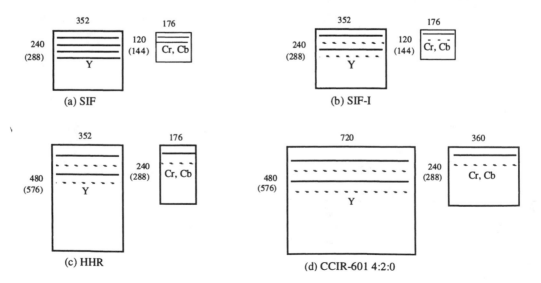

Fig. 5.12 SIF, SIF-1, HHR and CCIR-601 4:2:0 Video Formats

□ = Y samples   ● = Cr, Cb samples

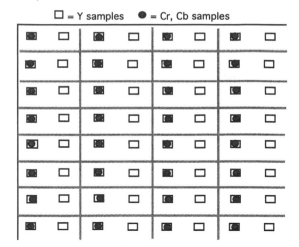

Fig. 5.13 Luminance and Chrominance samples in a 4:2:2 video frame.

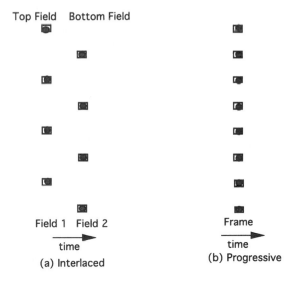

Top Field   Bottom Field

Field 1 Field 2   Frame
time   time
(a) Interlaced   (b) Progressive

Fig. 5.14 Interlaced and Progressive representation of a 4:2:2 frame.

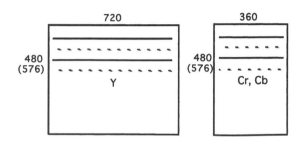

Fig. 5.15 CCIR-601 4:2:2 video format

As mentioned earlier, the 4:2:2 format can be used for MPEG-2 video coding. The main image resolution and format of interest in this case is CCIR-601 4:2:2 which uses an image size of either 720 × 480 with 30 frames/s or 720 × 576 with 25 frames/s. Of course, the chrominance resolution is half of this resolution horizontally. Figure 5.15 shows this image resolution and format.

5.3.8 The 4:4:4 Format

In a 4:4:4 video format, in each frame of video, the number of samples of each chrominance component, Cr or Cb, is the same as that of the number of samples of luminance, both horizontally and vertically. Alternatively, for every sample of luminance, two samples of chrominance, one a Cr sample and the other a Cb sample, exist. In a frame of 4:4:4 video the location of chrominance samples is the same as that of luminance samples as shown in Fig. 5.16.

The 4:4:4 format is useful mainly in computer graphics applications, where full chrominance resolution is required not only in the vertical but also in the horizontal direction. This format, although it is supported by MPEG-2, is not expected to find a wide use. In Fig. 5.17 we illustrate the vertical and temporal positions of samples in a 4:4:4 frame.

□ = Y samples   ● = Cr, Cb samples

Fig. 5.16 Luminance and Chrominance samples in a 4:4:4 video frame.

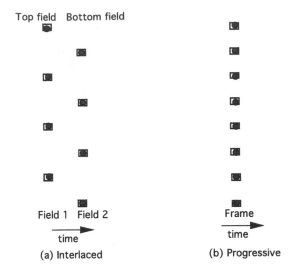

Top field   Bottom field

Field 1   Field 2   Frame

time   time

(a) Interlaced   (b) Progressive

Fig. 5.17   Inerlaced and Progressive representation of a 4:4:4 frame

5.4   SPATIAL CONVERSIONS

In this section we discuss conversion[10,11] between CCIR-601 4:2:2 video format and the various sizes and formats mentioned in the previous section that are typically employed in MPEG video coding. Often, owing to the interlaced nature of CCIR-601 video, some of the conversions that we are about to discuss impact not only the spatial resolution of video but jointly the spatiotemporal resolution; however, we identify these conversions as spatial conversions to differentiate them from mainly temporal conversions discussed in the next section.

5.4.1   CCIR-601 4:2:2 and CCIR-601 4:2:0

The conversion of CCIR-601 4:2:2 to 4:2:0 involves reduction of vertical resolution of each of the two chrominance signals by one-half. In principle, this can be accomplished by simply subsampling by 2 vertically, which is same as throwing away every other line.

This, however, results in two problems: one, that of jaggard edges introduced on significant chrominance edges within each frame, and the other, a more serious one caused by the fact that since CCIR-601 4:2:2 is interlaced, simple vertical subsampling causes loss of one complete field of

chrominance. These problems can be minimized by use of filtering although simply filtering across fields generates field to field artifacts due to motion in a scene. Thus, in each frame, each chrominance field should be filtered separately, field 1 by a filter of odd length and field 2 by a filter of even length (to generate lines at intermediate locations). Then each field is subsampled by a factor of 2 vertically. The filter used for filtering of chrominance field 1 is of length of 7 with coefficients $[-29, 0, 88, 138, 88, 0, -29]/256$ and that used for filtering chrominance field 2 is of length 4 with coefficients $[1, 7, 7, 1]/16$. The two resulting subsampled chrominance fields are merged to form a frame separately for each chrominance signal, which with the associated luminance frame is termed as the CCIR 601 4:2:0 frame. The operations involved in conversion of CCIR 601 4:2:2 to CCIR 601 4:2:0 are shown in Fig. 5.18.

The inverse operation, that is, conversion of CCIR 601 4:2:0 to CCIR 601 4:2:2, consists of upsampling the chrominance signal vertically by a factor of 2. Of course, upsampling can not really increase the resolution lost due to downsampling, but only increase the number of samples to that required for display purposes. Again, the simplest upsampling is pel repetition, which produces artifacts such as jaggy edges. Better upsampling involves use of an interpolation filter. Since we are dealing with interlaced 4:2:0 chrominance frames, interpolation needs to be performed separately on each field of chrominance samples. Referring to the two fields of a frame as field 1 and field 2, chrominance samples of field 1 of the 4:2:0 format are copied to form every other line of field 1 of the 4:2:2 format; the in-between lines of field 1 are generated by simple averaging using the line above and line below, which is same as using an interpolation filter of length 2 and coefficients $[^{1}/_{2}, {}^{1}/_{2}]$.

In field 2 of 4:2:2 format chrominance no samples of field 2 of 4:2:0 chrominance can be copied directly. All samples have to be calculated by using two different interpolation filters, each of length 2. The first interpolation filter has coefficients $[^{3}/_{4}, {}^{1}/_{4}]$ and the second interpolation filter has coefficients $[^{1}/_{4}, {}^{3}/_{4}]$. Each of the interpolation filters is used to compute alternate lines of field 2 of 4:2:2 chrominance. The various steps involved in con-

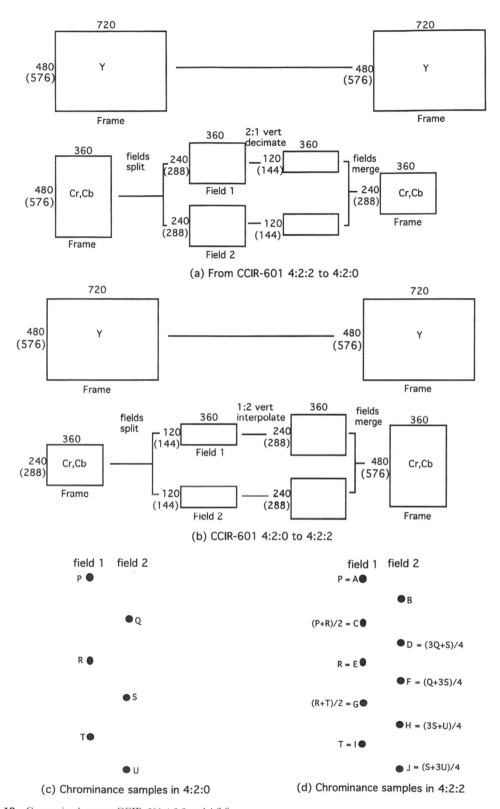

(a) From CCIR-601 4:2:2 to 4:2:0

(b) CCIR-601 4:2:0 to 4:2:2

(c) Chrominance samples in 4:2:0

(d) Chrominance samples in 4:2:2

Fig. 5.18 Conversion between CCIR-601 4:2:2 and 4:2:0

version of CCIR 601 4:2:0 to CCIR 601 4:2:2, including details of chrominance filtering, are shown in Figure 5.18.

5.4.2 CCIR-601 4:2:2 and SIF

The CCIR-601 4:2:2 format is converted to the SIF format by first discarding field 2 of each frame of luminance and chrominance signals, retaining field 1 only. Next, the luminance and chrominance signals are horizontally subsampled by one-half it is filtered. This operation, known as horizontal decimation, consists of applying a decimation filter and retaining every other sample per line of luminance and chrominance signals. As an example, a filter of length 7 with coefficients $[-29, 0, 88, 138, 88, 0, -29]/256$ can be applied horizontally to every pel in every line of the original image to obtain a horizontally filtered image, followed by 2:1 subsampling horizontally. Eight luminance pels and four chrominance pels per line are then discarded to obtain 352 and 176 pels, respectively. The chrominance signals in the field are also vertically decimated by first filtering vertically, followed by 2:1 subsampling. The filter used for vertical decimation can be the same as that used for horizontal decimation. The various steps involved in conversion of CCIR-601 4:2:2 to SIF are shown in Fig. 5.19.

The inverse operation of conversion of SIF to CCIR-601 4:2:2 consists of upsampling of the luminance and each of the chrominance signals. The SIF luminance signal is first upsampled horizontally to twice the resolution, and then it is also vertically upsampled to twice the resolution. The SIF chrominance signals are first upsampled vertically to twice the resolution and they are then further upsampled horizontally and vertically again to twice the resolution to obtain the CCIR 601 4:2:2 format. As an example, the horizontal upsampling operation consists of first copying every pel of source resolution, say horizontal resolution of SIF, to every other pel location in the destination horizontal resolution of CCIR-601. Next, the intermediate missing pels are generated by applying filter coefficients to the pels copied earlier. The operation just described applies to horizontal upsampling of luminance and chrominance signals. An

identical operation is applied for vertical upsampling. Both in horizontal and vertical upsampling the interpolation filter used is $[-12, 140, 140, -12]/256$. Note that although SIF can be upsampled to the same resolution as CCIR-601 4:2:2, the resulting signal is not quite same, because SIF is noninterlaced, whereas CCIR-601 4:2:2 is interlaced. The steps involved in the conversion of SIF to CCIR-601 4:2:2 are shown in Fig. 5.19.

5.4.3 CCIR-601 4:2:2 and HHR

The conversion of CCIR-601 4:2:2 to HHR consists of reducing the horizontal resolution of the luminance signal by one-half, whereas for chrominance signals, both horizontal and vertical resolution is reduced by one-half. The process of reducing luminance and chrominance resolution horizontally to one-half is identical to the one performed for reducing the horizontal resolution of CCIR 601 4:2:2 to SIF, described earlier. However, as compared to the conversion to SIF, no vertical subsampling of luminance is needed and thus no field dropping is necessary. The process of conversion of the chrominance signal to half of its resolution vertically is identical to the process followed in the conversion of CCIR-601 4:2:2 to 4:2:0. Eight luminance pels and four chrominance pels per line are then discarded to obtain 352 and 176 pels, respectively. An example filter that can be used in all the decimation operations uses a length of 7 with coefficients $[-29, 0, 88, 138, 88, 0, -29]/256$ and is used to compute each pel in source resolution before subsampling by a factor of 2 to achieve half-resolution. The steps involved in the conversion of CCIR-601 4:2:2 to HHR are shown in Fig. 5.20.

The conversion of HHR to CCIR-601 4:2:2 consists of upsampling the luminance and chrominance by a factor of 2 horizontally. Since full luminance resolution is present vertically in the HHR format, no upsampling is needed vertically. The chrominance signals are then upsampled vertically. Here, the upsampling operation is the same as that described for upsampling of 4:2:0 to CCIR-601 4:2:2, and involves use of interpolation filters. The sequence of steps in conversion of HHR to CCIR-601 4:2:2 is also shown in Fig. 5.20.

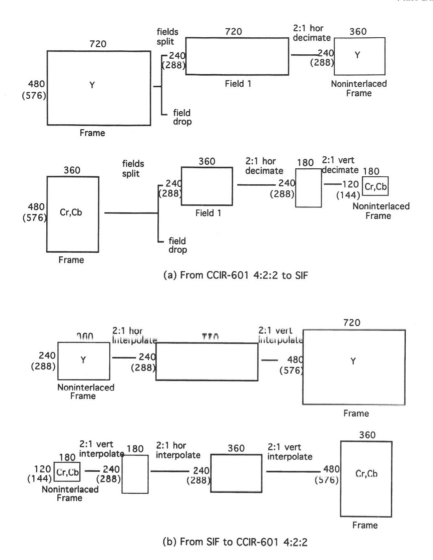

Fig. 5.19 Conversion between CCIR-601 4:2:2 and SIF. 8 out of 360 pels are discarded to obtain 352 luminance pels. Four out of 180 pels are discarded to obtain 176 chrominance pels.

5.4.4 CCIR-601 4:2:0 and SIF-I

In the conversion of CCIR-601 4:2:2 to SIF-I, although downsampling is performed to derive a resolution equal to that of SIF, the format of the resulting SIF is interlaced rather than the normal noninterlaced. A simple method is suggested here and is known to produce somewhat blurry images. For luminance downsampling, each frame of luminance is separated into individual fields and downsampled to one-half resolution horizontally, fol-lowed by further downsampling to one-half resolution vertically. The downsampling of chrominance is accomplished also by first separating chrominance to fields and then downsampling each field horizontally and vertically respectively, and finally again downsampling vertically prior to merging of the two fields of chrominance. Eight luminance pels and four chrominance pels per line are then discarded to obtain 352 and 176 pels, respectively. As an example, the filter specified by $[-29, 0, 88, 138, 88, 0, -29]/256$ can be used for horizontal

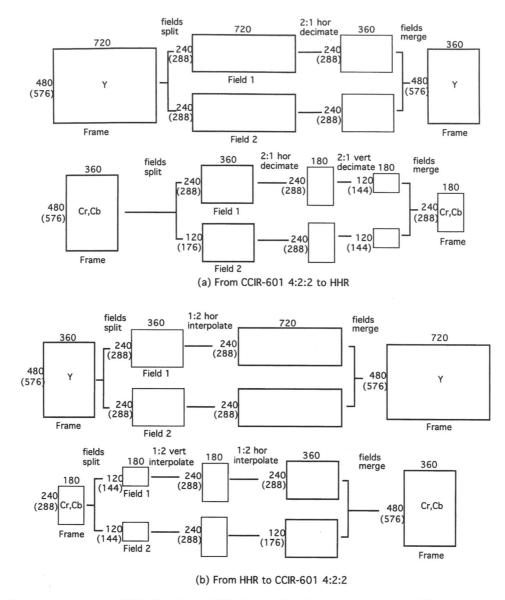

Fig. 5.20 Conversion between CCIR-601 4:2:2 and HHR. 8 out of 360 pels are discarded to obtain 352 luminance pels. Four out of 180 pels are discarded to obtain 176 chrominance pels.

filtering before subsampling. For vertical subsampling of field 1, the same filter can be used, whereas for vertical subsampling of field 2, a filter with an even number of taps such as $[-4, 23, 109, 109, 23, -4]/256$ can be used. The sequence of steps in the conversion of CCIR-601 4:2:2 to SIF-I is shown in Fig. 5.21.

The conversion of SIF-I to CCIR-601 4:2:2 is the inverse operation. For luminance, it consists of separating each frame into fields and upsampling each field horizontally and vertically, followed by field merging. Next, the SIF-I chrominance is separated into individual fields and upsampled to twice the vertical resolution two

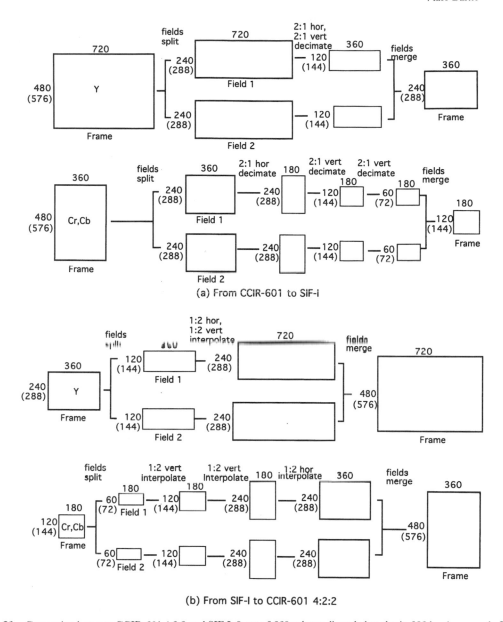

Fig. 5.21 Conversion between CCIR-601 4:2:2 and SIF-I. 8 out of 360 pels are discarded to obtain 352 luminance pels. Four out of 180 pels are discarded to obtain 176 chrominance pels.

times, followed by horizontal upsampling to twice the size. The resulting fields are merged. For upsampling, interpolation filters are used; in particular, horizontal upsampling can be performed using a $[-12, 140, 140, -12]/256$ filter. For vertical upsampling of field 1 the same filter as that used for horizontal upsampling can be used, whereas for vertical upsampling of field 2, an interpolation filter such as $[-2, 20, 110, 110, 20, -2]/256$ can be used. The sequence of steps in conversion of SIF-I to CCIR-601 4:2:2 are also shown in Fig. 5.21.

5.5 TEMPORAL CONVERSIONS

In the previous section we discussed some basic approaches for conversion between the source of CCIR-601 4:2:2 format to the various resolutions and formats used in MPEG-1 and MPEG-2 video coding and back to display resolutions and formats. Conversions of horizontal sizes and some conversions of vertical sizes are called spatial conversions, while other vertical size conversions also change field rates and are thus referred to as vertical/temporal or here simply as temporal conversions.[1,4,10,11] We have chosen to include only those conversions that may be directly relevant within the context of MPEG-2 video coding and display.

As a typical example, interlaced video of 60 fields/s normal rate, coded by MPEG-2, may have to be displayed on a computer workstation operating at 72 progressive frames/s. Another example is that of 24 frames/s film coded by MPEG-1 or MPEG-2 which needs conversion to 60 fields/s for display on a TV monitor. Yet another example may involve a PAL video format CD employing 50 fields/s compressed by MPEG-2 to be played on NTSC video format equipment using a display rate of 60 frames/s. In discussing the following general techniques for conversion, we do not assume a specific spatial resolution; our main purpose is to show temporal resolution conversions only.

5.5.1 Interlaced and NonInterlaced

We first consider the case of conversion[2–4] of interlaced video of 50 or 60 fields/s to noninterlaced video with the same number of noninterlaced frames/s. This operation is also called deinterlacing or line-doubling and involves generation of a noninterlaced frame corresponding to each field of interlaced video, effectively doubling the number of samples in the vertical direction.

A simple method to convert interlaced to noninterlaced is based on *line interpolation* in each field, which involves generating intermediate lines between each pair of consecutive lines in each field. A simple method to achieve line interpolation is by straightforward averaging of two lines to pro-

duce a line in between, which is the same as saying the interpolation filter has length 2 with coefficients of $[\frac{1}{2}, \frac{1}{2}]$. The complete noninterlaced frame can be now created by alternately copying lines of a field and interpolating lines, generating a frame with twice as many lines as a field.

Yet another technique to accomplish interlaced to noninterlaced conversion is by *field merging*, which consists of generating a frame by combining lines of two consecutive fields to produce a frame of twice the number of lines per field. It turns out that although in stationary areas, field merging results in full vertical resolution, in the areas with motion, field merging produces judder artifacts. For improved results, line interpolation and field merging could be combined, such that line interpolation would be used for moving areas and field merging for stationary areas. However, such adaptivity itself can introduce visible artifacts owing to switching between methods, and so using a weighting function to combine line interpolation and field merging methods may be a feasible solution.

As an alternative to all this complexity, a simple method for conversion of interlaced to noninterlaced uses a *frame based line interpolation* and works similar to line interpolation; however, the interpolated line is generated by applying the interpolation filter on an interlaced frame. The noninterlaced frame is generated by alternating between copying a line from a field of an interlaced frame and interpolating the next line using the interlaced frame. The interpolation filter applied on frame lines has length 5 and coefficients of $[-\frac{1}{8}, \frac{1}{2}, \frac{1}{4}, \frac{1}{2}, -\frac{1}{8}]$. The details of this method are shown in Fig. 5.22.

We now consider the case of conversion of noninterlaced to interlaced video, in which, given full resolution for each frame at say 60 frames/s, we reduce the vertical resolution by a factor of 2 to obtain 60 fields/s such that fields can be paired to form interlaced frames for display.

The simplest technique for noninterlaced to interlaced conversion is based on *line subsampling* vertically of each frame to form a field. On consecutive frames, line subsampling is implemented to yield fields of opposite parities that can be interleaved to form interlaced frames. Thus the lines of alternate fields are at identical locations but the

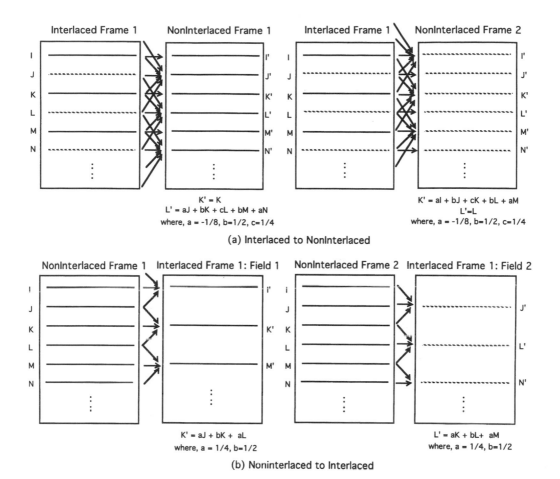

Fig. 5.22   Temporal Conversion between Interlaced and Noninterlaced

lines of consecutive fields are offset by half of the line distance between lines in a field.

Further improvements in noninterlaced to interlaced conversion is possible by *line weighting*, in which each line of a field is formed by a weighted combination of consecutive lines in the noninterlaced frame. This method, employing a simple example filter of length 3 with coefficients [$^1/_4$, $^1/_2$, $^1/_4$], is shown in Fig. 5.22. For better results, a filter of length 5 or 7 may be necessary.

5.5.2   Field Rate Conversion

We now discuss basic techniques for conversion[2-4] of field rates between NTSC and PAL TV systems. To convert 60 fields/s to 50 fields/s, for every six fields of input, five fields are output. A simple way of doing that is to discard one out of every six fields, and from the 60 fields/s source rate, the fields closest in timing to the 50 fields/s destination rate are selected. However, this simple method does not work very well, as with fast motion, glitches in time become apparent. A somewhat better solution is to interpolate fields in time by a weighted combination of fields at the source rate of 60 fields/s to produce destination fields at 50 fields/s. Figure 5.23 shows this operation, including the weighting factors needed for 60 fields/s to 50 fields/s conversion.

The inverse operation, that is, conversion of 50 fields/s video to 60 fields/s video, can also be done similarly using temporal interpolation. This is also

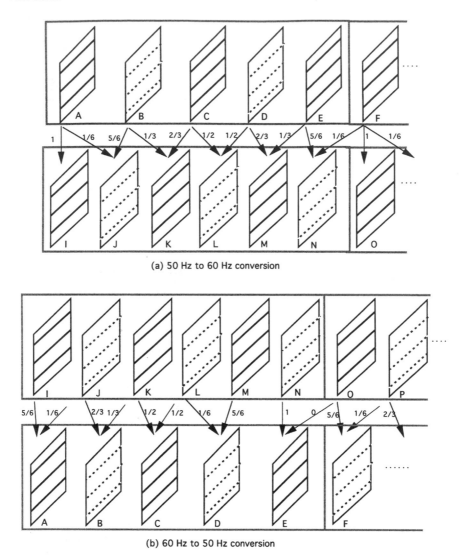

(a) 50 Hz to 60 Hz conversion

(b) 60 Hz to 50 Hz conversion

Fig. 5.23   Temporal Conversion between 50 and 60Hz Field rates

shown in Figure 5.23, including the weighting factors needed for the conversion.

It should be noted that the temporal interpolations just described have a problem when there is motion in a scene. In principle a much better method is to compensate for motion before interpolation. In this method each *target field* is segmented into nonoverlapping small blocks (say 8 × 8 or 16 × 16) and motion vectors are computed by block matching with respect to a *reference field*. These motion vectors are scaled by a factor reflecting the relative distance of the field to be interpo-

lated from the target field as compared to the total distance between the target and the reference field. To clarify, let us assume we want to interpolate a field located at a distance of two-thirds from a target field and one-third from a reference field. The motion vectors would then have to be scaled by a factor of two-thirds and motion-compensated blocks would be obtained from the corresponding reference field. There are still some problems with this approach, for example, block matching motion vectors may not reflect actual motion and thus may not be meaningfully scaled. Also, truncation or rounding errors in motion vectors affect the accuracy of motion compensation. The following two chapters will discuss motion compensation for compression in much more detail, but here our primary purpose was to mention that motion compensation can be used in field rate conversion. Incidentally, commercially available standards converters that perform conversion between NTSC and PAL TV formats use the same basic principles discussed here, including those that use motion compensation.

5.5.3 Film and Interlaced Video Conversions

As discussed earlier, film is usually recorded at the rate of 24 frames/s. For transfer of film to PAL or SECAM, it is often played back a little faster, at 25 frames/s. This results in a slight distortion of pitch of the accompanying sound; however, this distortion is not often very noticeable. For transfer of film to NTSC, which uses 30 interlaced frames/s, 3:2 pulldown is necessary. This involves generating five fields of video per pair of film frames, resulting in 60 fields/s. The process of conversion[1-4] of film to interlaced NTSC video is shown in Figure 5.24. As shown by solid arrows in the figure, a film frame may be used to generate three fields, two of which are essentially identical, whereas the next film frame may be used to generate two fields and so on. Depending on the method used for film to video conversion, the three fields generated from a film frame may be exactly identical. For scenes with high-speed motion of objects, the selection of a film frame to be used for generating a field may be performed manually to minimize motion artifacts.

Sometimes film may already be available in the TV format at 30 interlaced frames/s, because film may have been converted to video using a Telecine

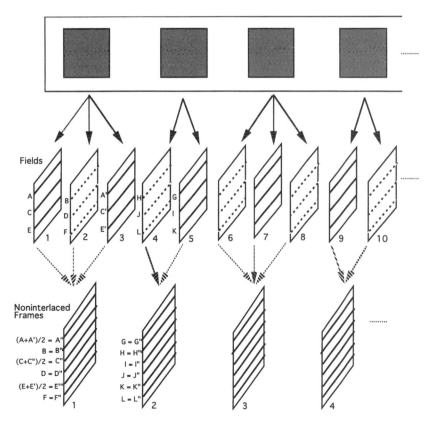

Fig. 5.24   Temporal Conversion between Film and Interlaced video, also referred to as 3:2 pulldown.

camera which can output interlaced video by performing automatic 3:2 pulldown to yield 60 fields/s. If so, it can be more efficiently encoded by MPEG video coding by first undoing the 3:2 pull down to regenerate 24 frames/s noninterlaced video. In undoing the 3:2 pulldown before coding, it may be a good idea to generate the noninterlaced frame by first averaging the first and third out of every group of 3 fields to reduce noise and then combining it with the second field to form a noninterlaced frame. The operation of undoing 3:2 pulldown is shown by dotted arrows in Figure 5.24.

However, since NTSC television uses 60 fields/s for display, after MPEG video decoding, a 3:2 pulldown would be necessary to convert 24 frames/s to 60 fields/s rate for display. On the other hand, if film is available at noninterlaced 24 frames/s, it can be coded directly by MPEG video coding and the only conversion needed is a 3:2 pulldown to produce 60 fields/s for display on TV.

5.6   MIXING AND KEYING

We now discuss combining of video signals by two methods, one called *mixing*[2-4] and the other called *keying*[2-4]

Mixing is used to overlay text or graphics on a video signal or fade to or from a color, say black. Mixing may consist of something as simple as switching between two video signals or something more such as a blended transition between two video signals.

Keying is used to generate special effects such as inserting a background from one scene into another scene containing foreground objects. Keying usually happens within video frames. It basically involves substituting a background signal in portions of the foreground signal based on some attribute such as a luminance or a chrominance value, or a specified region.

As a matter of interest, morphing involves smooth transition over time from an object of interest in one video sequence to an object of interest in another scene. It involves defining key points between the two objects and blending of shape as well as blending of amplitudes of signals.

5.6.1   Mixing

Mixing can be defined as the addition of two or more input video signals, each in some proportion *alpha* to generate a single mixed output video signal. The alpha information, α, refers to the degree of opacity of each scene and each video signal may carry its own α, a video signal with higher α contributing proportionally higher value to the output. Generalized alpha mixing can be represented by the following:

$$Video = \alpha_1 \times Video1 + \alpha_2 \times Video2 + \ldots$$

A specific situation occurs when a crossfade is implemented between two scenes. Typically, in the case of crossfade, alpha values and up to 1 and a single value of alpha is sufficient, resulting in the following operation:

$$Vmix = \alpha \times Video1 + (1 - \alpha) \times Video2$$

Here, when α is 1 it implies the absence of Video2, whereas when α is 0, only Video2 is present. In between these two extremes, a weighted combination of the two scenes results. By increasing the α values as a function of time, a crossfade from one scene to another, say from Video2 to Video1, can be implemented. If the values are increased gradually the crossfade is slow, otherwise, a fast crossfade is accomplished.

5.6.2   Luma Keying

In luma (short for luminance) keying, a level of luminance signal is specified for the scene containing the foreground such that all pels above that value are replaced by the corresponding pels of the scene containing the background. Luma keying can also be implemented by specifying two threshold values Y_l and Y_h for the luminance of the foreground, Y_{fg}. More specifically, the background can be keyed into the brightest areas in the foreground signal by using a keying signal k. The foreground can be retained in the low-brightness areas, and a linear blended signal can be introduced in other regions.

This is done by defining the keying signal in terms of luminance values as follows:

$k = 1$ when $Y_h < Y_{fg}$ use background

$k = 0$ when $Y_{fg} < Y_l$ use foreground

$k = (Y_{fg} - Y_l)/(Y_h - Y_l)$ else $Y_l \leq Y_{fg} \leq Y_h$ use blended

Alternatively, it is also possible to introduce the background into the darkest areas of the foreground signal.

5.6.3 Chroma Keying

Chroma keying consists of substituting a background signal in place of the foreground signal based on some key color identified in the foreground video signal. Chroma keying is also popularly known as the "blue screen," due to frequent choice of a bright blue color for keying in the background signal.

If chroma keying is implemented as a switch within a frame between foreground and background sources, problems occur in areas where the foreground is transparent since the background is not visible through these areas; also, shadows from the foreground are not present in areas containing the background. By implementing chroma keying as a linear combination of foreground and background, the problems due to transparent foreground can be resolved. However, the problem of shadows is not fully resolved. By using a uniformly lit blue background on which foreground objects cast shadows, and using this signal for mixing with the background, the effect of foreground shadows can be transferred to the areas where the background is selected. This type of chroma keying is called *shadow chroma keying* and also results in improvements in areas where the foreground is transparent. Chroma keying also has some problems when foreground colors are close to the key color. It also appears that although a lighted blue background resolves many problems, it also introduces a new one due to color leakage between the foreground signal and the blue background. The steps in chroma keying of two video signals are shown in Figure 5.25.

A good method to resolve problems associated with chroma keying involves processing foreground and background scenes separately and using two keying signals, k_{fg} for foreground and k_{bg} for background, each with values in the 0 to 1 range.

5.7 SUMMARY

In this chapter, we first learned about the raster scanning of video, human visual system, color spaces, and various formats used by MPEG video. Next, we discussed spatial conversions used in preprocessing prior to MPEG video encoding and postprocessing after video decoding for display purposes. Then, we described several temporal conversions used in preprocessing and postprocessing. Finally, we discussed video mixing, a topic of general interest in many applications. We now summarize the highlights of this chapter as follows:

• Raster scanning can be either interlaced or noninterlaced (progressive). Both TV and motion pictures, although fairly different, use the same basic principle of portraying motion in a scene by showing slightly different images taken over time to replicate motion in a scene. The image aspect ratio (IAR) controls the aesthetic appearance of displayed image frames and generally refers to the ratio of width to height.

• Color spaces of composite TV systems— NTSC, PAL, and SECAM—are all derived from YUV space. However, the component system, and in particular the CCIR-601 video format, is of primary interest in MPEG coding.

• Analog to digital conversion involves the steps of filtering, sampling, quantization, and PCM coding.

• MPEG coding employs various formats derived from the CCIR-601 digital video format. MPEG-1 video coding was optimized for the SIF noninterlaced format. MPEG-2 video coding was optimized for the CCIR-601 4:2:0 interlaced formats. MPEG-2 video coding often also employs HHR interlaced formats.

• Spatial conversions are either through decimation or interpolation, where decimation means subsampling to reduce resolution and interpolation means upsampling to produce a higher number of samples. To avoid aliasing, filtering is necessary before subsampling. To avoid upsampling jaggies, an interpolation filter is used instead of simple pel or line repetition. When performing the decimation or interpolation on interlaced frames, these operations are performed on individual fields rather than frames.

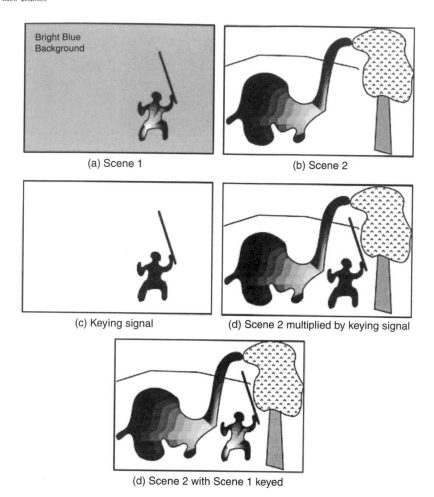

Fig. 5.25   Chroma Keying

• Temporal conversions involve conversions between interlaced and noninterlaced frames, conversion of field rate between 50 and 60 field/s TV systems, and 3:2 pull down between film and interlaced video. In field-rate conversion, sometimes simple methods such as weighted field interpolation are not sufficient and scene motion based weighted combination may be necessary. If film material is available as 60-Hz interlaced video, it may be converted to 24-Hz noninterlaced (progressive) format before MPEG coding.

• Mixing of several video signals to produce a new mixed signal can be done by hard-keying or by soft-keying (alpha blending). Keying can be done by using selected thresholds on brightness values or by using a selected color; the corresponding methods are referred to as luma-keying or chroma-keying. Chroma-keying is often referred to as blue-screen, because of the usual selection of blue for keying.

REFERENCES

1. A. N. Netravali and B. G. Haskell, *Digital Pictures: Representation, Compression and Standards,* Second Edition, Plenum Press, New York (1995).

2. A. F. Inglis, *Video Engineering,* McGraw-Hill, New York (1993).

3. K. B. Benson and D. G. Fink, *HDTV: Advanced Television for the 1990s,* McGraw-Hill, New York (1991).

4. K. JACK, *Video Demystified: A Handbook for the Digital Engineer,* Brooktree, (1993).

5. T. RZESZEWSKI (Ed.), *Color Television,* IEEE Press, New York (1983).

6. A. N. NETRAVALI and B. PRASADA (Ed.), *Visual Communication Systems,* IEEE Press, New York (1989).

7. W. K. PRATT, *Digital Image Processing,* John Wiley & Sons, New York (1978).

8. A. C. LUTHER, *Digital Video in the PC Environment,* McGraw-Hill, New York (1989).

9. A. ROSENFELD and A. C. KAK, *Digital Picture Processing,* Academic Press, New York (1982).

10. Video Simulation Model Editing Committee, MPEG Video Simulation Model 3 (SM3) (July 1990).

11. Test Model Editing Committee, "Test Model 5," *ISO/IEC JTC1/S29/WG11/N0400* (April 1993).

6

Digital Compression: Fundamentals

The world seen through the lens of a television camera is far more complex than a two-dimensional projection of a finite size that is imaged on the camera target. In this chapter we examine in more detail the limitations of a picture sequence captured by a television camera and its numerical representation in a finite bitrate. Using the previous chapter as a starting point, we will discuss the details of various compression algorithms that remove the superfluous information so as to reduce the data rate.

6.1 LIMITATIONS OF THE TELEVISION CAMERA

A television camera projects the three-dimensional world onto a two-dimensional camera plane by perspective projection. Thus, the depth information is largely lost. The finiteness and the resolution of the raster limit the extent of the scene viewed by the camera. Also, the radiation emanating from the scene contains a variety of wavelengths, only a fraction of which are visible to the human eye. Moreover, the radiated intensity as a function of wavelengths is reduced to only three primary colors by the camera based on the trichro-

macy of the human visual system. The motion rendered by the television camera also has limitations resulting from the frame rates at which the camera operates. Cameras that operate at a 30-Hz frame rate handle slow motion without any artifacts. But for very rapid motion, images are either blurred (as the result of integration in the camera) or the motion appears jerky owing to aliasing effects. In spite of these limitations, television produces a very usable replica of the world for many applications. However, these limitations must be factored into the design of any compression algorithm.

6.2 NUMERICAL REPRESENTATION

We will assume for the purpose of this discussion that a camera has captured a two-dimensional intensity signal $B(x, y)$, representing the gray level information as a function of the two spatial coordinates (x, y) of the raster. Our task is to study methods of representing $B(x, y)$ by a finite amount of data. If $B(x, y)$ is a function of time, then a finite data rate is desired.

The first step in this process is to discretize the continuous function $B(x, y)$ by spatial sampling.* If

---

*Later, we will deal with $B(x, y, t)$ by first sampling in time and then in space.

a finite set of points, $[(x_j, y_i)j = 1, \ldots, N; i = 1, \ldots, M]$ is chosen as the sampling grid and the brightness of each sample is represented by a finite number of bits (i.e., quantized), then we have succeeded in representing $B(x, y)$ by a finite number of bits. This procedure of sampling and quantization is often called *pulse code modulation* (PCM). The set of sampling points is usually a rectangular array consisting of rows and columns of samples, called *pels* (or *pixels*). The pels should be spatially close to each other to capture accurately the variation in intensity but not too close so as to be superfluous. In fact, quality permitting, the samples should be as far from each other as possible to reduce the redundancy. The density of the array of samples determines the spatial resolution of the picture. Some spatial variations in intensity that are too rapid for a chosen sampling density will be distorted in the discretization process, and such distortions are called *prealiasing* artifacts. To avoid prealiasing, the analog signal $B(x, y)$ is first filtered to remove variations that are too rapid. The filtered output $B(x, y)$ can be represented as a weighted average over a small neighborhood of each sampling point:

$$\overline{B}(x, y) = \int_{y-\delta}^{y+\delta} \int_{x-\delta}^{x+\delta} B(w, z) h_c(x - w, y - z) dw\, dz \qquad (6.1)$$

Depending on the weighting function, $h_c(x, y)$, and the spatial window, $[-\delta, +\delta]$, the filter can remove rapid variations in intensity.

Instead of thinking of the filter as a weighted average, we can also think of it as removing higher spatial frequencies by low-pass filtering which can be implemented as a convolution by an appropriate filter kernel, $h_c(x, y)$.

$$\overline{B}(x, y) = \int_0^X \int_0^Y B(w, z) h_c(x - w, y - z) dz\, dw \qquad (6.2)$$

Once again, the shape of the filter is determined by the filter kernel $h_c(\cdot, \cdot)$. This can be easily extended to time varying visual information, $B(x, y, t)$, by filtering in the space (x, y) and time (t) domains:

$$\overline{B}(x, y, t) = \int_0^T \int_0^X \int_0^Y B(w, z, \tau) h(x - w, y - z, t - \tau) dz\, dw\, d\tau \qquad (6.3)$$

Such a filtered signal is then sampled at a given frame rate in the time domain. For applications involving human viewers, temporal bandwidths of the order of 10 to 15 Hz are usually adequate for scenes involving slow to moderate motion, implying temporal sampling at frame rates of 20 to 30 Hz. Higher frame rates are required for accurate rendition of rapid motion, particularly if the moving object is tracked by the viewer's eye.

Given the discretized intensity at sampled locations (x_k, y_k), the replica image can be reconstructed by:

$$\tilde{B}(x, y) = \sum_k \overline{B}(x_k, y_k) h_d(x - x_k, y - y_k) \qquad (6.4)$$

By the proper choice of $h_d(\cdot, \cdot)$, called the interpolation filter, artifacts that may otherwise occur in the reconstruction process (called *postaliasing*) can be reduced. The post-aliasing filter essentially smoothes the spatially discrete pels to form a continuous image. Thus, not only is the filtering, as shown in Figure 6.1, generally required at the camera before sampling, but it is also needed before the display for reconstructing the image using the discrete samples of intensity.

The above method of reconstruction and the corresponding equations can be easily extended for intensities that are functions of (x, y, t).

An example of the pre- and postfilter is given below. As discussed above, the prefilter can be implemented as a weighted average over a small neighborhood of each sampling point. For this purpose a weighting function $h_c(u, v)$ can be used that peaks at $(u = 0, v = 0)$ and falls off in all directions away from the origin. For example, the Gaussian function

$$K \exp(-au^2 - bv^2) \qquad |u|, |v| < \delta$$

where K is a normalizing constant, is quite suitable for many applications. Gaussian weighting is also characteristic of many imaging devices. Using this

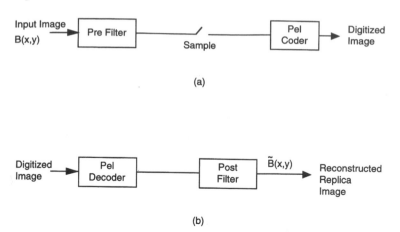

Fig. 6.1   Spatial sampling of images generally requires prefiltering prior to sampling, as shown in (a), and postfiltering following reconstruction, as shown in (b). Inadequate prefiltering leads to prealiasing, whereas inadequate post-filtering causes post-aliasing.

function, the filtered image can be computed as

$$\overline{B}(x, y) = \int\limits_{y-\delta}^{y+\delta} \int\limits_{x-\delta}^{x+\delta} KB(w, z)\exp\Big(-a(x-w)^2$$
$$-b(y-z)^2\Big)dw\,dz$$

where the intensity at each point is replaced by a weighted average over a small surrounding neighborhood. Such prefiltering could also be implemented simply by proper design of the optical components in the camera imaging system.

The postfilter is an interpolation filter. A commonly used interpolation filter is the so-called cubic B-spline $h_d(\mu, v) = S(\mu, \delta_x)S(v, \delta_y)$ where the horizontal and vertical pel spacings are δ_x and δ_y, respectively, and

$$S(z, \Delta) = \begin{cases} \frac{2}{3} - \left|\frac{z}{\Delta}\right|^2 + \frac{1}{2}\left|\frac{z}{\Delta}\right|^3 & \text{for } \left|\frac{z}{\Delta}\right| \le 1 \\ \frac{4}{3} - 2\left|\frac{z}{\Delta}\right| + \left|\frac{z}{\Delta}\right|^2 - \frac{1}{6}\left|\frac{z}{\Delta}\right|^3 & \text{for } 1 \le \left|\frac{z}{\Delta}\right| \le 2 \end{cases}$$

For positive pel values, the B-spline interpolation is always positive. The B-spline filter essentially smoothes the spatially discrete pels of an image and preserves the continuity of the first and the second derivatives in both x and y dimensions.

6.2.1   Raster Scanning

Raster scanning is a process used to convert a three-dimensional image intensity, $B(x, y, t)$, into a one-dimensional television signal waveform. First, the intensity, $B(x, y, t)$, is sampled many times per second to create a sequence of frames. Then, within each frame, sampling is done vertically to create scan lines. Scanning proceeds sequentially, left to right and from top to bottom. In a tube-type television camera, an electron beam scans across an electrically photosensitive target upon which the image is focused. At the other end of the television chain there is usually a cathode ray tube (CRT) display, in which the electronic beam scans and lights up the picture elements in proportion to the intensity, $B(x, y, t)$. While it is convenient to think of all the samples of a frame as occurring at the same time (similar to film), the scanning in a camera or display results in every sample corresponding to a different point in time.

Since the electron beam has some thickness, the scanning itself is partly responsible for local spatial averaging of the image over the area of the electron beam. This occurs both at the camera as well as the display. At the camera, as the beam scans the camera target, it converts the photons integrated (over a frame time) at each target point into an electrical signal. This results in both the spatial and temporal averaging of light and is commonly referred to as *camera integration*. Higher resolution camera systems have a much finer electron beam, but in that case, owing to averaging over a smaller area, the amount of light intensity as well as signal-to-noise ratio is reduced. At the display, the beam thickness may simply blur the image or a thinner beam may result in visible scan lines.

There are two types of scanning: *progressive* (also called *sequential*) and *interlaced*. In progressive scanning, within each frame all the raster lines are scanned from top to bottom. Therefore, all the vertically adjacent scan lines are also temporally adjacent and are highly correlated even in the presence of rapid motion in the scene. Film can be thought of as naturally progressively scanned, since all the lines were originally exposed simultaneously, so a high correlation between adjacent lines is guaranteed. Almost all computer displays are sequentially scanned.

In analog interlaced scanning (see Fig. 6.2) all the odd-numbered lines in the entire frame are scanned first, and then the even-numbered lines are scanned. This process results in two distinct images per frame, representing two distinct samples of the image sequence at different points in time. The set of odd-numbered lines constitute the *odd-field*, and the even numbered lines make up the *even-field*.* All the current analog TV systems (NTSC, PAL, SECAM) use interlaced scanning.

One of the principal benefits of interlaced scanning is to reduce the scan rate (or the bandwidth) in comparison to the bandwidth required for sequential scan. Interlace allows for a relatively high field rate (a lower field rate would cause flicker), while maintaining a high total number of scan lines in a frame (a lower number of lines per frame would reduce resolution on static images). Interlace cleverly preserves the high detailed visual information and at the same time avoids visible large area flicker at the display due to insufficient temporal postfiltering by the human eye. Interlace scanning often shows small flickering artifacts in scenes with sharp detail and may have poor motion rendition for fast vertical motion of small objects.

If the height/width ratio of the TV raster is equal to the number of scan lines/number of samples per line, the array is referred to as having *square pels*, that is, the pel centers all lie on the corners of squares. This facilitates digital image processing as well as computer synthesis of images. In particular, geometrical operations such as rotation, or filters for antialiasing can be implemented without shape distortions.

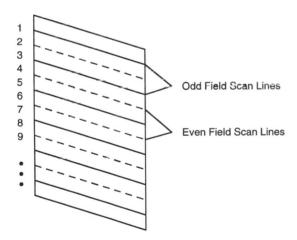

Fig. 6.2 Scanning pattern for 2:1 interlace. Odd-numbered lines (field 1) are scanned and displayed first, followed by the even-numbered lines (field 2). In this way, low spatial frequencies are displayed twice as often as high spatial frequencies, thus better matching the characteristics of human vision. Compared to scanning both the odd and even lines at the field rate, this also reduces the line rate (and bandwidth) by a factor of two.

6.2.2 Color Images

Light is a subset of the electromagnetic energy. The visible spectrum wavelengths range from 380 to 780 nanometers. Thus, visible light can be specified completely at a pel by its wavelength distribution $\{S(\lambda)\}$. This radiation, on impinging on the human retina, excites predominantly three different receptors that are sensitive to wavelengths at or near 445 (called blue), 535 (called green), and 570 (called red) nanometers. Each type of receptor measures the energy in the incident light at wavelengths near its dominant wavelength. The three resulting energy values uniquely specify each visually distinct color, **C**.

---

*MPEG has a different nomenclature, as we shall see.

This is the basis of the *trichromatic* theory of color which states that for human perception, any color can be synthesized by an appropriate mixture of three properly chosen primary colors **R, G,** and **B.** For video, the primaries are usually red, green, and blue. The amounts of each primary required are called the tristimulus values. If a color **C** has tristimulus values R_C, G_C, and B_C, then $\mathbf{C} = R_C\mathbf{R} + G_C\mathbf{G} + B_C\mathbf{B}$. The tristimulus values of a wavelength distribution $S(\lambda)$ are given by

$$R_S = \int S(\lambda)r(\lambda)d\lambda$$

$$G_S = \int S(\lambda)g(\lambda)d\lambda \qquad (6.5)$$

$$B_S = \int S(\lambda)b(\lambda)d\lambda$$

where $\{r(\lambda), g(\lambda), b(\lambda)\}$ are called the *color matching functions* for primaries **R, G,** and **B.** These are also the tristimulus values of unit intensity monochromatic light of wavelength λ. Figure 6.3 shows color matching functions with the primary colors chosen to be spectral (light of a single wavelength) colors of wavelengths 700.0, 546.1, and 435.8 nanometers. Equation (6.5) allows us to compute the tri-

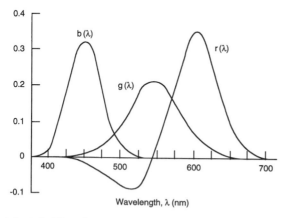

0.4

0.3

0.2

0.1

0

-0.1

Wavelength, λ (nm)

Fig. 6.3   The color-matching functions for the 2° Standard Observer, based on primaries of wavelengths 700 (red), 546.1 (green), and 435.8 nm (blue), with units such that equal quantities of the three primaries are needed to match the equal energy white.

stimulus values of any color with a given spectral distribution, $S(\lambda)$, using color matching functions.

One consequence of this is that any two colors with spectral distributions $S_1(\lambda)$ and $S_2(\lambda)$ match if and only if

$$R_1 = \int S_1(\lambda)r(\lambda)d\lambda = \int S_2(\lambda)r(\lambda)d\lambda = R_2$$

$$G_1 = \int S_1(\lambda)g(\lambda)d\lambda = \int S_2(\lambda)g(\lambda)d\lambda = G_2 \quad (6.6)$$

$$B_1 = \int S_1(\lambda)b(\lambda)d\lambda = \int S_2(\lambda)b(\lambda)d\lambda = B_2$$

where $\{R_1, G_1, B_1\}$ and $\{R_2, G_2, B_2\}$ are the tristimulus values of the two distributions $S_1(\lambda)$ and $S_2(\lambda)$, respectively. This could happen even if $S_1(\lambda)$ were not equal to $S_2(\lambda)$, for all the wavelengths in the visible region.

Instead of specifying a color, **C,** by its tristimulus values* $\{R, G, B\}$, normalized quantities called *chromaticity coordinates* $\{r, g, b\}$ are often used:

$$r = \frac{R}{R+G+B}$$

$$g = \frac{G}{R+G+B} \qquad (6.7)$$

$$b = \frac{B}{R+G+B}$$

Since $r + g + b = 1$, any two chromaticity coordinates are sufficient. However, for complete specification a third dimension is required. It is usually chosen to be the *luminance* (Y).

Luminance is an objective measure of brightness. Different contributions of wavelengths to the sensation of brightness are represented by $y(\lambda)$, *the relative luminance efficiency.* The luminance of any given spectral distribution $S(\lambda)$ is then given by

$$Y = k_m \int S(\lambda)y(\lambda)d\lambda \qquad (6.8)$$

where k_m is a normalizing constant. For any given

---

*We drop the subscripts C from now on.

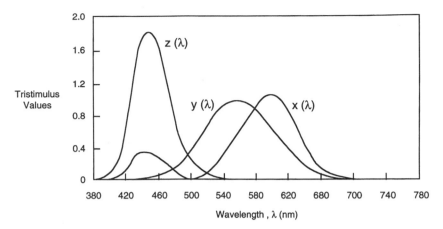

Fig. 6.4 Color matching functions $x(\lambda)$, $y(\lambda)$, $z(\lambda)$ for the 2° Standard Observer. Note that $y(\lambda)$ is the same as the relative luminous efficiency curve and therefore tristimulus value Y is the luminance.

choice of primaries and their corresponding color matching functions, luminance can be written as a linear combination of the tristimulus values, $\{R, G, B\}$. Thus, a complete specification of color is given either by the tristimulus values or by the luminance and two chromaticities. A color image can then be specified by luminance and chromaticities at each pel.

The above specification of color depends on the choice of the primary colors. To help precise measurements as well as cooperation and communication between colorimetrists, the CIE (Commission Internationale de L'Eclairge) has defined a standard observer and three new primaries, **X, Y,** and **Z,** whose color matching functions are given in Fig. 6.4. This CIE coordinate system has two useful properties: first, the Y tristimulus value corresponds to the definition of luminance normalized to equal energy white, and second, since the color matching functions are always positive, the X, Y, Z tristimulus values are also always positive. Also, of course, a color **C** can be specified in the CIE system by the two chromaticities (x, y) and the luminance Y.

In most color TV systems, computation of chromaticities is deemed too complicated. Instead, colors are usually described by luminance Y and two *chrominances*, which are other linear combinations of R, G, and B. A common choice for chrominances are $(B-Y)$ and $(R-Y)$, with suitable normalization to constrain their amplitudes.

If two colors $S_1(=X_1, Y_1, Z_1)$ and $S_2(=X_2, Y_2, Z_2)$ are mixed to obtain a color $S(=X, Y, Z)$, then

$$X = X_1 + X_2$$

$$Y = Y_1 + Y_2 \tag{6.9}$$

$$Z = Z_1 + Z_2$$

Since the chromaticities are ratios, they are given by

$$x = \frac{X_1 + X_2}{X + Y + Z} = \frac{x_1(Y_1/y_1) + x_2(Y_2/y_2)}{(Y_1/y_1) + (Y_2/y_2)}$$

and $\tag{6.10}$

$$y = \frac{Y_1 + Y_2}{(Y_1/y_1) + (Y_2/y_2)}$$

and the luminance is given by

$$Y = Y_1 + Y_2 \tag{6.11}$$

In summary, a picture can be represented as a two-dimensional array of pels. For video, the array is repeated at the frame rate. Proper filtering is required before sampling the intensity to create this array. Each pel in the array can be specified by three quantities, either tristimulus values $\{R, G, B\}$ or the two chromaticity coordinates plus luminance {*x, y, Y*} or simply luminance and chrominance, for example, {*Y, B–Y, R–Y*}. Based on the trichromacy of the human visual system, any of these triplets completely specifies any color.

6.3   BASIC COMPRESSION TECHNIQUES

We have so far represented a continuous image intensity function $B(x, y, t)$ as an array of samples on a rectangular spatial grid sampled at a frame rate. For color images, B is a vector expressed either as tristimulus values or chromaticity coordinates/luminance, measured according to some primary colors.

Considering for now only the luminance, the next step in representing image data with finite bits (or bitrates) is to quantize each sample of the array using 2^K levels. This generates K bits for each pel and is known as PCM encoding. The time discreteness is provided by sampling the signal at least at the Nyquist rate, whereas the amplitude discreteness is provided by using a sufficient number of quantization levels so that degradation due to quantization error is tolerable. PCM does not remove any statistical or perceptual redundancy from the signal, and therefore results in large bitrates; however, it provides the first input stage for most of the compression system.

Starting with a PCM coded bitstream, a number of compression techniques have been developed to reduce the bitrate. A compression system typically consists of a combination of these techniques to satisfy the requirements of an application.

The first step in compression usually consists of decorrelation, that is, reducing the spatial and temporal redundancy in the signal. The candidates for this step are:

1. Predicting the next pel using values of the "previously transmitted" pels, and then transmitting the quantized prediction error. This is called Differential PCM (DPCM). A variety of previously transmitted pels can be used for prediction. If previous pels are restricted to the current frame, it is called intraframe prediction; if previous frame pels

are used, it is called interframe prediction. Such an approach can be made adaptive by changing the prediction algorithm based on local picture statistics. In the case of interframe prediction, a pel in the current frame can be predicted by a pel in the previous frame either at the same spatial location (in case of no motion) or at a displaced location in case it is a part of the moving object in the scene.

2. Transforming (taking a linear combination of) a block of pels to convert them into a set of transform coefficients. The transform often packs most of the energy into a few coefficients, and makes the other coefficients close to zero so they can then be dropped from transmission. One-, two-, and three-dimensional blocks (i.e., two spatial dimensions and one time dimension) have been studied for transformation. Recently, however, two-dimensional transforms have become more popular based on complexity considerations. They remove the correlation only in the two spatial dimensions. Temporal correlation is removed by computing displaced prediction errors as above.

The second step in compression is selection of the important coefficients or pels and then quantizing them with just enough accuracy. For DPCM, the prediction error may be quantized a sample at a time or as a vector of many samples. As long as the quantization step size is sufficiently small and the quantizer range is large enough to cover occasional large prediction errors, a good picture quality can be maintained. Alternatively, for transform coding only the significant coefficients must be selected and quantized. For good picture quality it is important that all the transform coefficients that have significant amplitude are quantized with sufficient accuracy.

3. The final step is entropy coding. It recognizes that the different values of the quantized signal occur with different frequencies, and therefore representing them with unequal length binary codes reduces the average bitrates. As we will see later, both the prediction error in the case of predictive coding and the coefficients in the case of transform coding have highly nonuniform probability density functions. In such cases, entropy coding results in substantial benefits. Unfortunately, such a variable length code generates a variable data rate, requiring a buffer to smooth it out for transmission over a

fixed bitrate transmission channel. The queues in the buffer must be managed to prevent either overflow or underflow. In the remainder of this chapter we give more details of the following techniques, since they have formed the basis of most modern compression systems:

Predictive coding (DPCM) (including motion compensation)

Transform coding

Vector quantization

Entropy coding

Incorporation of perceptual factors.

6.4 PREDICTIVE CODING (DPCM)

In basic predictive coders (see Fig. 6.5), a prediction of the pel to be encoded is made from previously coded information that has been transmitted. The error (or differential signal) resulting from the

subtraction of the prediction from the actual value of the pel is quantized into one of L discrete amplitude levels. These levels are then represented as binary words of fixed or variable word lengths and sent to the channel coder for transmission. The three main parts of the predictive coder are (1) predictor, (2) quantizer, and (3) entropy coder.

6.4.1 Predictors

In its simplest form, DPCM uses the coded value of the horizontally previous pel as a prediction. This is equivalent to quantization of a horizontal slope of the intensity signal. However, to exploit all the correlation, more pels (than just the previous pel) in the horizontal direction, as well as a few pels in the previous line (in the same field for interlace scanning), are often used. Coding systems based on such two-dimensional predictors have been traditionally called intrafield codecs. Interframe coders use pels in the previous field/frame and therefore require field/frame memory. As the cost of memo-

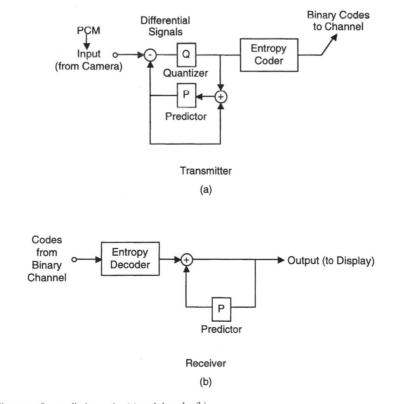

Fig. 6.5 Block diagram of a predictive coder (a) and decoder (b).

Table 6.1 Prediction coefficients and resulting variances of prediction errors for luminance and chrominance signals. Pel positions are shown in Fig. 6.6.

Video Signal	Prediction Coefficients			Variance
	α_A	α_B	α_C	σ_E^2
Y	1	—	—	53.1
	1	$-1/2$	$1/2$	29.8
	3/4	$-1/2$	3/4	27.9
	7/8	$-5/8$	3.4	26.3
$R-Y$	1	—	—	22.6
	—	—	1	5.6
	1/2	$-1/2$	1	4.9
	5/8	$-1/2$	7/8	4.7
$B-Y$	1	—	—	13.3
	—	—	1	3.2
	1/2	$-1/2$	1	2.5
	3/8	$-1/4$	7/8	2.5

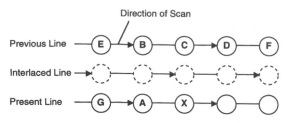

Fig. 6.6 Configuration of picture elements used in Table 6.1 for prediction of pel X. All the solid scan lines are from the same field, making these infrafield predictions.

is to render the sharp edges much better. In addition, keeping the predictor coefficients as powers of two leads to simpler implementation. It is worth noting that prediction performance is not improved by adding more pels. In fact, adding more pels to the configuration of Fig. 6.6 would give only marginal improvement.

6.4.2 Motion Estimation

Interframe predictors could also be constructed by using previously encoded pels from the previous fields/frames. If there are no changes in the image intensity from frame to frame, then temporally most adjacent pels are the best predictors. For small motion of objects with only moderately sharp edges, a combination of pels from current and two previous fields work well.

Figure 6.7 shows the relative position of pels in the previous field/frame that can be used for inter-

ry chips continues to drop, the distinction between intrafield and interframe coders is blurring.

Table 6.1 shows several practical predictors and their performance (in terms of mean square prediction error) as decorrelators of both the luminance and chrominance components of a color TV signal. The configuration of pels referred to in this table is shown in Fig. 6.6.

It is clear from the table that the two-dimensional predictors do a far better job of removing the correlation than the one-dimensional predictors. Practical experience also shows that the subjective effect of using a two-dimensional prediction

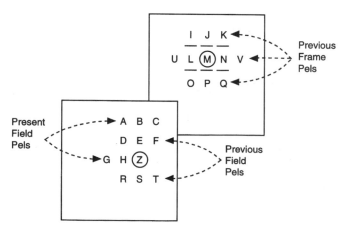

Fig. 6.7 Relative configuration of pels used for interframe prediction. Pel Z in the moving area is to be predicted from a combination of pels in the three successive fields. Pels Z and M are at the same spatial location but one frame period apart.

Table 6.2 Typical differential signal entropies of moving-area pels for some fixed linear predictors when the differential signal is quantized to 35 levels.

Transmitted Signal $Z - \hat{Z}$	Prediction $\hat{Z}$	Entropies in Bits per Moving-Area Pel (35 Level Quantization)
Frame-difference	M	≈ 2.1 to 3.9
Element-difference	H	≈ 2.0 to 3.7
Element-difference of frame-difference	$M + H - L$	≈ 1.8 to 3.1
Line-difference of frame-difference	$M + B - \hat{J}$	≈ 1.5 to 3.5
Field-difference	$(E + S)/2$	≈ 1.8 to 3.2
Element-difference of field-difference	$H + (E + S)/2 - (D + R)/2$	≈ 1.5 to 2.5

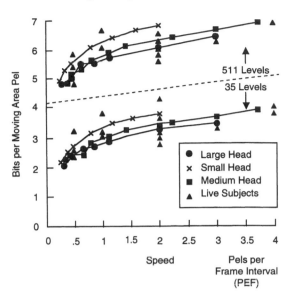

Fig. 6.8 Moving-area frame-difference entropy as a function of speed of movement. Fixed interframe prediction loses efficiency as the speed increases. This result is independent of the quantization of prediction and as the two scales used in the figure show.

frame prediction. Table 6.2 shows the compression efficiency (as measured by the entropy of the differential signal) of the various predictors.

Compared to a pure frame difference predictor, element-difference of field-difference prediction does quite well, particularly when frame differences starts to become large owing to motion. Such fixed predictors, which are based on temporal correlation, become more and more inefficient as the motion in the scene increases. Any object motion larger than $1 \approx 2$ pels per frame, which is common in many applications, results in loss of correlation between pel Z and M in Fig. 6.7 that are spatially adjacent to Z.

Figure 6.8 shows the predictor efficiency as a function of motion. Because of the loss of the prediction efficiency by motion, it is important to adapt the prediction when the scene contains large motion.

If a television scene contains moving objects and if an estimate of their translation in one frame time is available, then the temporal prediction can be adapted using pels in the previous frame that are appropriately spatially displaced. Such prediction is called motion compensated prediction. In real scenes, motion can be a complex combination of three-dimensional translation and rotation. Such motion is difficult to estimate even with a large amount of processing. In addition, three-dimensional motion can expose new features on the object's surface that were not visible in the previous frame, and therefore cannot be predicted from the previous frame. Fortunately, practical experience shows that translational motion occurs quite often in real scenes. In addition, translational motion in a plane perpendicular to the camera axis can be estimated much more easily and therefore has been used successfully for motion compensated predictive coding.

The crucial requirement is the algorithm for motion estimation. Two types of motion estimation algorithms are described below that have been successfully used in coding applications. These algorithms are derived under the following assumptions: (1) object motion is in a plane perpendicular to the camera axis, (2) illumination is spatially and temporally uniform, and (3) occlusion of one object by another and uncovered background are relatively small. Although these assumptions may appear unrealistic and quite constraining, practical experience suggests that these algorithms and their benefits are quite robust for a variety of real situations.

Under these assumptions, the monochrome intensities $b(\mathbf{Z}, t)$ and $b(\mathbf{Z}, t - \tau)$ of two consecutive frames are related by

$$b(\mathbf{Z}, t) = b(\mathbf{Z} - \mathbf{D}, t - \tau) \qquad (6.12)$$

where τ is the time between two frames, $\mathbf{D}$ is the two-dimensional translation vector of the object during the time interval, $[t - \tau, t]$ and $\mathbf{Z}$ is the two-dimensional vector representing spatial position. Thus, in real scenes a very good prediction of $b(\mathbf{Z}, t)$ is given by $b(\mathbf{Z} - \mathbf{D}, t - \tau)$. The problem then is to estimate $\mathbf{D}$ from the intensities of the present and previous frame. Two broad types of methods have been used: (1) Block Matching Methods, in which a block of pels in current (some yet to be transmitted) and previous frames are compared to estimate a $\mathbf{D}$, but because some "future" pels are used, $\mathbf{D}$ is generally coded and transmitted as side information, (2) Recursive Methods, in which $\mathbf{D}$ is not transmitted, but instead estimated only from previously transmitted pels. It is also possible to combine these two methods in a variety of ways.

6.4.2a   Block Matching Methods

Block Matching Methods assume that the object displacement is constant within a small two-dimensional $M \times N$ block of pels. The displacement $\mathbf{D}$ can be estimated by minimizing a measure of the prediction error

$$\text{PE}(\mathbf{D}) = \sum_{M \times N} G\Big(b(\mathbf{Z}, t) - b(\mathbf{Z} - \mathbf{D}, t - \tau)\Big) \quad (6.13)$$

where $G(\cdot)$ is a distance metric such as the magnitude or the square function. The minimization is done over a region of pels, $(M + 2d_{\max})(N + 2d_{\max})$, centered around pel $\mathbf{Z}_0$. $d_{\max}$ is the maximum horizontal and vertical displacement of pels. Thus, if the block size is 9×9 and a maximum displacement of $d_{\max} = 10$ is used, the search region in the previous frame would be an area containing 29×29 pels. An exhaustive search method would be to evaluate Eq. (6.13) for every pel shift in the horizontal and vertical direction and choose the minimum. This would require $(2d_{\max} + 1)^2$ evaluations of Eq. (6.13). A simple magnitude distance criterion $G(\cdot)$ for Eq. (6.13) results in:

$$\text{PE}(\mathbf{Z}_0, i, j) = \frac{1}{MN} \sum_{|m| \le \frac{M}{2}} \sum_{|n| \le \frac{N}{2}} \Big| b(\mathbf{Z}_{mn}, t)$$
$$- b(\mathbf{Z}_{m+i, n+j}, t - \tau) \Big| \quad (6.14)$$

where

$$-d_{\max} \le i, j \le +d_{\max}$$
$$\mathbf{Z}_{mn} = \mathbf{Z}_0 + [m, n]' \quad (6.15)$$

Several heuristics have been developed for the search procedure. The goal is to require as few shifts as possible or equivalently as few evaluations of PE. The number of shifts are reduced under the assumption that $\text{PE}(\mathbf{Z}_0, i, j)$ increases monotonically as the shift (i, j) moves away from the direction of minimum distortion. An example is the two-dimensional logarithmic search where the algorithm follows the direction of minimum distortion.

Figure 6.9 shows that at each step five shifts are checked. The distance between the search points is decreased if the minimum is at the center of the search location or at the boundary of each search area. For the example shown in Fig. 6.9, five steps were required to get the displacement vector at the point $(i, j) = (2, 6)$. Although the example shows the search process for pel displacements that are integers, many applications use fractional pel accuracy of either 0.25 pels or 0.5 pels. The effect of

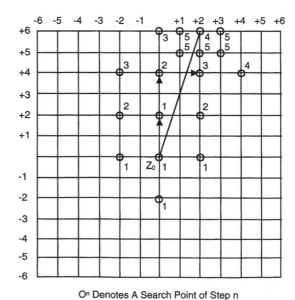

On Denotes A Search Point of Step n

Fig. 6.9   Illustration of two-dimensional logarithmic search procedure. The shifts in the search area of the previous frame are shown with respect to a pel $(\mathbf{Z}_0)$ in the present frame. Here the approximate displacement vectors $(0, 2)'$, $(0, 4)'$, $(2, 4)'$, $(2, 5)'$, $(2, 6)'$ are found in steps 1, 2, 3, 4, and 5. $d_{\max}$ is 6 pels.

this is simply to make a search grid finer and larger. Since block matching uses pels that are not yet transmitted to the receiver, displacement values need to be coded and sent separately as overhead. This increases the bitrate, but also has the potential for better motion estimation. As an example, if a 5 × 5 block is used, and if d_{max} is ±8 pels, then for a displacement accuracy of $1/2$ pel, for every block, 12 bits specifying the motion vector have to be transmitted, resulting in a bitrate of 0.48 bits/pel.

6.4.2b Recursive Methods

Recursive displacement estimation algorithms have also been developed to minimize local prediction errors. If a pel at location $\mathbf{Z}_n$ in the present frame is predicted with displacement $\hat{\mathbf{D}}_{k-1}$, that is, intensity $b(\mathbf{Z}_n - \hat{\mathbf{D}}_{k-1}, t - \tau)$, resulting in a certain amount of prediction error, the estimator should try to decrease the prediction error for $\mathbf{Z}_{n+1}$ by proper choice of $\hat{\mathbf{D}}_k$. If the displacement is updated at each pel, then $k = n$. Otherwise k is related to n in some predetermined manner. It is assumed that the pel at location $\mathbf{Z}_n$ is being predicted and displacement estimate $\hat{\mathbf{D}}_{k-1}$ is computed only with information transmitted prior to pel $\mathbf{Z}_n$.

The prediction error can be called displaced frame difference (DFD):

$$DFD(\mathbf{Z}, \hat{\mathbf{D}}) = b(\mathbf{Z}, t) - b(\mathbf{Z} - \hat{\mathbf{D}}, t - \tau) \quad (6.16)$$

DFD is defined in terms of two variables: (1) the spatial location $\mathbf{Z}$ at which DFD is evaluated and (2) the estimated displacement $\hat{\mathbf{D}}$ used to compute an intensity in the previous frame. In the case of a two-dimensional grid of discrete samples, an interpolation process must be used to evaluate $b(\mathbf{Z} - \hat{\mathbf{D}}, t - \tau)$ for values of $\hat{\mathbf{D}}$ that are a noninteger number of pel spacings. In the ideal case, DFD converges to zero as $\hat{\mathbf{D}}$ converges to the actual displacement, $\mathbf{D}$.

Pel recursive displacement estimators minimize recursively $[DFD(\mathbf{Z}, \hat{\mathbf{D}})]^2$ at each moving area pel using a steepest descent algorithm. Thus,

$$\hat{\mathbf{D}}_k = \hat{\mathbf{D}}_{k-1} - \frac{\varepsilon}{2} \nabla_{\hat{\mathbf{D}}_{k-1}} \left[DFD(\mathbf{Z}_n, \hat{\mathbf{D}}_{k-1}) \right]^2 \quad (6.17)$$

where ∇ is the two-dimensional gradient operator with respect to $\hat{\mathbf{D}}$. Carrying out the above operation using Eq. (6.16):

$$\hat{\mathbf{D}}_k = \hat{\mathbf{D}}_{k-1} - \varepsilon DFD(\mathbf{Z}_n, \hat{\mathbf{D}}_{k-1})$$
$$\cdot \nabla_z b(\mathbf{Z}_n - \hat{\mathbf{D}}_{k-1}, t - \tau) \quad (6.18)$$

Figure 6.10 shows that at every iteration we add to the old displacement estimate a vector quantity that is either parallel to or opposite to the direction of the spatial gradient of image intensity depending on the sign of DFD, the motion compensated prediction error. If the actual displacement error $(\mathbf{D} - \hat{\mathbf{D}}_k)$ is orthogonal to the intensity gradient $\nabla_z b$, the displaced frame difference is zero giving a zero update. This may happen even though the object has moved. Also if the gradient of intensity is zero, that is, absence of edges, even if there is an error in the displacement estimation, the algorithm will do no updating. It is only through the occurrence of edges with differing orientations in real television scenes that convergence of $\hat{\mathbf{D}}$ to the actual displacement $\mathbf{D}$ is possible.

Figure 6.11 shows the efficiency of motion compensated prediction in terms of reducing the bitrate. Compared to frame difference, motion compensated prediction can result in a factor of two to three improvement in compression. However, the improvement depends on the scene, the type of motion, and the type of objects in motion.

6.4.3 Quantization

DPCM systems achieve compression by quantizing the prediction error more coarsely than the original signal itself. A large number of properly placed quantization levels usually result in excellent picture quality. However, to reduce the bitrate to a minimum, the placement of levels has to be optimized. Three types of degradations may be seen from improper placement of quantizer levels. These are referred to as *granular noise, slope overload,* and *edge business.*

As shown in Fig. 6.12, which uses previous pel prediction, coarse inner levels lead to granular noise in flat or low detail areas of the picture. The result is similar to a picture upon which white noise

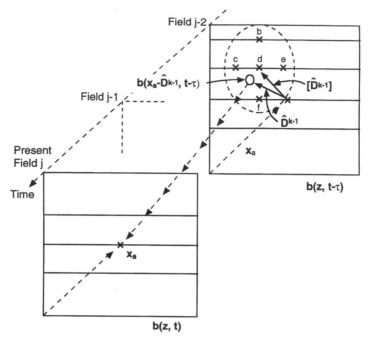

Fig. 6.10 Recursive motion estimation. Displacement estimate $\mathbf{D}_{k-1}$ is updated at pel a. Gradient of intensity $\nabla_z b(z - D_k, t - \tau)$ is obtained by using intensities at pels b, c, d, e, f in the field $(j - 2)$. [ ] represents rounding.

is superimposed. If the largest representative level is too small, then the decoder takes several samples to catch up with a high-contrast edge, resulting in slope overload. The result is a selective blurring of the picture. Finally, for edges whose contrast changes somewhat gradually, the quantizer output oscillates around the signal value which may change from line to line, or frame to frame, giving the appearance of a discontinuous jittery or "busy" edge. Thus, quantizer design must make tradeoffs between these three rather different subjective effects, and at the same time optimize the bitrate. This is a difficult task and is usually done by trial and error. However, two systematic approaches have emerged to guide the trial and error: (1) statistical design and (2) subjective design.

6.4.3a Statistical Design

Quantizers can be designed purely on the basis of statistical criteria to minimize the mean square

quantization error. Considering Fig. 6.13, if x is the input to the DPCM quantizer and has probability density $p(x)$, then quantizer parameters can be obtained to minimize the following measure of quantization error:

$$D = \sum_{k=1}^{L} \int_{t_{k-1}}^{t_k} f(x - \ell_k) \cdot p(x)dx \qquad (6.20)$$

where $t_0 < t_1 < \ldots < t_L$ and $\ell_1 < \ell_2 < \ldots < \ell_L$ are decision and representative levels, respectively, and $f(\cdot)$ is a non-negative error measure. We assume, as shown in Fig. 6.13, that all inputs, $t_{k-1} < x \leq t_k$ to the quantizer are represented* as ℓ_k. Differentiating Eq. (6.20) with respect to t_k and ℓ_k (for a fixed number of levels L), necessary conditions for optimality are given by:

$$f(t_{k-1} - \ell_{k-1}) = f(t_{k-1} - \ell_k), \qquad k = 2 \ldots L \qquad (6.21)$$

---

*MPEG uses a different quantizer nomenclature. MPEG quantizers output the integer k.

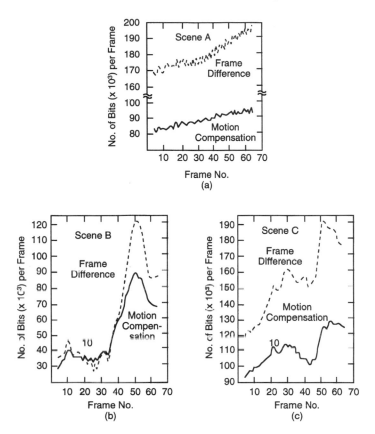

Fig. 6.11 Comparison of bitrates for frame difference and motion compensated prediction for color component signals from Scene A, Scene B, and Scene C. The three scenes contain varying amount of motion and spatial detail.

and

$$\int_{t_{k-1}}^{t_k} \frac{df(x - \ell_k)}{dx} \cdot p(x)dx = 0, \qquad k = 1 \dots L \quad (6.22)$$

assuming that $f(\cdot)$ is differentiable. If the mean square error criterion is to be used, that is, $f(z) = z^2$, the above equations reduce to the *Lloyd–Max* quantizer

$$t_k = \frac{1}{2}(\ell_k + \ell_{k+1}) \qquad (6.23)$$

and

$$\ell_k = \int_{t_{k-1}}^{t_k} xp(x)dx \; / \int_{t_{k-1}}^{t_k} p(x)dx \qquad (6.24)$$

Thus, given the probability density, statistically optimum quantizers are simple to compute. However, $p(x)$ is often not available. Fortunately, for the case of previous pel and other differential coders, $p(x)$ can be approximated by a Laplacian density, and therefore, other than a few parameters, the shape of $p(x)$ is known. The optimum quantizer in this case is *companded*, that is, the step size $(t_k - t_{k-1})$ increases with k. For highly peaked Laplacian densities, this increase with k is even more rapid. Since for much of the picture, the intensity signal has no sharp edges, the prediction error x is close to zero and therefore $p(x)$ is highly peaked around zero. This results in a close spacing of levels near $x = 0$ and a rather coarse spacing of levels as $|x|$ increases. For images, this generally results in over-specification of the low-detailed and low-motion areas of the picture, and consequently only a small amount of granular noise is visible, if at all. How-

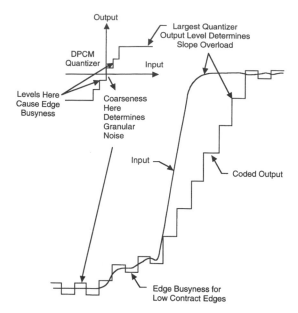

Fig. 6.12 An intuitive classification of distortion due to coarse DPCM quantization. Three classes of quantization noise are identified: granular noise, edge business, and slope overload.

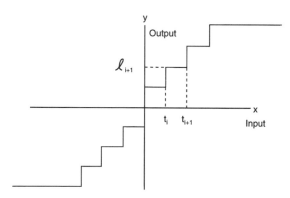

Fig. 6.13 Characteristics of a quantizer. x is the input and y is the output. (t_i) and (ℓ_i) are decision and representative levels, respectively. Inputs between decision levels t_i and t_{i+1} are represented by representative level $\ell_i + 1$.

ever, areas with detail and edges are poorly reproduced. Owing to the small dynamic range of such a quantizer, this is often visible as slope overload and edge business for high-contrast edges. Thus for subjective reasons, the quantizers are not designed based solely on the *mean square error* criterion.

Minimization of a quantization error metric, for a fixed number of levels, is relevant for DPCM systems that code and transmit the quantizer outputs using fixed-length binary words. However, since the probability of occurrence of different quantizer levels is highly variable, there is considerable advantage in using variable length words to represent quantizer outputs. In such a case, since the average bitrate (in bits/pel) is lower bounded by the entropy of the quantizer output, it is more relevant to minimize the quantization distortion subject to a fixed entropy, rather than a fixed number of levels. For the Laplacian density that is typical of prediction errors, it can be shown that such optimization leads to uniform quantizers. Although such entropy coders may provide adequate picture quality for sufficiently small step size, they may often be too complex to implement because of the very long code-word lengths required for some of the outer levels, which are very infrequent.

6.4.2b Subjective Design

For better picture quality, quantizers should be designed on the basis of psychovisual criteria. Since the human visual system is very complex, the optimum psychovisual criterion has eluded designers. However, several simple algorithmic models of the visual systems have emerged over the past decade, which can be used for optimization of quantizers. One straightforward method is to extend the Lloyd–Max technique by minimizing a weighted square quantization error, where the weighting is derived from subjective tests. This is similar to statistical design, where the probability density is replaced by a subjective weighting function in Eq. (6.20).

A typical subjective weighting function for a head and shoulders type of picture is shown in Fig. 6.14, where the error to be quantized is from a previous pel prediction. These subjective weights decrease with increasing magnitude of the prediction error. This is due to two factors: (1) quantization error is subjectively less visible at pels near or at large edges, which also correspond to pels where large prediction error usually occurs; and (2) the number of pels with large prediction error is few and isolated.

Such weighting functions are obtained empirically by subjective tests directly on individual pic-

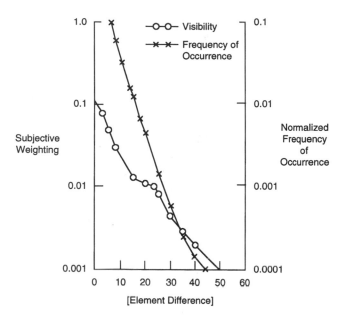

Fig. 6.14 Quantization noise visibility as a function of magnitude of the element difference for a typical head and shoulders view. This can be used as a weighing function for quantizer design. Also shown is a histogram of the magnitude of the element differences.

tures and are therefore picture dependent. However, experiments show that the variation for a class of pictures is not significant. As in the case of statistical design, such subjective quantizers can also be optimized for a fixed entropy instead of a fixed number of levels. In general, subjective quantizers are much less companded compared to statistical quantizers and therefore reproduce edges more faithfully. They do give more importance to better reproduction of edges even though, in statistical terms, edges do not occur as often. However, if proper care is not exercised, subjective quantizers may result in larger granular noise. Figures 6.15 and 6.16 give characteristics of quantizers optimized under different criteria and their performance in terms of the entropy of the quantizer output and a subjective measure of picture quality.

Quantizers can also be designed such that the quantization error is at or below the threshold of visibility, while minimizing either the number of quantizer levels or the entropy of the quantizer output. The threshold of visibility can be determined by a simple subjective experiment, in which subjects are shown both a perturbed (vary-

ing amount of perturbation ΔI.) and an unperturbed edge (see Fig. 6.17). The assumption here is that the perturbation corresponds to the quantization noise. The purpose of the experiment is to derive a relationship between the magnitude of the perturbation that is just visible and the edge contrast.

The perturbation is varied until it becomes just noticeable. In this way, the quantization error visibility threshold is obtained as a function of the amplitude of the edge. This relationship can be used in the design of the quantizer by constraining the quantization error to be below the threshold for all prediction errors of the picture signal. The algorithm for defining the characteristics of the quantizer satisfying this constraint is illustrated in Fig. 6.18, for an even L (number of quantizer output levels). In this case, $x = 0$ is a decision level, and the geometric procedure starts at the corresponding point on the visibility threshold function with a dashed line of inclination $-45°$. The intersection of this line with the abscissa gives the first reconstruction level. From this point, a dashed line of inclination $+45°$ is drawn. Its intersection with the visibility threshold func-

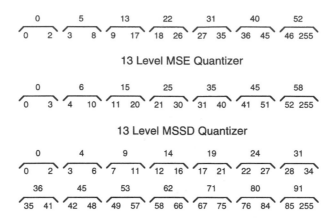

13 Level MSE Quantizer

13 Level MSSD Quantizer

MSSD Quantizer with Entropy 2.6 Bits/Pel

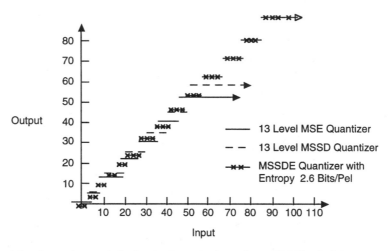

Fig. 6.15 Characteristics of quantizers optimized under three criteria are shown: (1) 13-level minimum mean square error (MSE); (2) 13-level minimum subjective weighted mean square error (MSSD); (3) MSSD with a constraint on the entropy (MSSDE) of the quantizer output. For (2) and (3), weights are derived from subjective weighting functions such as in Fig. 6.14.

tion gives the next decision level, and so on. This procedure is continued until the last +45° line exceeds the amplitude range of x. The number of points at which this path meets the abscissa is the minimum number of levels of the desired quantizer. The resulting dashed zigzag line represents the magnitude of the quantization error as a function of the quantizer input signal x. If L is odd, then the procedure starts with the reconstruction level at $x = 0$.

Experimental evidence indicates that quantizers satisfying the above threshold constraint turn out

to be the same whether the number of levels is minimized or the entropy is minimized. For monochrome videotelephone (i.e., 256 × 256 pels with 1 MHz bandwidth) signals with head and shoulders type of pictures, 27 levels are required for no visible error at normal (six times the picture height) viewing distance if previous element prediction is used. A larger number of levels is required for pictures containing a lot of sharp detail, for example, the resolution charts. Using two-dimensional prediction, this number may be reduced to 13 levels for natural (e.g., head and shoulders) pictures and

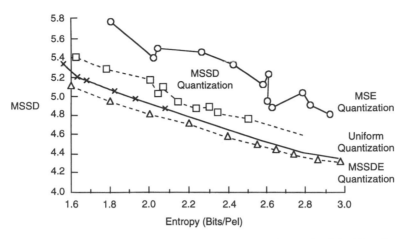

Fig. 6.16 Performance of quantizers optimized under different criteria for a typical head and shoulders picture. Mean square subjectively weighted quantization noise (MSSD) is plotted with respect to entropy. It is assumed that MSSD is a reasonable measure of picture quality.

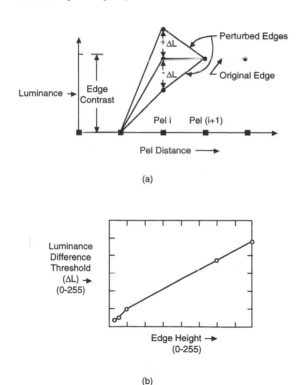

Fig. 6.17 (a) Stimuli for the determination of visibility thresholds (i.e., just noticeable perturbation in edge contrast). (b) Visibility threshold as a function of edge contrast.

21 levels for resolution charts. Chrominance components usually require fewer levels, for example, nine levels for the $(R-Y)$ component and five for the $(B-Y)$ component.

6.4.3c Adaptive Quantization

Because of the variation of the image statistics and the required fidelity of reproduction in different regions of the picture, adapting the DPCM quantizer usually leads to higher compression for the same quality. In general, one would like to divide a picture into several subpictures such that within each subpicture both the visibility of quantization noise and the statistical properties of the quantized differential signal are relatively similar. However, this goal is difficult or impossible to achieve since visibility depends on many factors besides the statistics. Several approximations of this idealized situation have been made, some purely statistical and some based on certain psychovisual criteria.

Pel domain approaches to adaptive quantization start out with a measure of spatial detail (called the *Masking Function*) and then obtain experimentally the relationship between the visibility of noise and the measure of spatial detail. In nonadaptive quantization, the prediction error is usually used as the measure of spatial detail. However, for adaptive quantization more complex measures are used.

A typical relationship for head and shoulders pictures is shown in Fig. 6.19, where a weighted sum of slopes in a 3×3 neighborhood is used as a measure of local spatial detail. If Δ_{ij}^H and Δ_{ij}^V are the horizontal and vertical slopes of the luminance

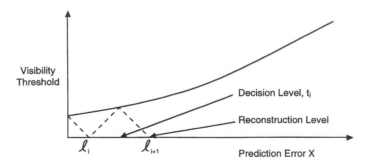

Fig. 6.18   Illustration of the subjective quantizer design procedure for an even number *L* of reconstruction levels. Note that by proper placement of the decision and representative levels, the procedure forces the quantization error (dotted line) to remain always below the visibility threshold function.

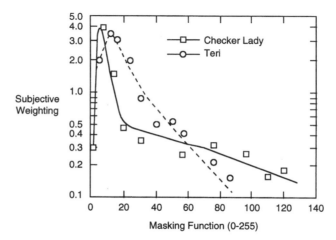

Fig. 6.19   Relationship between subjective visibility of noise and a measure of spatial detail that uses a combination of element and line differences in a 3 × 3 neighborhood. Dependence of the visibility on the picture content is illustrated for two different head and shoulders views.

at location (i, j) then the weighted sum of slopes is given by

$$M(i, j) = \sum_{l,m=-1}^{+1} \alpha^{\sqrt{\ell^2+m^2}} \left[|\, \Delta_{i-l,j-m}^H \,| + |\, \Delta_{i-l,j-m}^V \,| \right]$$

$$(6.25)$$

where (i, j) are the coordinates relative to the pel being coded, and $\alpha < 1$ is a factor based on psychovisual threshold data. Typically, α is taken to be 0.35, and therefore exponentially decreasing weights are assigned to pels of greater distance from the current (to be coded) pel.

Using the noise weighting function of Fig. 6.19, which relates the visibility of additive noise to the masking function, a picture is divided into (not necessarily contiguous) subpictures. Within each

segment, noise visibility is approximately constant, and a different quantizer is designed for each one based on the techniques of Sections 6.4.3a and 6.4.3b. The segmentation is controlled by the value of the masking function *M*, as indicated in Fig. 6.20. Figure 6.21 shows a head and shoulders picture divided into four segments. Figure 6.21a shows the original picture, and Figs. 6.21b–e show four subpictures (pels in these subpictures are shown white), which are in the order of increasing spatial detail, that is, decreasing noise visibility. Thus, flat, low detail areas of the picture constitute subpicture one (i.e., Fig. 6.21b), and sharp, high-contrast edges are included in subpicture four (i.e., Fig. 6.21e). In such adaptive quantization, the transmitter must signal to the receiver the subpicture each pel belongs to. Of course, for these

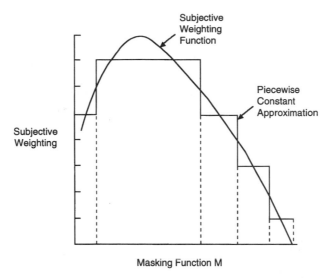

Subjective
Weighting
Function

Piecewise
Constant
Approximation

Subjective
Weighting

Masking Function M

Fig. 6.20 Segmentation of a picture by approximating the subjective weighting function by a piecewise constant function. Using this approximation, images are segmented into four subpictures, I to IV, according to the value of the masking function M at each pel.

methods to be worthwhile this signaling information must be substantially smaller than the decrease in bitrate due to adaptive quantization.

Although the optimum rules for adaptation are not known, rules based on intuition and trial-and-error have shown significant reduction in bitrate compared with nonadaptive systems.

6.4.4 Code Assignment

The probability distribution of the quantizer output levels is highly nonuniform for intraframe as well as interframe predictive coders. This lends itself to variable length codes (e.g., Huffman codes). The average bitrate for such a code is usually very close to and is lower bounded by the entropy of the quantizer output. A simple example of variable length codes is the Huffman Code. If the individual code words of a Huffman Code are concatenated to form a stream of bits, then correct decoding by a receiver requires that every combination of concatenated code words be uniquely decipherable. A sufficient condition for this is that the code satisfy the so-called *prefix rule*, which states that no code word may be the prefix of any other code word.

Construction of a Huffman Code, which follows the prefix rule, involves the use of a tree shown in Fig. 6.22 for 8-level quantization. Levels

$\{b(i)\}$ to be coded are first arranged in Section I of the tree according to decreasing order of probability. Construction of the tree then proceeds right to left as follows: the two bottom-most branches of Section I are combined via a *node* and their probabilities added to form a new branch in Section II of the tree. The branches of Section II are then rearranged according to decreasing order of probability. The procedure is repeated using the two bottom-most branches of Section II to form the branches of Section III which are then reordered according to decreasing probability, as with previous sections. The procedure continues until in Section VIII only one branch remains whose probability is one.

The entire tree may then be reordered to eliminate crossovers as shown in Fig. 6.23. Coding is then accomplished for each sample by moving from the left of the tree toward the level $b(i)$ to be coded and, at each node, transmitting a binary 0 if an upward step is required and a binary 1 to indicate a downward step. The receiver decodes using the same tree in exactly the same way, that is, moving upward for a received 0 and downward for a received 1. Resulting code words are shown at the right of Fig. 6.23. It can be shown that the Huffman Code minimizes the average word length.

A typical variable length code for a previous pel DPCM coder with 16 quantizer levels is shown in

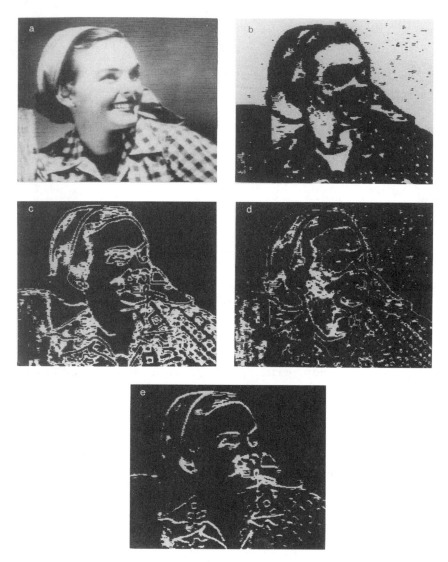

Fig. 6.21 Original picture (a) and its four segments (b), (c), (d), (e) designed on the basis of spatial detail (two-dimensional) and noise visibility. Segment (b) has the highest noise visibility, whereas segment (e) has the lease noise visibility.

Fig. 6.24. Inner levels occur much more often and therefore are represented by a smaller length word. A Huffman code requires a fairly reliable knowledge of the probability distribution of the quantized output. In the absence of such knowledge, the performance of the Huffman code may vary by a large amount. However, the probability distributions for DPCM are approximately Laplacian, and in practice, it is found that the Huffman Code efficiency is not too sensitive to small changes in the probability distributions that take place for different pictures. Therefore, Huffman codes based on the Laplacian distribution computed as an average of many pictures remain quite efficient. Typically, the average bitrate using Huffman coding is about one bit/pel less than the corresponding bitrate for a fixed length code.

One of the problems with the use of variable length codes is that the output bitrate from the source coder changes with local picture content.

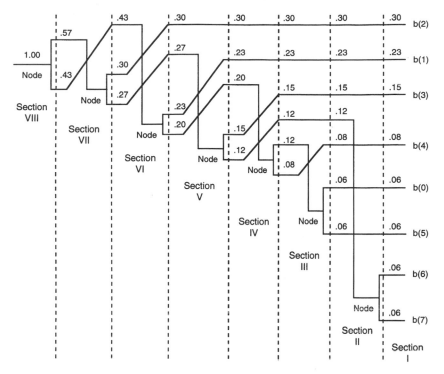

Fig. 6.22 Construction of a binary Huffman code proceeds from right to left. At each section the two bottom-most branches are combined to form a node and followed by a reordering of probabilities into descending order. These probabilities are then used to start the next section.

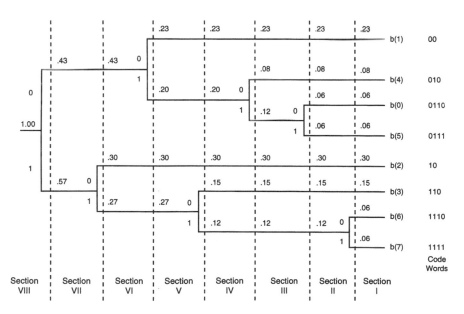

Fig. 6.23 After code construction, the tree is rearranged to eliminate crossovers, and coding proceeds from left to right. At each node a stepup produces a zero and a stepdown a one. Resulting code words are shown at the right for each of the eight levels. Note that no code word is a prefix of any other code word.

Level No.	Code Word Length	Code
1	12	100101010101
2	10	1001010100
3	8	10010100
4	6	100100
5	4	1000
6	4	1111
7	3	110
8	2	01
9	2	00
10	3	101
11	4	1110
12	5	10011
13	7	1001011
14	9	100101011
15	11	10010101011
16	12	100101010100

Fig. 6.24 A typical variable-length code for a DPCM coded signal with 16 quantizer levels.

Busy regions (or large motion) generate a large amount of data, whereas flat regions (or no motion) generate very little data. To transmit a variable bitrate digital signal over a constant bitrate channel, the source coder output has to be held temporarily in a first-in–first-out buffer that can accept input at a nonuniform rate, but whose output is emptied into the channel at a uniform rate. Since, in a practical system, a buffer must be of finite size, buffer queues, including buffer overflow and underflow, must be managed. This depends on the size of the buffer, the variable length code used, the image statistics, and the channel bitrate. Buffer underflow is usually not a problem since PCM values could be inserted to increase the coder output whenever required, which can only improve the picture quality.

By using a channel whose rate is somewhat higher than the entropy of the quantizer output, the probability of buffer overflow is markedly reduced. In addition, the likelihood of buffer overflow can be reduced by designing codes that minimize the probability of code words exceeding a certain length. However, since buffer overflow cannot always be prevented, strategies must be designed to gradually reduce the output bitrate of

the coder as the buffer begins to fill. This is particularly important in the case of interframe coders since the motion in a television scene is relatively bursty. Several techniques for reducing the input data rate to the buffer as it begins to fill have been studied, for example, reducing the sampling rate, coarser quantization, and so forth. Such methods allow graceful degradation of picture quality when the source coder overloads the channel.

6.5   BLOCK TRANSFORM CODING

In transform coding, blocks of pels are converted into another domain called the transform domain prior to coding and transmission. As discussed before, in the pel domain there is substantial correlation between neighboring pels, and therefore, transmitting each of them is wasteful. By transforming a block of pels, correlation between different transform coefficients is reduced substantially. Transform coding is relatively efficient for low bitrates, and achieves this mainly from three mechanisms. First, not all of the transform domain coefficients need to be transmitted to maintain good image quality; second, the coefficients that are coded need not be represented with full accuracy; and finally, coefficients have a very nonuniform probability distribution and therefore can be entropy coded for efficiency. Loosely speaking, transform coding is preferable to DPCM for compression below 2 to 3 bits per pel for single images, and is also used for coding of motion compensated prediction errors for video.

6.5.1   Discrete Linear Orthonormal Transforms

With one-dimensional, orthonormal linear transforms each block of pels to be transformed is first arranged into a column vector $\mathbf{b}$ of length N, and then transformed using an *orthonormal* (also known as unitary) linear transform by*

$$\mathbf{c} = \mathbf{Tb} \qquad (6.26)$$

---

*Superscript ′ on a vector or matrix denotes its transpose and complex conjugate. Superscript * indicates complex conjugate only.

$$\mathbf{b} = \mathbf{T}'\mathbf{c} \tag{6.27}$$

where $\mathbf{T}$ is the $\mathcal{N} \times \mathcal{N}$ transform matrix, and $\mathbf{c}$ is the column vector of transform coefficients. If the mth column of matrix $\mathbf{T}'$ is denoted by $\mathbf{t}_m$ then for orthonormal transforms

$$\mathbf{t}'_m \mathbf{t}_n = \delta_{mn} \tag{6.28}$$

where δ_{mn} is a delta function, equal to 1 if $m = n$; otherwise it is zero. The vectors $\mathbf{t}_m$ are *basis vectors* of the unitary transform $\mathbf{T}$, and based on Eq. (6.28), they are an orthonormal set. Using them as the basis functions, Eq. (6.27) can be written as

$$\mathbf{b} = \sum_{m=1}^{N} c_m \mathbf{t}_m \tag{6.29}$$

where c_m is the mth element of $\mathbf{c}$, and is given by

$$c_m = \mathbf{t}'_m \mathbf{b} \tag{6.30}$$

The first basis vector $\mathbf{t}_1$ consists of constant values for practically all transforms of interest, that is, it corresponds to zero spatial frequency and is given by

$$\mathbf{t}'_1 = \frac{1}{\sqrt{\mathcal{N}}}(1,1,1\ldots1) \tag{6.31}$$

For this reason, c_1 is often called the DC coefficient. Thus, if b_{max} is the largest possible pel value, then $\sqrt{\mathcal{N}}\, b_{max}$ is the largest possible value for c_1.

The vector $\mathbf{b}$ could be constructed from a two-dimensional $L \times L$ array of image pels by simply laying the L-pel columns end to end to form a single column vector of length $\mathcal{N} = L^2$.

Another possibility is to denote the $L \times L$ array of image pels by the square matrix $\mathbf{B} = [b_{ij}]$ and separate the transform into two steps. The first step is to transform the *rows* of $\mathbf{B}$ with a length L transform to exploit the horizontal correlation in the image. Next, the *columns* of $\mathbf{B}$ are transformed to exploit the vertical correlation in the image. The combined operation can be written

$$c_{mn} = \sum_{i=1}^{L}\sum_{j=1}^{L} b_{ij}t_{nj}t_{mi} \tag{6.32}$$

where the ts are the elements of the $L \times L$ transform matrix $\mathbf{T}$ and $\mathbf{C} = [c_{mn}]$ is the $L \times L$ matrix of transform coefficients. Assuming the one-dimensional basis vectors are indexed in order of increasing spatial frequency, then subscript m indexes the vertical spatial frequency, and n indexes the horizontal spatial frequency. If f_m and f_n are the vertical and horizontal spatial frequencies of the respective basis vectors, then the spatial frequency corresponding to coefficient c_{mn} is given by

$$f_{mn} = \sqrt{f_m^2 + f_n^2}\ \frac{\text{cycles}}{\text{distance}} \tag{6.33}$$

For unitary transforms, the inverse operation is simply

$$b_{ij} = \sum_{m-1}^{L}\sum_{n-1}^{L} c_{mn}t_{nj}^*t_{mi}^* \tag{6.34}$$

Matrices $\mathbf{C}$, $\mathbf{T}$, and $\mathbf{B}$ are related by[†]

$$\mathbf{C} = \mathbf{T}\mathbf{B}\mathbf{T}^t \tag{6.35}$$

and for unitary $\mathbf{T}$ the inverse operation is simply

$$\mathbf{B} = \mathbf{T}'\mathbf{C}\mathbf{T}^* \tag{6.36}$$

If $\{\mathbf{t}_m\}$ are the basis vectors of the Lth order transform $\mathbf{T}$, then Eq. (6.36) can be written

$$\mathbf{B} = \sum_{m=1}^{L}\sum_{n=1}^{L} c_{mn}\mathbf{t}_m^*\mathbf{t}'_n \tag{6.37}$$

That is, $\mathbf{B}$ can be thought of as a linear combination of $L \times L$ *basis images*

$$\{\mathbf{t}_m^*\mathbf{t}'_n\} \tag{6.38}$$

Eq. (6.37) is also known as an *outer product* expansion of $\mathbf{B}$.

[†]Superscript t indicates transpose only.

6.5.2 Discrete Cosine Transform (DCT)

In recent years, the DCT has become the most widely used of the unitary transforms, and under certain circumstances, as we shall see later, its performance can come close to that of the optimum transform. The elements of the DCT transform matrix **T** are given by

$$t_{mi} = \sqrt{\frac{2 - \delta_{m-1}}{\mathcal{N}}} \cos\left\{\frac{\pi}{N}\left(i - \frac{1}{2}\right)(m-1)\right\}$$

$$i, m = 1 \ldots \mathcal{N} \tag{6.39}$$

where $[\delta_0 = 1, \delta_p = 0$ for $p > 0]$. Thus, DCT basis vectors $\mathbf{t}_m$ are sinusoids with frequency indexed by m.

For $\mathcal{N} = 16$ the DCT basis vectors are shown in Fig. 6.25. One of the important factors in the choice of transforms, in addition to its ability to compress, is fast, inexpensive computation. The DCT has a fast computational algorithm along the lines of the FFT (Fast Fourier Transform).

A Fast DCT Algorithm

1. Reorder the pels to form the complex vector **w,** having length $\mathcal{N}/2$, as follows:

$$\text{Re}(\mathbf{w}) = (b_1 b_5 b_9 \ldots b_{12} b_8 b_4)'$$
$$\text{Im}(\mathbf{w}) = (b_3 b_7 b_{11} \ldots b_{10} b_6 b_2)' \tag{6.40}$$

2. Take the FFT of **w** to obtain the complex coefficients $\{W_m, m = 1 \ldots \mathcal{N}/2\}$.

3. Form the complex values X_m, Y_m, and G_m using Eqs. (6.41), and compute the complex DFT coefficients $\{V_m, 1 \le m \le 1 + \mathcal{N}/2\}$ using Eqs. (6.42).

$$X_m = \frac{1}{\sqrt{8}}\left[W_m + W'_{\frac{\mathcal{N}}{2}+2-m}\right]$$

$$Y_m = \frac{\sqrt{-1}}{\sqrt{8}}\left[W_m - W'_{\frac{\mathcal{N}}{2}+2-m}\right]\exp\left[-\frac{2\pi\sqrt{-1}(m-1)}{\mathcal{N}}\right]$$

$$G_m = \exp\left[-\frac{\pi\sqrt{-1}(m-1)}{2\mathcal{N}}\right] \tag{6.41}$$

$$V_1 = \frac{1}{\sqrt{2}}\left[\text{Re}(W_1) + \text{Im}(W_1)\right]$$

$$\left\{V_m = X_m - Y_m, \qquad 2 \le m < \frac{\mathcal{N}+5}{4}\right\}$$

$$\left\{V_{\frac{\mathcal{N}}{2}+2-m} = X'_m + Y'_m, \qquad 2 \le m < \frac{\mathcal{N}+5}{4}\right\} \tag{6.42}$$

$$V_{\frac{\mathcal{N}}{2}+2-m} = \frac{1}{\sqrt{2}}\left[\text{Re}(W_1) + \text{Im}(W_1)\right]$$

4. Form the complex products $\{U_m = G_m V_m, 2 \le m \le \mathcal{N}/2\}$.

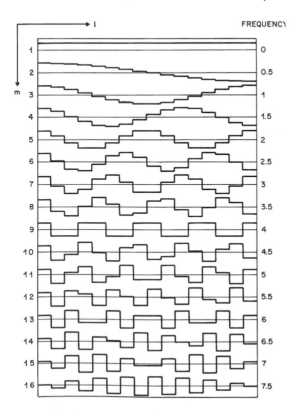

Fig. 6.25   Basis vectors for the DCT with $\mathcal{N} = 16$.

5. The DCT coefficients are then given by

$$c_1 = V_1$$

$$c_m = \sqrt{2}\,\mathrm{Re}(U_m), \qquad 2 \le m \le \frac{N}{2}$$

$$\tag{6.43}$$

$$c_{\frac{N}{2}+1} = V_{\frac{N}{2}+1}$$

$$c_{N-p} = -\sqrt{2}\,\mathrm{Im}(U_{2+p}), \qquad 0 \le p \le \frac{N}{2} - 2$$

The inverse DCT is the reverse of the above procedure. That is, we solve for U_m, then V_m, then X_m and Y_m, then W_m. An inverse $N/2$-point FFT then gives **w,** and a reordering produces **b.** The computation time for the Fast DCTs increases as $N \log_2 N$, as opposed to the N^2 increase for direct matrix multiplication. For sizable $N > 8$ the computational advantage of using the Fast DCT Algorithm can be enormous, as shown in Fig. 6.26. An alternative procedure for $N = 8$ is given in the Appendix.

An indication of the compaction capabilities of the DCT can be obtained by examining the spectral energy distribution in the transform domain. Figure 6.27 shows the mean square values (MSV) of the DCT coefficients for the image "Karen," for various one- and two-dimensional block sizes, arranged according to increasing spatial frequency as well as according to decreasing MSV. Good transforms compact the largest amount of energy into the smallest number of coefficients.

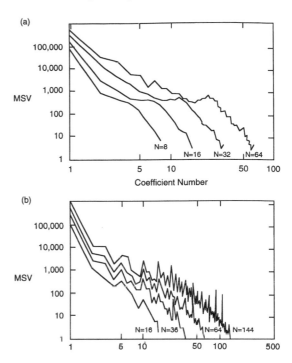

Fig. 6.27 DCT coefficient mean square values for the image "Karen" with various block sizes. (a) One-dimensional, arranged according to increasing frequency. (b) Separable two-dimensional, arranged according to increasing spatial frequency.

The transform block size and shape are important parameters affecting the compaction capability of the transforms, although the picture itself is, of course, the overriding factor. Generally, the larger the block size, the higher the energy compaction achieved by the transform. Also, two-dimensional blocks achieve more compaction than one-dimensional blocks. Experience has shown that over a wide range of pictures there is not much improvement in average energy compaction for two-dimensional blocks of sizes above 8×8 pels. Also, three-dimensional blocks have not found much use, owing to complexity (multiple frame stores required) and inability to incorporate motion compensation to capture correlation in the presence of motion.

6.5.3 Quantization of Transform Coefficients

Transform coding does a reasonably good job of exploiting *statistical* redundancy in images by

N	Number of Computations for		Speedup Factor
	Matrix DCT	Fast DCT	
4	16	8	2.0
8	64	24	2.7
16	256	64	4.0
32	1024	160	6.4
64	4096	384	10.7
138	16384	896	18.3
256	65536	2048	32.0

Fig. 6.26 Computational advantage of using the Fast DCT Algorithm compared with simple matrix multiplication.

decorrelating the pel data and by compacting information into the lower order coefficients. We now present a framework for the exploitation of *subjective* redundancy in transform coded pictures.

Unlike DPCM, where the quantization of a particular sample affects relatively few pels in the reconstructed picture, quantization (or dropping from transmission) of a particular transform coefficient affects every pel in the transformed block. This *spreading* of the quantization error often makes the design of quantizers for transform coefficient quite different than that of DPCM quantizers.

If a certain percentage of the low-energy, higher order transform coefficients are set to zero, then the effect is a blurring or loss of sharpness in regions of rapid luminance variation. Since each coefficient corresponds to a certain type of pattern (or frequency) of the signal, setting it to zero implies that the pattern or frequency is removed. The transform coefficient c_1 corresponds to a spatial frequency of zero (called the DC coefficient), and coarse quantization of c_1 causes visible discontinuity between adjacent blocks, especially in the low-detail areas of the image. Quantization error in other low-order coefficients may also contribute to these block boundary effects. The quantization error in higher order coefficients appears as granular noise; the higher the order, the finer the granularity.

The quantization problem thus involves: (1) which coefficients are retained, based on visibility of the truncation error, (2) how many bits (number of levels) are assigned for quantization of each retained coefficient, and (3) what is the shape of each quantizer? The tradeoff between the visibility of the truncation error versus quantization error is not well understood and therefore must be done experimentally. It can be shown under fairly general assumptions that the number of bits assigned for each coefficient is proportional to the log of the energy in that coefficient. The DC coefficient generally has a uniform distribution and, therefore, uniform quantizer with fixed length code words. Because of the perceptual importance of this coefficient, it is always given sufficient levels. For an 8 × 8 spatial transform, the DC coefficient is typically quantized using more than 64 levels.

The higher order coefficients are quantized either by a Lloyd–Max quantizer optimized for the Laplacian probability distribution and fixed-length code words, or a uniform quantizer optimized for variable length code words. The higher the order of the coefficient, the more skewed toward zero is the probability distribution. Therefore, quantizers for higher order coefficients have more widely spaced representation levels.

6.5.4 Adaptive Coding of Transform Coefficients

Since the image statistics vary widely, adapting the quantization strategy based on local areas of the picture is beneficial. In a technique called *threshold sampling*, only those coefficients in each block whose magnitude exceeds a prescribed threshold are transmitted. The thresholds are carefully determined based on subjective consideration. Therefore, low-detail blocks having a large number of insignificant coefficients are coded with far fewer bits than the higher detail blocks, resulting in a significantly lower overall bitrate. Also, to exploit the subjective effects, high-detail blocks are coded more coarsely than low-detail blocks.

A more systematic procedure is to classify each transform block according to its spatial activity into one of say L classes $\{C_\ell, \ell = 1 \ldots L\}$. Within each class C_ℓ the coefficient MSVs $\{_\ell\sigma^2_m, m = 1 \ldots N\}$ and the activity function should be about the same for each block. Optimum quantizers can then be designed for each class such that the distortion is equalized over all classes. This includes determining which coefficients are retained, the number of bits assigned for quantization of each retained coefficient, and the shape of the quantizer. Class information (i.e., which block belongs to which class) has to be signalled to the receiver. The reduction in bitrate due to adaptation is usually far higher than the overhead in the signalling information, thus justifying adaptive quantization.

6.6 HYBRID TRANSFORM CODING

Transform coding techniques described in the previous section involve high complexity, both in terms of storage and computational requirements. For sources with stationary statistics, transform coding with large block size removes essentially all of the statistical redundancy. However, real-world picture sources are not stationary and, therefore, small block sizes (e.g., 8 × 8) must be used so that techniques that change based on local area statistics can be accommodated. Small block sizes are helpful in taking advantage of psychovisual properties since, in that case, the quantization error spreads to a smaller area and therefore it can be better controlled to match the masking effects. Similarly, smaller blocks are better for motion compensation since there is a larger likelihood of motion being uniform over a smaller block. However, with smaller block sizes, considerable block-to-block redundancy may exist even after the transform operation. In addition, smaller blocks require larger adaptation information to be signalled and higher overhead due to more frequent transmission of motion vectors. It is this dilemma that is partially alleviated by hybrid transform coding.

Hybrid transform coding combines transform and predictive coding to reduce storage and com-

putations and at the same time give a higher degree of compression and adaptability. Intra-frame hybrid coding, for example, may involve a one-dimensional transform followed by line-by-line DPCM of each transform coefficient. However, by far the most popular design is the hybrid coder shown in Fig. 6.28. It performs DPCM in the temporal dimension using motion compensation and transform coding with two-dimensional blocks in the spatial dimensions.

The first step in the hybrid interframe coder is to partition each image frame into two-dimensional blocks of $L \times L$ pels. Each block can be transformed by a two-dimensional transform. However, the transform is not performed on the pel values. Since motion compensation is difficult to perform in the transform domain, the spatial transform is performed on motion compensated prediction error as shown in Fig. 6.28. This arrangement requires only a single frame store in the receiver. Just as in the case of transform coding, interframe hybrid transform coding benefits a great deal by adapting the truncation and quantization strategy based on local spatial and temporal detail. Owing

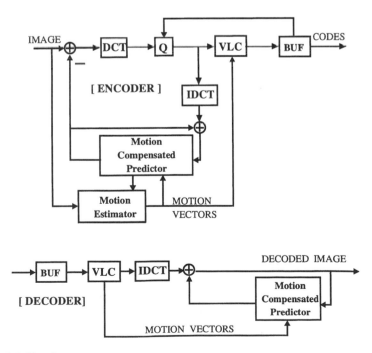

Fig. 6.28   Interframe hybrid coding.

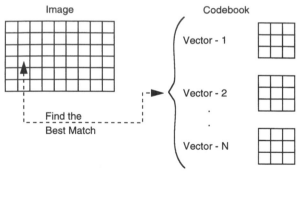

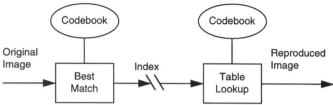

Fig. 6.29   Vector quantization.

to the compression efficiency and smaller computational and storage requirement, this scheme has become part of the MPEG compression standard described in detail later in this book.

6.7   VECTOR QUANTIZATION

In predictive coding, described in the previous section, each pel was quantized separately using a scalar quantizer. The concept of scalar quantization can be generalized to vector quantization in which a group of pels is quantized at the same time by representing them as a code vector. Such a vector quantization can be applied to a vector of prediction errors, or pels, or transform coefficients. Figure 6.29 shows a group of nine pels from a 3 $\times$ 3 block represented by one of the vectors from a codebook of k vectors. The problem of vector quantization is then to design the codebook and an algorithm to determine the vector from the codebook that offers the best match to the input data.

The design of the codebook usually requires a set of training pictures. The codebook can grow to a large size for a large block of pels. Thus, for an 8 $\times$ 8 block compressed to two bits per pel, one would need a codebook of size 2^{128} vectors! Matching the original pel blocks with each vector of such a large size codebook normally requires some short cuts. However, such matching is done only at the transmitter, and the receiver is considerably simpler since it does only a simple table lookup.

This is in contrast to transform coding in which a block of pels is first statistically decorrelated by taking a linear transform, and then each coefficient is separately quantized as a scaler. The main advantage of vector quantization is the simple receiver structure, which consists only of a lookup table or codebook containing 2^{NR} code vectors. The disadvantages include encoder complexity and the fact that images mismatched to the codebook may not be well represented. Over the years, substantial progress has been made in codebook design, development of perceptually matched error criteria, and motion compensation. The performance of vector quantization after these improvements approaches, but does not exceed, that of interframe hybrid transform coding described in the previous section. Therefore, vector quantization has not been able to displace interframe hybrid transform coding, which has already become the basis for the MPEG standards.

6.8 MOTION ADAPTIVE INTERPOLATION

In many instances, individual fields or frames must be dropped from transmission. One case of this is when the buffer at the transmitter starts to become full, in which case fields or frames may be dropped from transmission to reduce the chance of buffer overflow. In such instances, the dropped fields or frames must be recreated at the receiver for display. For example, the dropped fields or frames might be recreated simply by repeating the previous one or future ones by delaying the display, or by averaging the transmitted neighboring fields/frames. Simply

repeating fields/frames creates jerkiness in scenes even with small motion. On the other hand, averaging produces visible motion blur for moderate to high motion.

In some cases, it is possible to improve the quality and reduce both of these two artifacts by using estimates of motion of objects in the scene. Figure 6.30 shows an example where frame $(k - 1)$ is skipped and is reconstructed at the receiver by using the frames transmitted at instants $(k - 2)$ and (k). In this case, linear interpolation for a pel at location (x_1, y_1) in frame $(k - 1)$ would be a simple average of pels at the same location in frames $(k - 2)$ and (k). This would blur the picture since pels from the background and moving objects are averaged together. If, however, the moving object were undergoing pure displacement as shown in the figure, then a pel at position $(x_1 + dx/2, y_1 + dy/2)$ in the skipped frame should be interpolated using pels (x_1, y_1) and $(x_1 + dx, y_1 + dy)$ in frames $(k - 2)$ and (k), respectively. This is indicated by arrows in the figure. Thus, the schemes can be successful if

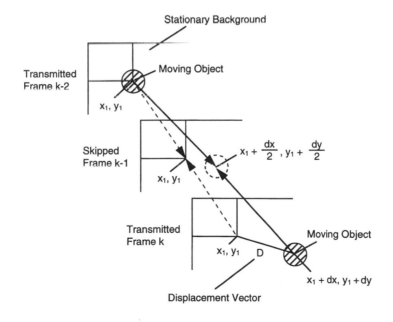

Fig. 6.30 Illustration of linear and motion interpolation.[1]

motion-based interpolation is applied for appropriate pels.

In addition to an accurate motion estimator, this requires segmentation of the skipped frame into moving areas, uncovered areas in the present frame relative to the previous frame, and areas to be uncovered in the next frame, so that proper interpolation can be applied. Obviously, pels as well as weighting values used for interpolation of these different segments have to be different. This is a subject of current research. Simulations indicate that under moderate motion, a 3:1 frame dropping (i.e., skipping two out of three frames) and motion adaptive interpolation often gives reasonable motion rendition.

In this as well as other interpolation schemes, since sometimes the interpolation may be inaccurate, techniques have been devised where the quality of interpolation is checked at the transmitter, and if the interpolation error is larger than a threshold, correction information is transmitted to the receiver. Owing to unavoidable inaccuracies of the displacement estimator (e.g., complex translational and rotational motion) and the segmentation process, such side information is necessary to reduce artifacts that may otherwise be introduced due to faulty interpolation.

6.9 INCORPORATION OF PERCEPTUAL FACTORS

One of the most important objectives in the design of visual communication systems for human viewing is that it represent, transmit, and display only the information that the human eye can see. To transmit and display characteristics of images that a human observer cannot perceive is a waste of channel resources and display media. Therefore, considerable work has been done to study those properties of human vision that are helpful in evaluating the quality of the coded picture and that help us to optimize the coder to achieve the lowest transmission rate for a given picture quality. In this section we summarize some of the major deficiencies of the human visual system and how they have been exploited in the design of compression algo-

rithms. Although our understanding of the human visual system has increased significantly over the past century, it has also revealed many complexities of the visual system, and we are far from developing a simple yet powerful model for visual processing that can be used for optimization of coders. Therefore, the major findings listed below should not be interpreted as very quantitative and must be engineered for the application at hand.

6.9.1 Chrominance Bandwidth

The trichromatic theory of color allows us to represent any color by using a linear combination of three primary colors for the purpose of visual matches. This already results in a substantial compression of color information since the wavelength distribution of the matched color may not be the same as the actual color, although for visual matches they are identical. In addition, if one of the primaries is chosen to be the luminance component of full bandwidth, then the other two chrominance components can be reduced in their spatial bandwidth without much noticeable blur. This is due to viewers being relatively insensitive to sharpness of the chrominance signal transitions in high-detail areas of a picture and at edges of objects. If luminance transitions are kept sharp by allocating higher bandwidths to them, they tend to mask the chrominance blurring. In NTSC television, for example, very good picture rendition can be obtained in most cases with chrominance components bandlimited to about 1.5 MHz and 0.6 MHz, respectively, (luminance bandwidth is approximately 4MHz). In the future, such bandlimiting of chrominance components could take place in the vertical as well as in the temporal dimension by using more sophisticated spatiotemporal filters.

In analog systems, bandlimiting the chrominance signal allowed a clever frequency multiplexing of the luminance and chrominance signals to create a composite color television signal that was upward compatible with the older monochrome TV signal. For digital television, the lower bandwidth of chrominance signals allows a lower rate sampling (usually by a factor of 2 to 4) in two spatial dimensions. While such bandlimiting of chrominance does not degrade the quality of most natural pictures, some pictures synthesized by computers have very sharp color transitions, and in such cases, the effects of chrominance bandlimiting may be visible.

6.9.2 Weber's Law and PCM Coding

Placement of quantization levels for PCM coding can be based on psychovisual criteria. For gray scale pictures, the principal effect used is Weber's law, which states that the visibility threshold of a perturbation in luminance increases almost linearly with increasing background luminance. This implies that the visibility of a given amount of quantization noise decreases with luminance level, and therefore the coarseness of the PCM quantizer can be increased for higher luminance levels. Such a nonlinear companding of the quantizer improves the picture quality for a given bitrate. Some of this improvement is offset by a nonlinearity in cameras and cathode ray tubes (called *gamma*). Gamma characteristics increase the amplitude of the quantization error for larger values of the luminance to some extent, and therefore partially compensates for the Weber's law effects.

The visibility of the quantization noise in PCM-coded pictures can be reduced by observing that the human eye is more sensitive to noise that has strong structure than to random noise. Therefore, to improve picture quality by breaking down the quantization noise structure, some high-frequency noise (called *dither*) may be added to the original signal before quantizing. This noise causes the coded signal to oscillate between the quantizing levels, thereby increasing the high-frequency content of the quantizing noise. The contouring patterns ordinarily arising from coarse quantization are broken and randomized, making them less visible.

For PCM coding of color components, the nonlinearity of the perceptual color metrics can be used by developing a nonlinear quantization scale. This could be done by choosing a perceptually uniform color space and mapping the uniform quantization of this space to the quantization of the original chrominance signal space. Since the perceptually uniform color space is a nonlinear

mapping of the chrominance signals, its uniform quantization creates a nonlinear quantization of the chrominance signals. Such nonlinear quantization can be easily implemented in the form of lookup tables. Simultaneous quantization of the chrominance components using perceptual redundancy and constraints on the possible range lead to a significant reduction in the PCM bitrate. However, the total bits required for PCM quantization of the chrominance signals is already low, usually ten to twenty percent of the total bitrate. This is a combined result of lower bandwidth and fewer quantization levels used for the chrominance components.

6.9.3 Visual Masking

It is known that there is a reduction in the visibility of stimuli (or an increase in the visibility threshold of quantization noise) superimposed on spatially and temporally nonuniform background. This is referred to as masking of the quantization noise by a nonuniform background. Spatial masking has been effectively used in designing DPCM quantizers. For the nonadaptive quantizers, the quantization step size is increased for large prediction errors because these correspond to large transitions (edges) of the luminance signal. Section 6.4.3b gives procedures for systematically incorporating subjective effects in DPCM quantizer designs. For adaptive quantization, as we discussed in previous sections, pels are classified depending on how much quantization error can be masked at their location. Coarser quantizers are used when larger errors can be masked.

In transform coding, higher order coefficients that correspond to higher frequency or detail are quantized more coarsely owing to the reduced visibility of high-frequency noise. In addition, if in a particular transform block, high-order coefficients have large amplitudes, then the block can be classified as busy, leading to coarse quantization of all the coefficients. Such adaptive quantization leads to significant success in bitrate reduction, as discussed in Section 6.5.4.

6.9.4 Temporal Effects

The television signal is sampled in time. The required sampling rate is not only a function of how rapid the motion is in the scene, but also the ability of the eye to discern rapidly changing stimuli. At normal levels of brightness, 50 fields/s is about the lower limit of the required temporal rate. At this rate, most typical motion of objects appears smooth and not jerky. However, the perceptible spatial resolution decreases as the speed of object motion increases, particularly if the moving object cannot be tracked by the eye. Thus, during nontrackable chaotic motion, spatial resolution can be reduced. Fast trackable object motion requires that higher spatial resolution be maintained, even during rapid motion. Moreover, rapid trackable motion may require higher temporal rate (e.g., > 60 Hz) to prevent jerkiness.

In interframe coding, particularly when variable length codes are used, the coder may generate a variable data rate depending on detail and motion in the scene. A buffer is usually required to smooth out the data to interface to a constant bitrate channel. One of the more effective techniques to pre-vent buffer overload is to dynamically reduce the temporal resolution by dropping fields or frames. Motion compensated interpolation of the dropped fields/frames, along with reduced visual perception during buffer overload, continues to give good picture quality during the overload.

During a scene change, a large increase in data may be generated. However, during the scene change our perception of both spatial and temporal detail is low for a short period of time (approximately 4 fields or $1/15$ s) immediately after the scene change. Therefore, an effective strategy to reduce buffer overflow during scene change is to increase the coarseness of the quantization, drop a larger number of transform coefficients, as well as drop some fields/frames from transmission. As long as the resolution is restored quickly, the effect of reduction of these resolutions is not visible. All the interframe coders use this strategy very effectively.

APPENDIX: FAST DCT FOR 1 × 8 BLOCKS

Here we provide fast one-dimensional DCT and IDCT algorithms for blocksize 8. They are based on the theory of rotations[2] and minimize the number of multiplications.

```
                /* 1D DCT block size 8*/
                /*Computed by rotations */
#define PI      3.14159265358979323846264338327950288419717169
#include <math.h>
dct(pels,coefs)
short   pels[8];        /* input values */
double  coefs[8];       /* output values */
{
       static int i;
       static double   c0,c1,c2,c3,c4,c5,c6,c7,c8,c9,c10,c11,c12;
       static double   alpha;
       static double   ts,m1,m2,m3;                    /* temp variables */
       static double   s1[8], s2[8], s3[8], s4[8], s5[8];
       static int first=1;
       double cos( ),sin( ),sqrt( );
       if(first){
          first=0;
          alpha = 2*PI/(4*8);
```

```
        c0 = sin(4.0*alpha) - cos(4.0*alpha);
        c1 = cos(4.0*alpha);
        c2 = -(sin(4.0*alpha) + cos(4.0*alpha));
        c3 = sin(1.0*alpha) - cos(1.0*alpha);
        c4 = cos(1.0*alpha);
        c5 = -(sin(1.0*alpha) + cos(1.0*alpha));
        c6 = sin(2.0*alpha) - cos(2.0*alpha);
        c7 = cos(2.0*alpha);
        c8 = -(sin(2.0*alpha) + cos(2.0*alpha));
        c9 = sin(3.0*alpha) - cos(3.0*alpha);
        c10 = cos(3.0*alpha);
        c11 = -(sin(3.0*alpha) + cos(3.0*alpha));
        c12 = (double)1 / sqrt((double)2);
    }
    s1[0] = pels[0] + pels[7]; s1[1] = pels[0] - pels[7];
    s1[2] = pels[2] + pels[1]; s1[3] = pels[2] - pels[1];
    s1[4] = pels[4] + pels[3]; s1[5] = pels[4] - pels[3];
    s1[6] = pels[5] + pels[6]; s1[7] = pels[5] - pels[6];

    s2[0] = s1[1]; s2[1] = s1[5];
    s2[2] = s1[0] + s1[4]; s2[3] = s1[0] - s1[4];
    s2[4] = s1[2] + s1[6]; s2[5] = s1[2] - s1[6];
    s2[6] = s1[3] + s1[7]; s2[7] = s1[3] - s1[7];

    s3[0] = s2[0]; s3[1] = s2[1];
    s3[2] = s2[2]; s3[3] = s2[5]*c12;
    s3[4] = s2[3]; s3[5] = s2[4];
    s3[6] = s2[6]; s3[7] = s2[7]*c12;

    /*The following can be skipped and incorporated
      into the quantization process by doubling
      the quantizer step size                      */
    for (i = 0; i < 8; ++i) s3[i] = s3[i]/2.0;

    s4[0] = s3[2]; s4[1] = s3[4];
    s4[2] = s3[5]; s4[3] = s3[6];
    s4[4] = s3[0] + s3[3]; s4[5] = s3[0] - s3[3];
    s4[6] = s3[7] + s3[1]; s4[7] = s3[7] - s3[1];

ts = s4[0] + s4[2]; m1 = s4[0] * c0; m2 = ts * c1;
m3 = s4[2] * c2; s5[0] = m2 + m3; s5[1] = m1 + m2;

ts = s4[4] + s4[6]; m1 = s4[4] * c3; m2 = ts * c4;
m3 = s4[6] * c5; s5[2] = m2 + m3; s5[3] = m1 + m2;
```

```
          ts = s4[1] + s4[3]; m1 = s4[1] * c6; m2 = ts * c7;
          m3 = s4[3] * c8; s5[4] = m2 + m3; s5[5] = m1 + m2;


          ts = s4[5] + s4[7]; m1 = s4[5] * c9; m2 = ts * c10;
          m3 = s4[7] * c11; s5[6] = m2 + m3; s5[7] = m1 + m2;


          coefs[0] = s5[1]; coefs[1] = s5[2]; coefs[2] = s5[4];
          coefs[3] = s5[6]; coefs[4] = s5[0]; coefs[5] = s5[7];
          coefs[6] = s5[5]; coefs[7] = s5[3];
return(0);
}
                    /*1d idct blocksize 8*/
#define PI      3.14159265358979323846264338327950288419716 9
#include <math.h>
idct(coefs,pels)
double coefs[8];
short pels[8];
{
          static double dout[8];
          static double   c1,c2,c3,c4,c5,c6,c7,c8,c9;
          static double   s1[8],s2[8],s3[8],s4[8],s5[8];
          static double   nmax2, alpha;
          static int i;
          static int first=1;

          if(first){
          first=0;
          nmax2 = 4;
          alpha = 2*PI/(4*8) ;

          c1 = cos(nmax2*alpha); c2 = sin(nmax2*alpha);
          c3 = cos(alpha); c4 = sin(alpha);
          c5 = cos(2*alpha); c6 = sin(2*alpha);
          c7 = cos(3*alpha); c8 = sin(3*alpha);
          c9 = sqrt((double)2);
          }
          s1[0] = coefs[4]; s1[1] = coefs[0]; s1[2] = coefs[1];
          s1[3] = coefs[7]; s1[4] = coefs[2]; s1[5] = coefs[6];
          s1[6] = coefs[3]; s1[7] = coefs[5];

          s2[0] = c1*2*s1[0] + c2*2*s1[1]; s2[1] = c5*2*s1[4] + c6*2*s1[5];
          s2[2] = c1*2*s1[1] - c2*2*s1[0]; s2[3] = c5*2*s1[5] - c6*2*s1[4];
          s2[4] = c3*2*s1[2] + c4*2*s1[3]; s2[5] = c7*2*s1[6] + c8*2*s1[7];
          s2[6] = c3*2*s1[3] - c4*2*s1[2]; s2[7] = c7*2*s1[7] - c8*2*s1[6];
```

```
s3[0] = 0.5*s2[4] + 0.5*s2[5]; s3[1] = 0.5*s2[6] - 0.5*s2[7];
s3[2] = s2[0]; s3[3] = 0.5*s2[4] - 0.5*s2[5];
s3[4] = s2[1]; s3[5] = s2[2]; s3[6] = s2[3];
s3[7] = 0.5*s2[6] + 0.5*s2[7];

s4[0] = s3[0]; s4[1] = s3[1]; s4[2] = s3[2];
s4[3] = s3[4]; s4[4] = s3[5]; s4[5] = s3[3]*c9;
s4[6] = s3[6]; s4[7] = s3[7]*c9;

s5[0] = 0.5*s4[2] + 0.5*s4[3]; s5[1] = s4[0];
s5[2] = 0.5*s4[4] + 0.5*s4[5]; s5[3] = 0.5*s4[6] + 0.5*s4[7];
s5[4] = 0.5*s4[2] - 0.5*s4[3]; s5[5] = s4[1];
s5[6] = 0.5*s4[4] - 0.5*s4[5]; s5[7] = 0.5*s4[6] - 0.5*s4[7];

dout[0] = 0.5*s5[0] + 0.5*s5[1]; dout[1] = 0.5*s5[2] - 0.5*s5[3];
dout[2] = 0.5*s5[2] + 0.5*s5[3]; dout[3] = 0.5*s5[4] - 0.5*s5[5];
dout[4] = 0.5*s5[4] + 0.5*s5[5]; dout[5] = 0.5*s5[6] + 0.5*s5[7];
dout[6] = 0.5*s5[6] - 0.5*s5[7]; dout[7] = 0.5*s5[0] - 0.5*s5[1];

for(i=0; i<=7; i++){
        if(dout[i]<0.0)pels[i] = dout[i] - 0.5;
        else                    pels[i] = dout[i] + 0.5;
}
return(0);
}
```

REFERENCES

1. H. G. MUSMANN et al., "Advances in Picture Coding," *Proc. IEEE*, April 1985.

2. C. LOEFFLER et al., "Practical Fast 1D DCT Algorithm with 11 Multiplications," *Proc. IEEE ICASSP-89*, 2:988–991 (February 1989).

7

Motion Compensation Modes in MPEG

We saw in the last chapter that exploitation of temporal redundancy commonly uses motion compensation. After the removal of temporal redundancy, spatial redundancy is reduced by transformation to the frequency domain, for example, by the Discrete Cosine Transform (DCT), followed by quantization to force many of the transform coefficients to zero.

MPEG-1[2] was developed specifically for coding progressive scanned video in the range of 1 to 2 Mbits/s. MPEG-2[4] aims at higher bitrates and contains all of the progressive coding features of MPEG-1. In addition, MPEG-2 has a number of techniques for coding interlaced video.

7.1 CODING OF PROGRESSIVE SCANNED VIDEO

The primary requirement of the MPEG video standards is that for a given bitrate they should achieve the highest possible quality of decoded video. As well as producing good picture quality under normal play conditions, some applications have additional requirements. For instance, multimedia applications may require the ability to randomly access and decode any single video picture*

in the bitstream. Also, the ability to perform a fast search of video in the bitstream, both forward and backward, is extremely desirable. It is also useful to be able to edit compressed bitstreams directly while maintaining decodability. Finally, a variety of video formats and image sizes should be supported.

Both spatial and temporal redundancy reduction are needed for the high compression requirements of MPEG. Some techniques used by MPEG were described in the previous chapter.

7.1.1 Exploiting Spatial Redundancy of Progressive Video

Because video is a sequence of still images, it is possible to achieve some compression using techniques similar to JPEG.[1] Such methods of compression are called *Intraframe* coding techniques, wherein each picture of the video is individually and independently compressed or encoded. Intraframe coding exploits the spatial redundancy that exists between adjacent pels of a picture. Pictures coded using only intraframe coding are called *I-pictures.*

As in JPEG, the MPEG intraframe video coding algorithms employ a Block-based two-dimensional DCT. An I-picture is first divided into 8×8

---

*Frames and pictures are synonymous in MPEG-1, whereas MPEG-2 has both frame-pictures and field-pictures.

Blocks of pels, and the two-dimensional DCT is then applied independently on each Block. This operation results in an 8×8 Block of DCT coefficients in which most of the energy in the original (pel) Block is typically concentrated in a few low-frequency coefficients. The coefficients are then quantized and transmitted as in JPEG.

7.1.2 Exploiting Temporal Redundancy of Progressive Video

Many of the interactive requirements can be satisfied by intraframe coding. However, for typical video signals the image quality that can be achieved by intraframe coding alone at the low bitrates desired is usually not sufficient.

Temporal redundancy results from a high degree of correlation between adjacent pictures. The H.261 and MPEG algorithms[3,5–7] exploit this redundancy by computing an interframe difference signal called the *Prediction Error*. In computing the prediction error, the technique of motion compensation is employed to correct the prediction for motion. As in H.261, the *Macroblock* (MB) approach is adopted for motion compensation in MPEG. In unidirectional motion estimation, called *Forward Prediction*, a *Target MB* in the picture to be encoded is matched with a set of displaced macroblocks of the same size in a past picture called the *Reference* picture. As in H.261, the Macroblock in the Reference picture that *best matches* the Target Macroblock is used as the *Prediction MB*.* The prediction error is then computed as the difference between the Target Macroblock and the Prediction Macroblock. This operation is illustrated in Fig. 7.1.

The position of this best matching Prediction Macroblock is indicated by a *Motion Vector* (MV) that describes the horizontal and vertical displacement[†] from the Target MB to the Prediction MB. Unlike H.261, MPEG MVs are computed to the nearest half pel of displacement, using bilinear interpolation to obtain brightness values between pels. Also, using values of past transmitted MVs, a *Prediction Motion Vector* (PMV) is computed. The difference $MV - PMV$ is then encoded and transmitted along with the pel prediction error signal. Pictures coded using Forward Prediction are called *P-pictures*.

MOTION-COMPENSATED PREDICTION

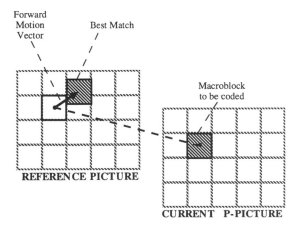

Fig. 7.1 Motion-Compensation using Unidirectional prediction. A displaced macroblock in a previous reference picture is used as the prediction of the moving-area macroblock in the current P-picture. The prediction error signal is then coded and transmitted.

---

*Prediction MBs do not, in general, align with coded MB boundaries in the Reference picture.

[†]Positive horizontal and vertical displacements imply a rightward and downward shift, respectively, from the Target MB to the Prediction MB.

MOTION-COMPENSATED INTERPOLATION

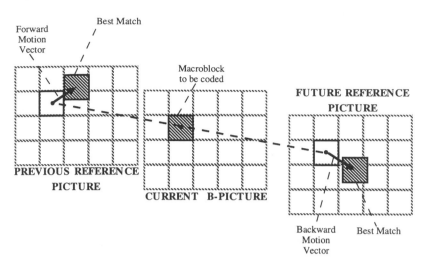

Fig. 7.2   Motion-Compensated Interpolation using Bidirectional prediction. A displaced macroblock in the previous picture is used as one prediction of the moving-area macroblock in the current B-picture. A displaced macroblock in the future picture is used as another prediction of the moving-area macroblock in the current B-picture. One, or the other, or an average of both can be used as the final prediction. The prediction error signal is then coded and transmitted.

The pel prediction error signal itself is transmitted using the DCT-based intraframe encoding technique as described in Chapter 6 and summarized above. In MPEG-1 video (as in H.621), the Macroblock size for motion compensation is chosen to be 16 × 16 luminance pels, representing a reasonable tradeoff between the compression provided by motion compensation and the cost associated with transmitting the motion vectors.

7.1.3   Bidirectional Temporal Prediction for Progressive Video

Bidirectional temporal prediction,[2,3,7] also called Motion-Compensated Interpolation, is a key feature of MPEG video. Pictures coded with Bidirectional prediction use two Reference pictures, one in the past and one in the future. A Target Macroblock in bidirectionally coded pictures can be predicted by a Prediction Macroblock from the past Reference picture (*Forward Prediction*), or one from the future Reference picture (*Backward Prediction*), or by an average of two Prediction Macroblocks, one from each Reference picture (*Interpolation*). In every case, a Prediction Macroblock from

a Reference picture is associated with a motion vector, so that up to two motion vectors per MB may be used with Bidirectional prediction. Motion-Compensated Interpolation for a Macroblock in a Bidirectionally predicted picture is illustrated in Fig. 7.2.

Pictures coded using Bidirectional Prediction are called *B-pictures*. Pictures that are Bidirectionally predicted are never themselves used as Reference pictures, that is, Reference pictures for B-pictures must be either P-pictures or I-pictures. Similarly, Reference pictures for P-pictures must also be either P-pictures or I-pictures.

Bidirectional prediction provides a number of advantages. The primary one is that the compression obtained is often higher than can be obtained using Forward (unidirectional) prediction alone. To obtain the same subjective picture quality, Bidirectionally predicted pictures can often be encoded with fewer bits than pictures using only Forward prediction.

However, Bidirectional prediction does introduce extra delay in the encoding process, because pictures must be encoded and transmitted out of sequence as shown in Fig. 7.3. Further, it entails extra encoding complexity because Macroblock matching (the most

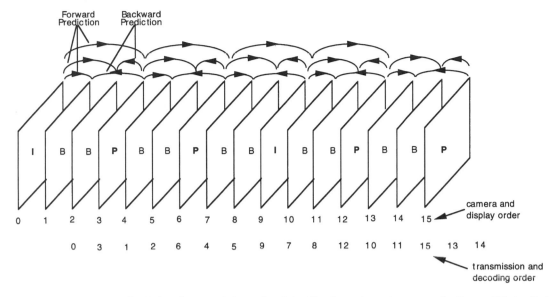

Fig. 7.3   With Bidirectional Prediction, the transmission order of the video frames is not the same as the Camera/Display Order. Note that the coded sequence starts with an I-picture followed by a P-picture.

computationally intensive encoding procedure) has to be performed twice for each Target Macroblock, once with the past Reference picture and once with the future Reference picture.

Bidirectional prediction is most useful in normal scenes containing slow to moderate motion or during camera panning over a relatively stationary background. It is less useful if there are numerous and frequent scene changes or rapid, erratic motion of large areas of the picture. For this type of material, where coding artifacts are much more difficult for viewers to detect, P-pictures by themselves offer a reasonable tradeoff between image quality and updating the picture as quickly as possible.

7.2   CODING OF INTERLACED VIDEO

The original objective of MPEG-2 was to code interlaced CCIR* 601 video at a bitrate that would serve a large number of consumer applications[6]. In fact, one of the main differences between MPEG-1 and MPEG-2 is that MPEG-2 handles interlace efficiently.[4–5,8–10] Since the picture resolution of CCIR 601 is approximately four times that of the SIF of MPEG-1, the bitrate chosen for MPEG-2 optimization was 4 Mbits/s. However, MPEG-2 allows rates as high as 429 Gbits/s.

For interlace, MPEG-2 provides a choice of two *Picture Structures*, as shown in Fig. 7.4. *Field-pictures* consist of individual fields that are each divided into macroblocks and coded separately.[†] With *Frame-pictures*, on the other hand, each interlaced field pair is interleaved together into a frame that is then divided into macroblocks and coded. MPEG-2 requires interlaced video to be displayed as alternate top and bottom fields.[‡] However, within a frame either the top or bottom field can be temporally first and sent as the first Field-picture of the frame.[§]

MPEG-2 also allows for progressive coded pictures, but interlaced display. In fact, coding of a 24 frame/s progressive film source with 3:2 pulldown interlaced display (see Section 5.6.3) is also supported.

---

[*]CCIR has been renamed to International Telecommunications Union—Radio Sector (ITU-R).

[†]Each pair of fields still belongs to a unique frame, however.

[‡]The top field contains the top line of the frame. The bottom field contains the second (and bottom) line of the frame.

[§]If the temporal ordering of the fields in a frame differs from the previous frame, due perhaps to editing, then typically one field must be displayed twice to maintain the top/bottom alternation. See **repeat_first_field** in Chapter 10.

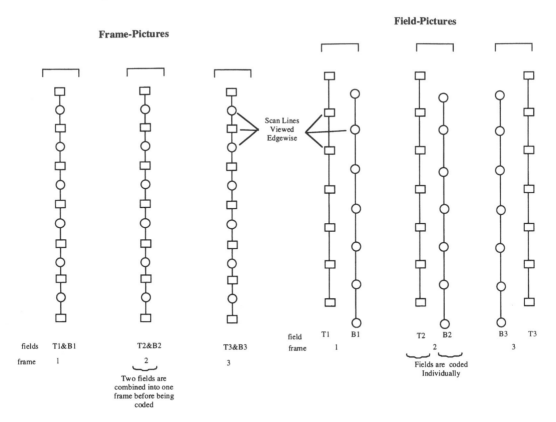

Fig. 7.4 MPEG-2 provides a choice of two Picture Structures: Frame-pictures and Field-pictures. Each frame consists of a top field and a bottom field, either of which may be temporally first. Note that the temporal order of the fields in frame 3 is different than in frames 1 and 2. Thus, field T2 must be displayed again after field B2 to maintain the required top/bottom alternation at the display.

The main effect of interlace in Frame-pictures is that since alternate scan lines come from different fields, motion in the scene causes a reduction of correlation between vertically adjacent pels. MPEG-2 provides several motion compensation features[4,5,9] for dealing with this.

As in MPEG-1, the quality achieved by intraframe coding alone is not sufficient for typical MPEG-2 video signals at bitrates around 4 Mbits/s. Thus, MPEG-2 uses all the temporal redundancy reduction techniques of MPEG-1, plus other methods to deal with interlace.

7.2.1 Frame Prediction for Frame-Pictures

This mode of prediction is used only for Frame-pictures and is exactly the same[3,7] as in MPEG-1. It exploits temporal redundancy by means of

motion compensated Macroblocks and the use of P-Frame-pictures and B-Frame-pictures. We refer to these methods as *Frame Prediction for Frame-pictures*, and, as in MPEG-1, we assign up to one motion vector to P-frame Target MBs and up to two motion vectors to B-frame Target MBs. Prediction MBs are taken from previously coded Reference frames, which must be either P-pictures or I-pictures. Frame Prediction works well for slow to moderate motion, as well as panning over a detailed background. Frame Prediction cannot be used in Field-pictures.

7.2.2 Field Prediction for Field-Pictures

A second mode of prediction provided by MPEG-2 for interlaced video is called *Field Prediction for Field-pictures* and is used only for Field-

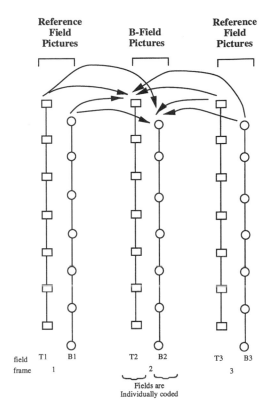 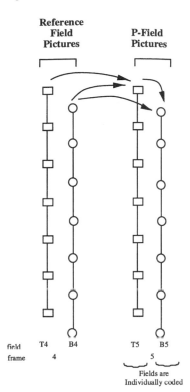

Fig. 7.5 With Field Prediction for Field-pictures, Target MB pels all come from the same field. Up to one motion vector is assigned to each P-field Target MB and up to two motion vectors to B-field Target MBs. Each Prediction MB also comes from one field, which may or may not be of the same parity as the Target MB.

pictures. This mode is conceptually similar to Frame Prediction, except that Target MB pels all come from the same Field-picture, as shown in Fig. 7.5. Prediction MB pels also come from one field, which may or may not be of the same *parity* (top or bottom) as the Target MB field.

For P-Field-pictures, the Prediction MBs may come from either of the two most recently coded I- or P-fields.* Thus in Fig. 7.5, P-field T5 may take its Prediction MBs from either field T4 or B4. P-field B5 may take its Prediction MBs from either field T5 or B4. Note that if the field display order in the frame were reversed, the prediction MBs for field B5 would still be taken from fields T5 and B4.

For B-Field-pictures, the Prediction MBs are taken from the two most recently coded I- or P-frames. Thus, in Fig. 7.5, each Target MB in B-field T2 has its Forward Prediction MB taken from either field T1 or B1 and its Backward Prediction MB taken from either field T3 or B3. The same applies for Target MBs in B-field B2.

Up to one motion vector is assigned to each P-field Target MB and up to two motion vectors to B-field Target MBs. This mode is not used for Frame-pictures and is useful mainly for its simplicity.

7.2.3 Field Prediction for Frame-Pictures

A third mode of prediction[5,8,9] for interlaced video is called *Field Prediction for Frame-pictures*. With this mode, the Target MB in a Frame-picture is first split into top-field pels and bottom-field pels,

---

*However, the two prediction fields MUST be of opposite parity. If the previous I- or P-frame was a progressive Frame-picture the temporal order of the fields might not be indicated.

Frame Macroblock

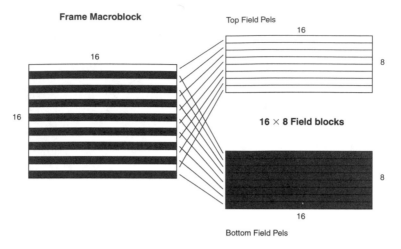

Fig. 7.6 With Field Prediction for Frame-pictures, Target MBs are first split into top field pels and bottom field pels. Two motion vectors are assigned to each P-frame Target MB and two or four motion vectors to B-frame Target MBs. A 16 × 8 prediction cannot come from the same frame as its Target MB.

as shown in Fig. 7.6. Field prediction, similar to but not exactly as defined above, is then carried out independently on each of the two 16 × 8 parts of the Target MB.

For P-frames, two motion vectors* are assigned to each Target MB. The 16 × 8 predictions may be taken from either field, of the most recently decoded I- or P-frame. Thus, a 16 × 8 prediction cannot come from the same frame as its Target MB.

For B-frames, two or four motion vectors are assigned to each Target MB. The 16 × 8 predictions may be taken from either field of the two most recently decoded I- or P-frames.

Field Prediction for Frame-Pictures is not used for Field-pictures and is useful mainly for rapid motion.

7.2.4 Dual-Prime for P-Pictures

A fourth mode of prediction,[4,5,9] called *Dual-Prime for P-pictures*, transmits one motion vector per MB. It can be used either for Field-pictures or Frame-pictures. This mode cannot be used if the previously coded picture was a B-picture or the first I-Field-picture of a frame.

From the transmitted motion vector, two preliminary predictions are computed, which are then averaged together to form the final prediction. An

example of Dual-Prime for the case of Frame-pictures is shown in Fig. 7.7.

The first preliminary prediction is identical to Field Prediction, except that the Reference pels must all come from the previously coded fields having the *same parity* (top or bottom) as the Target pels. Reference pels, which are obtained using the *transmitted motion vector*, are taken from one field for Field-pictures and from two fields for Frame-pictures. In Fig. 7.7 the transmitted motion vector has vertical displacement value 3. For Target pels in field T2, prediction pels are taken from field T1. For Target pels in field B2, prediction pels are taken from field B1.

The second preliminary prediction is derived using a computed motion vector plus a small transmitted *Differential Motion Vector* (dmvector) correction. For this prediction, Reference pels are taken from the opposite parity field as the first preliminary prediction. The computed motion vectors are obtained by a temporal scaling of the transmitted motion vector to match the field in which the Reference pels lie, as shown in Fig. 7.7. The dmvector motion vector correction of up to one-half pel is transmitted in the MB Header and when added to the computed motion vectors gives the final *corrected motion vector*. In Fig. 7.7 the dmvector motion vector correction has vertical displace-

---

*If both MVs = 0, they need not be sent. However, in this case the prediction mode is Frame Prediction for Frame-picture

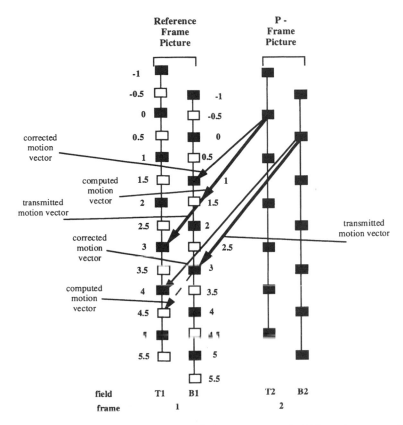

Fig. 7.7   With Dual-Prime for P-pictures, two preliminary predictions are computed, which are then averaged together to form the final prediction. The first preliminary prediction is identical to Field Prediction, while the second preliminary prediction is derived using computed motion vectors plus a small Differential Motion Vector (dmvector) correction. White rectangles represent interpolated scan lines corresponding to half-pel motion vectors.

ment value −0.5. For Target pels in field T2, prediction pels are taken from field B1. For Target pels in field B2, prediction pels are taken from field T1.

For interlaced video the performance of Dual-Prime prediction can rival that of B-picture prediction under some circumstances and has the advantage of lower encoding delay. However, memory bandwidth requirements are usually higher for Dual-Prime since, unlike B-pictures, the decoded pels must be written into a picture store for future use as reference pels.

7.2.5   16 × 8 MC for Field-Pictures

A fifth mode of prediction is called *16 × 8 MC for Field-pictures*. Basically, this mode splits the Field-

picture MB into an upper half and lower half, and performs a separate Field Prediction for each half. Two motion vectors are transmitted per P-picture MB, and two or four motion vectors per B-picture MB. This mode of motion compensation may be useful in field pictures that contain a lot of irregular motion.

Thus, five modes of motion compensation are provided by MPEG-2, as shown in Table 7.1.

7.2.6   Restrictions and Other Features

Field-pictures have some further restrictions on I-, P-, B-picture coding type and motion compensation mode. Normally, the second Field-picture of a frame must be of the same coding type as the first field. However, if the first Field-picture of a

Table 7.1   The five modes of motion compensation that can be used in MPEG-2.

Motion Compensation Mode	Use in Field Pictures?	Use in Frame Pictures?
Frame Prediction for Frame Pictures	No	Yes
Field Prediction for Field Pictures	Yes	No
Field Prediction for Frame Pictures	No	Yes
Dual-Prime for P-Pictures	Yes	Yes
16 × 8 MC for Field-Pictures	Yes	No

frame is an I-picture, then the second field can be either I or P. If it is a P-picture, the Prediction MBs must all come from the previous I-picture, and Dual-Prime cannot be used.

Motion vectors for chrominance pels are derived from the luminance MVs by a scaling that depends on the chrominance sampling density. For 4:2:0 video, both horizontal and vertical components must be divided by 2. For 4:2:2 video, only the horizontal component of the luminance MVs must be divided by two.

MPEG-2 also allows motion vectors to be sent in intra Macroblocks for the purpose of *Error Concealment*. Basically the idea is to use them to estimate intracoded pels that are lost owing to transmission errors. For these MVs, field prediction for Field-pictures or frame prediction for Frame-pictures is all that is allowed.

7.3 CHOOSING THE MODE OF MOTION COMPENSATION

It is the responsibility of the Encoder to decide which picture coding type and which prediction mode[9,10] is best. Needless to say, this can be an extremely complex process if done with full optimality. Thus, in practice numerous shortcuts are usually employed for economical implementation.

For example, during the process of calculating motion vectors, it is possible to save partial sums of prediction errors. Using these, one can then compare the prediction error of field prediction with frame prediction and choose the best mode.

7.4 CODING THE MOTION VECTORS

As mentioned previously, motion vectors (MVs) are coded differentially with respect to a previously transmitted MV called the *Prediction Motion Vector* (PMV). At the start of a Slice* we set PMV = 0.

For I- and P-pictures, PMV is the motion vector of the previous Macroblock if it has one. Otherwise, PMV = 0. If two motion vectors exist in the current MB, but only one in the previous MB, then both MVs of the current MB use the same PMV.

For B-pictures, the forward PMV is the most recently transmitted forward MV in the Slice, if there was one; similarly for the backward PMV. Of course, forward MVs are predicted from forward PMVs, and backward MVs are predicted from backward PMVs. As before, if two forward MVs exist in the current MB, but only one in a previous MB, then both forward MVs of the current MB use the same forward PMV; similarly for backward MVs.

The first step in coding a horizontal or vertical MV component is to compute

$$\text{delta} = 2 \times (\text{MV} - \text{PMV})$$

which is an integer since MVs have half-pel accuracy. The absolute value of delta $\neq$ 0 is then coded using three parameters, namely ***r_size, motion_code,*** and ***motion_residual*** through the following formula:

$$|\text{delta}| = (\textbf{\textit{motion\_code}} - 1) \times 2^{\textbf{\textit{r\_size}}} + \textbf{\textit{motion\_residual}} + 1$$

---

*A slice is a group of Macroblocks. Slice will be defined precisely in the next chapter.

Table 7.2   Variable Length Code (VLC) for sending the sign of delta and the value of **motion_code.**

VLC Code Word	value of motion_code
1	0
01s	1
001s	2
0001s	3
4*0 11s	4
4*0 101s	5
4*0 100s	6
5*0 011s	7
5*0 1011s	8
5*0 1010s	9
5*0 1001s	10
5*0 10001s	11
5*0 10000s	12
6*0 1111s	13
6*0 1110s	14
6*0 1101s	15
6*0 1100s	16

N*0 means N zeros. The sign bit s = 0 for positive, 1 for negative.

Table 7.3   Variable Length Code (VLC) for sending the Dual-Prime Differential Motion Vector (dmvector).

Code Word	dmvector
11	−0.5
0	0
10	+0.5

If delta = 0, it is signalled simply by setting **motion_code** = 0. **r_size** is sent as a 4-bit variable* having values ranging from 0 to 8, which can be changed only once per field or frame.

The parameters **motion_code** and **motion_residual** may be sent in each MB. The sign of delta and the value of **motion_code** are sent using the Variable Length Code of Table 7.2. The value of **motion_residual** is sent as simple binary using **r_size** bits per code word.

For Dual-Prime, the Differential Motion Vector components are sent using the VLC of Table 7.3.

REFERENCES

1. R. Aravind etal, "Image and Video Coding Standards" *AT&T Technical J.* pp. 67–89 (January/February 1993).

2. "Coding of Moving Pictures and Associated Audio for Digital Storage Media at up to about 1.5 Mbit/s," ISO/IEC11172-2: Video (November 1991).

3. Video Simulation Model Editing Committee, "MPEG-I Video Simulation Model" (July 1990).

4. "Generic Coding of Moving Pictures and Associated Audio Information: Video" ISO/IEC 13818-2: Draft International Standard (November 1994).

5. Test Model Editing Committee, "MPEG-2 Video Test Model 5," ISO/IEC JTC1/SC29/WGII/NO400 (April 1993).

6. D. J. LeGall, "MPEG Video: The Second Phase of Work," Proc. Int. Symposium of Society for Information Display, pp. 113–116, (October 1992).

7. A. Puri, "Video Coding Using the MPEG-1, Compression Standard," Proc. Int. Symposium of Society for Information Display, pp. 123–126, (May 1992).

8. A. Puri, R. Aravind, and B. G. Haskell, "Adaptive Frame/Field Motion Compensated Video Coding," *Signal Processing: Image Commun.*, Vol. 1–5, pp. 39–58 (February 1993).

9. A. Puri, "Video Coding Using the MPEG-2 Compression Standard," Proc. SPIE Visual Commun. and Image Proc., SPIE Vol. R199, pp. 1701–17 (November 1993).

10. R. L. Schmidt, A. Puri, and B. G. Haskell, "Performance Evaluation of Non scalable MPEG-2 Video Coding," Proc. SPIE Visual Commun. and Image Proc., SPIE Vol. 12308, pp. 296–310 (September 1994).

*In MPEG parlance, what is actually sent is **f_code** = **r_size** + 1

8

MPEG-2 Video Coding and Compression

The MPEG group was chartered by the International Organization for Standardization (ISO) to standardize coded representations of video and audio suitable for digital storage and transmission media. Digital storage media include magnetic computer disks, optical compact disk read-only-memory (CD-ROM), digital audio tape (DAT), and so forth. Transmission media include telecommunications networks, home coaxial cable TV (CATV), and over-the-air broadcast. The group's goal has been to develop a *generic* coding standard that can be used in many digital video implementations. Thus, some applications will typically require further specification and refinement. As of this writing, MPEG has produced two standards, known colloquially as MPEG-1[1] and MPEG-2[2]. Work also continues on methods and applications for future coding standards and is known colloquially as MPEG-4.

MPEG-1 is an International Standard[1] for the coded representation of digital video and associated audio at bitrates up to about 1.5 Mbits/s. Its official name is ISO/IEC 11172. If video is coded at about 1.1 Mbits/s and stereo audio is coded at 128 kbits/s per channel, then the total audio/video digital signal will fit onto the CD-ROM bitrate of approximately 1.4 Mbits/s as well as the North American ISDN Primary Rate (23 B-channels) of 1.47 Mbits/s. The specified bitrate of 1.5 Mbits/s

is not a hard upper limit. In fact, MPEG-1 allows rates as high as 100 Mbits/s. However, during the course of MPEG-1 algorithm development, coded image quality was optimized at a rate of 1.1 Mbits/s using progressive* scanned color images. Two *Source Input Formats* (SIFs) were used for optimization[6,8,10] of MPEG-1. One, corresponding to NTSC, was 352 pels, 240 lines, 29.97 frames/s. The other, corresponding to PAL, was 352 pels, 288 lines, 25 frames/s. SIF uses 2:1 chrominance subsampling, both horizontally and vertically, in the same 4:2:0 format as H.261.

Originally, MPEG-2 video[2] (ISO/IEC 13818-2) was meant to code interlaced CCIR 601 video for a large number of consumer applications. One of the main differences between MPEG-1 and MPEG-2 is that MPEG-2 handles interlace efficiently.[5,9,11,12] Since the picture resolution of CCIR 601 is about four times the SIF of MPEG-1, the bitrate chosen for MPEG-2 optimization was 4 Mbits/s. However, MPEG-2 allows rates as high as 429 Gbits/s.

8.1 MPEG-2 CHROMINANCE SAMPLING

A bitrate of 4 Mbits/s was deemed too low to enable high-quality transmission of every CCIR

---

*The term *Noninterlaced* is also used for progressively scanned pictures.

601 chrominance sample. Thus, an MPEG-2 4:2:0 format was defined to allow for 2:1 vertical sub-sampling of the chrominance, in addition to the normal 2:1 horizontal chrominance subsampling of CCIR 601.

For interlace, the temporal integrity of the 4:2:0 chrominance samples must be maintained. Thus, MPEG-2 normally defines the first, third, etc. rows of 4:2:0 chrominance *CbCr* samples to be from the same field as the first, third, etc. rows of luminance *Y* samples. The second, fourth, etc. rows of chrominance *CbCr* samples are from the same field as the second, fourth, etc. rows of luminance *Y* samples. However, an override capability is also available to indicate that the 4:2:0 chrominance samples are all temporally the same as the temporally first field of the frame. The MPEG-2 4:2:0 chrominance sampling is shown in Fig. 8.1.

At higher bitrates the full 4:2:2 chrominance format[7] of CCIR 601 may be used, in which the chrominance is subsampled 2:1 horizontally only. In 4:2:2 video, the first luminance and chrominance samples of each line are geometrically cosit-ed. MPEG-2 also allows for a 4:4:4 chrominance format,[7] in which the luminance and chrominance samplings are identical.

8.2 REQUIREMENTS OF THE MPEG-2 VIDEO STANDARD

The main requirement of MPEG-2 video is that it should achieve the highest possible quality of the decoded video during normal play. In addition to obtaining excellent picture quality, we need to randomly display any single frame in the video stream. Also, the capability of performing fast searches directly on the video stream, both forward and backward, is extremely desirable if the storage medium has seek capabilities. It is also useful to be able to edit compressed video streams directly while maintaining decodability. Also, multipoint network communications may require the ability to communicate simultaneously with SIF and CCIR 601 decoders. Communication over packet networks may require prioritization so that the net-

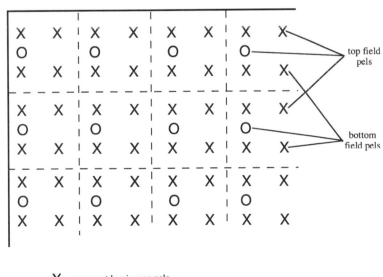

X   represent luminance pels

O   represent chrominance pels

Fig. 8.1   MPEG-2 4:2:0 chrominance format for interlaced video. Chrominance is subsampled 2:1 both horizontally and vertically. Alternate lines of chrominance are temporally aligned with alternate fields.

work can drop low-priority packets in case of congestion. Broadcasters may wish to send a progressive scanned HDTV program to CCIR 601 interlace receivers as well as to progressive HDTV receivers.

To satisfy all these requirements MPEG-2 has defined a large number of capabilities. However, not all applications will require all the features of MPEG-2. Thus, to promote interoperability among applications, MPEG-2 has designated several sets of constrained parameters using a two-dimensional rank ordering. One of the dimensions, called *Profile*, specifies the coding features supported. The other dimension, called *Level*, specifies the picture resolutions, bitrates, and so forth that can be handled. A number of Profile-Level combinations have been defined (see Chapter 11), the most important of which is called *Main Profile at Main Level*, or MP@ML for short. Parameter constraints for MP@ML are shown in Table 8.1.

8.3 MAIN PROFILE ALGORITHM OVERVIEW

Uncompressed digital video requires an extremely high transmission bandwidth. Digitized NTSC resolution video, for example, has a bitrate of approximately 100 Mbits/s. With digital video, compression is necessary to reduce the bitrate to suit most applications. The required degree of compression is achieved by exploiting the spatial and temporal

Table 8.1 Parameter Bounds for MPEG-2 Main Profile at Main Level (MP@ML) Video Streams.

Parameter	Bound
Samples/line	720
Lines/frame	576
Frames/second	30
Samples/second	10 368 000
Bitrate	15 Mbits/s
Buffer size	1 835 008 bits
Chroma format	4:2:0
Image aspect ratio	4:3, 16:9 and square pels

redundancy present in a video signal. However, the compression process is inherently lossy, and the signal reconstructed from the compressed video stream is not identical to the input video signal. Compression sometimes causes some visible artifacts in the decoded pictures.

For progressive scanned video there is very little difference between MPEG-1 and MPEG-2 compression capabilities. However, interlace presents complications in removing both types of redundancy, and many features have been added to deal specifically with it.

As we saw in Chapter 7, MPEG-2 specifies a choice of two picture structures. *Field-pictures* consist of fields that are coded independently. In *Frame-pictures*, on the other hand, field pairs are merged into frames before coding. MPEG-2 requires interlace to be displayed as alternate top and bottom fields.* However, either field can be displayed first within a frame. The Main Profile allows only 4:2:0 chrominance sampling.

Both spatial and temporal redundancy reductions are needed for the high-compression requirements of MPEG-2. Many techniques used by MPEG-2 have been described in previous chapters.

8.3.1 Exploiting Spatial Redundancy

As in JPEG and H.261,[3] the MPEG-1 and MPEG-2 video-coding algorithms employ a Block-based two-dimensional Discrete Cosine Transform (DCT). A picture is first divided into 8×8 Blocks of pels. The two-dimensional DCT is then applied independently on each Block. This operation results in an 8×8 Block of DCT coefficients in which most of the energy in the original (pel) Block is typically concentrated in a few low-frequency coefficients. The coefficients of each Block may be scanned and transmitted in the same zigzag order as JPEG and H.261 (see Fig. 8.2a).

The main effect of interlace in Frame-pictures is that since adjacent scan lines come from different fields, vertical correlation is reduced when there is motion in the scene. MPEG-2 provides two features for dealing with this.

---

*The top field contains the top line of the frame. The bottom field contains the second (and bottom) line of the frame.

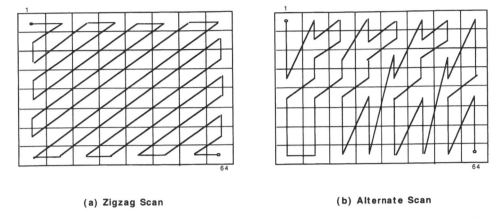

(a) Zigzag Scan   **(b) Alternate Scan**

Fig. 8.2 Two methods for scanning DCT coefficients are available in MPEG-2. The zigzag order (a) is used in JPEG, H.261, and MPEG-1. The alternate scan (b) often gives better compression for interlaced video when there is significant motion.

First, with reduced vertical correlation, the zigzag scanning order for DCT coefficients shown in Fig. 8.2a may not be optimum. Thus, MPEG-2 has an *Alternate Scan,* shown in Fig. 8.2b, that may be specified by the encoder on a picture-by-picture basis to allow the significant bottom-left frequencies to be sent earlier.

Second, a capability for *field_DCT* coding within in a Frame-picture Macroblock (MB) is provided. That is, just prior to performing the DCT, the encoder may reorder the luminance lines within a MB so that the first eight lines come from the top field, and the last eight lines come from the bottom field. This reordering is undone after the IDCT in the encoder and the decoder. The effect of this reordering is to increase the vertical correlation within the luminance blocks and thus increase the energy compaction in the DCT domain. Again, a comparison of the sum of absolute vertical line differences is often sufficient for deciding when to use field_DCT coding and when to use frame_DCT coding in a MB. Chrominance MBs are not reordered in Main Profile field_DCT coding.

After scanning, a quantizer is applied to the DCT coefficients, which results in many of them being set to zero. As with JPEG, a different quantization step size may be applied to each DCT coefficient. This is specified by a *Quantizer Matrix* that is sent in the video stream. This quantization is responsible for the lossy nature of the compression algorithms in JPEG, H.261, and MPEG video.

Compression is achieved by transmitting only the nonzero quantized coefficients and by entropy-coding their locations and amplitudes.

8.3.2 Exploiting Temporal Redundancy

Temporal redundancy results from similarity between adjacent pictures. MPEG-2 exploits this redundancy by computing and transmitting an interframe difference signal called the *prediction error.* In computing the prediction error, the technique of Macroblock (MB) motion compensation is employed to correct for motion, as described in Section 7.1.2. Pictures coded using Forward Prediction are called *P-pictures.* Pictures coded using Bidirectional Prediction are called *B-pictures.*

Pictures that are Bidirectionally predicted are never themselves used as Reference pictures, that is, Reference pictures for B-pictures must be either P-pictures or I-pictures. Similarly, Reference pictures for P-pictures must also be either P-pictures or I-pictures.

The positions of the best-matching *Prediction Macroblocks* are indicated by motion vectors that describe the displacement between them and the *Target Macroblocks.* The motion vector information is also encoded and transmitted along with the prediction error.

The prediction error itself is transmitted using the DCT-based intraframe encoding technique

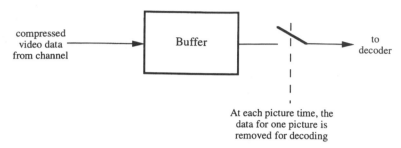

At each picture time, the
data for one picture is
removed for decoding

Fig. 8.3   Video Buffer Verifier (VBV). The buffer must never overflow, and underflow is allowed only under restricted circumstances. The buffer size is specified in the video stream. For MP@ML the maximum allowable VBV buffer size is 1 835 008 bits.

summarized previously. In MPEG-2 video, as in H.261 and MPEG-1, the MB size is chosen to be 16×16 pels, representing a reasonable tradeoff between the compression provided by motion compensation and the cost of transmitting the motion vectors.

Many encoder procedures are not specified by the standard, so that different algorithms may be employed at the encoder as long as the resulting video stream is consistent with the specified syntax. For example, the details of the motion estimation procedure are not part of the standard. Neither are the decision methods used to determine the various coding modes and structures. It is the responsibility of the encoder to decide which prediction mode is best. Needless to say, this can be an extremely complex process if done with full optimality. Thus, in practice numerous shortcuts are usually employed for economical implementation (see Section 7.3).

8.3.3   Video Buffer Verifier

As with H.261 and MPEG-1, MPEG-2 has one important encoder restriction, namely a limitation on the variation in bits/picture, especially in the case of constant bitrate operation. This limitation is enforced through a *Video Buffer Verifier* (VBV), which corresponds to the Hypothetical Reference Decoder of H.261.

The contents of the transmitted video stream must meet the requirements of the VBV, which is shown in Fig. 8.3. Data enter the VBV at a piecewise constant rate for each picture. At each picture decode time the data for one picture are removed instantaneously from the VBV buffer and decoded. At no time must the VBV buffer overflow, and underflow is allowed only under restricted circumstances as described later.

If the VBV input data rate is the same for each picture, then the video is said to be coded at *Constant Bitrate* (CBR). Otherwise, the video is said to be coded at *Variable Bitrate* (VBR). The VBV video bitrate is not specified per se. What is specified instead is the amount of time, called **vbv_delay**,* that the picture header[†] resides in the VBV before being extracted for decoding. Thus, the picture header for picture n arrives at Systems Time

$$t(n) = \text{DTS}(n) - \text{vbv\_delay}(n) \qquad (8.1)$$

where $\text{DTS}(n)$ is the Systems Decoding Time Stamp for picture n. The VBV input bitrate for picture n is then given by

$$R(n) = \text{\textit{N}bits}(n) / \left[t(n + 1) - t(n) \right] \qquad (8.2)$$

where $\text{\textit{N}bits}(n)$ is the number of bits[‡] in picture n. Of course, the actual instantaneous transmission bitrate after Systems packetization and multiplexing may be very bursty and far from this value.

In normal operation, the interval between picture decoding times is determined by the specified picture rate and structure. However, during

---

*See Section 10.2.8 for the specification of vbv_delay.
[†]Actually the last byte of the Picture Start Code.
[‡]More precisely, the number of bits up to the next Picture Start-Code.

low_delay operation, a departure from this is allowed in case the encoder requires a very large number of bits to code a particular picture. Such a picture is called a *Big Picture*. In this case the VBV decoder must wait longer than one picture interval, perhaps several picture intervals, for all the bits of the Big Picture to enter the VBV buffer before instantaneous decoding can take place.

For Big Pictures, the values of vbv_delay and DTS may not be correct, since most encoders are well into the coding of a picture before they realize it is a Big Picture. However, Eqs. (8.1) and (8.2) are still valid for most encodings of Big Pictures.

Note that this definition of VBV allows for varying the delay during the video coding. If the encoder requires a relatively large number of bits for a Big Picture, the decoder display must repeat the previously decoded picture a few times while waiting for the arrival of those bits, thus increasing the delay between camera input and display output.

If the encoder then resumes normal operation with the frame following the Big Picture, the total delay will remain at the new higher value. However, in many applications the users would like the delay to be reduced to its lower nominal value as soon as possible. Thus, to reduce the delay, the encoder normally skips a few frames following the Big Picture and does not send them.

8.4 OVERVIEW OF THE MPEG-2 VIDEO STREAM SYNTAX

The MPEG-2 video standard specifies the *syntax* and *semantics* of the compressed video stream produced by the video encoder. The standard also specifies how this video stream is to be parsed and decoded to produce a decompressed video signal. Most of MPEG-2[2] consists of additions to MPEG-1.[1] However, unlike MPEG-1, Big Pictures, as in H.261, are allowed.

The video stream syntax is flexible to support the variety of applications envisaged for the MPEG-2 video standard. To this end, the overall

Table 8.2 Six Headers of MPEG-2 Video Stream Syntax

Syntax Header	Functionality
Sequence	Definition of entire video sequence
Group of Pictures	Enables random access in video stream
Picture	Primary coding unit
Slice	Resynchronization, refresh, and error recovery
Macroblock	Motion compensation unit
Block	Transform and compression unit

syntax is constructed in a hierarchy of several *Headers,* each performing a different logical function. The different Headers in the syntax and their use are illustrated in Table 8.2.

8.4.1 Video Sequence Header

The *Video Sequence* Header and its extensions contains basic parameters such as the size of the coded video pictures, size of the displayed video pictures if different, Image Aspect Ratio (IAR), picture rate, maximum bitrate (R_{max}), VBV buffer size, low_delay indication, Profile and Level identification, Interlace or Progressive sequence indication, private user data, plus certain other global parameters.

This Header also allows for the optional transmission of JPEG style Quantizer Matrices, one for Intra-coded MBs and one for NonIntra coded MBs.* Unlike JPEG, if one or both quantizer matrices are not sent, default values are defined. These are shown in Fig. 8.4.

Private *user data* can also be sent in the Sequence Header extension as long as they do not contain a *Start Code Prefix (Psc)*, which MPEG-2 defines as a string of 23 or more binary zeros followed by a binary one.

8.4.2 Group of Pictures (GOP) Header

Below the Video Sequence Header is the *Group of Pictures* (GOP) Header, which provides support for random access, fast search, and editing. The GOP Header contains a time code (hours, minutes, seconds, frames) used by certain recording

---

*With 4:2:2 video, two additional quantizer matrices are sent for Intra chrominance and NonIntra chrominance.

```
 8  16  19  22  26  27  29  34

16  16  22  24  47  49  34  37

19  22  26  27  29  34  34  38

22  22  26  27  29  34  37  40

22  26  27  29  32  35  40  48

26  27  29  32  35  40  48  58

26  27  29  34  38  46  56  69

27  29  35  38  46  56  69  83
```
Intra

```
16  16  16  16  16  16  16  16

16  16  16  16  16  16  16  16

16  16  16  16  16  16  16  16

16  16  16  16  16  16  16  16

16  16  16  16  16  16  16  16

16  16  16  16  16  16  16  16

16  16  16  16  16  16  16  16

16  16  16  16  16  16  16  16
```
Inter

Fig. 8.4 Default Quantizer Matrices for Intra and NonIntra Pictures.

devices. It also contains editing flags to indicate whether the B-pictures following the first I-picture of the GOP can be decoded following a random access.

In MPEG, a sequence of transmitted video pictures is typically divided into a series of GOPs, where each GOP begins with an Intra-coded picture (I-picture) followed by an arrangement of For-

ward Predictive-coded pictures (P-pictures) and Bidirectionally Predicted pictures (B-pictures).* Figure 8.5 shows examples of MPEG GOPs.

The top GOP is comprised of pictures 0 to 14, and since there are no B-pictures the encoding/transmission order is the same as the camera/display order.

The bottom GOP contains pictures 1 to 12, consisting of one I-picture, three P-pictures, and eight B-pictures. The encoding/transmission order of the pictures in this GOP is shown at the bottom of Fig. 8.5. B-pictures 1 and 2 are encoded after I-picture 3, using P-picture 0 and I-picture 3 as reference. Note that B-pictures 13 and 14 are part of the next GOP because they are encoded after I-picture 15.

Pictures are displayed in their camera order 0, 1, 2, 3, 4 . . . If B-pictures are to appear in the sequence, then a *Reordering Picture Delay* must be used for all I- and P-pictures to produce the correct display order, as shown in Fig. 8.6.

Random access and fast search are enabled by the availability of the I-pictures, which can be decoded independently and serve as starting points for further decoding. The MPEG-2 video standard allows GOPs to be of arbitrary structure and length. The GOP Header may be used as the basic unit for editing an MPEG-2 video stream.

8.4.3   Picture Header

Below the GOP is the *Picture Header*, which contains the type of picture that is present, for example, I, P, or B, as well as a *Temporal Reference* indicating the position of the picture in camera/display order within the GOP.† It also contains the parameter ***vbv_delay*** that indicates how long to wait after a random access before starting to decode. Without this information, a decoder buffer could underflow or overflow following a random access.

Within the Picture Header several picture coding extensions are allowed. For example, the quantization accuracy of the Intra DC coefficients may be increased from the 8 bits of MPEG-1 to as much as 10 bits for MP@ML. A 3:2 pulldown flag,

---

*A GOP usually contains only one I-picture. However, more than one are allowed.

†The Temporal Reference is reset to zero in the first picture to be displayed of each GOP.

Group of Pictures with an I-Picture every N Pictures and a P-Picture every M Pictures

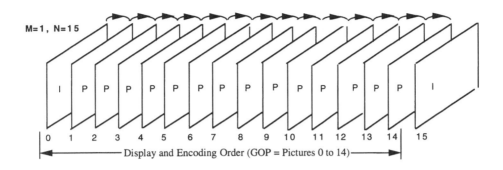

M=1, N=15

I	P	P	P	P	P	P	P	P	P	P	P	P	P	P	I
0	1	2	3	4	5	6	7	8	9	10	11	12	13	14	15

Display and Encoding Order (GOP = Pictures 0 to 14)

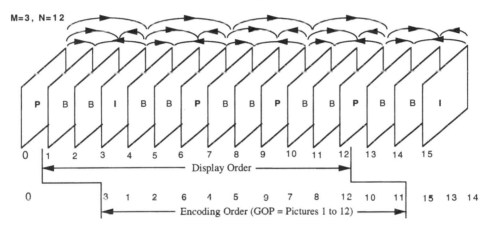

M=3, N=12

P	B	B	I	B	B	P	B	B	P	B	B	P	B	B	I
0	1	2	3	4	5	6	7	8	9	10	11	12	13	14	15

Display Order

0   3   1   2   6   4   5   9   7   8   12   10   11   15   13   14

Encoding Order (GOP = Pictures 1 to 12)

Fig. 8.5 Examples of MPEG Group of pictures.

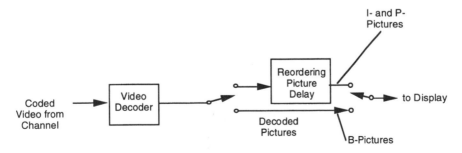

Fig. 8.6 Decoder for MPEG video stream containing I-, P-, and B-pictures. B-pictures are displayed immediately through the bottom path, whereas each decoded I- or P-picture passes first via the top path to a Reordering Picture Delay, to await display after the ensuing B-pictures. The switches are either both up or both down.

called ***repeat_first_field,*** indicates, for Frame-pictures, that the first field of the picture should be displayed one more time following the display of the second field. An alternative scan to the DCT zigzag scan may be specified. Also, the presence of error concealment motion vectors in I-pictures may be indicated. Other information includes Picture Structure (field or frame), field temporal order (for Frame-pictures), progressive frame indicator, and information for reconstruction of a composite NTSC or PAL analog waveform.

Within the Picture Header a picture display extension allows for the position of a *display rectangle* to be defined for each picture. This feature is useful, for example, when coded pictures having IAR 16:9 are to be also received by conventional TVs having IAR 4:3. This capability is also known as *Pan and Scan.*

8.4.4 Slice Header

A *Slice* is a string of consecutive MBs of arbitrary length running from left to right across the picture, for example, as shown in Fig. 8.7. In I-pictures, all MBs are *transmitted.* In P-pictures and B-pictures, typically some MBs of a slice are transmitted and some are not, that is, they are *skipped.* However, the first and last MBs of a Slice must always be transmitted. A Slice is not allowed to extend beyond the right edge of the picture, and Slices must not overlap.

The *Slice Header* is intended to be used for resynchronization in the event of transmission bit errors. It is the responsibility of the encoder to choose the length of each Slice depending on the expected bit error conditions. Prediction registers used in the differential encoding of motion vectors and DC Intra coefficients are reset at the start of a Slice.

The Slice Header contains the vertical position of the Slice within the picture, as well as a ***quantizer_scale_code*** parameter used to define the quantizer step size until such time as a new step size is optionally sent at the MB level. The Slice Header may also contain an indicator for Slices that contain only Intra MBs. These may be used in certain fast forward and fast reverse display applications.

All Profiles defined so far have the *restricted Slice structure,* for which all MBs in the picture must belong to a Slice, that is, the Slices cover the entire picture with no gaps in between, as shown in Fig. 8.7.

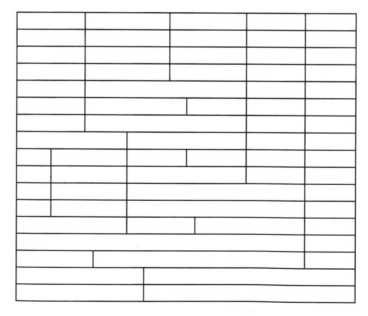

Fig. 8.7 Possible arrangement of Slices in which slice lengths vary throughout the picture. In MPEG-2 the left edge of the picture always starts a new slice.

8.4.5 Macroblock Header

The MacroBlock (MB) is the 16×16 motion compensation unit, and each MB begins with a *MacroBlock Header.* For the first MB of each Slice, the horizontal position with respect to the left edge of the picture (in MBs) is coded using the **macroblock_address-increment** VLC shown in Table 8.3. The positions of additional transmitted MBs are then coded differentially with respect to the most recently transmitted MB, also using the macroblock_address-increment VLC.

In P-pictures, skipped MBs are assumed NonIntra with zero DCT coefficients and zero motion vectors. In B-pictures, skipped MBs are assumed NonIntra with zero DCT coefficients and motion vectors the same as the previous MB, which cannot be Intra.

Also included in the MacroBlock Header are **Macroblock_type** (Intra, NonIntra, etc.), Motion Vector Type, DCT_type (field_DCT or frame_DCT), **quantizer_scale_code,** motion vectors, and a **coded block pattern** indicating which blocks in the MB are coded. As with other Headers, many parameters may or may not be present, depending on Macroblock_type, as shown in Fig. 8.8. The VLC for coded block pattern, which is present if macroblock_pattern = 1, is given for Main Profile in Fig. 8.9.

MPEG-2 has many more MB Types than MPEG-1, owing to the additional features provided as well as to the complexities of coding interlaced video. Some of these are discussed later.

8.4.6 Block

A *Block* consists of the data for the quantized DCT coefficients of an 8×8 Block in the MB. It is VLC coded as described in the next Sections. MP@ML has six blocks per MB. For noncoded Blocks, the DCT coefficients are assumed to be zero.

8.4.6.1 Quantization of DCT Coefficients

As with JPEG, Intra-coded blocks have their DC coefficients coded differentially with respect to

Table 8.3 **Macroblock_address-increment** Variable Length Code (VLC). The escape value is used for addresses larger than 33. N*0 means N zeros.

VLC Codeword	macroblock_ address- increment	VLC Codeword	macroblock_ address- increment
1	1	5*0 101 01	18
011	2	5*0 101 00	19
010	3	5*0 100 11	20
0011	4	5*0 100 10	21
0010	5	5*0 100 011	22
0001 1	6	5*0 100 010	23
0001 0	7	5*0 100 001	24
4*0 111	8	5*0 100 000	25
4*0 110	9	6*0 11 111	26
4*0 1011	10	6*0 11 110	27
4*0 1010	11	6*0 11 101	28
4*0 1001	12	6*0 11 100	29
4*0 1000	13	6*0 11 011	30
5*0 111	14	6*0 11 010	31
5*0 110	15	6*0 11 001	32
5*0 101 11	16	6*0 11 000	33
5*0 101 10	17	7*0 1 000	escape_word

the previous block of the same *YCbCr* type, unless the previous block is NonIntra, belongs to a skipped MB, or belongs to another Slice. In any of these cases the prediction value is reset to the midrange value of 1024. The range of unquantized Intra DC coefficients is 0 to 8×255, which means the range of differential values is -2040 to 2040. Intra DC Differentials are quantized with a uniform midstep quantizer with stepsize* 8, 4, or 2, as specified in the Picture Header by the parameter **intra_dc_precision** = $(3 - \log_2$ stepsize). Quantized Intra Differential DC Levels for MP@ML have a range of -1020 to 1020.

Intra AC coefficients are quantized with a uniform midstep quantizer having variable step size that can change every MB if need be under control of the parameter **quantizer_scale,** which in turn is determined by the two transmitted parameters **q_scale_type** and **quantizer_scale_code.** If q_scale_type = 0, then we have simply **quantizer_scale** = $2 \times$ **quantizer_scale_code.** If q_scale_type = 1, then **quantizer_scale** is given by Table 8.4.

---

*A step size 1 is also available in the High Profile.

• VLC Table for macroblock_type modes in I-pictures

macroblock _type	macroblock _quant	macroblock _motion _forward	macroblock _motion _backward	macroblock _pattern	macroblock _intra	VLC
Intra					1	1
Intra+macroblock_quant	1				1	01

• macroblock_type modes in P-pictures

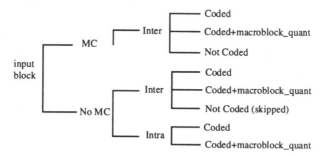

• VLC Table for macroblock_type modes in P-pictures

macroblock _type	macroblock _quant	macroblock _motion _forward	macroblock _motion _backward	macroblock _pattern	macroblock _intra	VLC
MC, Coded		1		1		1
No MC, Coded				1		01
MC, Not Coded		1				001
Intra					1	00011
MC, Coded+macroblock_quant	1	1		1		00010
No MC, Coded+macroblock_quant	1			1		00001
Intra+macroblock_quant	1				1	000001

• macroblock_type modes in B- pictures

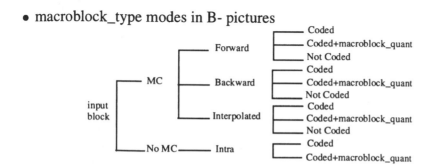

Fig. 8.8 **Macroblock_type** VLCs and the variables that depend on the value of **macroblock_type** for nonscalable or base layer video streams. A 1 indicates a *true* value for the variable. Otherwise, the variable has value 0 = *false*. (*Continued*)

● VLC Table for macroblock_type modes in B- pictures

macroblock _type	macroblock _quant	macroblock _motion _forward	macroblock _motion _backward	macroblock _pattern	macroblock _intra	VLC
MC Interpolated, Not Coded		1	1			10
MC Interpolated, Coded		1	1	1		11
MC Backward, Not Coded			1			010
MC Backward, Coded			1	1		011
MC Forward, Not Coded		1				0010
MC Forward, Coded		1		1		0011
Intra					1	00011
MC Interpolated, Coded+macroblock_quant	1	1	1	1		00010
MC Forward, Coded+macroblock_quant	1	1		1		000011
MC Backward, Coded+macroblock_quant	1		1	1		000010
Intra+macroblock_quant	1				1	000001

Fig. 8.8 (*Continued*)

For Intra AC coefficient $c[m][n]$ the quantization step size is given by

$$(2 \times \boldsymbol{quantizer\_scale} \times \text{Intra Qmatrix } [m][n])/32$$
$$\text{for } m + n \neq 0 \tag{8.3}$$

where $\boldsymbol{quantizer\_scale}$ is set by a Slice or MB Header as described previously, and Intra Qmatrix$[m][n]$ is the Intra quantizer matrix value for DCT frequency $[m][n]$. Intra quantization is usually performed simply by division and rounding to produce the Quantized Levels.

All NonIntra coefficients are quantized in a way similar to H.261 and MPEG-1, except that MPEG-2 does not require odd representative levels. The NonIntra quantization step size is given by

$$(2 \times \boldsymbol{quantizer\_scale}$$
$$\times \text{NonIntra Qmatrix}[m][n])/32$$
$$= 2 \times \text{QUANT} \tag{8.4}$$

where NonIntra Qmatrix$[m][n]$ is the NonIntra quantizer matrix value for DCT frequency $[m][n]$, and QUANT is half the stepsize. The quantizer is nonuniform with a *dead zone* around zero, as shown

in Fig. 8.10. Thus, coefficient values should be shifted toward zero by at least QUANT prior to dividing by $2 \times$ QUANT when computing quantized Levels.

MPEG-2 allows for Quantized Levels to have a range of -2047 to 2047. This is enforced by clipping after division by the step size. However, clipping may cause unacceptable picture distortion, in which case usually a better solution is to increase the step size.

8.4.6.2 Variable Length Coding of Quantized DCT Coefficients

The Intra DC VLC is similar to JPEG and MPEG-1, that is, Differential DC levels DIFF are coded as a VLC size followed by a Variable Length Integer (VLI), where the VLC indicates the size in bits of the VLI. The VLCs and VLIs are given in Table 8.5 for quantized Intra DC DIFFs, which have a range* of -1020 to 1020 for MP@ML.

The remaining AC coefficients are coded either in zigzag order or alternate scan order using one of two available VLCs (see Tables 8.6 and 8.7). Intra AC coefficients may use either VLC as specified by Picture Header parameter $\boldsymbol{intra\_vlc\_format}$.

---

*Assuming the quantizer simply divides by the step size and rounds.

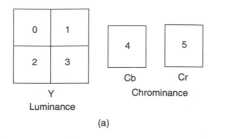

0 | 1
2 | 3

Y
Luminance

4 | 5

Cb | Cr
Chrominance

(a)

VLC codeword	cbp	VLC codeword (N*0 means N zeros)	cbp
111	60	0001 1100	35
1101	4	0001 1011	13
1100	8	0001 1010	49
1011	16	0001 1001	21
1010	32	0001 1000	41
1001 1	12	0001 0111	14
1001 0	48	0001 0110	50
1000 1	20	0001 0101	22
1000 0	40	0001 0100	42
0111 1	28	0001 0011	15
0111 0	44	0001 0010	51
0110 1	52	0001 0001	23
0110 0	56	0001 0000	43
0101 1	1	4*0 1111	25
0101 0	61	4*0 1110	37
0100 1	2	4*0 1101	26
0100 0	62	4*0 1100	38
0011 11	24	4*0 1011	29
0011 10	36	4*0 1010	45
0011 01	3	4*0 1001	53
0011 00	63	4*0 1000	57
0010 111	5	5*0 111	30
0010 110	9	5*0 110	46
0010 101	17	5*0 101	54
0010 100	33	5*0 100	58
0010 011	6	6*0 11 1	31
0010 010	10	6*0 11 0	47
0010 001	18	6*0 10 1	55
0010 000	34	6*0 10 0	59
0001 1111	7	7*0 1 1	27
0001 1110	11	7*0 1 0	39
0001 1101	19		

(b)

Table 8.4 Relation between *quantizer_scale* and *quantizer_scale_code* when q_scale_type = 1.

quantizer_ scale_code	quantizer_ scale	quantizer_ scale_code	quantizer_ scale
0	reserved	16	24
1	1	17	28
2	2	18	32
3	3	19	36
4	4	20	40
5	5	21	44
6	6	22	48
7	7	23	52
8	8	24	56
9	10	25	64
10	12	26	72
11	14	27	80
12	16	28	88
13	18	29	96
14	20	30	104
15	22	31	112

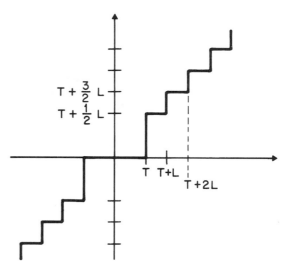

Fig. 8.9 (a) Macroblock with numbered blocks and (b) coded block pattern VLC. The coded block pattern (cbp) is computed by adding the values 2^i for each coded block, where i is the coded block number. For example, if only the top two luminance blocks are to be coded, then cbp = 3. The cbp is then coded according to the VLC shown. The variable named "pattern_code[i]" will be used later. pattern_code[i] = 1 if block i is coded, and = 0 otherwise.

Fig. 8.10 Quantizer used for all NonIntra transform coefficients. The characteristic is uniform except for the dead zone around zero (see Eq. 8.6)

Table 8.5 Variable Length Codes (VLC) and Variable Length Integers (VLI) for quantized Intra DC differential values, which have a range of −1020 to 1020.

Range of Differential DC (DIFFs)	Size	Size VLC Luminance	Size VLC Chrominance	VLIs
−2047 to −1024	11	9*1	9*1 1	9*0 00 to 0 9*1 1
−1023 to −512	10	8*1 0	9*1 0	9*0 0 to 0 9*1
−511 to −256	9	7*1 0	8*1 0	9*0 to 0 8*1
−255 to −128	8	6*1 0	7*1 0	8*0 to 0 7*1
−127 to −64	7	5*1 0	6*1 0	7*0 to 0 6*1
−63 to −32	6	4*1 0	5*1 0	6*0 to 0 5*1
−31 to −16	5	1110	4*1 0	5*0 to 0 4*1
−15 to −8	4	110	1110	4*0 to 0111
−7 to −4	3	101	110	000 to 011
−3 to −2	2	01	10	00 to 01
−1	1	00	01	0
0	0	100	00	
1	1	00	01	1
2 to 3	2	01	10	10 to 11
4 to 7	3	101	110	100 to 111
8 to 15	4	110	1110	1000 to 4*1
16 to 31	5	1110	4*1 0	1 4*0 to 5*1
32 to 63	6	4*1 0	5*1 0	1 5*0 to 6*1
64 to 127	7	5*1 0	6*1 0	1 6*0 to 7*1
128 to 255	8	6*1 0	7*1 0	1 7*0 to 8*1
256 to 511	9	7*1 0	8*1 0	1 8*0 to 9*1
512 to 1023	10	8*1 0	9*1 0	1 9*0 to 9*1 1
1024 to 2048	11	9*1	9*1 1	1 9*0 0 to 9*1 11

N*0 means N zeros; N*1 means N ones

NonIntra coefficients always use the VLC of Table 8.6. Runlength–Level combinations that are not in the VLCs are coded using a a 6-bit Escape, 6-bit FLC for Runlength, and a 12-bit FLC for Level as shown in Table 8.8. Quantized Levels have a range of −2047 to 2047. Every transmitted block ends with End-of-Block (EOB).

8.4.6.3 Reconstruction of Quantized DCT Coefficients

Intra DC Differential values are obtained by simple multiplication of received Levels by the step size specified by ***intra_dc_precision***. Intra AC coefficient reconstruction is simply

$$c[m][n] = (2 \times \text{Level}[m][n] \times \textbf{\textit{quantizer\_scale}} \times \text{Intra Qmatrix}[m][n])/32 \qquad (8.5)$$

All NonIntra coefficients are reconstructed (using the dead zone) by

$$c[m][n] = ((2 \times \text{Level}[m][n] + \text{Sign}\,[m][n]) \times \textbf{\textit{quantizer\_scale}} \times \text{NonIntra Qmatrix}[m][n])/32 \qquad (8.6)$$

where $\text{Sign}[m][n] = -1, 0, +1$ if Level $[m][n] < 0$, $= 0, > 0$, respectively, and the/implies integer division with truncation toward zero.

All reconstructed coefficients are then clipped to the range −2048 to 2047, followed by IDCT mismatch control to try to alleviate the effects on picture quality when the encoder and decoder have slightly different IDCT implementations. Mismatch control is very different than for MPEG-1 or H.261. MPEG-2 basically adds all the reconstructed coefficients of each block, and if the sum is even, the least significant bit of the highest frequency coefficient is changed.*

---

*In practice, if all coefficients are zero, mismatch control should be bypassed.

Table 8.6a   VLC Table zero for RUNS of zero levels and nonzero LEVEL. It is used for NonIntra coefficients and for Intra AC coefficients if intra_vlc_format = 0. This VLC is the same as MPEG-1. The sign bit s is 0 for positive, and 1 for negative.

Run	Level	Code Word	Run	Level	Code Word
0	1	1s (if first inter coefficient)	4	1	0011 0s
0	1	11s (otherwise)	4	2	0000 0011 11s
0	2	0100 s	4	3	0000 0001 0010s
0	3	0010 1s	5	1	0001 11s
0	4	0000 110s	5	2	0000 0010 01s
0	5	0010 0110s	5	3	0000 0000 1001 0s
0	6	0010 0001s	6	1	0001 01s
0	7	0000 0010 10s	6	2	0000 0001 1110s
0	8	0000 0001 1101s	7	1	0001 00s
0	9	0000 0001 1000s	7	2	0000 0001 0101s
0	10	0000 0001 0011s	8	1	0000 111s
0	11	0000 0001 0000s	8	2	0000 0001 0001s
0	12	0000 0000 1101s	9	1	0000 101s
0	13	0000 0000 1100s	9	2	0000 0000 1000 1s
0	14	0000 0000 1100s	10	1	0010 0111 s
0	15	0000 0000 1011s	10	2	0000 0000 1000 0s
1	1	011s	11	1	0010 0011 s
1	2	0001 10s	12	1	0010 0010 s
1	3	0010 0101s	13	1	0010 0000 s
1	4	0000 0011 00s	14	1	0000 0011 10s
1	5	0000 0001 1011s	15	1	0000 0011 01s
1	6	0000 0000 1011 0s	16	1	0000 0010 00s
1	7	0000 0000 1010 1s	17	1	0000 0000 1111s
			18	1	0000 0001 1010s
2	1	0101 s			
2	2	0000 100s	19	1	0000 0001 1001s
2	3	0000 0010 11s	20	1	0000 0001 0111s
2	4	0000 0001 0100s	21	1	0000 0001 0110s
2	5	0000 0000 1010 0s	22	1	0000 0000 1111 1s
3	1	0011 1s	23	1	0000 0000 1111 0s
3	2	0010 0100s	24	1	0000 0000 1110 1s
3	3	0000 0001 1100s	25	1	0000 0000 1110 0s
3	4	0000 0000 1001 1s	26	1	0000 0000 1101 1s
EOB	10	Escape 0000 01			

8.5 TYPICAL ENCODER ARCHITECTURE

Figure 8.11 shows a typical MPEG-2 video encoder. It is assumed that frame reordering takes place before encoding, that is, I- or P-pictures used for B-picture prediction must be coded and transmitted before any of the corresponding B-pictures, as shown in Fig. 7.3.

Input video is fed to a Motion Compensation Estimator/Predictor that feeds a prediction to the minus input of the Subtractor. For each MB, the Inter/Intra Classifier then compares the input pels with the prediction error output of the Subtractor. Typically, if the mean square prediction error exceeds the mean square pel value, an Intra MB is decided. More complicated comparisons involving DCT of both the pels and the prediction error yield somewhat better performance, but are not usually deemed worth the cost.

A thresholding on the sum of absolute vertical line differences is often sufficient for deciding when to use the alternate scan. Alternatively, DCT coef-

Table 8.6b   VLC Table zero for RUNS and LEVELS. It is used for NonIntra coefficients and for Intra AC coefficients if intra_vlc_format = 0. This VLC is the same as MPEG-1. The sign bit s is 0 for positive, and 1 for negative.

Run	Level	Code Word	Run	Level	Code Word
0	16	8*0 0111 11s	1	8	8*0 0011 111s
0	17	8*0 0111 10s	1	9	8*0 0011 110s
0	18	8*0 0111 01s	1	10	8*0 0011 101s
0	19	8*0 0111 00s	1	11	8*0 0011 100s
0	20	8*0 0110 11s	1	12	8*0 0011 011s
0	21	8*0 0110 10s	1	13	8*0 0011 010s
0	22	8*0 0110 01s	1	14	8*0 0011 001s
0	23	8*0 0110 00s	1	15	8*0 0001 0011s
0	24	8*0 0101 11s	1	16	8*0 0001 0010s
0	25	8*0 0101 10s	1	17	8*0 0001 0001s
0	26	8*0 0101 01s	1	18	8*0 0001 0000s
0	27	8*0 0101 00s	6	3	8*0 0001 0100s
0	28	8*0 0100 11s	11	2	8*0 0001 1010s
0	29	8*0 0100 10s	12	2	8*0 0001 1001s
0	30	8*0 0100 01s	13	2	8*0 0001 1000s
0	31	8*0 0100 00s	14	2	8*0 0001 0111s
0	32	8*0 0011 000s	15	2	8*0 0001 0110s
0	33	8*0 0010 111s	16	2	8*0 0001 0101s
0	34	8*0 0010 110s	27	1	8*0 0001 1111s
0	35	8*0 0010 101s	28	1	8*0 0001 1110s
0	36	8*0 0010 100s	29	1	8*0 0001 1101s
0	37	8*0 0010 011s	30	1	8*0 0001 1100s
0	38	8*0 0010 010s	31	1	8*0 0001 1011s
0	39	8*0 0010 001s			
0	40	8*0 0010 000s			

8*0 = 0000 0000

ficient amplitudes after transformation and quantization could be examined.

For Intra MBs the prediction is then switched off. Otherwise, it comes from the Predictor, as described previously. The prediction error MB is then formatted if necessary for field_DCT, passed through the DCT and Quantizer* before being coded, multiplexed, and sent to the Buffer.

Quantized Levels are converted to reconstructed DCT coefficients by the Inverse Quantizer, inverse transformed by the IDCT, and then unformatted if necessary for field_DCT to produce a coded prediction error MB. The Adder adds the prediction to the prediction error and clips the result to the range 0 to 255 to produce coded pel values.

For B-pictures the Motion Compensation Estimator/Predictor uses both the Previous Picture and the Future Picture. These are kept in picture stores and remain unchanged during B-picture coding. Thus, the Inverse Quantizer IDCT Frame/Field Unformatter, and Adder may be disabled during B-picture coding if desired to improve implementation efficiency.

For I- and P-pictures the coded pels output by the Adder are written to the Future Picture Store, while at the same time the old pels are copied from the Future Picture Store to the Previous Picture Store. In practice this is usually accomplished by a simple change of memory addresses.

The Coding Statistics Processor in conjunction with the Quantization Adapter adjusts the output bitrate to conform to the Video Buffer Verifier (VBV) and to optimize the picture quality as much as possible. For constant bitrate (CBR) operation at constant frame-rate and bitrate R, buffer control is

---

*The output of an MPEG quantizer is a signed integer.

Table 8.7a   VLC Table one for RUNS and LEVELS. It is used for Intra AC coefficients if intra_vlc_format = 1. The escape code plus the FLC of Table 8.8 are used for values not in the table.

VLC Codeword	Run	Level	VLC Codeword	Run	Level
0110	End of Block		0000 01	Escape	
10s	0	1	8*0 0010 111 s	0	33
110 s	0	2	8*0 0010 110 s	0	34
0111 s	0	3	8*0 0010 101 s	0	35
1110 0 s	0	4	8*0 0010 100 s	0	36
1110 1 s	0	5	8*0 0010 011 s	0	37
0001 01 s	0	6	8*0 0010 010 s	0	38
0001 00 s	0	7	8*0 0010 001 s	0	39
1111 011 s	0	8	8*0 0010 000 s	0	40
1111 100 s	0	9	010 s	1	1
0010 0011 s	0	10	0011 0 s	1	2
0010 0010 s	0	11	1111 001 s	1	3
1111 1010 s	0	12	0010 0111 s	1	4
1111 1011 s	0	13	0010 0000 s	1	5
1111 1110 s	0	14	8*0 1011 0 s	1	6
1111 1111 s	0	15	8*0 1010 1 s	1	7
8*0 0111 11 s	0	16	8*0 0011 111 s	1	8
8*0 0111 10 s	0	17	8*0 0011 110 s	1	9
8*0 0111 01 s	0	18	8*0 0011 101 s	1	10
8*0 0111 00 s	0	19	8*0 0011 100 s	1	11
8*0 0110 11 s	0	20	8*0 0011 011 s	1	12
8*0 0110 10 s	0	21	8*0 0011 010 s	1	13
8*0 0110 01 s	0	22	8*0 0011 001 s	1	14
8*0 0110 00 s	0	23	8*0 0001 0011 s	1	15
8*0 0101 11 s	0	24	8*0 0001 0010 s	1	16
8*0 0101 10 s	0	25	8*0 0001 0001 s	1	17
8*0 0101 01 s	0	26	8*0 0001 0000 s	1	18
8*0 0101 00 s	0	27	0010 1 s	2	1
8*0 0100 11 s	0	28	0000 111 s	2	2
8*0 0100 10 s	0	29	1111 1100 s	2	3
8*0 0100 01 s	0	30	0000 0011 00 s	2	4
8*0 0100 00 s	0	31	8*0 1010 0 s	2	5
8*0 0011 000 s	0	32			

reasonably easy, since the encoder and decoder buffers track each other. More specifically, at start-up the VBV is allowed to fill to some value vbv_size before decoding the first picture. Thereafter, we have the following relationship:

$$B_e(n) = \text{vbv\_size} - B_d(n) \qquad (8.7)$$

where $B_e(n)$ is the encoder buffer fullness just before encoding picture n and $B_d(n)$ is the decoder (VBV) buffer fullness just before decoding picture n. Thus, as long as there is no underflow or overflow beyond vbv_size of the encoder buffer, there will

be no underflow/overflow of the decoder buffer. For CBR, the overall delay* is given by

$$\text{Delay} = \text{vbv\_delay}(1) = \text{vbv\_size}/R \qquad (8.8)$$

and the vbv_delay values may be computed from

$$\text{vbv\_delay}(n) = B_d(n)/R \qquad (8.9)$$

A simple control that works reasonably well is to define a target buffer fullness for each picture in the GOP. For each picture the quantizer_scale value is then adjusted periodically to try to make the

---

*Assuming no Big Pictures.

Table 8.7b VLC Table one for RUNS and LEVELS. It is used for Intra AC coefficients if intra_vlc_format = 1. The escape code plus the FLC of Table 8.8 are used for values not in the table.

VLC Codeword	Run	Level	VLC Codeword	Run	Level
0011 1 s	3	1	8*0 0001 1001 s	12	2
0010 0110 s	3	2	0010 0100 s	13	1
0000 0001 1100 s	3	3	8*0 0001 1000 s	13	2
8*0 1001 1 s	3	4	0000 0010 1 s	14	1
0001 10 s	4	1	8*0 0001 0111 s	14	2
1111 1101 s	4	2	0000 0011 1 s	15	1
0000 0001 0010 s	4	3	8*0 0001 0110 s	15	2
0001 11 s	5	1	0000 0011 01 s	16	1
0000 0010 0 s	5	2	8*0 0001 0101 s	16	2
8*0 1001 0 s	5	3	0000 0001 1111 s	17	1
0000 110 s	6	1	0000 0001 1010 s	18	1
0000 0001 1110 s	6	2	0000 0001 1001 s	19	1
8*0 0001 0100 s	6	3	0000 0001 0111 s	20	1
0000 100 s	7	1	0000 0001 0110 s	21	1
0000 0001 0101 s	7	2	8*0 1111 1 s	22	1
0000 101 s	8	1	8*0 1111 0 s	23	1
0000 0001 0001 s	8	2	8*0 1110 1 s	24	1
1111 000 s	9	1	8*0 1110 0 s	25	1
8*0 1000 1 s	9	2	8*0 1101 1 s	26	1
1111 010 s	10	1	8*0 0001 1111 s	27	1
8*0 1000 0 s	10	2	8*0 0001 1110 s	28	1
0010 0001 s	11	1	8*0 0001 1101 s	29	1
8*0 0001 1010 s	11	2	8*0 0001 1100 s	30	1
0010 0101 s	12	1	8*0 0001 1011 s	31	1

Table 8.8 FLC Table for RUNS and LEVELS. It is used following the escape code of a VLC.

FLC Codeword	Run	FLC Codeword	signed_level
0000 00	0	1000 0000 0000	reserved
0000 01	1	1000 0000 0001	−2047
0000 10	2	1000 0000 0010	−2046
...	...	...	...
...	...	1111 1111 1111	−1
...	...	0000 0000 0000	not allowed
...	...	0000 0000 0001	+1
...	...	...	...
1111 11	63	0111 1111 1111	+2047

actual buffer fullness meet the assigned value. More complicated controls could, in addition, exploit spatiotemporal masking in choosing the quantizer_scale parameter for each MB.

In systems where the transmission facilities are shared among many users, variable bitrate (VBR) operation is often advantageous. With VBR, the video bitrate can be small when there is little motion is the picture and large when motion is

rapid. The net effect, especially in video conferencing systems, is often that the transmission channel can handle more users. However, with VBR the control of buffers may be much more complicated than with CBR.[4]

8.6 TYPICAL DECODER ARCHITECTURE

Figure 8.12 shows a typical MPEG-2 video decoder. It is basically identical to the pel reconstruction portion of the encoder. It is assumed that frame reordering takes place after decoding and video output. However, extra memory for reordering can often be avoided if during a write of the Previous Picture Store the pels are routed also to the display.

The decoder cannot tell the size of the GOP from the video stream parameters. Indeed it does not know until the Picture Header whether the picture is I, P, or B Type. This could present problems

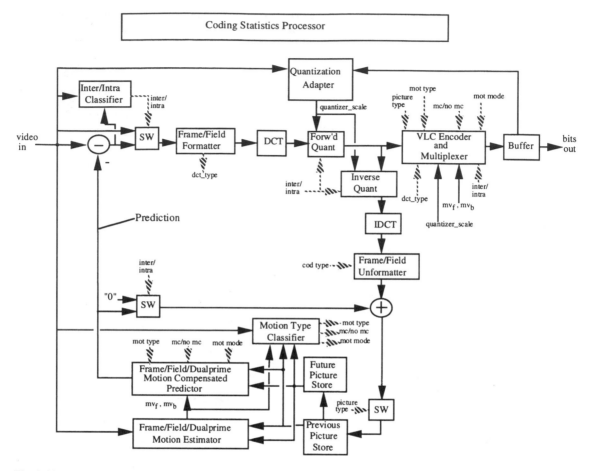

Fig. 8.11   A typical MPEG-2 encoder.

in synchronizing audio and video were it not for the Systems part of MPEG-2, which provides Time Stamps for audio, video, and ancillary data. By presenting decoded information at the proper time as indicated by the Time Stamps, synchronization is assured.

8.7 MPEG-2 ERROR CONCEALMENT

Decoders often must accommodate occasional transmission bit errors. Typically, when errors are detected either through external means or internally through the arrival of illegal data, the decoder replaces the coded data with skipped MBs until the next Slice is detected. The visibility of

errors is then rather low unless they occur in an I-picture, in which case they may be very visible throughout the GOP. A cure for this problem was developed for MPEG-2 and is described next.

MPEG-2 has added an important feature for concealing the effects of transmission bit errors. Intra pictures may optionally contain coded motion vectors, which are used only for error concealment. Then in P-pictures, B-pictures, or I-pictures with motion vectors, a Slice that is known to be erroneous may be replaced by motion-compensated pels from the previous I- or P-picture. Motion vectors from the slice above could be used for this process.

In demonstrations of MPEG-2 video at a bitrate of 4.0 Mbits/s, interlaced CCIR 601 pictures have been used, with 4:2:0 chrominance sub-

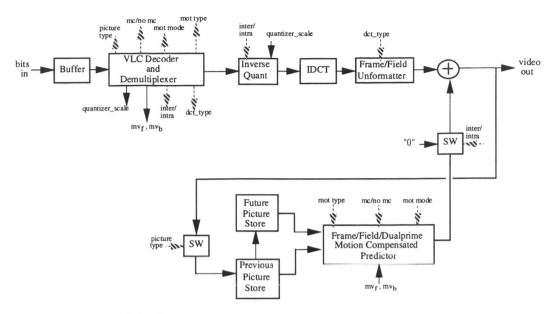

Fig. 8.12 A typical MPEG-2 decoder.

sampling. The quality achieved by the MPEG-2 video encoder is good to excellent, depending on scene content. Error resilience features have been found to be very effective at bit error rates up to 10^{-4} and usable up to 10^{-3}.

8.8 PERFORMANCE OF MPEG-2

Here we evaluate the performance[13] of various encoding options and tools in motion compensation, DCT coding, DCT coefficient quantization, DCT coefficient scanning, variable length coding, B-pictures, and frame- or field-picture structure. Simulations on several MPEG-2 standard test sequences and several other test sequences are performed. Although our simulations primarily report results on Peak-Signal to RMS-Noise Ratio (PSNR), we are aware that a subjective measure of performance is usually needed to truly evaluate the video coding quality. Also, the simulation results presented here are for illustration purposes and do not represent the results of an optimized MPEG-2 video encoder. Generally, they use the various MB mode decisions and rate control of the MPEG-2 Test model.[5] Within the context of MPEG-1 video

coding a number of alternate rate control techniques[14,15] have also been developed.

In our simulations we employ the following picture formats and source sequences for coding:

—4:2:0 and 4:2:2 interlaced sequences of 704 pel × 480 line active resolution derived from 720 pel × 486 line CCIR-601 originals. From the sequences Flowergarden, Mobile&Calendar, Football, Bus, and Carousel, 150 frames are employed.

—Progressive sequences of 704 × 480 active resolution at a frame rate of 60 Hz were derived by windowing on progressive HDTV sources. The progressive sequences employed are Raft (60 frames), Robots (60 frames), and Basketball2 (40 frames).

We have tried to keep performance comparisons as meaningful as possible. Whenever certain comparisons were difficult owing to lack of availability of certain options, we performed the simulations without those options for more accurate comparisons. For example, in comparisons of field-picture and frame-picture coding, results do not employ concealment motion vectors. Also, during the simulations we found various macroblock mode decisions of the Test Model[5] when applied to field-pictures to be insufficient. Thus, for consistency with other results we did not use better mode decisions. We now present results of individual experiments.

8.8.1 Zigzag versus Alternate Scan Performance Comparison

We compare the performance[13] of the DCT coefficient Zigzag scan of MPEG-1 with the MPEG-2 Alternate scan[5,19] on interlaced video. For this experiment we employ 4:2:0 format sequences, P-picture prediction distance $M = 3$ (i.e., two B-pictures), a GOP length $N = 15$, and a bitrate of 4 Mbits/s. Motion compensation and DCT coding is fully adaptive, and concealment motion vectors are enabled.

The results in Table 8.9 show an improvement of up to 0.3 dB by using the alternate scan for interlaced video. Sequences such as "Mobile & Cal" with little movement between fields behave as progressive sequences and do not result in an improvement over the zigzag scan, whereas larger gain is possible on faster moving sequences such as "Football."

8.8.2 MPEG-1 VLC Zero versus MPEG-2 Intra VLC One on Intra Macroblocks

We compare the performance[13] of MPEG-1 VLC zero with Intra VLC one[5–18,19] on interlaced video. For this experiment[5–18,19] we employ 4:2:0 format sequences, all Intra picture coding at bitrates of 12 and 25 Mbits/s. Concealment motion vectors are not used in these simulations. Results are shown in Tables 8.10a and b.

Table 8.9 Luminance PSNR (dB) for DCT coefficient scans for frame-pictures. $M = 3$, $N = 15$, bitrate $= 4$ Mbits/s

Sequence	Zigzag Scan	Alternate Scan
Flowergarden	29.36	29.61 (+0.25)
Mobile & Cal	28.20	28.24 (+0.04)
Football	34.77	35.07 (+0.30)
Bus	31.35	31.57 (+0.22)
Carousel	29.57	29.68 (+0.11)

Table 8.10a Luminance PSNR (dB) for DCT coefficient VLC Tables for frame-pictures. Intra pictures, bitrate = 12 Mbits/s

Sequence	MPEG-1 VLC	Intra VLC
Flowergarden	28.46	28.55 (+0.09)
Mobile & Cal	25.85	25.88 (+0.03)
Football	39.25	39.35 (+0.10)
Bus	32.10	32.21 (+0.11)
Carousel	34.05	34.18 (+0.13)

Table 8.10b Luminance PSNR (dB) for DCT coefficient VLC Tables for frame-pictures. Intra pictures, bitrate = 25 Mbits/s

Sequence	MPEG-1 VLC	Intra VLC
Flowergarden	34.42	35.20 (+0.78)
Mobile & Cal	31.01	31.81 (+0.80)
Football	42.18	42.34 (+0.16)
Bus	37.76	38.54 (+0.78)
Carousel	38.83	39.50 (+0.67)

At higher bitrates for I-pictures (or Intra MBs anywhere) significant improvement in PSNR performance can be obtained for nearly all scenes with Intra VLC one. Although overall improvement is less significant for $M = 1$ or $M = 3$ coding, it is to be noted that Intra VLC is only used for Intra MBs, and is highly dependent on their frequency of occurrence.

8.8.3 Performance Comparison of Motion Compensation Modes

Next, we compare the performance[5,16,17] of various options for motion compensation modes on interlaced video. For this experiment we employ 4:2:0 format sequences, P-picture prediction distance $M = 1$ and $M = 3$, a GOP length $N = 15$, and a bitrate of 4 Mbits/s. For frame-pictures, DCT coding is frame/field adaptive and concealment motion vectors are enabled. Field-pictures do not use concealment motion vectors. The performance of various motion compensation modes is given in Tables 8.11a to 8.11c. All simulations use zigzag scan and MPEG-1 VLC for Intra as well as other MB types.

Table 8.11a Luminance PSNR (dB) for Motion-Compensated Prediction modes for frame-pictures. $M = 1, N = 15$, bitrate = 4 Mbits/s

Sequence	Frame MC	Field MC	Frame/Field MC	Dualp MC	Frame/Field/Dualp MC
Flowergarden	27.72	28.06 (+0.34)	28.22 (+0.50)	28.39 (+0.67)	29.38 (+1.66)
Mobile & Cal	25.69	25.86 (+0.17)	26.04 (+0.35)	25.51 (−0.18)	26.63 (+0.94)
Football	34.20	35.60 (+1.40)	35.69 (+1.49)	35.69 (+1.49)	36.04 (+1.84)
Bus	28.99	30.26 (+1.27)	30.43 (+1.44)	30.70 (+1.71)	31.31 (+2.32)
Carousel	28.67	29.97 (+1.30)	30.07 (+1.40)	29.99 (+1.32)	30.53 (+1.86)

Table 8.11b Luminance PSNR (dB) for Motion-Compensated Prediction modes for frame-pictures. $M = 3, N = 15$, bitrate = 4 Mbits/s

Sequence	Frame MC	Field MC	Frame/Field MC
Flowergarden	29.07	29.20 (+0.13)	29.63 (+0.56)
Mobile & Cal	28.11	27.86 (−0.25)	28.27 (+0.16)
Football	34.54	35.01 (+0.47)	35.12 (+0.58)
Bus	30.79	31.32 (+0.53)	31.60 (+0.81)
Carousel	29.22	29.54 (+0.32)	29.73 (+0.51)

Table 8.11c Luminance PSNR (dB) for Motion-Compensated Prediction modes for field-pictures. $M = 1, N = 15$, bitrate = 4 Mbits\s

Sequence	Field MC	16×8 MC	Field/16×8 MC
Flowergarden	26.99	25.94 (−1.05)	27.18 (+0.19)
Mobile & Cal	25.02	23.61 (−1.41)	25.21 (+0.19)
Football	36.07	35.07 (−1.00)	35.89 (−0.18)
Bus	29.63	28.76 (−0.87)	29.83 (+0.20)
Carousel	30.31	29.30 (−1.01)	30.29 (+0.12)

Dual-prime motion compensation (allowed only when prediction distance is $M = 1$) provides significant improvement for $M = 1$ and often makes the performance of $M = 1$ coding comparable to $M = 3$ in a PSNR sense. However, dual-prime requires computational complexity and bandwidth higher than the $M = 3$ case and is meaningful for low-delay applications where high performance is required. Also, for frame-pictures, the $M = 1$ case provides a bigger relative improvement in performance for frame/field adaptive motion compensation when compared to frame motion compensation only. From field-picture PSNR results it appears that field/16×8 adaptive motion compensation requires a better decision criterion than that of the Test Model; further use of smaller block sizes at the expense of extra motion vectors has potential for better subjective quality not reflected by the PSNR.

8.8.4 DCT Coding Modes Performance Comparison

We now compare the performance of various DCT coding modes on interlaced frame-pictures. We employ 4:2:0 format sequences, prediction distance $M = 1$ and $M = 3$, a group-of-picture length $N = 15$, concealment motion vectors, and a bitrate of 4 Mbits/s. All simulations employ zigzag scan and MPEG-1 VLC for Intra as well as other MB types.

The PSNR results, shown in Tables 8.12a and b, show an improvement of up to 0.85 dB with typical improvement of about 0.35 dB by using frame/field adaptive DCT over frame DCT only.

Table 8.12a Luminance PSNR (dB) for DCT modes for frame-pictures with frame/field/dual MC. $M = 1$, $N = 15$, bitrate = 4 Mbits/s

Sequence	Frame DCT	Field DCT	Frame/Field DCT
Flowergarden	29.36	29.04 (−0.32)	29.38 (+0.02)
Mobile & Cal	26.66	25.87 (−0.79)	26.63 (−0.03)
Football	35.54	35.95 (+0.41)	36.04 (+0.50)
Bus	31.05	31.00 (−0.05)	31.31 (+0.26)
Carousel	29.68	30.36 (+0.68)	30.53 (+0.85)

Table 8.12b Luminance PSNR (dB) for DCT modes for frame-pictures with frame/field MC. $M = 3$, $N = 15$, bitrate = 4 Mbits/s

Sequence	Frame DCT	Field DCT	Frame/Field DCT
Flowergarden	29.61	29.46 (−0.15)	29.63 (+0.02)
Mobile & Cal	28.34	27.74 (−0.60)	28.27 (−0.07)
Football	34.67	35.04 (+0.37)	35.12 (+0.45)
Bus	31.34	31.41 (+0.07)	31.60 (+0.26)
Carousel	29.04	29.59 (+0.55)	29.73 (+0.69)

Further improvements are possible with optimized DCT mode decisions.

8.8.5 Motion Compensation and DCT Coding Modes Performance Comparison

We now present results[13] of PSNR performance comparisons[13] of various motion compensation and DCT mode combinations to illustrate degradation in quality if fully adaptive motion compensation and DCT coding modes cannot be afforded because of complexity. All simulations use zigzag scan, MPEG-1 VLC, and concealment motion vectors.

Table 8.13a illustrates that for M = 1 coding the combination of frame/field adaptive motion compensation with frame/field adaptive DCT out-

performs the combination of frame motion compensation and frame DCT in a PSNR sense by up to 2.36 dB with a possibility of another 1 dB or so improvement when adaptive motion compensation also includes dual-prime. Table 8.13b shows that for M = 3 coding the PSNR improvement with a combination of frame/field adaptive motion compensation and frame/field adaptive DCT can be as much as 1.3 dB.

8.8.6 Frame-versus Field-Picture Coding Performance Comparison

We now show PSNR results[3] of experiments of frame-picture and field-picture coding for prediction distances M = 1 and M = 3 and group-of-

Table 8.13a Luminance PSNR (dB) for MC and DCT mode comparisons for frame-picture $M = 1$, $N = 15$, bitrate = 4 Mbits/s

Sequence	Frame MC, Frame DCT	Frame MC, Frame/Field DCT	Field MC, Field DCT	Frame/Field MC, Frame/Field DCT	Frame/Field/DualpMC Frame/Field DCT
Flowergarden	27.67	27.72 (+0.05)	27.76 (+0.09)	28.22 (+0.55)	29.38 (+1.71)
Mobile & Cal	25.72	25.69 (−0.03)	25.12 (−0.60)	26.04 (+0.32)	26.63 (+0.91)
Football	33.57	34.20 (+0.63)	35.51 (+1.94)	35.69 (+2.08)	36.04 (+2.47)
Bus	28.65	28.99 (+0.34)	29.99 (+1.34)	30.43 (+1.78)	31.31 (+2.66)
Carousel	27.71	28.67 (−0.04)	29.80 (+2.09)	30.07 (+2.36)	30.53 (+2.82)

Table 8.13b   Luminance PSNR (dB) for MC and DCT mode comparison for frame-pictures $M = 3$, $N = 15$, bitrate = 4 Mbits/s

Sequence	Frame MC, Frame DCT	Frame MC, Frame/Field DCT	Field MC, Field DCT	Frame/Field MC, Frame/Field DCT	Frame/Field/DualpMC Frame/Field DCT
Flowergarden	29.03	29.07 (+0.04)	28.99 (−0.04)	29.20 (+0.17)	29.63 (+0.60)
Mobile & Cal	28.13	28.11 (−0.02)	27.27 (−0.86)	27.86 (−0.27)	28.27 (+0.14)
Football	33.99	34.54 (+0.55)	34.92 (+0.93)	35.01 (+1.02)	35.12 (+1.13)
Bus	30.46	30.79 (+0.33)	31.10 (+0.64)	31.32 (+0.86)	31.60 (+1.14)
Carousel	28.35	29.22 (+0.87)	29.38 (+1.03)	29.54 (+1.19)	29.73 (+1.38)

pictures size $N = 15$. Our simulations do not use dual-prime motion compensation or concealment motion vectors. Adaptive motion compensation modes are allowed both for frame- and field-pictures, and adaptive DCT is used for frame-pictures. Zigzag scan and MPEG-1 VLCs are used in these simulations.

The results in Table 8.14 show that for $M = 1$ frame-pictures can give up to 0.05 dB improvement in PSNR over field pictures. For fast motion sequences, there may be no improvement. However, for $M = 3$, frame pictures are usually better than field-pictures in all cases.

group-of-pictures size $N = 15$, frame-field adaptive motion compensation, frame/field adaptive DCT, zigzag scan, and MPEG-1 VLC. Concealment motion vectors are not used.

It is clear from Table 8.15 that luminance PSNR performance of 4:2:0 is better by about 0.2 to 0.75 dB as compared to the 4:2:2 format at moderate bitrates. Thus, better chrominance rendition is accomplished at some loss in luminance quality. Although chrominance PSNRs are specified for both cases they should not be compared because of different references used for computing PSNR in the two cases.

8.8.7   4:2:0 versus 4:2:2 Format Video Coding Performance Comparison

Next, we experiment[13] with coding of 4:2:2 format* video and evaluate the loss in luminance quality due to coding of increased number of chrominance samples (because of 4:2:2 format). Our coder uses a prediction distance $M = 3$,

8.8.8   No B-pictures versus with B-pictures Comparison for Progressive Video

We now present results[13] of progressive 60-Hz video coding at a bitrate of 6 Mbits/s using prediction distance $M = 1$ and 3 and GOP size $N = 30$. It is our observation that for $M = 3$, the coding quality follows that of P-pictures and although the

Table 8.14   Luminance PSNR (dB) for frame-pictures versus field-pictures comparison. $M = 1$ and $M = 3$, $N = 15$, bitrate = 4 Mbits/s

Sequence	$M = 1$		$M = 3$
	Frame-picture	Field-picture	Frame-picture
Flowergarden	27.93	27.18 (−0.75)	29.43
Mobile & Cal	26.06	25.21 (−0.85)	28.31
Football	35.37	35.89 (+0.52)	34.90
Bus	30.19	29.83 (−0.36)	31.45
Carousel	30.00	30.29 (+0.29)	29.71

*Main Profile does not support 4:2:2.

Table 8.15 Luminance and chrominance PSNR (dB) for 4:2:0 versus 4:2:2 format comparison. $M = 3, N = 15$, bitrate = 4 Mbits/s

Sequence	4:2:0 Format			4:2:2 Format		
	Y	Ch	Cr	Y	Cb	Cr
Flowergarden	29.25	32.67	34.43	28.93 (−0.32)	32.35	33.40
Mobile & Cal	28.10	34.19	34.20	27.43 (−0.67)	33.54	33.99
Football	33.87	36.79	38.78	33.54 (−0.33)	36.87	38.75
Bus	30.78	37.88	40.27	30.57 (−0.21)	37.72	40.34
Carousel	28.48	33.32	33.89	28.22 (−0.26)	33.89	34.38

average PSNR of I-, P-, and B-pictures does not show a big difference, the coding quality with M = 3 is better than the M = 1 case. Thus we specify not only average PSNR for all picture types (as in previous tables) but also compare PSNR performance of P-pictures.

Our results in Table 8.16 indicate that using P-picture PSNR as prediction of quality is consistent with the subjective quality obtained from these coded sequences, that is, M = 3 results in less noise in reconstructed pictures. In a PSNR sense, comparing average picture results shows that M = 3 provides up to 0.57 dB better PSNR.

8.8.9 Observations and Comments

Due to the large number of tables in our simulations we summarize our key observations as follows. However, the reader is encouraged to look at details of individual experiments to draw specific conclussions.

1. For interlaced video, use of the alternate scan can typically provide between 0.2 and 0.3 dB improvement over the zigzag scan. This improvement is important as it comes at a relatively small increase in complexity.

2. In all Intracoding, the use of the Intra VLC can provide 0.1 to 0.9 dB PSNR improvement over the MPEG-1 VLC, depending on the bitrate. In all other cases, the overall benefit of the Intra VLC obviously depends on the frequency of Intra MBs and will not be as large. However, the additional implementation complexity of Intra VLC is relatively small.

3. For interlaced video and frame-pictures, the frame/field adaptive motion compensation of MPEG-2 provides an improvement of 0.50 to 1.50 dB in PSNR over frame motion compensation (of MPEG-1). Moreover, the use of dual-prime motion compensation (allowed only for M = 1) can provide an additional 0.5 to 1.0 dB improvement, but requires extra bandwidth and computational complexity.

4. For interlaced video and frame-pictures, the frame/field adaptive DCT can provide up to 0.9 dB PSNR improvement over frame DCT alone. The improvements with M = 1 are slightly higher than for M = 3. For scenes with faster motion, improvement is higher than for scenes with slow to medium motion.

5. For interlaced video and frame-pictures, the combination of frame/field adaptive motion compensation and frame/field adaptive DCT can provide 0.3 to 2.5 dB improvement over the combination of frame motion compensation and

Table 8.16 Luminance PSNR (dB) for progressive 60-Hz video. M = 1 versus $M = 3$ comparison, $N = 30$, bitrate = 6 Mbits/s

Sequence	P-Pictures Only		Avg. Picture	
	$M = 1$	$M = 3$	$M = 1$	$M = 3$
Raft	25.43	25.87 (+0.44)	25.50	25.64 (+0.14)
Robots	30.66	31.26 (+0.60)	30.67	31.24 (+0.57)
Basketball2	27.06	27.45 (+0.39)	27.10	27.33 (+0.23)

frame DCT. With M = 1 the improvements are larger than for M = 3. Moreover, the use of dual-prime (e.g., M = 1) can provide an additional 0.5 to 1 dB improvement.

6. For interlaced video, in comparing frame- and field-pictures, frame-pictures provide up to 0.9 dB better PSNR than field-pictures. For sequences with fast motion, field-pictures can provide up to 0.5 dB improvement over frame-pictures. Overall, frame-pictures outperform field-pictures more frequently.

7. For coding of interlaced video at bitrates of 4 Mbits/s, it is preferable to use the 4:2:0 format rather than the 4:2:2 format. In coding the 4:2:2 format, better chrominance quality can be retained at the expense of about 0.25 to 0.7 dB in luminance PSNR. However, the improvements in chrominance quality are visible only for critical graphics scenes.

8. Progressive 60-Hz video of normal TV resolution requires about 6 Mbits/s to achieve quality comparable to that obtained for interlaced video at 4 Mbits/s.

9. The use of GOP of length N = 9 instead of N = 15 incurs only a small penalty in luminance PSNR (0.05 to 0.25 dB).

10. The use of B-pictures provides an increase of overall PSNR by 0.3 to 0.6 dB over the case of no B-pictures and no dual-prime. However, the average PSNR for B-picture coding is highly influenced by the high occurrence of B-pictures (which have a lower PSNR). Thus, the subjective quality is much higher with B-pictures than reflected by the PSNR improvement.

11. The use of a nonuniform quantizer does not provide a statistical gain. However, its use is useful for maintaining a steady picture quality, especially in cases of extreme coding where either very many or very few bits are available.

8.8.10 Discussion

Within the framework of the MPEG-2 standard, we have studied various encoding configurations that allow a wide range of tradeoffs in performance versus complexity. While the decoder specification is mandated by the profile/level chosen, considerable flexibility in encoder design is possible. Our studies have also included the performance expected with various coding parameters. Some general conclusions regarding our experi-

ence in evaluating the performance of the MPEG-2 video standard are drawn next.

Frame-pictures generally outperform field-picture coding, although field picture coding offers a smaller delay, which may be significant in low-delay applications. In fast motion sequences, field-pictures can nearly equal the performance of frame-pictures.

For interlaced video, coding with B-pictures clearly outperforms coding without B-pictures as long as the dual-prime motion compensation mode is not used. When the dual-prime mode is used in coding without B-pictures, the difference with respect to coding with B-pictures is not very big. The use of B-pictures requires two prediction stores, whereas in coding without B-pictures, only one frame store is sufficient. However, memory speed requirements of dual-prime are 50% greater than that of B-pictures. For low-delay applications, the use of dual-prime with other motion compensation modes can result in good performance. In the coding of progressive-pictures, since the dual-prime mode is not available, B-pictures are needed to obtain good performance. Although the PSNR difference between with and without B-pictures for progressive video does not appear very large, the subjective quality with B-pictures is distinctly better. If both interlaced and progressive video are required to be coded, the need for B-pictures becomes crucial.

The adaptations for interlace, including frame/field motion compensation, frame/field DCT, and alternative scan, considerably improve the coding quality over MPEG-1 motion compensation and DCT. Intra VLC can improve the quality for Intra coding.

At lower bitrate coding of difficult sequences, if there is a choice between 4:2:0 and 4:2:2 formats, 4:2:0 offers the best overall tradeoff between better chrominance rendition and quality of luminance. At 9 Mbits/s, sequences with saturated chrominance can visibly benefit from the better chrominance rendition of the 4:2:2 format.

REFERENCES

1. ISO/IEC 11172 (MPEG-1). "Information Technology—Coding of Moving Pictures and Associated

Audio for Digital Storage Media up to About 1.5 Mbits/s."

2. ISO/IEC 13818 (MPEG-2). "Information Technology—Generic Coding of Moving Pictures and Associated Audio Information."

3. R. Aravind etal., "Image and Video Coding Standards" *AT&T Technical J.*, pp. 67–89 (January/February 1993).

4. A.R. Reibman and B. G. Haskell, "Constraints on Variable Bit-Rate Video for ATM Networks," *IEEE Trans. Circuits Syst. Video Technol.* 2 (4) 361–372 (December 1992).

5. Test Model Editing Committee, "MPEG-2 Video Test Model 5," ISO/IEC JTC1/SC29/WG11 Doc. N0400 (April 1993).

6. D. J. LeGall, "MPEG: A Video Compression Standard for Multimedia Applications," *Commun. of the ACM*, Vol. 34, No. 4, pp. 47–58 (April 1991).

7. "Digital Video," *IEEE Spectrum Magazine*, Vol. 29, No. 3 (March 1992).

8. Video Simulation Model Editing Committee, "MPEG-1 Video Simulation Model" (July 1990).

9. D. J. LeGall, "MPEG Video: The Second Phase of Work," Proc. Int. Symposium of Society for Information Display, pp. 113–116, (October 1992).

10. A. Puri, "Video Coding Using the MPEG-1 Compression Standard," Proc. Int. Symposium of Society for Information Display, pp. 123–126. (May 1992).

11. A. Puri, R. Aravind, and B. G. Haskell, "Adaptive Frame/Field Motion Compensated Video Coding," *Signal Processing: Image Commun.*, Vol. 5, pp. 39–58. (February 1993).

12. A. Puri, "Video Coding Using the MPEG-2 Compression Standard," Proc. SPIE Visual Commun. and Image Proc., SPIE Vol. 1199, pp. 1701–17 (November 1993).

13. R. L. Schmidt, A. Puri, and B. G. Haskell, "Performance Evaluation of Nonscalable MPEG-2 Video Coding," Proc. SPIE Visual Commun. and Image Proc. SPIE Vol 2308, pp. 296–310 (September 1994).

14. E. Viscito and C. Gonzales, "A Video Compression Algorithm with Adaptive Bit Allocation and Quantization," Proc. SPIE Visual Commun. and Image Proc., Vol 1605, pp. 205–215 (November 1991).

15. A. Puri and R. Aravind, "Motion Compensated Video Coding with Adaptive Perceptual Quantization," *IEEE Trans. on Circuits and Systems for Video Technology*, Vol. CSVT-1, No. 4, pp. 351–361 (December 1991).

16. Y. Nakajima, "Results of Prediction Core Experiment: L-14 Special Prediction Mode," ISO/IEC JTC1/SC29/WG11 MPEG 93/035 (January 1993).

17. T. Savatier, "Simulation Results on Modified L-14 and Dual-prime," ISO/IEC JTC1/SC29/WG11 MPEG 93/303 (March 1993).

18. Y. Yagasaki and J. Yonemitsu, "Introduction of the second VLC table for Intra Coding," ISO/IEC JTC1/SC29/WG11 MPEG92/452 Rev., (November 1992).

19. A. Puri, "Scan and VLC Experiment Results," ISO/IEC JTC1/SC29/WG11 MPEG 93/388 (March 1993).

9

MPEG-2 Scalability Techniques

Realizing that many applications require video to be simultaneously available for decoding at a variety of resolutions or qualities, MPEG-2 includes several efficient techniques that allow video coding to be performed in a manner suitable to accomplish this, these are loosely referred to as scalability techniques.[2,19] Scalable coding can be useful in many applications. Further, there are many types of scalabilities, each allowing a different set of tradeoffs in bandwidth partitioning, video resolution, or quality in each layer and overall implementation complexity.

Simply speaking, scalability of video means the ability to achieve video of more than one resolution or quality simultaneously. Scalable video coding involves generating a coded representation (bitstream) in a manner that facilitates the derivation of video of more than one resolution or qualities from a single bitstream.

Bitstream Scalability is the property of a bitstream that allows decoding of appropriate subsets to generate complete pictures of resolution or quality commensurate with the proportion of the bitstream decoded. In principle, if a bitstream is truly scalable, decoders of different complexities, from inexpensive, low-performance decoders to expensive, high-performance decoders can coexist, and whereas inexpensive decoders would be expected to decode only small portions of the same bitstream producing basic quality pictures expensive decoders would be able to decode much more and produce higher quality pictures.

Scalable video coding is also referred to as layered video coding or hierarchical video coding. MPEG-2 scalable coding can be thought of as based on or related to two popular forms of scalable coding methods known as pyramid coding and sub-band coding. However, as you are about to find out, it is much more than that. Before finalizing scalability in MPEG-2, a very exhaustive analysis of many scalability techniques was performed. Only the most essential and practical techniques that had the potential of addressing real-world applications were included. Redundant techniques or techniques that did not seem to meet the performance objectives of anticipated applications were not included.

For now we list a general set of applications where scalability is useful; a little later we will point out which type of scalability can best address which applications.

- Error Resilience on noisy channels
- Multipoint videoconferencing
- Video database browsing
- Multiquality video services on ATM, Internet, and other networks
- Windowed video on computer workstations
- Compatibility with existing standards and equipment

- Compatible digital HDTV and digital TV
- Migration to high temporal resolution progressive HDTV
- Video on computer workstations and LANs.

In this chapter we first discuss basics of scalability including scalability approaches, scalability types, and their applications. We then discuss details of each type of scalability including codec structures, specific operations, the encoding process, the decoding process, and example results for each type of scalability included in the MPEG-2 standard. Finally, we discuss combinations of individual scalabilities referred to as hybrid scalability and provide example codec structures involving hybrid scalability.

9.1   SCALABILITY BASICS

Now that we have clarified the significance of scalability, it is time to discuss how to achieve scalability and the different types of scalable coding techniques. We can then try to determine which type of scalable coding can be used for which applications.

9.1.1   Scalability Approaches

We now discuss the two basic approaches to achieve scalability.

In *Simulcast Coding*, each layer of video representing a resolution or quality is coded independently. Thus any layer can be decoded by a single-layer (nonscalable) decoder. In simulcast coding, total available bandwidth is simply partitioned depending on the quality desired for each independent layer that needs to be coded. In simulcast coding, it is assumed that independent decoders would be used to decode each layer.

We now summarize attributes of simulcast video coding:

—All layers are independently coded.
—Total Bandwidth is proportionately partitioned between all layers.

In *Scalable Video Coding*, one layer of video is coded independently whereas other layers are coded

dependently, each following layer coded with respect to the previous layer. The independently coded layer can be decoded by single-layer decoders. If this layer is coded with another video coding standard, compatibility is said to be achieved. For example, an independent layer may be coded by MPEG-1, thus providing compatibility with MPEG-1.

Generally, scalable coding is more efficient than simulcast coding. Except for the independently coded layer, each layer is able to reuse some of the bandwidth assigned to the previous layer. The exact amount of increased efficiency is dependent on the specific technique used, the number of layers, and the bandwidth partitioning used. Also, this increase in efficiency comes at the expense of some increase in complexity compared to simulcast coding. Different scalabilities offer different tradeoffs, and in general some are more suited for one set of applications whereas others are better suited for a second set of applications and so forth. We are now ready to briefly introduce various types of scalabilities.

Summarizing the attributes of scalable video coding:

—The lowest layer is independently coded; each following layer is coded with respect to the previously decoded layer.
—The bandwidth is efficient because of the ability to reuse the portion of bandwidth spent on coding the previous layer.
—There is ease of networking with other standards and integration of multiple video services.
—Higher layer encoders/decoders are more complex.

9.1.2   Scalable Coding Types

We now list the various types of scalable coding techniques[2,6,19,20] supported by the MPEG-2 video standard and provide a very short explanation of each technique.

- *Data Partitioning*
  Single-coded video bitstream is artificially partitioned into two or more layers.
- *Signal-to-Noise Ratio (SNR) Scalability*
  More than one layer; each layer is at different quality, but at the same spatial resolution.

- *Spatial Scalability*
  More than one layer; each layer has the same or different spatial resolution.
- *Temporal Scalability*
  More than one layer; each layer has the same or different temporal resolution but is at the same spatial resolution.

Although we have listed data partitioning among scalabilities, it is strictly not a scalability coding technique but a way of partitioning a coded single-layer bitstream into two parts such that error-resilient transmission or storage can be accomplished. SNR, Spatial, and Temporal scalabilities, on the other hand, are truly scalability coding techniques.

Besides the above types of scalability, the MPEG-2 standard also supports Hybrid scalability, which involves using two different type of scalabilities from among SNR, Spatial, and Temporal scalabilities.

9.1.3   Applications of Each Scalability Type

- *Data Partitioning*
  —Error-resilient video over ATM and other networks
- *SNR Scalability*
  —Digital broadcast TV (or HDTV) with two quality layers
  —Error-resilient video over ATM and other networks
  —Multiquality video-on-demand services
- *Spatial Scalability*
  —Interworking between two different video standards
  —Digital HDTV with compatible digital TV as a two-layer broadcast system
  —Video on LANs and computer networks for computer workstations
  —Error-resilient video over ATM and other networks
- *Temporal Scalability*
  —Migration to high temporal resolution progressive digital HDTV from first generation interlaced HDTV
  —Communication services requiring progressive video format and interworking with existing equipments

—Video on LANs and computer networks for computer workstations
—Error-resilient video over ATM and other networks.

9.2   DATA PARTITIONING

Data partitioning, although not a true form of scalable coding, does provide a means of partitioning coded video data into two priority classes: essential data and additional data. In data partitioning the distinction between essential data and additional data is not very strict since both types of data are simply derived by partitioning a single stream of coded MPEG-2 video data, depending on the amount of data desired for each data type.

To be more precise, in data partitioning,[2,3,19,20] coded video bitstream is partitioned into two layers, called partitions, such that partition 0 contains address and control information as well as lower order DCT coefficients while partition 1 contains high-frequency DCT coefficients. Partition 0 is referred to as the base partition (or the high-priority partition) whereas partition 1 is called the enhancement partition (or the low-priority partition). The syntax elements included in partition 0 are indicated by priority breakpoint (PBP); the remaining elements are placed in partition 1. Some syntax elements belonging to partition 0 are redundantly included in partition 1 to facilitate error recovery.

9.2.1   The High-Level Codec

A high-level codec employing data partitioning is shown in Fig. 9.1. It consists of a data partitioning encoder, corresponding decoder, and a System multiplexer and demultiplexer. Further, the data partitioning encoder consists of a motion-compensated DCT encoder followed by a data partitioner, whereas the data partitioning decoder consists of a data combiner and a motion-compensated DCT decoder, which perform functions complementary to those of the data partitioner and motion-compensated DCT encoder, respectively.

Although, for generality, we have not discussed any specific motion-compensated DCT encoder and corresponding decoder employed in the data

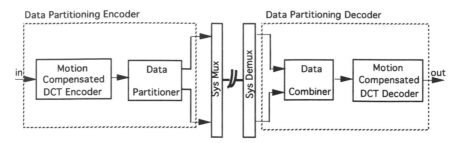

Figure 9.1   Data partitioning Codec.

partitioning codec, in the context of MPEG-2, a nonscalable MPEG-2 video encoder/decoder, with or without B-picture, is intended. Further, for clarity we have shown data partitioning to be an independent operation; however, in reality it may actually take place within the motion-compensated DCT encoder. Likewise, data combining may either be an independent operation or take place within the motion-compensated DCT decoder.

9.2.2 The Partitioner and Combiner

The functional details of the data partitioner are shown in Fig. 9.2. A single stream of MPEG-2 coded video input to the partitioner is applied to a bitstream demultiplexer along with the same bitstream variable length decoded by VLD. The outputs of the video bitstream demultiplexer are two bitstreams, the base partition bitstream, bits_b, and the enhancement partition bitstream, bits_e. Depending on the overall bitrate target for each

partition, the value of PBP can be adjusted in the priority bitrate controller at every picture slice.

The data combiner performs the inverse function of the data partitioner; thus, the data combiner, as its name suggests, combines two bitstreams into one. This operation takes place within a video bitstream multiplexer which performs actual combining of appropriate portions of the two bitstreams as indicated by the value of PBP, which is decoded for each slice of every picture. In Fig. 9.3 we show a functional block diagram of a data combiner.

The base partition bitstream bits_b and the enhancement partition bitstream bits_e are variable length decoded in their respective VLDs, which output the length of each of the variable length codes in each bitstream. The actual bitstreams and information about length of each variable length code are provided to the bitstream loss concealer, whose function is to regenerate syntactically correct bitstreams from bitstreams that have undergone losses owing to errors. A simple way to regenerate correct bitstreams from bitstreams with

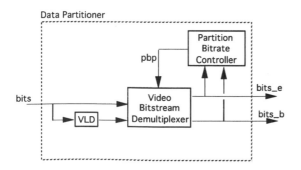

Figure 9.2   Data Partitioner used in Data Partitioning Encoder.

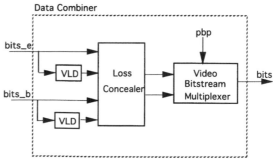

Figure 9.3   Data Combiner used in Data Partitioning Decoder.

losses owing to channel errors involves replacing errored parts of bitstreams by codes that correspond to zeros for lost DCT coefficients.

9.2.3 The Encoding Process

Strictly speaking, what we are about to describe is not encoding but the process of generating two partitions of an already encoded video bitstream. However, for generality it is referred to as the encoding process.

Data partition encoding consists of determining the value of PBP, which can be changed as often as on a slice basis and helps identify portions of the bitstream belonging to the base partition, whereas the enhancement partition, partition 1, carries all the remaining parts of the bitstream not belonging to the base partition, partition 0, as well as some redundant information to facilitate error recovery. The redundant information consists of several headers, such as sequence_header( ), group_of_pictures_header( ), picture_header( ), and sequence_end_code; the values in these redundant headers in partition 1 are exactly the same as those in the original headers in partition 0. In terms of syntax for partition 1, the only extensions expected are sequence_extension( ) and picture_extension( ), which are normal for every MPEG-2 video bitstream, and sequence_scalable_extension( ), which is normal for scalable MPEG-2 video bitstreams and among other things identifies the scalability mode (data partitioning) employed.

Earlier, we had mentioned that the PBP value identifies which portions of a bitstream belong to which partition. A detailed list of different values that PBP can have and their meaning in terms of syntax elements to be included in partition 0 is shown in Table 9.1. From Table 9.1 it should be noted that a PBP right after the DC coefficient is not allowed because of the potential to cause emulation of start codes.

In general PBP can take valid values in the range of 0 to 127, where a value of 0 is used to signal partition 1 and thus is not available for partition 0. To take a few examples, a PBP of 1 includes various syntax elements at sequence, a group of pictures, and a slice level. A PBP of 2 includes all syntax elements in PBP of 1 along with a few Macroblock (MB) syntax elements (up to macroblock_address_increment), whereas a PBP of 3 includes all syntax elements included in PBP of 2 and additional MB syntax elements just before coded_block_pattern( ). Values of PBP in the range of 4 to 63 are reserved. Further, PBP values in the range between 64 and 127 include all syntax elements included in the PBP of 3 and coded_block_pattern( ) or DC coefficient (dct_dc_differential), along with one (for PBP of 64) through 64 (for PBP of 127) run/level pairs of DCT coefficients. Put another way, depending on whether a PBP value is equal to 64 or higher, one

Table 9.1   Syntax elements in Data Partition 0

priority_break point	Syntax elements included in Partition 0
0	Value reserved for partition 1; all slices in partition 1 shall have priority_breakpoint=0
1	All data at the sequence, GOP, picture and slice( ) down to extra_bit_slice in slice( )
2	All data included above, and macroblock syntax elements up to and including macroblock_address_increment
3	All data included above, and macroblock syntax elements up to but not including coded_block_pattern( )
4 . . . 63	Reserved
64	All syntax elements up to and including coded_block_pattern( ) or DC coefficient (dct_dc_differential), and the first (run, level) DCT event (or EOB). Note that priority breakpoint immediately following DC coefficient is disallowed due to potential of start code emulation
65	All syntax elements above, and up to 2 (run, level) DCT events
. . . .	
65+j	All syntax elements above, and up to j (run, level) DCT events
. . . .	
127	All syntax elements above, and up to 64 (run, level) DCT events

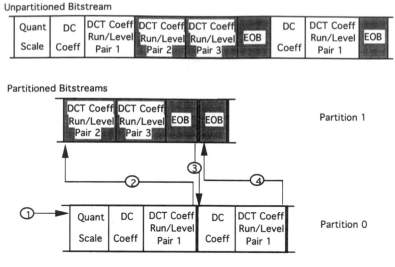

Figure 9.4   Data partitioned bitstream example with priority breakpoint of 64.

or more run/level DCT coefficient pairs is included in the base partition. Data partitioning retains the DCT coefficient run/level structure intact except that run/level pairs get partitioned between two partitions. The end-of-block (EOB) codes, indicating the last nonzero coefficient along the (zigzag or alternate) scan path in DCT-coded blocks, is preserved; it is included only in partition 1. Since partition 0 does not include EOB codes, partition 1 by itself cannot be decoded by a nonscalable MPEG-2 decoder.

We now consider a specific example assuming PBP of 64 and in Fig. 9.4 show how a single-layer bitstream is partitioned into two layers, precisely showing which syntax elements are included in which partition.

We have chosen to focus on a portion of an unpartitioned bitstream showing a quantizer scale used for a MB followed by DCT coefficient data and EOB of the first block, followed by DCT coefficient data and EOB of the next block within an MB and so forth. Since PBP is assumed to be 64 in this example, syntax elements up to and including the first DCT coefficient run/level pair are assigned to partition 0, whereas the remaining elements belong to partition 1. Thus, besides the quantizer scale, block level data such as DC coeff followed by the first run/level pair of each coded block, is included in partition 0, and the remaining

data for each block such as higher order DCT coefficient run/level pairs and EOB codes are included in partition 1.

9.2.4   The Decoding Process

The decoding process in data partitioning is the inverse of encoding and constitutes the following steps:

- Set current_partition to 0, start decoding the bitstream from partition 0 (which contains sequence_scalable_extension).
- If current_partition is 0, check if the current point in the bitstream is a PBP.
  —If yes, set current_partition to 1; decode next item from partition 1.
  —If no, continue decoding from partition 0.
- If current_partition is 1, check PBP to determine if the next item to be decoded should be from partition 1.
  —If yes, set current_partition to 0; decode next item from partition 0.
  —If no, continue decoding from partition 1.

9.2.5   Coding Issues and Constraints

Data partitioning does not standardize rate-partitioning control between layers since this is an encoding operation. Typically, many scalability applications require a base layer to carry one-half or less of the total bitrate. Typical rate partitioning control methods to achieve such a split require buffering two or three frames of data for base and enhancement layers followed by examining buffer fullness at every slice to determine if the PBP value needs to be adjusted. If possible, a value of PBP higher than 63 that does not cause buffer to overflow or underflow should be selected. As you may have already envisaged, if PBP is smaller than 64, no DCT coefficient data are included in the base layer and if channel errors occur causing losses in the enhancement layer, the base layer quality appears very poor. Thus only when it is unavoidable, that is, when the base layer buffer is quite full, should PBP be allowed to have a value lower than 64.

Although we mentioned that PBP should not be allowed to go to under 63, there is, however, one possible exception. In the case of B-pictures, it may be even desirable to allow PBP to take smaller values, even as low as 1, because typically in B-pictures, DCT coefficients arc only a fraction of the overall bitrate, with motion vectors consuming a significant portion. Since B-pictures do not feed back into the coding loop, and if bitrate partitioning constraints are severe, they could be poorly represented or even ignored.

One of the shortcomings[3] of data partitioning is the implicit restriction on the range of partitions it allows. This restriction is really implicit as it is caused by use of the same quantizer value (besides scaling produced by the quantization matrix) for both base and enhancement layers. Therefore, if data partitioning on a bitstream at a higher coded bitrate is employed, it still does not offer sufficient flexibility in rate partitioning since, with higher bitrates, the use of finer quantizer values proportionally increases bits not only for the enhancement layer but also for the base layer. Thus, the base layer for the same PBP ends up including finely quantized low-frequency DCT coefficients which place constraints on the lowest bitrate that can be achieved.

Yet another shortcoming of data partitioning is the presence of drift errors. In general, this is caused when only partial information is available to generate prediction, and thus this prediction differs from what would be generated if full information were available. Here, by partial information we mean that prediction is generated by using past pictures reconstructed using a few low-order coefficients instead of all coefficients. Drift can cause mismatch between encoder and decoder prediction loops and has a tendency to accumulate over P-pictures. B-pictures, as they are outside the predictive feedback loop, do not contribute to drift. Since I-pictures reset feedback prediction, increasing the frequency of I-pictures can in fact reduce or even minimize the effect of drift; however, I-pictures are not as efficient in the coding sense and thus require extra bits which can be regarded as overhead.

9.2.6 Results

In Table 9.2 we show an example of coding results comparison[1] of the base layer of data partitioning with a single-layer (unpartitioned) bitstream, each using 2.25 Mbits/s. Three sequences of CCIR 601 4:2:0 resolution are coded using nonscalable MPEG-2 with frame-picture coding.

The results show that in terms of SNR values, picture quality achieved for the base layer of data partitioning is indeed quite poor, and is significantly less than that achieved with nonscalable coding. This is also confirmed by visual comparison of the quality of two picture sequences.

Next, in Table 9.3 we compare results[4] of nonscalable coding at 4.5 Mbits/s to that of data partitioning with base and enhancement layer partitions combined.

The results show that for the case of no errors (packets or cell losses), with base and enhancement layers combined, there is really no significant dif-

Table 9.2 SNR comparison of Base layer in Data Partitioning with Single layer coding

Sequence	2.25 Mbits/s	
	Nonscalable	Data Partitioning Base 2.25 Mbit/s
Mobile	26.07	22.20 (−3.87)
Cheerleaders	26.41	22.54 (−3.87)
Bus	27.65	23.28 (−4.37)

Table 9.3   SNR comparison of Base+Enh layer coded by Data partitioning with Single layer coding

	4.5 Mbits/s	
Sequence	Nonscalable	Data Partitioning Enh 2.25 Mbit/s; Base 2.25 Mbit/s
Mobile	30.54	30.49 (−0.05)
Cheerleaders	30.10	30.03 (−0.07)
Bus	32.22	32.15 (−0.07)

Table 9.4   SNR comparison of Base layer in Data Partitioning with Single layer coding

	4.5 Mbits/s	
Sequence	Nonscalable	Data Partitioning Base 4.5 Mbits/s
Mobile	30.54	23.22 (−7.32)
Cheerleaders	30.10	24.73 (−5.37)
Bus	32.22	24.75 (−7.47)

ference between performance of data partitioning and nonscalable coding. In fact, these results just confirm what we said earlier, that data partitioning incurs practically close to zero overhead.

At this point an interesting question to ask is whether the performance of each of layers in Data partitioning will improve, remain the same, or worsen at higher bitrates. In Table 9.4 we show an SNR comparison[4] of the base layer of data partitioning with a single-layer (unpartitioned) bitstream, each using 4.5 Mbits/s with the same three sequences as before.

In comparing the results of Tables 9.2 and 9.4 we conclude that by doubling the bitrate, the SNR performance of nonscalable coding has improved by about 4 to 4.5 dB, whereas the SNR performance of the base layer of data partitioning has improved by only about 1 to 1.5 dB. At 4.5 Mbits/s, the performance of the base layer of data partitioning appears to be rather poor as compared to nonscalable coding.

With a base layer bitrate of 4.5 Mbits/s, for data partitioning, the performance of nonscalable coding is compared with combined base and enhancement layers at 9 Mbits/s.[4] This is shown in Table 9.5.

As in the case of Table 9.3, the SNR results of Table 9.5 again show that there is really no signifi-

Table 9.5   SNR comparison of Base+Enh layer coded by Data partitioning with Single layer coding

	9 Mbits/s	
Sequence	Nonscalable	Data Paritioning Enh 4.5 Mbits/s; Base 4.5 Mbits/s
Mobile	34.27	34.24 (−0.03)
Cheerleaders	34.27	34.24 (−0.03)
Bus	36.15	32.12 (−0.03)

cant difference between the performance of Data partitioning and nonscalable coding.

9.2.7   Summary of Data Partitioning

To draw some conclusions before we proceed to the next type of scalability, data partitioning is a simple and very low penalty (close to zero overhead) technique for generating layers of data. Such layering is primarily useful for applications requiring error resilience at little or no overhead cost. Data partitioning can place constraints on achieving desired bandwidth partitioning between two layers. In addition, it suffers from potential accumulation of drift (which can be reset at I-pictures). If total bandwidth is equally partitioned between the two layers, the picture quality of the base layer is significantly lower (by as much as 4 dB or more) as compared to that achieved with nonscalable coding, and may not be fit for viewing by itself. If the nature of errors is such that much more than half of the data can be assigned high priority for added protection and means for drift reduction are incorporated, tolerable picture quality may be obtained. However, this either reduces the coding efficiency or increases the complexity, making data partitioning less competitive. Because data partitioning incurs low overhead, the quality obtained by combining base and enhancement layers in the case of no errors is nearly the same as that obtained by nonscalable coding.

9.3   SNR SCALABILITY

SNR scalability[2,3,19,20] involves coding in a manner to generate separate bitstreams representing indi-

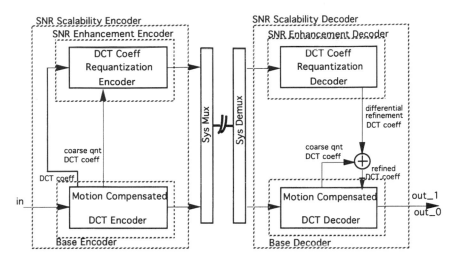

Figure 9.5 SNR scalability codec with drift in enhancement layer.

vidual layers such that, for example, an enhancement layer carries DCT refinement coefficients which are intended to enhance DCT coefficients included in the lower layer. Obviously, using coefficients from both layers can result in pictures with better quality than can be obtained by using the lower layer only. Since the enhancement layer is said to enhance the signal-to-noise ratio (or SNR) of the lower layer, this type of scalability is called SNR scalability. Alternatively, SNR scalability could have been called Coefficient Amplitude scalability or Quantization Noise scalability; these terms, although a bit wordy, are more precise.

The layers in SNR scalability are at the same spatial and temporal resolutions but produce different qualities. Among the important applications of SNR scalability are two layer TV[4,6] and two layer HDTV.[8,21]

9.3.1 The High-Level Codec

Figure 9.5 shows a high-level block diagram of an SNR scalability codec. It consists of an SNR Scalability Encoder, Sys Mux, Sys Demux, and SNR Scalability Decoder. The systems multiplexer Sys Mux and systems demultiplexer Sys Demux perform the same function as in the case of data partitioning, that is, Sys Mux packetizes video bitstreams representing various layers to generate a single stream of packets for transmission whereas Sys Demux separates packets

belonging to different video layers, unpacks, and provides appropriate bitstreams to decoders of corresponding layers.

Functionally, a two-layer SNR Scalability Encoder consists of a Base Encoder and an SNR Enhancement Encoder. Likewise, an SNR Scalability Decoder consists of corresponding complementary decoders, a Base Decoder, and an SNR Enhancement Decoder.

Considering two-layer SNR scalability, the Base Encoder/Decoder is a Motion-Compensated DCT Encoder/Decoder and the SNR Enhancement Encoder/Decoder is a DCT coefficient Requantization Encoder/Decoder. To be more precise, within the context of MPEG-2, the Base Encoder is a Nonscalable MPEG-2 Encoder with or without B-pictures and the Base Decoder is a Nonscalable MPEG-2 Decoder. The unquantized DCT coefficients as well as coarsely quantized DCT coefficients are input to the DCT Coeff Requantization Encoder which quantizes them finely and produces an enhancement layer bitstream that along with the base layer bitstream is multiplexed by Sys Mux. Sys Demux separates the two bitstreams and inputs the one corresponding to the enhancement layer to the DCT Coeff Requantization Decoder which decodes differential refinement coefficients, to which are added coarse quantized DCT coefficients decoded in the Base Decoder. The resulting refined coefficients are fed back to the Base Decoder.

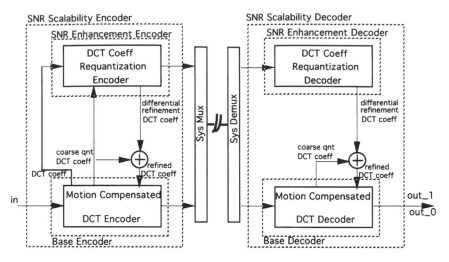

Figure 9.6   SNR Scalability codec without drift in enhancement layer.

In SNR scalability there is very tight coupling between the two layers; both base and enhancement layer outputs can appear at the same output, although not simultaneously. If the enhancement layer is temporarily unavailable as a result of channel errors or has incurred packet losses, the base layer is decoded and displayed by itself; that is, differential refinement coefficients of the enhancement layer are zeroed out.

Although it may not be entirely obvious from the discussion so far, the simple high-level codec structure just described suffers from drift problems since differential refinement coefficients do not feed back into the lower layer motion-compensated prediction loop at the encoder whereas they do so at the decoder. However, if there are channel errors or packet losses and only the base layer is to be decoded and displayed, no drift is expected.

9.3.2   High-Level Codec Without Drift

In Figure 9.6 we show an improved codec structure that prevents drift in the enhancement layer by feeding back the differential refinement DCT coefficients to the lower-layer via an Adder, at the other input of which are coarse quantized DCT coefficients.

It should become obvious soon that as long as there are no channel errors or packet losses, no drift is expected in the enhancement layer. Howev-

er, if there are channel errors or packet losses, drift will appear as the motion compensation loops at the encoder and decoder again become unsynchronized owing to the decoder using only lower layer DCT coefficients in its prediction loop.

9.3.3   The Encoder

Because of tight coupling between layers in SNR scalability, it is a bit difficult to clearly separate the two layers in the high-level codecs discussed thus far. Even worse, from a high-level codec structure, it is difficult to visualize where DCT coefficients are originating from and where they are being fed back to. However, the high-level codec did serve an important purpose in introducing the important fact that coarse DCT coefficients are coded in the lower layer and refinement coefficients are coded in the enhancement layer and that there is a possibility of drift between encoder and decoder, and it included a solution to prevent drift.

We are now ready to discuss the internals of the Base Encoder and SNR Enhancement Encoder to clearly show the paths of interchange of various DCT coefficient data. We have chosen to show an encoder configuration that along with corresponding decoders (to be discussed) operates in a drift-free manner as per the high-level codec of Fig. 9.6. The aforementioned encoders are shown in Fig. 9.7.

The base layer in Fig. 9.7 essentially uses a normal Motion-Compensated DCT Encoder with

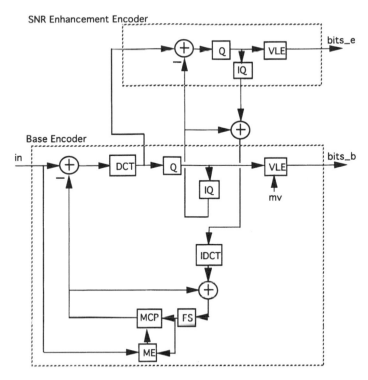

Figure 9.7 Details of SNR scalability encoder without drift in enhancement layer.

minor modifications. We briefly review the operation of such an encoder. In motion-compensated DCT encoding, video frames (or fields) are input at *in* and are partitioned into nonoverlapping macroblocks (16 × 16 luminance block and two corresponding 8 × 8 chrominance blocks) that are predicted using previous decoded frame(s); the resulting difference signal is DCT transformed using 8 × 8 blocks and the transform coefficients are quantized in a quantizer, Q. The quantization indices, motion vectors, and other data are variable length encoded in a VLE to generate a coded bitstream. The transform quantization indices are also fed back to a local decoding loop at the encoder consisting of an Inverse Quantizer (IQ), Inverse DCT (IDCT), an Adder, Frame Store(s) (FS), Motion Estimator (ME), and Motion-Compensated Predictor (MCP). Of course, in nonscalable MPEG-2 coding, there are other elements as well that permit efficient coding of interlace; they are, however, not shown for simplicity.

Now we discuss the SNR Enhancement Encoder and how it interacts with the Base Encoder. The DCT coefficients produced by the Base Encoder are input to the SNR Enhancement Encoder along with dequantized DCT coefficients from the local decoding loop of the Base Encoder. In the enhancement layer, these two sets of coefficients are differenced to generate refinement DCT coefficients which are requantized by a Quantizer, Q, which generates quantization indices encoded by a Variable Length Encoder (VLE). These quantization indices are also input to the Inverse Quantizer (IQ), which outputs a dequantized DCT coefficient refinement signal that is added back to the coarse dequantized DCT coefficient's produced by the Base Encoder. The resulting sum represents the high-quality DCT coefficients and is fed back to the base encoder and undergoes Inverse DCT (IDCT), and to which is added a motion-compensated prediction signal at the output of MCP. The output of the Adder is the high-quality decoded signal.

9.3.4 The Decoder

Having examined the details of the SNR Scalability Encoder you can probably guess what the

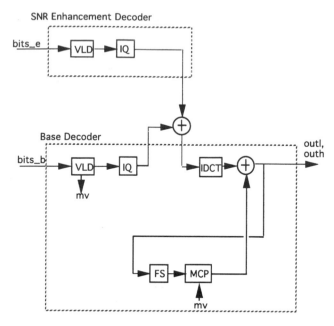

Figure 9.8   Details of a SNR scalability decoder without drift in enhancement layer.

SNR Scalability Decoder looks like. In any case, we are now ready to present a simplified yet meaningful block diagram of the SNR Scalability decoder, shown in Fig. 9.8.

The operation of the base layer decoder of the SNR Scalability Decoder shown in Fig. 9.8 is very similar to that of the normal Motion-Compensated DCT Decoder with a few exceptions. The operation of the Motion-Compensated DCT Decoder is similar to that of the local decoding loop in the Motion-Compensated DCT Encoder. The base layer bitstream, bits_b, and enhancement layer bitstream, bits_e, are provided by the Sys Demux (not shown) to the Base Decoder and SNR Enhancement Decoder, respectively. As mentioned earlier, the operation of the Base Decoder is similar to that of a Motion-Compensated DCT Decoder. The bits_b is first presented to a Variable Length Decoder (VLD), and then forwarded to an Inverse Quantizer (IQ), Inverse DCT (IDCT), and then to an Adder, the other input of which is a prediction signal generated by a Motion-Compensated Predictor (MCP), using frame(s) stored in a frame store (FS).

In an enhancement layer bitstream bits_e is input to a Variable Length Decoder (VLD), which decodes quantization indices of DCT refinement coefficients and forwards them to an Inverse Quan-

tizer (IQ), resulting in dequantizing of DCT refinement coefficients. These refinement coefficients are fed to an Adder; at the other input of the adder are dequantized coarse coefficients from the base layer. The output of this Adder, consisting of DCT refined coefficients, is fed back to the base layer, in particular at the input of the IDCT. Normally, the decoded enhancement layer pictures are output by the Adder in the base layer. In the case of channel errors, only the base layer is displayed and the enhancement layer is temporarily discarded.

The encoder and decoder presented prevent drift, as the same refined DCT coefficients used at the decoder are used in the feedback loop of the base layer encoder. However, in the case of errors, when the enhancement layer is temporarily unavailable at the decoder and only the base layer must be shown, there is a drift between encoder and decoder, as the encoder uses refined DCT coefficients not available at the decoder. As mentioned earlier, if refined DCT coefficients are not fed back into the base layer to generate a high-quality signal to be used for prediction, and since the normal output at the SNR Scalability Decoder is a high-quality signal, there is a consistent drift between the encoder and decoder prediction loops which is not desirable.

9.3.5 Coding Issues and Constraints

In SNR scalability encoding, the base layer bitstream carries, besides quantization indices for coarse quantized DCT coefficients, all motion compensation and MB type information. The enhancement layer bitstream carries quantization indices representing quantized DCT refinement coefficients and a small amount of overhead. In coding of the enhancement layer, not all MBs coded in the base layer may be coded. The standard allows for use of different quantization matrices in the enhancement layer, if needed. Further, the slice structure is required to be the same for both layers.

Although separate quantization of each of the two layers allows for flexibility in bitrates for each layer, because the base layer carries all motion vectors and other overhead there is an implicit constraint on the lowest bitrate that can be used for the base layer. The actual lowest bitrate that can be achieved depends on the resolution employed and the amount of motion activity in a scene. Other than the constraint mentioned, there is no real constraint on the quantization matrix employed in each layer and on the mechanism for quantization parameter selection to achieve desired bitrate partitioning between layers. Undoubtedly, SNR scalability offers much more flexibility[3] as compared to data partitioning.

SNR scalability, since it does not use completely independent coding loops for base and enhancement layers, has the potential for drift. When using an encoder that does not feed back DCT refinement coefficients into the prediction loop, drift can accumulate, affecting higher layer picture quality, which is the normal decoder output. On the other hand, if an encoder that feeds back DCT refinement coefficients is employed, and owing to channel errors the enhancement layer is lost, only the base layer output with drift effects will be available. The drift does get reset with I-pictures or Intra MBs and is less severe than that with data partitioning.

9.3.6 Results

In Table 9.6 we show example of coding results[4] of the base layer in SNR scalability to that of single-layer (nonscalable) coding, each using

Table 9.6 SNR comparison of Base layer in SNR scalability with Single layer coding

| | 2.25 Mbits/s | |
| | | |
Sequence	Nonscalable	SNR Scalability Base 2.25 Mbits/s
Mobile	26.07	25.50 (−0.57)
Cheerleaders	26.41	25.43 (−0.98)
Bus	27.65	26.92 (−0.73)

2.25 Mbits/s. The same three sequences as used for data partitioning are shown.

The results show that in terms of SNR values, with the bitrate used for the base layer the picture quality achieved in the base layer using SNR scalability is only slightly lower, by about 0.5 to 1.0 dB, than that using nonscalable coding. This is also confirmed by visual comparison of subjective quality.

Next, in Table 9.7, we compare results of nonscalable coding at 4.5 Mbits/s to that of SNR scalability with base and enhancement layers combined.

The results show that with base and enhancement layers combined, with the chosen bits partitioning, the performance of SNR salability is lower, by 0.5 to 1.0 dB, as compared to nonscalable coding. This means that SNR scalability incurs more overhead than data partitioning.

At this point an interesting question is whether the performance of each of the layers in SNR will improve, remain the same, or worsen at higher bitrates. In Table 9.8, we show an results[4] of comparison of the base layer of SNR scalability with a single-layer bitstream, each using 4.5 Mbits with the same three sequences as before.

The results in Table 9.8 show, with the bitrates selected for the base layer, the SNR performance

Table 9.7 SNR comparison of Base+Enh layer coded by SNR scalability with Single layer coding

| | 4.5 Mbits/s | |
| | | |
Sequence	Nonscalable	SNR Scalability Enh 2.25 Mbits/s; Base 2.25 Mbits/s
Mobile	30.54	29.59 (−0.95)
Cheerleaders	30.10	29.62 (−0.48)
Bus	32.22	31.39 (−0.83)

Table 9.8   SNR comparison of Base layer in SNR scalability with Single layer coding

	4.5 Mbits/s	
Sequence	Nonscalable	SNR Scalability Base 4.5 Mbits/s
Mobile	30.54	29.38 (−1.16)
Cheerleaders	30.10	29.09 (−1.01)
Bus	32.22	30.92 (−1.30)

of SNR scalability is 1 to 1.5 dB lower than that for nonscalable coding. Also, this is about 0.5 dB worse than that in Table 9.6, indicating that with increasing bitrates, performance of the base layer in SNR scalability decreases further.

With a base layer bitrate of 4.5 Mbits/s, the coding results of the combined base and enhancement layer at 9 Mbits can now be compared[4] with that obtained for single-layer coding. This is shown in Table 9.9.

The SNR results of Table 9.9 indicate, for the bitrate partitioning employed, the performance of the combined base and enhancement layer for SNR scalability is about 0.5 dB to 1.2 dB lower than for nonscalable coding. These results are very similar to the SNR scalability results in Table 9.7, and clearly lower than those of data partitioning with both layers combined.

9.3.7   Other Features

Besides error resilience and multiple quality video services, variations of SNR scalability can be used to provide other functionalities as well. We briefly discuss two such variations and the func-

Table 9.9   SNR comparison of Base+Enh layer coded by SNR scalability with Single layer coding

	9 Mbits/s	
Sequence	Nonscalable	SNR Scalability Enh 4.5 Mbits/s; Base 4.5 Mbits/s
Mobile	34.27	33.16 (−1.11)
Cheerleaders	34.27	33.74 (−0.53)
Bus	36.15	35.48 (−0.72)

tionalities they provide along with possible applications requiring these functionalities.

9.3.7.1   Chroma Simulcast

As mentioned elsewhere in this book, the MPEG-2 standard allows 4:2:0, 4:2:2, and 4:4:4 video formats. As a brief reminder, the 4:2:2 format uses twice the vertical resolution for each chrominance component as compared to the 4:2:0. The 4:4:4 format uses, both horizontally and vertically, twice the chroma resolution as compared to the 4:2:0 format. Whereas the 4:2:0 format is employed for distribution to the consumer, the 4:2:2 format is necessary for contribution and studio applications. The Chroma Simulcast scheme provides a means for simultaneous distribution of services that use 4:2:0 and 4:2:2 formats, with 4:2:0 as the lower layer which, in this case, undergoes SNR enhancement by simulcasting chrominance components of the 4:2:2 format as well as some enhancement for luminance of the 4:2:0 format. The high-quality signal is thus generated by using both lower and enhancement layer luminance, and 4:2:2 chrominance components (which are simulcast).

9.3.7.2   Frequency Scalability

SNR scalability can also be used as a frequency domain method to achieve scalability in spatial resolution. In this variation of SNR scalability called frequency scalability[9,11,18] lower layer video is intended for display at reduced spatial resolution. This may be useful if the display is incapable of showing full resolution or in software decoding where owing to lack of sufficient processing resources, only reduced spatial resolution can be decoded for display.

To extract reduced spatial resolution directly from an SNR scalability bitstream, SNR scalability syntax without any changes can be used; the only modifications required are those in decoding semantics; however, these do not need to be standardized. The lower layer decoding can be performed separately and its decoding loop can employ an inverse DCT of reduced size along with reduced size frame memory. If only a single motion compensation loop is used at the encoder as is normal for SNR scalability, the lower layer video would incur drift. This drift is caused by reduction in accuracy of motion compensation at the lower layer decoder owing to scaling of motion vectors computed at the encoder for motion compensation at full resolution. This drift can be reduced to some extent by a variety of means such as more frequent I-pictures and better interpolation filters.

SNR scalability encoding can be customized to support decoding using a smaller size DCT by using a quantization matrix that allows a corresponding subset of coefficients of the normal size DCT to be selected in the lower layer at the encoder. The enhancement layer can then carry the remaining coefficients. It is also possible to support drift-free decoding at lower spatial resolution by using independent motion compensation loops for each layer at the encoder. However, if independent motion compensation loops are required then other forms of scalability (such as Spatial Scalability, to be discussed) become even more attractive owing to increased flexibilities at comparable implementation complexity.

9.3.8 Summary SNR of Scalability

To draw some conclusions before moving on to the next type of scalability, SNR scalability is more complex than data partitioning and incurs some additional penalty (higher overhead) for generating layers of data. It produces video data layers that can be useful in applications such as multiquality video services in broadcast applications and in communication applications requiring error resilience that can tolerate some extra cost. SNR scalability, like data partitioning, also places constraints on achieving desired bandwidth partitioning between layers; this depends on picture sizes and amount of motion in scenes, primarily owing to the fact that all motion vectors have to be carried by the base layer. Further, like data partitioning, it also suffers from drift problems, although if the base encoder uses the locally decoded enhancement layer data in the feedback loop, drift may not take place in the decoded enhancement layer but in the base layer. Using a simpler configuration without feedback of locally decoded enhancement layer data in the base encoder, drift results in the

enhancement layer at the decoder also. The effects of drift can also be reduced by other methods. In SNR values and from the visual quality aspect, the picture quality produced by the base layer can be significantly higher (by more than 3 dB) as compared to that obtained with data partitioning, although it is still lower (by about 0.75 dB) than that of nonscalable coding. However, if drift can be kept low, satisfactory picture quality can be obtained for the base layer. The combined picture quality of base and enhancement layers is less than that obtained with nonscalable coding and data partitioning (by about 0.5 to 1.1 dB).

9.4 SPATIAL SCALABILITY

Spatial scalability refers to layered coding in a spatial domain which means that in this type of coding scheme, a lower layer uses spatial resolution equal to or less than the spatial resolution of the next higher layer. The layers are loosely coupled, allowing considerable freedom in resolutions and video formats to be used for each layer. A higher layer uses interlayer spatial prediction with respect to decoded pictures of its lower layer.

Spatial scalability in MPEG-2 owing to the flexibilities[1,2,5,19,20] it offers in choice of resolution and formats for each layer, allows interoperability between different video formats. Further, the lower layer may also be coded using a different standard, providing interoperability between standards. For example, spatial scalability offers backward compatibility with other standards such as H.261 and MPEG-1.

Spatial scalability can be considered related to and built on the framework of earlier layered video coding approaches commonly referred to as pyramid coding. However, spatial scalability clearly extends the pyramid coding in a significant way, resulting in higher coding efficiency and a higher degree of flexibility. By higher flexibility we mean the ability to deal with different picture formats in each layer. Thus, calling spatial scalability pyramid coding is like basically ignoring its subtleties and intricacies which were primarily responsible for getting this scheme accepted in the MPEG-2 video standard in the first place.

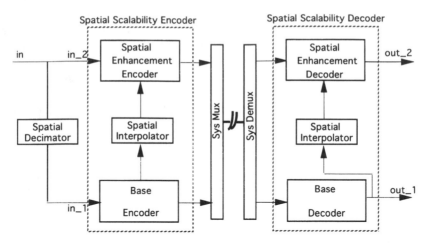

Figure 9.9   Spatial scalability codec.

Spatial scalability supports interoperability between applications using different video formats, interworking with other standards, interworking between broadcast and computer applications, interworking between various telecommunication applications, a two-layer system consisting of HDTV and normal TV,[8,13,21] database browsing, error resilince on ATM and other networks, and more.

9.4.1   The High-Level Codec

We now describe a high-level codec structure[1,2] for spatial scalability. This codec is only a functional representation of high-level operations. Although it shows separate encoders and decoders for each layer, an encoder/decoder for a high layer may not explicitly include a encoder/decoder for a lower layer, meaning that some of the operations may be implicit or combined within a single encoder/decoder. Further, we consider the case of two-layer spatial scalability, only in such a case, the lower layer is usually referred to as the base layer and a higher layer is referred to as the enhancement layer. Such a two-layer codec structure is shown in Fig. 9.9.

Video input at *in* is spatially downsampled by a Spatial Decimator and applied at *in_1* to a Base Encoder, which encodes it, producing an encoded bitstream that is fed to a systems multiplexer, Sys Mux. A locally decoded base layer video signal is also input to a spatial upsampler called Spatial Interpolator, which provides a spatially interpolated signal to a Spatial Enhancement Encoder, where it is combined with a temporal prediction signal generated within a Spatial Enhancement Encoder when coding signal at *in* (also called *in_2*). The coded bitstream produced by the Spatial Enhancement Encoder is also fed to Sys Mux, where both the base layer bitstream and enhancement layer bitstreams are packetized and a stream of packets is readied for transport to a network or a storage medium. At the decoder, Sys Demux separates packets of base and enhancement layers and regenerates base layer and enhancement layer bitstreams by unpacketizing and inputs them to corresponding decoders. The Base Decoder decodes the base layer bitstream, resulting in decoded video at *out_1;* this is also input to the Spatial Interpolator which feeds the spatially interpolated signal to be used for prediction by the Spatial Enhancement Decoder depending on the actual coding mode when decoding the spatial enhancement bitstream. The decoded enhancement layer video is output by the Spatial Enhancement Decoder at *out_2.*

9.4.2   Spatiotemporal Prediction

Earlier, we had mentioned that spatial scalability represents a significant advancement over traditional pyramid coding as it allows increased efficiency and higher flexibility. We are now ready to

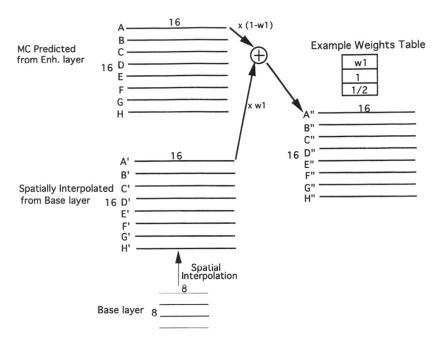

Figure 9.10   Principle of spatiotemporal weighted prediction in Spatial scalability.

discuss a major difference with respect to pyramid coding; it is called spatiotemporal weighted prediction.[1,17] In spatiotemporal weighted prediction, on a macroblock basis, a spatially interpolated decoded lower layer signal is adaptively combined with motion-compensated prediction using a set of predefined weights from a selected table. For each macroblock, corresponding to each weight entry in the table, a spatiotemporal weighted prediction is generated and used for computing prediction error; the weight entry corresponding to the smallest prediction error is selected and identified in the bitstream for the decoder to use. The process of generating spatiotemporal weighted prediction is explained in Fig. 9.10.

In the example shown in Fig. 9.10 to illustrate the principle of spatiotemporal weighted prediction, for every 16 × 16 block in the enhancement layer, the corresponding decoded 8 × 8 block of the base layer is upsampled spatially to 16 × 16 and combined with a 16 × 16 temporal prediction block (obtained by motion compensation) using a choice of weights from a predetermined table to generate 16 × 16 candidate blocks of weighted spatiotemporal prediction. On a macroblock basis,

the weighting factor producing the smallest prediction error is selected and identified in the encoded bitstream.

9.4.3 The Detailed High-Level Codec

Earlier in Fig. 9.9 we had shown a high-level codec structure for spatial scalability. We are now ready to revisit this high-level codec structure by clarifying the functions performed in each of the encoders and decoders and suggesting examples wherever appropriate.

First, we want to specify the base encoder/decoder pair; this is expected to be a Motion-Compensated Encoder/Decoder such as H.261, MPEG-1, and nonscalable MPEG-2. Next, we discuss the enhancement encoder/decode which is obviously an MPEG-2 Spatial Enhancement Encoder/Decoder consisting of a Motion-Compensated Encoder/Decoder modified to use weighted spatiotemporal prediction. An improved high-level codec structure is shown in Fig. 9.11.

In Fig. 9.11, the operation of Spatial Decimator, Spatial Interpolator, Sys Mux, and Sys Demux is

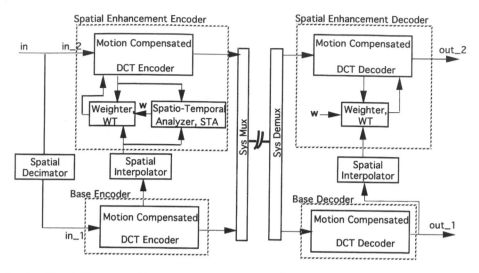

Figure 9.11   Spatial scalability codec with a few details.

exactly the same as explained for Fig. 9.9. The Spatial Decimator reduces spatial resolution of the input and offers this reduced spatial resolution signal to the Base Encoder, a Motion-Compensated Encoder, which produces a coded bitstream and sends it to Sys Mux for packetizing and multiplexing; this encoder also produces a locally decoded signal which is upsampled in the Spatial Interpolator and forwarded to a Spatial Enhancement Encoder. The Spatial Enhancement Encoder functionally consists of a Motion-Compensated DCT Encoder modified to accept weighted spatiotemporal prediction. On a macroblock basis, a SpatioTemporal Analyzer (STA) is used to determine the best prediction weights that result in the minimum prediction error. Using a spatially interpolated lower layer signal, a temporal prediction signal from the enhancement layer, and given weights, spatiotemporal weighting is performed by a Weighter (WT). At the receiver, Sys Demux separates and inputs base and enhancement bitstreams to corresponding decoders. Similar to the Spatial Interpolator at the encoder, the Spatial Interpolator at the decoder upsamples the base layer signal and presents it to WT, in the Spatial Enhancement Decoder. Using recently decoded pictures and motion vectors included in the bitstream, temporal prediction is also generated and input to WT. The WT also needs weights that are extracted from the coded bitstream. This completes the generation of spatiotemporal prediction which is used by the modified Motion-Compensated DCT Decoder to decode the enhancement layer.

9.4.4   The Encoder

Although the codec structure of Fig. 9.11 provides a significant amount of clarification over the codec structure of Fig. 9.9, it still leaves several important details hidden. We now describe the internals of a simplified Spatial Enhancement Encoder and show its interaction with other elements of a Spatial Scalability Encoder, thus offering a limited opportunity to uncover a few hidden details. We start by first presenting a simplified yet meaningful block diagram of a Spatial Scalability Encoder in Fig. 9.12 and then present an explanation of its operation.

In Fig. 9.12, the Spatial Enhancement Encoder shown is a Motion-Compensated DCT Encoder modified to compute weighted spatiotemporal prediction candidates and select the best candidate on a macroblock basis. As a reminder, in normal motion-compensated DCT encoding, video frames (or fields) are input at *in* and are partitioned into nonoverlapping macroblocks (16 × 16 luminance

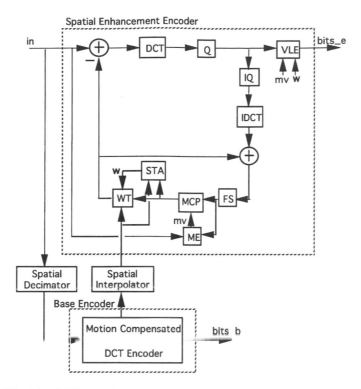

Figure 9.12 Details of Spatial scalability encoder.

block and two corresponding 8 × 8 chrominance blocks) which are predicted using previous decoded frame(s), the resulting difference signal is DCT transformed using 8 × 8 blocks, and the transform coefficients are quantized in a quantizer (Q). The quantization indices, motion vectors, and other data are variable length encoded in a VLE to generate a coded bitstream. The transform quantization indices are also fed back to a local decoding loop at the encoder consisting of an Inverse Quantizer (IQ), Inverse DCT (IDCT), and Adder, Frame Store(s) (FS), Motion Estimator (ME), and Motion-Compensated Predictor (MCP). Of course, in nonscalable MPEG-2 coding, there are other elements as well that permit efficient coding of interlace, but for simplicity they are not shown. Besides the blocks found in normal motion-compensated DCT coding, the encoder also shows new blocks such as STA and WT, as expected. The VLE encoder shown not only includes in the bitstream motion vectors, but also spatiotemporal weights.

As explained earlier, the lower layer signal upsampled by the Spatial Interpolator is provided both to WT and to STA. Also, temporal prediction from the MCP is fed both to STA and WT. The STA selects the weight that minimizes the prediction error and WT generates a weighted spatiotemporal prediction, which is differenced with a current frame on a block basis and the prediction error coded by DCT coding. The resulting enhancement layer bitstream is bits_e and is sent to Sys Mux (not shown here) which also receives the base layer bitstream bits_b.

9.4.5 The Decoder

Having examined the details of the Spatial Enhancement Encoder and its interaction with the remaining elements of the Spatial Scalability Encoder, you can probably guess what a Spatial Enhancement Decoder looks like and how it interacts with other elements of a Spatial Scalability

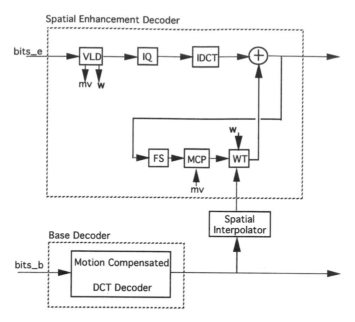

Figure 9.13 Details of Spatial scalability decoder.

Decoder. In any case, we present the internals of a simplified Spatial Enhancement Decoder and explain its interaction. A simplified yet meaningful block diagram of a Spatial Scalability Decoder is shown in Fig. 9.13.

The operation of a Spatial Enhancement Decoder shown in Fig. 9.13 is very similar to that of a normal Motion-Compensated DCT Decoder with a few exceptions. Also, the operation of this decoder is nearly identical to the operation of a local decoding loop in the Spatial Enhancement Encoder.

The enhancement layer bitstream, bits_e, from Sys Demux (not shown) is input to a Spatial Enhancement Decoder, where it is first presented to a Variable Length Decoder (VLD), and then forwarded to an Inverse Quantizer (IQ), Inverse DCT (DCT), and then to an Adder at the other input of which is a prediction signal. This prediction signal is generated by spatiotemporal weighted prediction, following a procedure similar to that at the encoder. The only difference is that the weighting to be used by WT is in fact decoded from the bitstream by VLD along with other data. The WT uses upsampled spatial prediction output by the Spatial Interpolator, temporal prediction from

MCP, and weights to generate weighted spatiotemporal prediction. The decoded enhancement layer pictures are output by the Adder.

9.4.6 Spatial Scalability Types and Their Performance

There are four basic types of spatial scalabilities:[1,19]

- Progressive–Progressive Spatial Scalability
- Progressive–Interlace Spatial Scalability
- Interlace–Interlace Spatial Scalability
- Interlace–Progressive Spatial Scalability

We now present details and coding performance results for each type of spatial scalability.

9.4.6.1 Progressive–Progressive Spatial Scalability

We now discuss the first type of spatial scalability, called progressive–progressive spatial scalability. In this type of spatial scalability both the lower layer and the enhancement layer use a progressive video format. The lower layer may use lower or the same spatial resolution as compared to the

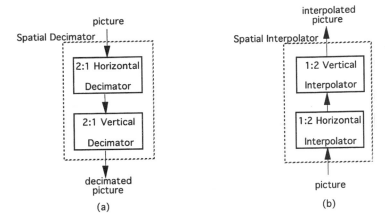

Figure 9.14   Spatial Decimator and interpolator examples for Progressive base layer.

enhancement layer. One potential application of this type of scalability is in a two-layer system, allowing interoperability between different standards, for example, using H.261 or MPEG-1 in the base layer and MPEG-2 spatial enhancement in the enhancement layer. Other potential applications are in scalable multimedia allowing windowing on computer workstations and multiquality service in telecommunications applications.

In spatial scalability, the spatial decimation operation, since it occurs only at the encoder, does not need to be standardized; however, the spatial interpolation operation which takes place both at the encoder and decoder does need to be standardized and is specified by the MPEG-2 video standard. As an example, assume that the base layer uses a picture resolution of one-quarter the size of input pictures (or enhancement layer pictures). Then the Spatial Decimator downsamples both the horizontal and vertical resolutions by a factor of 2 in each direction. The filters used prior to downsampling are not specified since spatial decimation is not standardized. The Spatial Interpolator performs the inverse function by upsampling the decoded base layer by a factor of 2, both horizontally and vertically. Corresponding block diagrams of Spatial Decimator and Spatial Interpolator are shown in Fig. 9.14a and b.

The next issue for discussion is the generation of spatiotemporal weighted prediction. Continuing with the example that we have been discussing, that is, the base layer has one-quarter res-

olution as compared to the enhancement layer, every 8×8 block of the base layer can be spatially interpolated to a 16×16 block and used for prediction of the corresponding 16×16 block of the enhancement layer. As discussed earlier, the spatiotemporal weighted prediction is generated by a weighted combination of a spatially interpolated decoded base layer frame and a previously decoded enhancement layer frame. Figure 9.15 shows the process of generating spatiotemporal weighted prediction as well as a sample table of weights that performs reasonably well. This table is one of several tables supported by the standard that allows table selection as frequently as once per picture.

As shown in Fig. 9.15, the weight table has only one component, $w1$ (as we shall see later, in other cases, there may be two components, $w1$ and $w2$). Lines of spatial interpolation blocks are weighted by $w1$, whereas lines of temporal prediction blocks are weighted by $(1 - w1)$. Using each of the entries in the weights table, spatiotemporal prediction error candidates are generated by differencing from the block to be coded the spatiotemporal weighted block. For each macroblock, the weight entry in the table for which prediction error is minimum is selected and an index identifying it is included in the encoded bitstream for the decoder to use.

We are now ready to discuss the performance aspects of progressive–progressive spatial scalability. Earlier, we had mentioned that interoperability

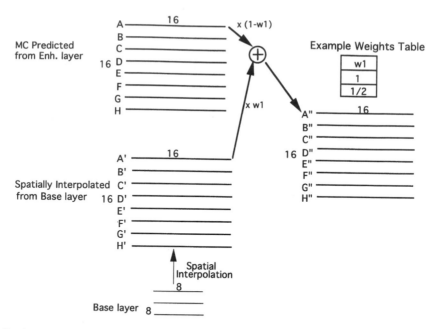

Figure 9.15   Spatiotemporal weighted prediction for Progressive–Interlace Spatial scalability.

of standards is a potential main application of this type of spatial scalability.

Thus the performance results that we will present will be based on simulations using two-layer progressive–progressive spatial scalability with MPEG-1 coding in the base layer and MPEG-2 spatial enhancement for enhancement layer coding.

In Table 9.10 we show coding results for a progressive base layer of SIF resolution coded at 1.15 Mbits/s using MPEG-1. Since, in Spatial scalability, the base layer is coded independently (loose coupling between layers), the results shown in Table 9.6 are the same as for nonscalable coding.

Next, in Table 9.11, we show a comparison of results for an enhancement layer coded simulcast, an enhancement layer coded with progressive–progressive spatial scalability, and single-layer cod-

ing. For the case of simulcast and spatial scalability, the base layer is assumed coded by 1.15 Mbits/s.

From Table 9.11, for the bitrate and bits partitioning selected, we observe that spatial scalability performs about 0.65 dB better than simulcast, while single-layer coding performs about 0.95 dB better, implying that for the enhancement layer,

Table 9.10   SNR for Progressive Base layer MPEG-1 coded at 1.15 Mbits/s used in Progressive-Progressive Spatial scalability

	Prog; 1.15 Mbits/s
	Spatial Scalability
Sequence	Base 1.15 Mbits/s
Raft	24.57

Table 9.11   SNR comparison of Progressive-Progressive Spatial scalability (using 1.15 Mbits/s for Base layer and 2.85 Mbits/s for Enhancement layer) with Simulcast and Single layer coding

	Prog-Prog; 4 Mbits/s		Prog; 4 Mbits/s
	Simulcast Enh 2.85 Mbits/s; Base 1.15 Mbits/s	Spatial scalability Enh 2.85 Mbits/s; Base 1.15 Mbits/s	
Sequence			Nonscalable
Raft	24.81	25.45 (+0.64)	25.75 (+0.94)

progressive–progressive spatial scalability incurs a penalty of about one-third of the total difference between simulcast and single-layer coding.

We now study the effect on SNR for each layer of using different bits partitioning between base and enhancement layers, in particular using a relatively higher percentage of bits for the base layer and a relatively lower percentage of bits for the enhancement layer.

In Table 9.12, we show the SNR performance of the progressive base layer of SIF resolution coded at 1.5 Mbits/s using MPEG-1.

In comparing Tables 9.10 and 9.12 we note that by increasing the bitrate of the base layer by about one-third of the total bitrate, an increase in SNR of about 0.9 dB is obtained.

Next, in Table 9.13, we show, a comparison of results for an enhancement layer coded as simulcast, an enhancement layer coded with progressive–progressive spatial scalability, and single-layer coding. For the case of simulcast and

spatial scalability, the base layer is assumed coded at 1.5 Mbits/s.

From Table 9.13, for the bitrate and bits partitioning selected, we observe that spatial scalability performs about 1 dB better than simulcast, while single-layer coding performs about 1.35 dB better, again implying that for the enhancement layer, progressive–progressive spatial scalability incurs a penalty of about one-third of the total difference between simulcast and single-layer coding.

For each of the two bits partitions selected, we now compare the percentage of times the spatial scalability mode is selected on a macroblock basis for each picture type. The results are shown in Table 9.14.

From Table 9.14 we observe that in progressive–progressive spatial scalability, in I-pictures the spatial scalability mode is selected 95% of the time, in P-pictures about 85% to 90% of the time, and for B-pictures, about 20% to 25% of the time. Of course, these results are for one sequence only; in general, results can be sequence dependent.

9.4.6.2 *Progressive–Interlace Spatial Scalability*

We now discuss the second type of spatial scalability, called progressive–interlace spatial scalability. In this type of spatial scalability the lower layer uses a progressive video format and the enhancement layer uses an interlaced video format. The lower layer may use lower or the same spatial reso-

Table 9.12   SNR for Progressive Base layer MPEG-1 coded at 1.50 Mbits/s used in Progressive-Progressive Spatial scalability

	Prog; 1.50 Mbits/s
Sequence	Spatial Scalability Base 1.50 Mbits/s
Raft	24.45

Table 9.13   SNR comparison of Progressive-Progressive Spatial scalability (using 1.50 Mbits/s for Base layer and 2.50 Mbits/s for Enhancement layer) with Simulcast and Single layer coding

	Prog-Prog; 4 Mbits/s		Prog; 4 Mbits/s
Sequence	Simulcast Enh 2.50 Mbits/s; Base 1.50 Mbits/s	Spatial Scalability Enh 2.50 Mbits/s; Base 1.50 Mbits/s	Nonscalable
Raft	24.39	25.32 (+0.93)	25.75 (+1.36)

Table 9.14   Percentage blocks in compatible mode in Progressive-Progressive Spatial scalability

	% age blocks in compatible mode with Enh bitrate 2.85 Mbits/s			% age blocks in compatible mode with Enh bitrate 2.50 Mbits/s		
Sequence	I-picture	P-picture	B-picture	I-picture	P-picture	B-picture
Raft	94.62	85.76	20.76	95.15	89.17	23.18

lution as compared to the enhancement layer. One potential application of this type of scalability is in a two-layer system, allowing interoperability between different standards, for example, using H.261 or MPEG-1 in the base layer, and MPEG-2 spatial enhancement for the enhancement layer. Another potential application is in scalable multimedia which can be simultaneously delivered as high-quality broadcast entertainment and as a window on computer workstations.

As mentioned earlier, in spatial scalability, the spatial decimation operation, since it occurs only at encoder, does not need to be standardized; however, the spatial interpolation operation, which takes place both at encoder and decoder, does need to be standardized and is specified by the MPEG-2 video standard. As an example assume that the base layer uses a picture resolution of one-quarter of the size of input pictures (or enhancement layer pictures). In the case of progressive–interlace spatial scalability, the Spatial Decimator simply performs horizontal decimation consisting of filtering and subsampling by a factor of 2, whereas vertical decimation, is performed by simply retaining one field per frame (either the first field in every frame or the second field of every frame). Thus the Spa-

tial Decimator downsamples both the horizontal and vertical resolutions by a factor of 2 in each direction. The filters used prior to horizontal downsampling are not specified since spatial decimation is not standardized. The Spatial Interpolar performs the inverse function of upsampling the decoded base layer by a factor of 2, both horizontally and vertically, by using linear interpolation. Corresponding block diagrams of the Spatial Decimator and Spatial Interpolator are basically the same as shown in Fig. 9.14a and b.

Next, we consider the issue of spatiotemporal weighted prediction in progressive–interlace spatial scalability. Continuing with the example that we have been discussing, that is the base layer has one-quarter resolution of the enhancement layer, every 8×8 block of the base layer can be spatially interpolated to a 16×16 block and used for prediction of the corresponding 16×16 block of the enhancement layer. As discussed earlier, the spatiotemporal weighted prediction is generated by a weighted combination of a spatially interpolated decoded base layer frame and a previously decoded enhancement layer frame. Figure 9.16 shows the process of generating a spatiotemporal weighted prediction as well as a sample table of weights

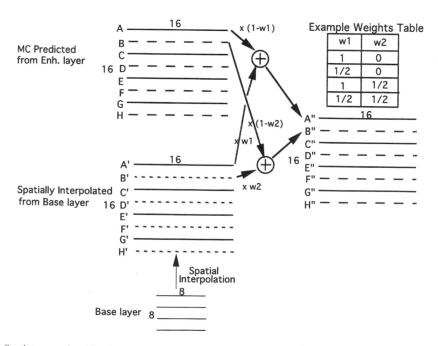

Figure 9.16   Spatiotemporal weighted prediction for Progressive–Interlace Spatial scalability.

that performs reasonably well. This table is one of several supported by the standard that allows table selection as frequently as once per picture.

As shown in Fig. 9.16, weight component *w1* is applied to liens of one field and weight component *w2* is applied to the lines of the other field of each frame. In our example, we have assumed the base layer to consist of the first field of each frame, which after spatial interpolation by 2 both horizontally and vertically can be used to spatially interpolate a frame. Lines of spatial interpolated blocks are weighted by either *w1* or *w2* as appropriate, whereas lines of temporal prediction block are weighted by either *(1 − w1)* or *(1 − w2)*. Using each of the entries in the weights table, spatiotemporal prediction error candidates are generated by differencing from the block to be coded the spatiotemporal weighted block. For each macroblock, the weight entry in the table for which prediction error is minimum is selected and an index identifying it is included in the encoded bitstream for the decoder to use.

We are now ready to discuss the performance aspects of progressive–interlace spatial scalability. Earlier, we had mentioned that interoperability of standards is a potential important application of this type of spatial scalability.

In Table 9.15 we show SNR results for a progressive base layer of SIF resolution coded at 1.15 Mbits/s using MPEG-1.

Since in spatial scalability, the base layer is coded independently (loose coupling between layers), the results shown in Table 9.15 are the same as for nonscalable coding.

Next in Table 9.16 we show as a comparison of results for an enhancements layer as simulcast, an enhancement layer coded with progressive–

Table 9.15 SNR for Progressive Base layer MPEG-1 coded at 1.15 Mbits/s used in Progressive-Interlace Spatial scalability

Sequence	Prog; 1.15 Mbits/s Spatial scalability Base 1.15 Mbits/s
Flowergarden	26.15
Football	34.58
Bus	28.05

interlace spatial scalability, and single-layer coding. For the case of simulcast and spatial scalability, the base layer is assumed coded by 1.15 Mbits/s.

From Table 9.16, for the bitrate and bits partitioning selected, we observed that spatial scalability performs about 0.5 to 1.5 dB better than simulcast, while single-layer coding performs about 2.0 to 2.25 dB better, implying that for the enhancement layer, progressive–interlace spatial scalability incurs a penalty ranging from slightly more than one-third to as much as two-thirds of the total difference between simulcast and single-layer coding.

We now study the effect for each layer of using different bits partitioning between base and enhancement layers, in particular using a relatively higher percentage of bits for the base layer and relatively lower percentage of bits for the enhancements layer.

In Table 9.17 we show the SNR performance for a progressive base layer of SIF resolution coded at 1.5 Mbits/s using MPEG-1.

In comparing Tables 9.15 and 9.17 we note that by increasing the bitrate of the base layer by about one-third of the total bitrate, an increase in SNR by about 1.25 dB is obtained.

Table 9.16 SNR comparison of Progressive-Interlace Spatial scalability (using 1.15 Mbits/s for Base layer and 2.85 Mbits/s for Enhancement layer) with Simulcast and Single layer coding

Sequence	Prog-Interl; 4 Mbits/s		Interl; 4 Mbits/s
	Simulcast Enh 2.85 Mbits/s; Base 1.15 Mbits/s	Spatial Scalability Enh 2.85 Mbits/s; Base 1.15 Mbits/s	Nonscalable
Flowergarden	28.03	28.72 (+0.69)	30.02 (+1.99)
Football	33.21	34.61 (+1.40)	35.50 (+2.29)
Bus	29.30	29.83 (+0.53)	31.26 (+1.96)

Table 9.17   SNR for Progressive Base layer MPEG-1 coded at 1.50 Mbits/s used in Progressive-Interlace Spatial scalability

	Prog; 1.50 Mbits/s

Sequence	Spatial scalability Base 1.50 Mbits/s
Flowergarden	27.52
Football	35.84
Bus	29.30

Next, in Table 9.18 we show an SNR comparison among an enhancement layer using simulcast, an enhancement layer using progressive–interlace spatial scalability, and single-layer coding. For the case of simulcast and spatial scalability, the base layer is assumed coded at 1.5 Mbits/s.

From Table 9.18 for the bitrate and bits partitioning selected, we observe that spatial scalability performs in the range of 1.0 to 2.25 dB better than simulcast, while single-layer coding performs in the range of 2.75 to 3.5 dB better, implying that for the enhancement layer, progressive–interlace spatial scalability incurs a penalty in the range of one-third to two-thirds of the total improvement in going from simulcast to single-layer coding.

For each of the two bits partitions selected, we now compare the percentage of times a spatial scalability mode is selected on a macroblock basis for each picture type. The results are shown in Table 9.19.

From Table 9.19 we observe that in progressive–interlace spatial scalability, in I-pictures the spatial scalability mode is selected 80% to 98% of the time, in P-pictures about 45% to 80% of the time, and for B-pictures, about 5% to 20% of the time. For progressive–interlace spatial scalability, as expected, the spatial scalability mode is selected less frequently as compared to progressive–progressive spatial scalability and thus this spatial scalability is not as effective as progressive–progressive spatial scalability. The reason has to do with the fact that since the progressive format is used for the base layer, only one corresponding field of the interlaced enhancement layer can be predicted well, which limits the coding efficiency of spatial scalability.

9.4.6.3   Interlace-Interlace Spatial Scalability

In interlace–interlace spatial scalability, the lower layer uses the interlaced video format and the enhancement layer also uses the interlaced

Table 9.18   SNR comparison for Progressive-Interlace Spatial scalability (using 1.50 Mbits/s for Base layer and 2.50 Mbits/s for Enhancement layer) with Simulcast and Single layer coding

	Prog-Interl; 4 Mbits/s		Interl; 4 Mbits/s
Sequence	Simulcast Enh 2.50 Mbits/s; Base 1.50 Mbits/s	Spatial Scalability Enh 2.50 Mbits/s Base 1.50 Mbits/s	Nonscalable
Flowergarden	27.22	28.49 (+1.27)	30.02 (+2.80)
Football	32.05	34.26 (+2.22)	35.50 (+3.46)
Bus	28.46	29.53 (+1.07)	31.26 (+2.80)

Table 9.19   Percentage blocks in compatible mode in Progressive-Interlace Spatial scalability

	%age blocks in compatible mode with Enh bitrate 2.85 Mbits/s			%age blocks in compatible mode with Enh bitrate 2.50 Mbits/s		
Sequence	I-picture	P-picture	B-picture	I-picture	P-picture	B-picture
Flowergarden	98.25	49.24	3.86	98.41	65.23	5.75
Football	80.45	80.45	15.91	80.83	89.39	21.51
Bus	82.57	43.10	4.09	83.25	56.81	9.39

video format. The lower layer may use lower or the same spatial resolution as compared to the enhancement layer. One potential application of this type of scalability is in two-layer scalable interlaced HDTV with normal interlaced TV as the base layer.

As previously mentioned, the spatial decimation operation, since it occurs only at the encoder, does not need to be standardized; however, the spatial interpolation operation, which takes place both at the encoder and decoder, does need to be standardized and is specified by the MPEG-2 standard. The decimation and interpolation operations can be best explained by taking specific examples of resolutions for base and enhancement layers. Assume that the base layer is required to have one-quarter the spatial resolution of the enhancement layer. Then the spatial decimation consists of deinterlacing, followed by horizontal decimation by a factor of 2, followed by vertical decimation by a factor of 4. The inverse operation, spatial interpolation, consists of deinterlacing, followed by horizontal upsampling by a factor of 2, followed by either outputting a field as is or resampling that field and outputting. Corresponding block diagrams of a Spatial Decimator and Spatial Interpolator are shown Fig. 9.17 a and b.

Thus, far, our explanation of a Spatial Decimator and Interpolator has been far from clear. This situation is remedied by choosing a specific input resolution of CCIR 601–4:2:0 and showing the various interim steps and resulting picture resolutions and formats after each step (Fig. 9.18).

Figures 9.18a and b correspondingly explain the spatial decimation and spatial interpolation operations for the luminance signal. In interlace-interlace spatial decimation[7,20] interlaced input frames comprising a pair of fields of 704×240 (for an NTSC system or 704×288 for the PAL system) undergo downsampling to produce pairs of video fields of 352×120 size (or 352×144) resolution. The process of spatial interpolation is the inverse of decimation and consists of upsampling pairs of video fields of 352×120 (or 352×144) resolution to pairs of video fields of 704×240 (or 704×288) resolution. Earlier, we had mentioned that the process of spatial interpolation needs standardization while spatial decimation does not need to be standardized. Well, the decoding process, which uses spatial interpolation, uses exactly the same steps as shown here using a linear interpolation filter to perform upsampling or resampling when required. Since spatial decimation does not need to be standardized, there is

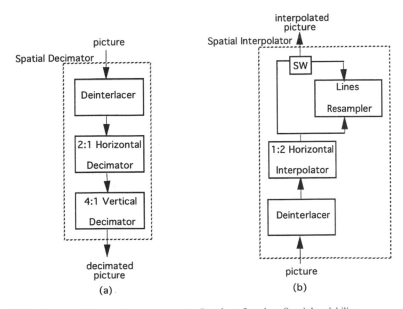

Figure 9.17   Spatial Decimator and Interpolator examples in Interlace–Interlace Spatial scalability.

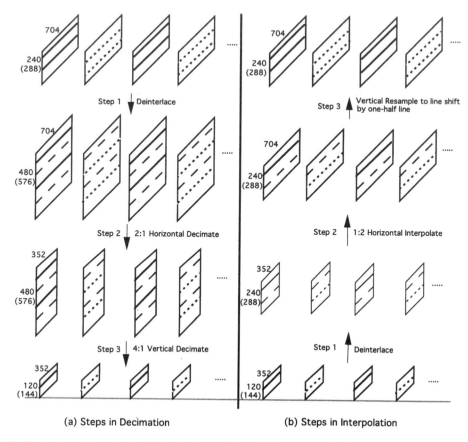

(a) Steps in Decimation   (b) Steps in Interpolation

Figure 9.18   Explanation of steps in Spatial Decimation and Interpolation examples in Interlace–Interlace Spatial scalability.

freedom in choosing filters for decimation and is purely an encoder operation not mandated by the standard.

The next issue for discussion is the generation of spatiotemporal weighted prediction. Continuing with the example that we have been discussing, that is, the base layer has one-quarter resolution of the enhancement layer, every 8 × 8 block of the base layer can be spatially interpolated to a 16 × 16 block and used for prediction of the corresponding 16 × 16 block of the enhancement layer. As for other types of spatial scalability discussed thus far, this prediction is generated by a weighted combination of a spatially interpolated decoded base layer frame and previously decoded enhancement layer frame. Figure 9.19 shows the process of generating spatiotemporal weighted prediction as well as a sample table of weights that performs reasonably well. This table is one of several sup-

ported by the standard that allows table selection as frequently as once per picture.

As shown in Fig. 9.19, weight component $w1$ is applied to lines of one field and weight component $w2$ is applied to the lines of the other field of each frame. Lines of the spatial interpolation block are weighted by $w1$ and $w2$ as appropriate, whereas lines of the temporal prediction block are weighted by either $(1 - w1)$ or $(1 - w2)$. Using each of the entries in the weights table, spatiotemporal prediction error candidates are generated by differencing from the block to be coded the spatiotemporal weighted block. For each macroblock, the weight entry in the table for which the prediction error is minimum is selected and an index identifying it is included in the encoded bitstream for the decoder to use.

We are now ready to discuss the performance aspects of interlace–interlace spatial scalability.

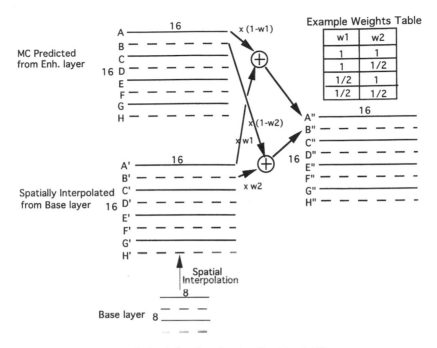

Figure 9.19 Spatiotemporal weighted prediction in Interlace–Interlace Spatial scalability.

Earlier we had mentioned that two-layer HDTV[8,13,21] is a potential major application of this type of scalability. The performance results that we will present[19] will be based on scaled down simulations using one-quarter of the resolution that may actually be used for two-layer spatial scalability HDTV application. These results are still reflective of performance tradeoffs expected in using a two-layer spatial scalability based HDTV system.

Table 9.20 shows SNR results of interlaced SIF resolution (SIFI) coded by nonscalable MPEG-2 at 1.5 Mbits/s. This is used as a base layer in interlace–interlace spatial scalability experiments.[14,19]

Next, Table 9.21 shows a comparison[14] of interlaced CCIR-601 4:2:0 resolution coded at a total of 4.0 Mbits/s, as a single layer, as an

Table 9.20 SNR for Interlace Base layer MPEG-2 coded at 1.50 Mbits/s

	Interl; 1.50 Mbits/s
Sequence	Spatial scalability Base 1.50 Mbits/s
Flowergarden	30.04
Cheerleaders	26.66
Bus	30.60

Table 9.21 SNR comparison of Interlace-Interlace Spatial scalability (using 1.50 Mbits/s for Base layer and 2.50 Mbits/s Enhancement layer) with Simulcast and Single layer coding

	Interl-Interl; 4 Mbits/s		Interl; 4 Mbits/s
Sequence	Simulcast Enh 2.50 Mbits/s; Base 1.50 Mbits/s	Spatial scalability Enh 2.50 Mbits/s; Base 1.50 Mbits/s	Nonscalable
Flowergarden	27.18	27.87 (+0.69)	30.10 (+2.92)
Cheerleaders	26.85	28.00 (+1.15)	29.12 (+2.27)
Bus	29.25	30.47 (+1.22)	31.43 (+2.18)

enhancement layer using simulcast, and as an enhancement layer using interlace–interlace spatial scalability. The bitrate for interlaced enhancement layer is 2.5 Mbits/s, using 1.5 Mbits/s for the interlaced base layer.

The SNR results of Table 9.21 show that, for the selected bitrate and bits partitioning, enhancement layer coding at 2 Mbits/s using interlace–interlace spatial scalability provides an improvement in the range of 0.75 to 1.25 dB over simulcast, whereas single-layer coding at 4.0 Mbits/s provides an improvement over simulcast enhancement (single-layer coding at 2.5 Mbits/s) by as much as 2 to 3 dB. More specifically, we can say that interlace–interlace spatial scalability incurs a penalty typically around one-half and as much as three-fourths of the total difference between single-layer coding and simulcast enhancement layer.

9.4.6.4 Interlace–Progressive Spatial Scalability

In interlace–progressive spatial scalability, as the name implies, the lower layer uses an interlaced video format and the enhancement layer uses a progressive video format. The lower layer may use lower or the same spatial resolution as compared to the enhancement layer. One potential application of this type of scalability is in two-layer scalable progressive HDTV with normal interlaced TV as the base layer.

As in the case of other spatial scalability types, the spatial decimation operation, which occurs only at the encoder, does not need to be standardized; however, the spatial interpolation operation, which takes place both at the encoder and the decoder, does need to be standardized and is specified by the MPEG-2 standard. There are various ways of decimating a progressive video to achieve an interlaced video of the same or lower spatial resolution and include simply line subsampling or prefiltering followed by line subsampling. In the next section on Temporal scalability, we will discuss the line subsampling method in more detail. As far as interpolation from an interlaced video to the same or higher spatial resolution progressive format is concerned, it can be achieved by deinterlacing followed by further spatial upsampling as required. During our discussion on interlace–

interlace spatial scalability, for interpolating the lower layer of the interlaced format to the higher spatial resolution interlaced video format used in the enhancement layer, the various steps in the example shown in Fig. 9.18 include generation of a progressive format by deinterlacing, than horizontal upsampling followed by vertical resampling. In interlace–progressive scalability, the first two steps are identical, that is, deinterlacing followed by upsampling, and the last step involving vertical resampling is not needed.

Using the decoded base layer interlaced video format, once the spatial interpolator produces an upsampled progressive format of the same spatial resolution as the enhancement layer, it is used to generate weighted spatiotemporal prediction. As in the case of other spatial scalability types, a spatiotemporal weight table is needed for computing the spatiotemporal weighted prediction and can be selected from available choices, as frequently as every picture, if desired. The process of spatiotemporal weighted prediction is similar to that for other spatial scalability types.

Although a very valid example application of this type of spatial scalability with interlaced TV as the base layer and progressive HDTV as the enhancement layer exists, it has not been experimented with sufficiently. For HDTV, other types of scalability, such as interlace–interlace spatial scalability, and progressive:interlace–interlace Temporal scalability (to be discussed) seem to have attracted more attention. Besides HDTV, other applications of interlace–progressive spatial scalability are somewhat limited. We will not be discussing detailed SNR results of this type of scalability because of availability of illustrative results. In any case, its performance is expected to be quite similar to that of interlace–interlace spatial scalability.

9.4.7 Summary of Spatial Scalability

Spatial scalability is useful for applications requiring interworking between standards, interworking of picture formats, broadcast digital HDTV with compatible digital TV, channel bandwidth and decoder complexity scaling, and high degree of error resilience. It offers considerable flexibility in achieving bandwidth partitioning between layers, as for the lower layer, picture size, same or lower than that of enhancement layers can be used. Since, in spatial scalability, two loosely coupled independent coding loops are employed, there are no drift problems. In Spatial scalability, the base layer is independently coded and thus its picture quality is exactly the same as that using nonscalable coding. The spatial scalability enhancement layer picture quality, although better than that for the simulcast enhancement layer (by 0.5 to 1.25 dB), is however also lower than that using nonscalable coding (by 0.75 to 1.5 dB). Since there is considerable flexibility between resolutions, formats, bitrate partitioning, and coding parameters, the SNR quality comparisons mentioned are not universal but merely examples. The differences in SNR values (whether higher or lower) in most cases range from being marginally visible to distinctly noticeable. In Spatial scalability, when compared to data partitioning and SNR scalability, the higher degree of flexibility, increased robustness, and improved performance (for the base layer) come at the expense of increased complexity.

9.5 TEMPORAL SCALABILITY

Temporal scalability[10,29] refers to layered coding that produces two or more layers, each with either the same or different temporal resolutions, which when combined provide full temporal resolution as available in the input video. Moreover, the spatial resolution of frames in each layer is assumed to be identical to that of the input video, although the picture formats used in the layers may be different from that of the input video. Each layer is encoded to generate separate bitstreams which are multiplexed, transmitted or stored, demultiplexed, and decoded to generate a separate layer of frames to be recombined for display.

In Temporal scalability, since the input frame rate is simply partitioned between the base layer and the enhancement layers, the Temporal scalability decoder does not need to be much more complex than a single-layer decoder. In fact, in MPEG-1 or in MPEG-2, B-pictures themselves

provide an inherent simpler form of Temporal scalability. In MPEG-2 video, Temporal scalability was a relatively late addition[10,12,15,16] compared to other scalability techniques and its potential is still largely untapped. The main use[8,16] of Temporal scalability is expected to be for HDTV application where migration to high temporal resolution progressive HDTV may be possible from the first generation interlaced HDTV in a compatible manner.

There are also other applications such as a new generation of conversational services using TV resolution video but with a progressive format at twice the frame rate, which would require interoperability with existing TV displays. Besides these applications, Temporal scalability is also useful in a software decoding environment where the decoding processor may not be powerful enough to decode video at the full frame rate or may be sharing its processing power among a number of different tasks.

9.5.1 The High-Level Codec

A two-layer Temporal scalability codec structure[12] consisting of a base layer encoder and decoder, a System Multiplexer, a System Demultiplexer, and an enhancement layer encoder and decoder is shown in Fig. 9.20. Consider input video sequence frames at full temporal rate fed to the Temporal Demultiplexer, which in our example splits up input video sequence frames into two video sequences, one of which is applied to the base encoder and the other is applied to the enhancement encoder. The base encoder is a single-layer encoder for efficiently compressing video, such as a motion-compensated DCT encoder and more specifically in MPEG-2, it is an MPEG-2 video encoder with or without B-pictures. The enhancement encoder is also a motion-compensated DCT encoder similar to the base encoder, and moreover can also exploit additional redundancies by using interlayer motion compensation with respect to locally decoded frames generated by the base encoder. In the example used in our discussion, both the base and the enhancement encoders each process the same frame rate, which is in fact half of the input video

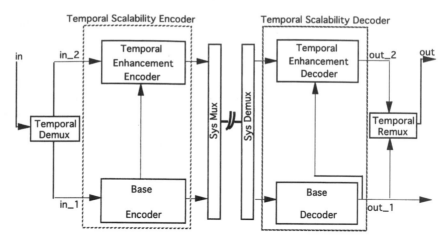

Figure 9.20   Temporal scalability codec.

frame rate, although, in general, Temporal scalability is quite flexible to allow different frame rates in each layer.

The encoded bitstreams generated by the base encoder and the enhancement encoder are multiplexed by the Systems Multiplexer into a single stream of packets allowing identification of the source of their origination. This stream can now be transmitted or stored and is eventually input to a Systems Demultiplexer which performs the complementary function of separating them back into two streams; the bitstream generated by the base encoder is input to the base decoder and the bitstream generated by the enhancement encoder is input to enhancement decoder for decoding. The base decoder is complementary to the base encoder and thus, depending on the encoder chosen, is an MPEG-2 video encoder with or without B-pictures. The enhancement decoder is complementary to the enhancement encoder and can exploit interlayer motion compensation; this decoder is referred to as the Temporal Enhancement decoder. Whereas the base layer bitstream can be decoded by the base decoder to produce decoded frames, the enhancement layer bitstream is decoded by the enhancement decoder which also uses base layer decoded frames. The decoded frames at the output of the base decoder can be shown by themselves at half the frame rate of the input video or can be temporally multiplexed in the Temporal Remultiplexer with the output of the enhancement decoder to provide full frame rate, the same as that of the input video.

9.5.2 Prediction Configurations

In Temporal scalability, the base layer is coded in exactly the same manner as in nonscalable coding with any of the prediction structures such as $M = 1$ (no B-pictures) or $M = 3$ (two B-pictures between a pair of reference pictures). There are, however, many possible ways of coding enhancement layer pictures using predictions from decoded base layer pictures and previous decoded enhancement layer pictures. The primary issues are how many pictures to use as reference at a time, and which pictures to use. MPEG-2 Temporal scalability syntax restricts the maximum number of reference pictures to two and these can be either both decoded base layer pictures or a decoded base layer picture and a decoded previous enhancement layer picture. Although enhancement layer coding allows use of I-, P-, or B-pictures, it is envisaged that for high coding efficiency mainly B-pictures would be used. However, enhancement layer B-pictures are a bit different than B-pictures in the base layer although there are similarities since both types of B-pictures use two references. We have already answered the question of how many pictures to use as references for coding an enhancement layer B-picture; we now only have to address which pictures to use as reference. The answer to

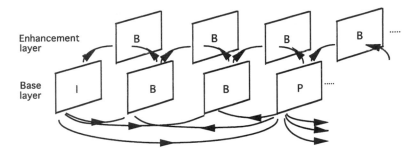

Figure 9.21   Interlayer Motion-Compensated Predictions.

this question is contained in two prediction configurations[10] which are explained next.

9.5.2.1 Configuration 1: Interlayer MC Predictions

This prediction configuration uses two interlayer motion-compensated predictions. In this configuration, the base layer although it can use any prediction structure, for the purposes of illustration, it
uses $M = 3$ structure. To maintain high coding
efficiency, the enhancement layer consists of modified B-pictures, although, if needed, additional P-
pictures can also be introduced. In this configuration, each B-picture uses two adjacent decoded
base layer pictures as references. This configuration is shown in Fig. 9.21.

In nonscalable coding, B-pictures use two references, a forward prediction reference and a backward prediction reference, to generate a choice of
three prediction modes. These are: forward prediction, backward prediction, and interpolated prediction and on a macroblock basis one of three
prediction modes can be chosen. Likewise,
enhancement layer B-pictures also use two refer-

ences, a forward prediction reference and a backward prediction reference, to generate three prediction modes. In this configuration, forward prediction implies the most recent lower layer picture
in display order and backward prediction implies
the next lower layer picture in display order.

9.5.2.2 Configuration 2: MC Prediction and Interlayer MC Prediction

This prediction configuration uses one motion-
compensated prediction and one interlayer motion-
compensated prediction. For the purpose of illustration, we employ a $M = 3$ prediction structure for
base layer coding, although other structures can
also be used. To maintain high coding efficiency,
the enhancement layer (with the exception of the
first picture, a P-picture) consists only of modified
B-pictures; if needed, additional P-pictures can be
introduced. In this configuration, each B-picture
uses a and decoded enhancement layer picture and
a decoded base layer picture as references. This
configuration is shown in Fig. 9.22.

Just as for the previous configuration and for
nonscalable coding, enhancement layer B-pictures

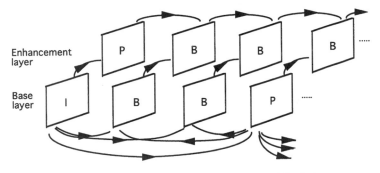

Figure 9.22   Motion-Compensated Prediction and Interlayer Motion-Compensated Prediction.

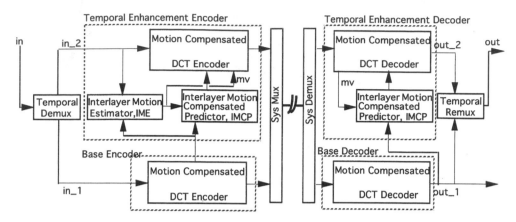

Figure 9.23   Temporal scalability codec with a few details.

use two references, a forward prediction reference and a backward prediction reference, to generate three prediction modes referred to as forward prediction, backward prediction, and interpolated prediction modes for selection on a macroblock basis. With the configuration of Fig. 9.22, forward prediction implies the most recent decoded enhancement picture(s) and backward prediction implies the most recent lower layer picture in display order. Another variation of this configuration is also supported. This variation uses the most recent decoded enhancement picture(s) as the forward reference and the next lower layer picture in display order as the backward reference.

9.5.3 The Detailed High-Level Codec

In principle, Temporal scalability can be looked upon as building on the motion-compensated DCT coding structure of nonscalable MPEG-2 coding by adding interlayer prediction between the layers. Again assuming two-layer coding consisting of a base layer and an enhancement layer, temporal scalability basically involves motion-compensated prediction such as that used in B-picture coding, with the addition that motion compensation can also be performed between the layers. This leads to an improved organization of bitstreams for flexible decoding while maintaining high coding efficiency.

To make the codec structure of Fig. 9.20 a bit more precise, we consider a specific class of

encoders/decoders that can be employed for base and enhancement layers; the revised codec is shown in Fig. 9.23.

For example, the base layer encoder/decoder is simply a motion-compensated encoder/decoder and more precisely, an MPEG-2 nonscalable encoder/decoder. The temporal enhancement layer encoder/decoder is a modified motion-compensated encoder/decoder that uses interlayer motion compensation, and primarily codes B-pictures. The temporal enhancement encoder includes a motion-compensated encoder like that used for base layer, an interlayer motion estimator (IME), and a corresponding interlayer motion-compensated predictor (IMCP). As its name suggests, IME computes motion vectors between blocks of enhancement layer pictures with respect to base layer pictures. These motion vectors are used by IMCP to perform motion compensation and are also included in the enhancement layer bitstream, depending on the mode selected, on a macroblock basis. The temporal enhancement decoder consists of a motion-compensated decoder and an interlayer motion-compensated predictor (MCP). From enhancement layer bitstream, depending on the mode of a coded macroblock, when available, interlayer motion vectors are extracted to perform motion-compensated prediction in IMCP. The IMCP at the enhancement encoder and enhancement decoder perform identically. The remaining operations in Fig. 9.23 are identical to that in Fig. 9.20.

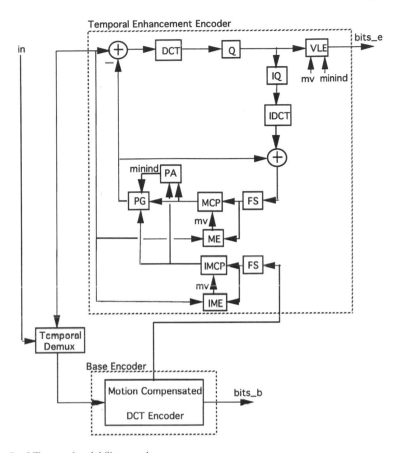

Figure 9.24   Details of Temporal scalability encoder.

9.5.4 The Encoder

Although the codec structure of Fig. 9.23 provides a significant amount of clarification over the codec structure of Fig. 9.20, it still leaves several important details hidden. We now describe the internals of a simplified Temporal Enhancement Encoder and show its interaction with other elements of a Temporal Scalability Encoder, thus offering a limited opportunity to uncover a few hidden details. We start by first presenting a simplified yet meaningful block diagram of a Temporal Scalability Encoder in Fig. 9.24 and then present an explanation of its operation.

In Fig. 9.24, the Temporal Enhancement Encoder shown is a motion-compensated DCT encoder with modifications to support basic prediction configurations of Fig. 9.21 and Fig. 9.22. This encoder can be considerably simplified to

support only one of the two prediction configurations, since it is the decoder that has to be able to support both structures, not the encoder. However, we show only one encoder that can support both basic configurations. These prediction configurations primarily employ B-pictures, which can use up to two references for prediction.

The basic coding loop in the Temporal Enhancement Encoder works like normal MPEG-1 or nonscalable MPEG-2 coding supporting B-pictures. Video frames (or fields) are input at *in* and are partitioned into nonoverlapping macroblocks (16 × 16 luminance block and two corresponding 8 × 8 chrominance blocks) which are predicted using previously decoded frame(s), the resulting difference signal is DCT transformed using 8 × 8 blocks and the transform coefficients are quantized in a Quantizer, Q. The quantization indices, motion vectors, and other data are variable length encoded in a VLE to generate a coded bitstream. The transform quantization indices are also fed back to a local decoding loop at the encoder consisting of an Inverse Quantizer (IQ), Inverse DCT (IDCT), an Adder, Frame Store(s) (FS) a Motion Estimator (ME), and Motion-Compensated Predictor (MCP). The FS contains one or two frame stores as needed. In each of the two configurations only two frame stores are needed; in the case of prediction configuration shown in Fig. 9.21, both frame stores are considered to be part of a single FS, shown at the output of a locally decoded output from the base layer while in the case of prediction configuration shown in Fig. 9.22, each FS contains one frame store. For both prediction configurations, in using decoded base layer picture(s) as reference, interlayer motion estimation (IME) and interlayer motion-compensated prediction (IMCP) are employed. In fact these operations are exactly the same as the corresponding operations of ME and MCP, and are shown to be separate only for clarity. We also notice two new operations of the Prediction Analyzer PA and prediction Generator (PG). In fact these operations are not new but also employed in MPEG-1 and nonscal-

able MPEG-2 encoding but were not shown earlier for simplicity. Here, however, they are used to emphasize the point that having selected a prediction configuration in B-pictures on a macroblock basis, one of the three modes–forward, backward, bidirectional–is selected. The modified meaning of these modes was explained earlier in the discussion of prediction configurations.

The resulting enhancement layer bitstream is bits_e and is sent to Sys Mux (not shown here) which also receives the base layer bitstream bits_b.

9.5.5   The Decoder

After having examined details of the Temporal Enhancement Encoder and its interaction with the remaining elements of a Temporal Scalability Encoder, you can probably guess what a Temporal Enhancement Decoder looks like and how it interacts with other elements of the Temporal Scalability Decoder. In any case, we present the internals of a simplified Temporal Enhancement Decoders and explain its interaction. A simplified yet meaningful block diagram of a Temporal Scalability Decoder is shown in Fig. 9.25.

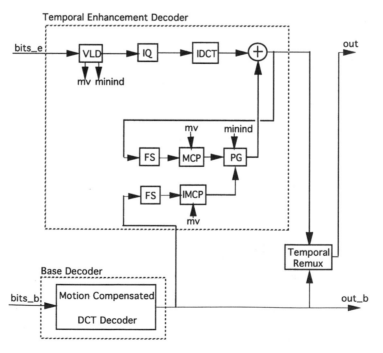

Figure 9.25   Details of Temporal scalability decoder.

The operation of a Temporal Enhancement Decoder shown in Fig. 9.25 is somewhat similar to that of normal motion-compensated DCT decoder besides the exceptions mentioned while discussing Temporal Enhancement Encoder. The operation of this decoder is also similar to that of the local decoding loop in the Temporal Enhancement Encoder.

The enhancement layer bitstream, bits_e, from Sys Demux (not shown) is input to Temporal Enhancement Decoder, where it is first presented to a Variable Length Decoder (VLD), and then forwarded to an Inverse Quantizer (IQ) Inverse DCT, (IDCT), and then to an Adder at the other input of which is a prediction signal.

First, the prediction signal is dependent on the prediction structure employed by the encoder, that is, whether the basic configuration of Fig. 9.21 or that of Fig. 9.22 is employed. Second, for B-pictures, on a macroblock basis, the prediction can use one of the three modes: forward, backward, or bidirectional. The prediction is generated by a Prediction Generator (PG), which uses a *minind* signal decoded from the enhancement layer bitstream and determines the prediction mode. With the prediction structure of Fig. 9.21, the feedback loop used is that connecting the base layer to the enhancement layer, that is, frame FS and IMCP, while with the prediction structure of Fig. 9.22, both feedback loops, the aforementioned loop from the base layer as well as the feedback loop from the enhancement layer. consist of FS and MCP. The decoded enhancement layer pictures are output by the Adder and are temporally multiplexed with decoded base layer pictures for display.

9.5.6 Temporal Scalability Types and Results

In this section, we discuss the various types[12] of of Temporal scalability and provide example results for each type. Temporal scalability can be classified into the following types.

- Progressive: Progressive–Progressive Temporal Scalability
- Progressive:Interlace–Interlace Temporal Scalability
- Interlace:Interlace–Interlace Temporal Scalability

9.5.6.1 *Progressive: Progressive-Temporal Scalability*

Progressive:Progressive–Progressive Temporal Scalability means that the video input is of a progressive format and is coded as two layers, both of which are also progressive.

Assuming input video uses progressive frames and is available at high temporal resolution (60 Hz in NTSC countries, 50 Hz in PAL or SECAM countries). This input progressive frames are temporally separated into two sequences, each of which are also of composed of progressive frames for example, each with half of the input frame rate. In the codec shown in Fig. 9.20, this operation of temporal separation actually takes place in Temporal Demux. To clarify further, in Fig. 9.26 we show by example how a progressive sequence can be temporally separated into two progressive sequences, each with half a frame rate.

If frames 1, 2, 3, 4,.. are consecutive frames of high temporal resolution progressive sequence, then simply selecting odd-numbered frames as the first sequence and even-numbered frames as the second sequence yields two progressive sequences, each with half a frame rate. By the way, Temporal scalability syntax allows considerable flexibility in separation of frames in the two layers, the equal separation of frames between two layers was intended as a simple example only.

Now following the high-level codec structure of Fig. 9.20, either one of the two separated progressive sequences can be encoded and decoded as the base layer by the Base Encoder/Decoder and the other one encoded and decoded as the enhancement layer by the Temporal Enhancement Encoder/Decoder. The two decoded progressive sequences are temporally combined to form a single sequence in Temporal Remux. This process is basically the reverse of the temporal separation process shown in Fig. 9.26.

We now provide example results for this type of Temporal scalability using bitrates of 4 and 6 Mbits/s for the base layer; the enhancement layer is always coded at 2 Mbits/s.

In Table 9.22 we show SNR results of the progressive base layer of CCIR-601 4:2:0 resolution coded by nonscalable MPEG-2 coding at 4 Mbits/s.

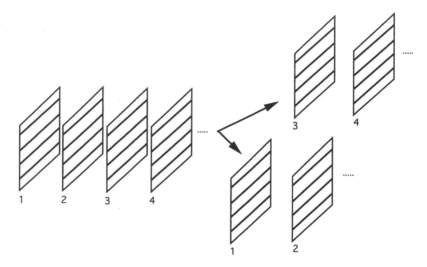

Figure 9.26   Demultiplexing of a progressive sequence into two progressive sequences.

Next, in Table 9.23 we show an SNR comparison of progressive enhancement layer of CCIR-601 4:2:0 resolution coded using the two Temporal scalability configurations and by simulcast at 2 Mbits/s using the progressive base layer coded at 4 Mbits/s.

With the bits partition selected, results indicate that for natural scenes, configuration 1 yields an improvement of 1 to 1.5 dB, while configuration 2

provides an improvement of 0.75 to 1.25 dB over simulcast. For synthetic scenes, the improvements can be significantly higher, in the 4 to 5 dB range.

Next, in Table 9.24 we show SNR results of the progressive base layer of CCIR-601 4:2:0 resolution coded by nonscalable MPEG-2 coding at 6 Mbits/s.

In Table 9.25 we show an SNR comparison of the progressive enhancement layer of CCIR-601 4:2:0 resolution coded using the two Temporal scalability configurations and by simulcast at 2 Mbits/s using the progressive base layer coded at 6 Mbits/s.

With the bits partition selected, results indicate that for natural scenes, configuration 1 yields an improvement of 1 to 2 dB, while configuration 2 provides an improvement of 1 to 1.5 dB over simulcast. For synthetic scenes, the improvements can be significantly higher, in the 5 to 6 dB range.

Table 9.22   SNR for Progressive Base layer of CCIR-601 4:2:0 resolution coded at 4 Mbits/s

	Prog; 4 Mbits/s
Sequence	Temporal Scalability Base 4 Mbits/s
Raft	25.62
Robots	31.36
Basketball2	27.69

Table 9.23   SNR for Progressive Enhancement layer of CCIR-601 4:2:0 resolution coded at 2 Mbits/s, Base layer at 4 Mbits/s

	Prog:Prog-Prog; 2 Mbits/s		
Sequence	Simulcast Enh 2 Mbits/s; Base 4 Mbits/s	Temporal scalability config 1 Enh 2 Mbits/s; Base 4 Mbits/s	Temporal scalability config 2 Enh 2 Mbits/s; Base 4 Mbits/s
Raft	23.72	25.25 (+1.53)	24.88 (+1.16)
Robots	26.98	31.72 (+4.74)	30.83 (+3.85)
Basketball2	26.58	27.47 (+0.89)	27.36 (+0.78)

Table 9.24 SNR for Progressive Base layer of CCIR-601 4:2:0 resolution coded at 6 Mbits/s

	Prog; 6 Mbits/s
Sequence	Temporal Scalability Base 6 Mbits/s
Raft	26.74
Robots	34.01
Basketball2	28.47

Finally, in Table 9.26 we present an SNR comparison of the progressive sequence at 60 Hz coded without scalability and with Temporal scalability, selecting the configuration that performs best for both bitrate combinations tested when both layers are progressive.

The SNR results indicate that for natural scenes, for both layers in the progressive format and for both bitrate combinations tested, the performance of Temporal scalability with both layers combined nearly equals (or even slightly exceeds) that of nonscalable coding. For a synthetic scene tested, Temporal scalability outperformed nonscalable coding by 1 to 2 dB. Informal visual tests confirm the conclusion from SNR results.

9.5.6.2 Progressive: Interlace–Interlace Temporal Scalability

Progressive:Interlace–Interlace Temporal Scalability means that video input is of the progressive format and is coded as two layers, both of which are interlaced.

Again, assume input video consists of progressive frames and is available at high temporal resolution. Now, input progressive frames are temporally separated into two sequences, each of which are interlaced, for example, each with half of the input frame rate but each having an interlaced format. This operation of temporal separation of a progressive sequence into two interlaced sequences takes place in Temporal Demux in the code shown in Fig. 9.20. In Fig. 9.27 we show details of a simple method to accomplish this operation.

As shown in Fig. 9.27, frames 1, 2, 3, 4,... are consecutive frames of a high temporal resolution progressive sequence. Consider each progressive frame to be composed of two sets of lines, with odd-numbered lines (assuming line numbers start from 1) in the first set and even-numbered lines in the second set. Now consider progressive frames in pairs such as frames 1 and 2, frames 3 and 4 and

Table 9.25 SNR for Progressive Enhancement layer of CCIR-601 4:2:0 resolution coded at 2 Mbits/s, Base layer at 6 Mbits/s

	Prog:Prog-Prog; 2 Mbits/s		
Sequence	Simulcast Enh 2 Mbits/s; Base 6 Mbits/s	Temporal scalability config 1 Enh 2 Mbits/s; Base 6 Mbits/s	Temporal scalability config 2 Enh 2 Mbits/s; Base 6 Mbits/s
Raft	23.72	25.53 (+1.81)	25.08 (+1.36)
Robots	26.98	33.39 (+6.41)	32.03 (+5.05)
Basketball2	26.58	27.60 (+1.02)	27.45 (+0.87)

Table 9.26 SNR comparison of best Progressive-Progressive Temporal Scalability configuration with Single layer coding

	Prog 60 Frames/s; 6 Mbits/s		Prog 60 Frames/s; 8 Mbits/s	
Sequence	Nonscalable	Temporal Scalability config 1: Mux Prog Base and Prog Enh	Nonscalable	Temporal Scalability config 1: Mux Prog Base and Prog Enh
Raft	25.25	25.44 (+0.19)	26.01	26.14 (+0.13)
Robots	30.20	31.54 (+1.24)	31.79	33.70 (+1.91)
Basketball2	27.43	27.58 (+0.15)	27.73	28.04 (+0.31)

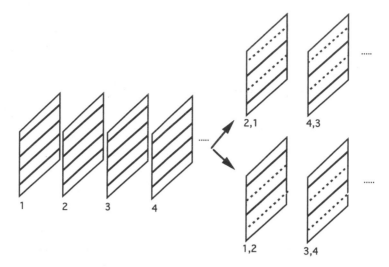

Figure 9.27   Demultiplexing of a progressive sequence to two interlaced sequences.

so on. Next, select odd-numbered lines from frame 1 of the progressive input and interleave, even-numbered lines from frame 2 of progressive input to create an interlaced frame referred to as 1, 2 in the first interlaced sequence. Conversely, to create the first interlaced frame referred to 2, 1, select odd-numbered lines from frame 2 and even-numbered lines from frame 1 of the progressive input. This procedure is repeated to create the second frame of each of the two interlaced sequences and uses frames 3 and 4 of the progressive input. Other interlaced frames are also created in the same manner.

Now, following the high-level codec structure shown in Fig. 9.20, either one of the two interlaced sequences can be encoded and decoded as the base layer by the Base Encoder/Decoder and the other one encoded and decoded as the enhancement layer by the Temporal Enhancement Encoder/Decoder. The two decoded interlaced sequences are combined to form a single sequence in Temporal Remux. This process is basically the reverse of the temporal separation process shown in Fig. 9.27.

We now provide example results for this type of Temporal scalability using bitrates of 4 and 6 Mbits/s for the base layer; the enhancement layer is always coded at 2 Mbits/s.

In Table 9.27 we show SNR results of interlaced base layer of CCIR-601 4:2:0 resolution

coded by nonscalable MPEG-2 coding at 4 Mbits/s.

Table 9.28 shows an SNR comparison of progressive enhancement layer of CCIR-601 4:2:0 resolution coded using the two Temporal scalability configurations and by simulcast at 2 Mbits/s using a progressive base layer coded at 4 Mbits/s.

With the bits partition selected, results indicate that for natural scenes, both configuration 1 and configuration 2 yield an improvement of 1 to 1.25 dB over simulcast. For synthetic scenes, the improvements can be significantly higher, as much as 4 dB.

Next, in Table 9.29 we show SNR results of an interlaced base layer of CCIR-601 4:2:0 resolution coded by nonscalable MPEG-2 coding at 6 Mbits/s.

Table 9.27   SNR for Interlace Base layer of CCIR-601 4:2:0 resolution coded at 4 Mbits/s

Sequence	Interl: 4 Mbits/s Temporal scalability Base 4 Mbits/s
Raft	24.29
Robots	29.54
Basketball2	26.98

Table 9.28   SNR for Interlace Enhancement layer of CCIR-601 4:2:0 resolution coded at 2 Mbits/s, Base layer at 4 Mbits/s

| Sequence | Prog:Interl-Interl; 2 Mbits/s | | |
	Simulcast Enh 2 Mbits/s; Base 4 Mbits/s	Temporal Scalability Enh 2 Mbits/s config 1 Base 4 Mbits/s	Temporal Scalability Enh 2 Mbits/config 2 Base 4 Mbits/s
Raft	22.95	24.21 (+1.26)	24.18 (+1.23)
Robots	25.81	29.84 (+4.03)	29.75 (+3.94)
Basketball2	25.70	26.79 (+1.09)	26.68 (+0.98)

In Table 9.30 we show an SNR comparison of an interlaced enhancement layer of CCIR-601 4:2:0 resolution coded using the two Temporal scalability configurations and by simulcast at 2 Mbits/s using an interlaced base layer coded at 6 Mbits/s.

Table 9.29   SNR for Interlace Base layer of CCIR-601 4:2:0 resolution coded at 6 Mbits/s

| Sequence | Interl, 6 Mbits/s |
	Temporal Scalability Base 6 Mbits/s
Raft	25.65
Robots	32.07
Basketball2	27.84

With the bits partition selected, results indicate that for natural scenes, both configurations 1 and 2 yield an improvement of 1.25 to 2 dB over simulcast. For synthetic scenes, the improvements can be significantly higher, in the 5 to 6 dB range.

Finally in Table 9.31 we present an SNR comparison of a progressive sequence at 60 Hz coded without scalability and with Temporal scalability, selecting the configuration that performs best for both bitrate combinations tested with both layers interlaced.

The SNR results indicate that for natural scenes, using the interlace format for both layers and for both bitrate combinations tested, the performance of Temporal scalability with both layers combined is lower than that using nonscalable coding by about 0.25 dB to 0.75 dB in one case and

Table 9.30   SNR for Interlace Enhancement layer of CCIR-601 4:2:0 resolution coded at 2 Mbits/s, Base layer at 6 Mbits/s

| Sequence | Prog:Interl-Interl; 2 Mbits/s | | |
	Simulcast Enh 2 Mbits/s; Base 6 Mbits/s	Temporal Scalability Enh 2 Mbits/s config 1; Base 6 Mbits/s	Temporal Scalability Enh 2 Mbits/s config 2; Base 6 Mbits/s
Raft	22.95	24.83 (+1.88)	24.18 (+1.83)
Robots	25.81	31.45 (+5.64)	29.75 (+5.37)
Basketball2	25.70	27.00 (+1.30)	26.68 (+1.16)

Table 9.31   SNR comparison of best Interlace-Interlace Temporal scalability configuration with Single layer coding

| Sequence | Prog 60 Frames/s; 6 Mbits/s | | Prog 60 Frames/s; 8 Mbits/s | |
	Nonscalable	Temporal Scalability config 1: Mux Interl Base and Interl Enh	Nonscalable	Temporal Scalability config 1: Mux Interl Base and Interl Enh
Raft	25.25	24.25 (−1.00)	26.01	25.24 (−0.77)
Robots	30.20	29.69 (−0.51)	31.79	31.76 (−0.03)
Basketball2	27.43	26.88 (−0.55)	27.73	28.04 (−0.31)

0.5 to 1 dB lower in another case. For a synthetic scene tested, the performance of Temporal scalability was lower than that using nonscalable coding by 0.5 dB. However, a word of caution is in order. The SNR results for nonscalable coding at 60 Hz are based on coding of progressive video, however, the results of the combined Temporal layer are based on coding of interlaced video in each layer, and thus there is an inherent difficulty in such a comparison. Informal visual tests confirm this conclusion and despite somewhat lower SNR, Temporal scalability results appear to be the same as that using nonscalable coding.

9.5.6.3 Interlace:Interlace–Interlace Temporal Scalability

Interlace:Interlace–Interlace Temporal Scalability means that video input is of the interlaced format and is coded as two layers, both of which are also interlaced.

Assume the input video is composed of interlaced frames and is to be temporally separated to form two interlaced sequences in Temporal Demux of Fig. 9.20. This operation can be accomplished as shown in Fig. 9.26 with the modified assumption that input video is interlaced. Thus, odd-numbered frames of the input video can be considered as forming one interlaced sequence and even-numbered frames as forming another interlaced sequence, each sequence having half a frame rate. Again, as before, Temporal scalability syntax is quite flexible and supports not only equal partitioning of frames between two layers but unequal partitioning as well.

Experiments have recently been conducted and results confirm that performance of this type of Temporal scalability is quite similar to the performance of Progressive:Progressive–Progressive Temporal scalability. In other words, its performance is nearly equal to that of nonscalable coding, and in some cases even exceeds it.

9.5.7 Summary of Temporal Scalability

Temporal scalability is useful for applications requiring migration to higher temporal resolution progressive video formats such as in HDTV or

telecommunications, interworking between standards, channel bandwidth and decoder complexity scaling, and a high degree of error resilience. It offers sufficient flexibility in bandwidth partitioning between layers; however, since the same picture size is used in each layer, the flexibility is somewhat less than that of spatial scalability. As in Spatial scalability, since it employs two loosely coupled independent coding loops, there are no drift problems. Further, as in Spatial scalability, the base layer is independently coded and thus its picture quality is exactly the same as that obtained using nonscalable coding. However, unlike Spatial scalability, which employs only spatial prediction between layers, Temporal scalability uses motion-compensated prediction between layers, resulting in very efficient coding. Thus, the enhancement layer picture quality is much better than simulcast (by 1.0 to 2.0 dB) and the quality obtained by temporal multiplexing of base and enhancement layers is nearly the same as that using nonscalable coding (in the range of 0.3 dB lower to 0.2 dB higher). Further, since there is sufficient flexibility in formats of base and enhancement layers compared to the input format, prediction structures, and bitrate partitioning, the SNR quality comparisons mentioned are not universal but merely examples. In terms of subjective quality, enhancement layer quality is significantly better than simulcast and indistinguishable from single layer. Lastly, the complexity of Temporal scalability, although it employs two independent coding loops, is less than that of Spatial scalability, since each loop operates at half of the input frame rate. In fact the complexity of temporal scalability is only slightly more than that of nonscalable coding.

9.6 HYBRID SCALABILITY

The MPEG-2 standard allows combining individual scalabilities such as SNR scalability, Spatial scalability, and Temporal scalability to form hybrid scalabilities[19] as may be needed for more demanding applications. If two scalabilities are combined to form hybrid scalability, three layers result and are called the base layer, enhancement layer 1, and enhancement layer 2. Here, enhancement layer 1 is the lower layer (and sort of functions as the base layer) for enhancement layer 2.

9.6.1   High-Level Codecs

As discussed earlier, with Spatial and Temporal scalability, coupling of layers is looser than with SNR scalability. This means that in Spatial and Temporal scalability, one or more fields or frames of the base layer are first decoded before decoding the enhancement layer frames which use these base layer fields or frames for prediction, and this process is repeated. Thus, for these scalabilities, we can generally say that the base layer is decoded before the enhancement layer. However, in SNR scalability, since the base and enhancement layers are coupled more tightly, we can say that both layers are decoded simultaneously. Our observation on decoding order in individual scalabilities can be easily extended to the case of Hybrid Scalabilities. Say, in the case of three layer Hybrid scalability, either Spatial or Temporal scalability is employed between base and enhancement layer 1 and SNR scalability is employed between enhancement layer 1 and enhancement layer 2. This implies that the base layer would be decoded first followed by simultaneous decoding of both enhancement layer 1 and enhancement layer 2.

Considering Hybrid scalability with three layers employing two different scalabilities at a time, three scalability pairs result, and depending on the order in which these scalabilities are applied, a total of six scalability combinations are possible. Next, we discuss example high-level codec struc-tures for three scalability combinations out of six, taking one combinations out of every pair of com-binations formed by taking two different scalabili-ties at a time.

9.6.1.1   *Spatial and Temporal Hybrid Scalability*

In Fig. 9.28 we show a three-layer Hybrid codec employing Spatial scalability and Temporal scala-bility. In this codec, Spatial scalability is used between the base and enhancement layer 1, while Temporal scalability is used between enhancement layer 1 and enhancement layer 2. Video sequence at "in" is temporally partitioned by Temporal Demux into two sequences which appear at "in_1" and "in_2"; further, sequence "in_1" is spatially downsampled in Spatial Decimator to yield "in_0." The sequence at "in_0" is fed to the base layer of the Spatial Scalability Encoder and the sequence at "in_1" is fed to its enhancement layer, which in the context of hybrid scalability is known as enhancement layer 1. The locally decoded enhancement layer from Spatial Scalability Encoder is used to form interlayer motion-compensated prediction in Temporal Enhance-ment Encoder, which uses the second sequence from the Temporal Demultiplexer at its input, "in_2." The Spatial Scalability Encoder generates two coded bitstreams representing two spatial lay-ers and the Temporal Enhancement Encoder gen-erates the third coded bitstream representing the third layer. The three coded bitstream are packe-tized in the systems multiplexer, Sys Mux, and are ready for transmission or storage.

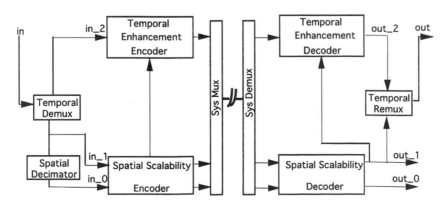

Figure 9.28   Spatial and Temporal Hybrid scalability codec.

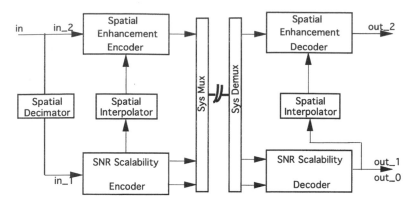

Figure 9.29 SNR and Spatial Hybrid scalability codec.

The systems demultiplexer, Sys Demux, unpacks packetized data and forwards coded bitstreams for decoding to one of the three decoders: Base Decoder or Enhancement Decoder of Spatial Scalability or Temporal Enhancement Decoder, as appropriate. Spatial Scalability Decoder performs a function complementary to that of the Spatial Scalability Encoder, decoding coded bitstreams to produce a base layer video sequence at "out_0." spatial enhancement layer at "out_1." The decoded video sequence at "out_1" is used to generate motion-compensated prediction as needed by the Temporal Enhancement Encoder while decoding its corresponding bitstream. The highest layer output is generated by temporally combining decoded video sequences at "out_1" and "out_2" in Temporal Remux, which produces the multiplexed video sequence at "out."

9.6.1.2 SNR and Spatial Hybrid Scalability

Next, we show another example of Hybrid scalability. Figure 9.29 shows a three-layer Hybrid codec employing SNR scalability and Spatial scalability. In this codec, SNR scalability is used between the base and enhancement layer 1, while Spatial scalability is used between enhancement layer 1 and enhancement layer 2. The video sequence at "in" is spatially decimated to lower resolution (the same resolution is also allowed if needed) and is available at "in_1" for input to the SNR scalability Encoder. The locally decoded higher quality video from the SNR Scalability Encoder is upsampled to full resolution in the Spatial Interpolator and is used by the Spatial Enhancement Encoder which encodes full spatial resolution video at "in_2" (same as "in"). The SNR Scalability Encoder generates two coded bitstreams representing two SNR layers and the Spatial Enhancement Encoder generates the third coded bitstream representing the spatial layer. The three coded bitstreams are packetized in the systems multiplexer, Sys Mux, and are ready for transmission or storage.

The systems demultiplexer, Sys Demux, unpacks packetized data and forwards coded bitstreams for decoding to one of the three decoders: Base Decoder and Enhancement Decoder of SNR Scalability or Spatial Enhancement Decoder, as appropriate. The SNR Scalability Decoder performs a function complementary to that of the SNR Scalability Encoder decoding coded bitstreams to produce either base layer video sequence at "out_0" or SNR enhancement layer at "out_1." Incidentally, both "out_0" and "out_1" appear on the same line, although not simultaneously. The highest layer output at "out_2" is generated by a Spatial Enhancement Decoder which decodes the highest layer bitstream and utilizes spatial prediction consisting of lower layer pictures interpolated by Spatial Interpolator as required.

9.6.1.3 SNR and Temporal Hybrid Scalability

In Fig. 9.30, we show another example of a three-layer Hybrid codec; this codec employs SNR

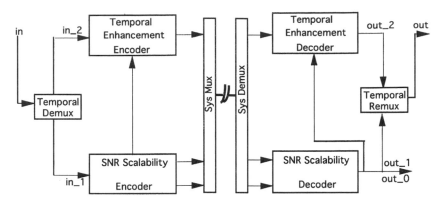

Figure 9.30　Temporal and SNR Hybrid scalability codec.

scalability and Temporal scalability. In this codec, SNR scalability is used between the base and enhancement layer 1, while Temporal scalability is used between enhancement layer 1 and enhancement layer 2. The video sequence at "in" is temporally partitioned by the Temporal Demux into two sequences which appear at "in_1" and "in_2." The sequence at "in_1" is fed to the SNR Scalability Encoder. The locally decoded higher quality video from the SNR Scalability Encoder is used by the Temporal Enhancement Encoder to form motion-compensated interlayer prediction. The SNR Scalability Encoder generates two coded bitstreams representing two SNR layers and the Temporal Enhancement Encoder generates the third coded bitstream representing the temporal layer. The three coded bitstreams are packetized in the systems multiplexer, Sys Mux, and are ready for transmission or storage.

The systems demultiplexer, Sys Demux, unpacks packetized data and forwards coded bitstreams for decoding to one of the three decoders: Base Decoder and Enhancement Decoder of SNR Scalability or Temporal Enhancement Decoder, as appropriate. The SNR Scalability Decoder performs a function complementary to that of SNR Scalability Encoder, decoding coded bitstreams to produce either base layer video sequence at "out_0" or SNR enhancement layer at "out_1." Incidentally, both "out_0" and "out_1" appear on the same line, although not simultaneously. The highest layer output at "out_2" is generated by the Temporal Enhancement Decoder which decodes

the highest layer bitstream and utilizes lower layer video for motion-compensated prediction as needed. The decoded lower layer video at "out_0" or "out_1," whichever is available, is multiplexed with the decoded higher layer video in Temporal Remux and output on "out" as the full temporal resolution signal.

9.6.2　Hybrid Scalability Applications

We have presented three examples of hybrid scalability codec structures each using three layers. As you may have noticed, it is easily possible to mix and match individual scalabilities to create hybrid scalability structures. Further, with an increase in number of layers, the number of combinations increases substantially and so does the hardware complexity. So which hybrid scalabilities are expected to be more useful than others? As in the case of individual scalabilities, the answer is not straightforward. It all depends on demands of applications where hybrid scalability may be deployed. That is, what type of picture quality is required for each layer, what type of bitrate partitioning is necessary, what implementation complexity can be afforded for each layer, etc. To stimulate further thinking and debate, we consider a few concrete examples of application scenarios where three-layer hybrid scalability may be employed.

Consider the case of digital HDTV and normal TV. From the aspect of efficient usage of available spectrum and integration of services, normal TV should migrate to digital format allowing digital TV to be a lower layer of HDTV. Further, HDTV itself has challenges to overcome owing to bandwidth constraints and receiver costs in the short term which appear to favor use of interlaced video format or progressive video format with lower spatial resolution. Undoubtedly, in the future or even now in nonbroadcast applications, higher resolution progressive video formats would be used to provide high-quality HDTV; such HDTV services would have to interoperate with less expensive HDTV receivers, or what may eventually be called first generation HDTV receivers. So what is the answer to this interoperability nightmare? Well, Spatial and and Temporal Hybrid Scalability certainly offers a solution that would be far more meaningful than simulcasting these three layers. Of course there are too many other factors so it is hard to say whether such a service can become practical or not.

Consider another case, again that of digital HDTV and digital TV. Now, assume that digital TV of two qualities is required, implying that consumers can either receive normal quality TV or high-quality TV, depending on the type of receivers (and even bandwidth) they can afford. Further, consider that there is yet another class of consumer who can afford HDTV receivers. So how can a three-layer service offering TV at two qualities and HDTV at one quality be possible? The answer is SNR and Spatial Hybrid Scalability, of course. Again, simulcasting three such layers would not make a lot of sense from the bandwidth efficiency standpoint. Again, there are too many potential factors that may or may not allow such a service, not only for terrestrial broadcast, but also on cable, satellite, or other means.

Consider yet another case, this time either that of digital TV or digital HDTV. If we consider the case of digital TV, in designing such a service, it is not difficult to anticipate the need for three layers. These three layers could very well be normal quality TV, enhanced quality TV, and progressive TV and there would be corresponding TV receivers to choose from. Viewers can then choose the quality of receiver (and cost of service) they can afford. How can such a three-layer service be possible in a

bandwidth efficient manner? The answer is SNR and Temporal Hybrid scalability. Again, simulcasting the three layers would not be efficient compared to scalable coding. The scenario just described applies equally well to digital HDTV where three layers may be used.

9.7 SCALABILITY SUMMARY

Scalability of video means the same video at different resolutions or qualities simultaneously. Simulcast coding is one way to achieve scalability of video. A more efficient way to achieve scalability of video is by scalable video coding which is also referred to as layered video coding.

We now summarize the various scalabilities as follows:

- Data partitioning is a simple technique that allows partitioning of encoded MPEG-2 video bitstreams into two partitions or layers for transmission or storage. At the decoder, the two such layers can be recombined to produce the same quality as produced by an unpartitioned bitstream. However, the quality of pictures produced by the lower layer can be quite poor and thus insufficient for standalone viewing in case of losses in the enhancement layer.

- SNR scalability is a layered coding method that allows two (or more) layers. It is a little bit more involved than data partitioning. Typically, the performance of recombined layers at the decoder produces lower quality as compared to data partitioning. However, the lower layer in SNR scalability can provide quality sufficient for standalone viewing. It can be used to provide multiquality services and in error-resilient applications.

- Spatial scalability is a layered coding method that allows two (or more) layers. It is somewhat more complex than SNR scalability but also offers the flexibilities of different resolution in each layer and compatibility with other standards. The lower layer can achieve quite good quality whereas the quality of combined layers can in some cases be notably lower than that obtained with nonscalable coding. It can be used for interworking between standards, HDTV with embedded TV, error resilience, and other applications.

• Temporal scalability is a layered coding method that allows two (or more) layers in time. It involves coding complexity similar to that obtained with nonscalable coding. In a simplified scenario, both the lower layer and the enhancement layer may have half the temporal resolution of the input video. The spatial quality of the lower layer and the combined lower and enhancement layers is usually quite high. It can be used to provide inter-operability between progressive and interlaced video in applications such as digital HDTV, digital TV, and others.

• Hybrid Scalability allows combining two scalabilities at a time from among Spatial scalability, the SNR scalability, and the Temporal scalability. It is useful in more demanding applications requiring three or more layers.

REFERENCES

1. A. PURI and A. WONG, "Spatial Domain Resolution Scalable Video Coding," *Proc. SPIE Visual Commun. Image Proc,* pp. 1701–1713 (November 1993).

2. A. PURI, "Video Coding Using the MPEG-2 Compression Standard," *Proc. SPIE Visual Commun. Image Proc,* pp. 718–729 (November 1993).

3. R. ARAVIND, M. R. CIVANLAR, and A. R. REIBMAN, "Packet Loss Resilience of MPEG-2 Scalable Video Coding Algorithms," Technical Memorandum, AT&T Bell Labs, (June 1995).

4. C. HERPEL, "SNR Scalability vs Data Partitioning for High Error-Rate Channels," ISO/IECJTC1/SC29/WG11 Doc. MPEG 93/658 (July 1993).

5. G. MORRISON and I. PARKE, "A Spatially Layered Hierarchical Approach to Video Coding," *Signal Proc. Image* 5 (5–6) 445–4623 (December 1995).

6. T. CHIANG and D. ANASTASSIOU, "Hierarchical Coding of Digital Television," *IEEE Commun. Magazine* 32 (5) 38–45 (May 1994).

7. Y. WANG and A. PURI, "Spatial-Temporal Adaptive Interlace–Interlace Extraction," ISO/IEC JTC1/SC29/WG11 Doc. MPEG 92/509 (September 1992).

8. D. ANASTASSIOU, "Scalability for HDTV," International Workshop on HDTV'92, Signal Processing of HDTV, IV, pp. 9–15, 1993.

9. M. R. CIVANLAR and A. PURI, "Scalable Video Coding in Frequency Domain," *Proc. SPIE Visual Commun. Image Proc.* 1818:1124–1134 (1992).

10. A. PURI, L. YAN, and B. G. HASKELL, "Temporal Resolution Scalable Video Coding," *Proc. IEEE Int. Conference on Image Processing,* 1994.

11. C. GONZALES and E. VISCITO, "Flexibly Scalable Digital Video Coding," Signal Processing: Image Commun. Vol. 5, Nos. 1–2, February 1993.

12. A. PURI, L. YAN, and B. G. HASKELL, "Syntax, Semantics and Description of Temporal Scalability," ISO/IEC JTC1/SC29/WG11 Doc. MPEG 93/795, (September 1993).

13. ETRI, "Comparison of CTV/HDTV Compatible Coding Schemes," ISO/IEC JTC1/SC29/WG11 Doc. MPEG93/569 (July 1993).

14. T. NAVEEN and S. C.-F HU, "Comparison of Spatial and Frequency Kies," ISO/IEC JTC1/SC29/WG11 Doc. MPEG93/615, (1993).

15. A. PURI and B. G. HASKELL, "Picture Format Scalable Coding Structures," ISO/IEC JTC1/SC29/WG11 Doc. MPEG93/673 (1993).

16. A. PURI, "Picture Format Scalable Coding for HDTV," ISO/IEC JTC1/SC29/WG11 Doc. MPEG93/390 (March 1993).

17. A. WONG, A. PURI, and D. ANASTASSIOU, "Multiplexing and Syntax Improvements in Spatially Scalable Coding," ISO/IEC JTC1/SC29/WG11 Doc. MPEG92/697 (November 1992).

18. T. SIKORA, T. K. TAN and K. N. NGAN, "A Performance Comparison of Frequency Domain Pyramid Scalable Coding Schemes within the MPEG Framework," Proc. Picture Coding Symposium, pp. 16.1–16.2, March 1993.

19. "Generic Coding of Moving Pictures and Associated Audio Information; Video," ISO/IEC 13818-2: Draft International Standard, November 1994.

20. Test Model Editing Committee, "Test Model 5," ISO/IEC JTC1/SC29/WG11/NO400, April 1993.

21. O. PONCIN et al., "New Results on Spatial Scalability," ISO/IEC JTC1/SC29/WG11 MPEG93/495, July 1993.

10

Video Stream Syntax and Semantics

In this chapter, we present the syntax of the MPEG-2 video stream plus the semantics of the variables that make up the video stream.* As mentioned elsewhere, the MPEG-2 video standard[3] is a syntactic superset[4] of the MPEG-1 video standard.[2] The video stream (in general) consists of a set of substreams called Layers. For a nonscalable video stream,[3,5] there is only one layer. If there are two layers or more, the video data is coded as a scalable hierarchy.[5] The first layer (zero) is called the base layer, and, except for data partitioning, it does not contain a sequence_scalable_extension( ) and can always be decoded independently. Other layers are referred to as enhancement layers, and as described in the previous chapter, they always contain sequence_scalable_extension( ) and can be decoded only after decoding all of the lower layers.

10.1 VIDEO STREAM SYNTAX

As mentioned in previous chapters, the video stream is organized as a hierarchy of headers and data that provide all the information necessary to decode and display pels. In this chapter we will use a number of shorthand notations to succinctly describe the stream syntax.

For example, each Start-Code consists of a Start-Code prefix *Psc* followed by a Start-Code ID. Psc is a string of 23 or more binary zeros followed by a binary one. The Start-Code ID consists of two hexadecimal characters that identify the type of Start-Code, as shown in Table 10.1.

A number of *Extensions* are also allowed for various purposes at specified places in the video stream. They all begin with the extension_start_code, followed by one of the allowable Extension_IDs shown in Table 10.2.

Figures 10.1 to 10.19 describe the video stream syntax using a hierarchy of flow charts that a decoder would use to extract the data elements needed for decoding. Elements containing ( ) indi-

Table 10.1 Video Start-Code IDs.

Start Code Type	start code ID (8 bits) (hexadecimal)
picture_start_code	00
slice_start_code	01 to AF
user_data_start_code	B2
sequence_header_code	B3
sequence_error_code	B4
extension_start_code	B5
sequence_end_code	B7
group_start_code	B8
reserved	B0, B1, B6

---

*Detailed study of this chapter is not necessary for the novice reader.

Table 10.2 Extension ID codes.

Extension_ID (4 bits) (hexadecimal)	Name
1	Sequence Extension
2	Sequence Display Extension
3	Quant Matrix Extension
4	Copyright Extension
5	Sequence Scalable Extension
7	Picture Display Extension
8	Picture Coding Extension
9	Picture Spatial Scalable Extension
A	Picture Temporal Scalable Extension
0, 6, B to F	reserved

cate procedures that will be described in later flow charts. Conditional steps are indicated inside oval shapes with question marks "??". If at any time a reserved Start-Code or reserved Extension_ID value is encountered, then all ensuing data in the stream is discarded until the next Start-Code. A boldface variable indicates that a value is read in from the stream. If a numerical value appears in parentheses with the boldface variable, then the data read in must have that value. The variable NextSC refers to the immediately upcoming Start-Code that must occur next in the stream. The variable NextEXT refers to the immediately upcoming Extension_ID that must occur next in the stream. NextHex refers to the next hexadecimal characters in the stream, and NextBits refers to the next bit pattern in the stream. Characters preceded by 0x are hexadecimal. A string of N zeros or ones may be denoted by $N*0$ or $N*1$, respectively. Powers of 10 are denoted by the letter "e," $27e6 = 27,000,000$. Flag and indicator variables are all one bit, unless otherwise indicated.

10.1.1 Error Codes Used in Syntax Diagrams

E1 In repeated Sequence and Extension Headers, only the quantizer matrix data can change.

E2 The picture following a Sequence_extension must not be a B-picture.

E3 The first and last Macroblocks of each slice must be coded, that is, not skipped. In all Profiles defined to date, slices must cover the entire picture with no gaps between them.

E4 In a B-picture, the MB following an Intra MB should not be skipped. In an I-picture no skipped MBs are allowed at all.

E5 Sequence Headers cannot appear before the second field picture in a frame.

E6 After a group_of_pictures header, the next picture must be an I-picture.

E7 Within a GOP, the temporal_reference repeats only for the second field picture of a frame. It never repeats for frame pictures.

E8 If the first field picture of a frame is not an I-picture, then the second field picture must be of the same picture_coding_type.

E9 If the first field picture of a frame is an I-picture, then the second field picture must not be a B-picture.

E10 Extension_IDs must occur in their specified places in the video stream.

E11 This Extension ID can occur only once per sequence_header.

E12 This Extension ID can occur only once per picture_header.

10.2 SEMANTICS FOR THE SYNTAX VARIABLES

This Section describes the semantic meaning of all variables in the video stream. Figure 10.1 illustrates the high-level structure of the video stream. In this figure we see that if sequence_extension( ) is missing, then the stream conforms to MPEG-1 and is not described in this syntax. Also, if a decoder encounters a Start-Code or extension ID that is "reserved" then the decoder discards all subsequent data until the next Start-Code. This allows future definition of compatible extensions.

10.2.1 Video Sequence (Fig. 10.1)

sequence_end_code The sequence_end_code terminates a video sequence. A new video sequence may then commence. However, decoder behavior during this transition is not defined. Thus, care must be taken in the video decoder to provide for a graceful transition. For example, buffers should probably be empty at the start of a new sequence.

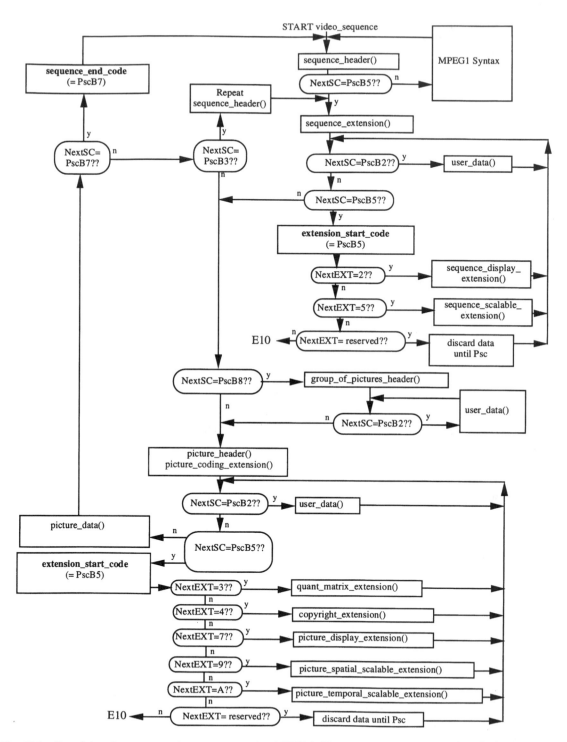

Fig. 10.1 Overall decoding procedure for a single layer of an MPEG-2 video sequence. Psc is the Start-Code Prefix, which is 23 or more binary zeros followed by a binary one.

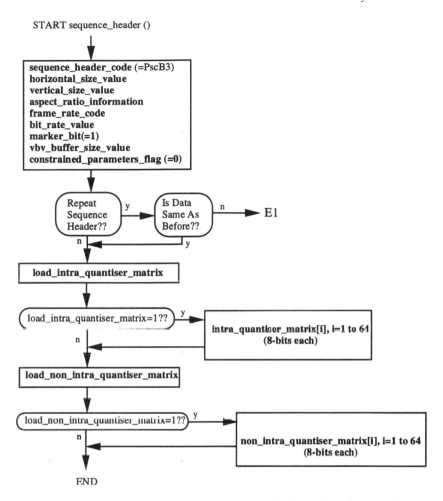

Fig. 10.2 MPEG-2 Sequence Header. The following picture must not be a B-picture. Most of the variables are used only for an MPEG-1 Sequence.

10.2.2 Sequence_header( ) (Fig. 10.2)

sequence_header_code Start-Code Prefix Psc followed by hexadecimal B3.

horizontal_size_value 12 least significant bits of horizontal_size.

vertical_size_value 12 least significant bits of vertical_size.

horizontal_size 14-bit unsigned integer (see also horizontal_size_extension). The horizontal_size is the width in pels of the visible part of the luminance frame. The width in Macroblocks (MBs) of the encoded luminance frame, MB_width, is given by (horizontal_size + 15)/16. The visible part is left-aligned.

vertical_size 14-bit unsigned integer (see also vertical_size_value). The vertical_size is the height in lines of the visible part of the luminance frame. If progressive_sequence =1, the height in MBs of the encoded luminance frame picture, MB_height, is given by (vertical_size + 15)/16. If progressive_sequence = 0, MB_height = 2 × ((vertical_size + 31)/32). For field pictures, MB_height = ((vertical_size + 31)/32). The visible part is top-aligned.

aspect_ratio_information (4 bits) See Table 10.3. Specifies either a square Pel Aspect Ratio (PAR) for the reconstructed frame, or alternatively it specifies the Image Aspect Ratio (IAR), in which case PAR may be calculated (see Section 5.1.4)

frame_rate_code (4 bits) Defines frame_rate_value as shown in Table 10.4. The frame_rate in Hertz is

Table 10.3 Pel and Image Aspect Ratio Information

aspect_ratio_information (hexadecimal)	PAR or IAR
0	forbidden
1	PAR = 1
2	IAR = 4:3
3	IAR = 16:9
4	IAR = 2.21:1
5 to F	reserved

Table 10.4 frame_rate_value.

frame_rate_code (hexadecimal)	frame_rate_value
0	forbidden
1	24 000 ÷ 1001 (23.976 . . .)
2	24
3	25
4	30 000 ÷ 1001 (29.97 . . .)
5	30
6	50
7	60 000 ÷ 1001 (59.94 . . .)
8	60
9 to F	reserved

then derived from the parameters frame_rate_value, frame_rate_extension_n and frame_rate_extension_d by frame_rate = frame_rate_value × (frame_rate_extension_n + 1) ÷ (frame_rate_extension_d + 1)

bit_rate_value The lower 18 bits of bit_rate. See also **bit_rate_extension** below for the upper 12 bits.

bit_rate A 30-bit integer that specifies the maximum rate of data delivery (in units of 400 bits/s) to the Video Buffer Verifier (VBV), that is $R_{max} = 400 \times$ **bit_rate** bits/s. Thus, $R(n) \leq R_{max}$ for all n, where $R(n)$ is the VBV input bit-rate for picture n. See Eq. (8.2).

marker_bit Set to 1 to prevent emulation of Start-Codes.

vbv_buffer_size_value The lower 10 bits of vbv_buffer_size.

vbv_buffer_size An 18-bit integer specifying the size (in units of 2048 bytes) of the VBV. See also vbv_buffer_size_extension.

constrained_parameters_flag Set to 0.

load_intra_quantizer_matrix, intra_quantizer_matrix, load_non_intra_quantizer_matrix, non_intra_quantizer_matrix See Quant matrix extension( )

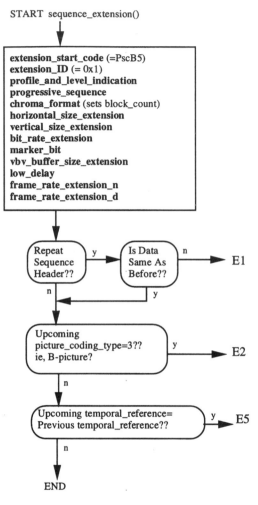

Fig. 10.3 Sequence Extension.

10.2.3 Sequence_extension( ) (Fig. 10.3)

extension_start_code Start-Code Prefix Psc followed by hexadecimal B5.

extension_ID 0x01

profile_and_level_indication 8-bit integer specifying the profile and level.

progressive_sequence 1 indicates that the sequence contains only progressive frame-pictures. This is the primary switch between interlaced and progressive video sources.

chroma_format A 2-bit variable indicating the chrominance formal as defined in Table 10.5. block_count-the number of blocks per MB. See Table 10.5.

Table 10.5 chroma_format.

chroma_format	Meaning	block_count
00	reserved	
01	4:2:0	6
10	4:2:2	8
11	4:4:4	12

horizontal_size_extension 2 most significant bits from horizontal_size.

vertical_size_extension 2 most significant bits from vertical_size.

bit_rate_extension 12 most significant bits from bit_rate.

vbv buffer size extension 8 most significant bits from vbv_buffer_size.

low_delay 1 indicates the sequence contains no B-pictures, that frame reordering is not present and that *big pictures* may be present. See Section 8.3.3.

frame_rate_extension_n 2-bit integer used to determine the frame_rate. See frame_rate_code.

frame_rate_extension_d 5-bit integer used to determine the frame_rate. See frame_rate_code.

10.2.4 User_data( ) (Fig. 10.4)

user_data_start_code Psc followed by hexadecimal B2. The user data continues until another Start-Code.

user_data An arbitrary string bytes defined by users for their specific applications. The data are not allowed to contain Psc.

10.2.5 Sequence_display_ extension( ) (Fig. 10.5)

extension_ID Hexadecimal 2.

video_format 3-bit integer indicating the original video format, as shown in Table 10.6.

color_description 1 indicates the presence of color_primaries, transfer_characteristics and color matrix_coefficients.

color_primaries 8-bit integer defining the chromaticity coordinates of the source primaries, as shown in Table 10.7.

transfer_characteristics 8-bit integer specifying the *gamma* of the original video, as shown in Table 10.8.

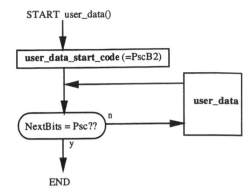

Fig. 10.4 User Data can be arbitrary, except that a Start-Code Prefix (Psc) is not allowed.

matrix_coefficients 8-bit integer describing the transformation from the red, green and blue primaries to integer values Y, Cb, and Cr, as shown in Table 10.9. We assume

E'_Y, E'_R, E'_G, and E'_B are analog with values between 0 and 1

E'_{PB} and E'_{PR} are analog with values between -0.5 and 0.5

$$Y = (219 \times E'_Y) + 16$$

$$Cb = (224 \times E'_{PB}) + 128$$

$$Cr = (224 \times E'_{PR}) + 128$$

display_horizontal_size, display_vertical_size (14 bits each) Define the display active region, which may be smaller or larger than the decoded frame size

10.2.6 Sequence_scalable_ extension( ) (Fig. 10.6)

extension_ID _Hexadecimal 5.

scalable_mode (2 bits) Indicates the type of scalability as shown in Table 10.10. Depending on the mode and presence of picture_spatial_scalable_ extension( ), different VLCs are used for macroblock_type. See Section 10.2.16.

layer_id (4 bits) Identifies the layers in a scalable hierarchy. Each higher layer has a layer_id one greater than the lowe layer. In the case of data partitioning, layer_id = 0 for partition zero and 1 partition one.

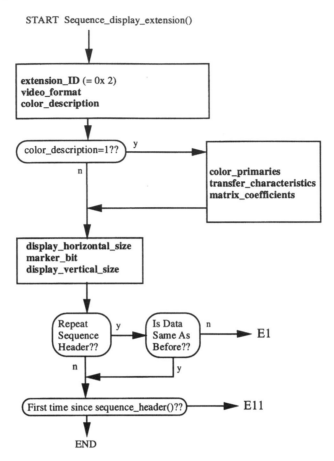

START Sequence_display_extension()

extension_ID (= 0x 2)
video_format
color_description

color_description=1??

color_primaries
transfer_characteristics
matrix_coefficients

display_horizontal_size
marker_bit
display_vertical_size

Repeat
Sequence
Header??

Is Data
Same As
Before??

E1

First time since sequence_header()??   E11

END

Fig. 10.5 Sequence Display Extension.

lower_layer_prediction_horizontal_size, lower_layer_prediction_vertical_size   14-bit integers equal to horizontal_size and vertical_size, respectively, in the lower layer.

horizontal_subsampling_factor_m, horizontal_subsampling_factor_n, vertical_subsampling_factor_m, vertical_subsampling_factor_n   (5 bits each) Affect the spatial scalable upsampling process, as defined in Chapter 9.

picture_mux_enable   1 indicates picture_mux_order and picture_mux_factor are used for remultiplexing prior to display.

mux_to_progressive_sequence   1 indicates the decoded pictures in the two layers are temporally multiplexed to generate a progressive sequence for display. Otherwise, an interlaced sequence is generated.

Table 10.6   video_format.

video_format	Original video
000	component
001	PAL
010	NTSC
011	SECAM
100	MAC
101	Unspecified video format
110	Reserved
111	Reserved

picture_mux_order   (3 bits) The number of enhancement layer pictures prior to the first base layer picture.

picture_mux_factor   (3 bits) The number of enhancement layer pictures between consecutive base layer pictures.

Table 10.7 color_primaries.

Value	Color Standard	Reference White (x,y)	Red, green, blue (x,y)
0	(forbidden)		
1	Recommendation ITU-R BT.70	0.3127, 0.3290	0.640, 0.330 0.300, 0.600 0.150, 0.060
2	Unspecified Video		
3	Reserved		
4	Recommendation ITU-R BT.470–2 System M	0.310, 0.316	0.67, 0.33 0.21, 0.71 0.14, 0.08
5	Recommendation ITU-R BT.470–2 System B, G	0.313, 0.329	0.64, 0.33 0.29, 0.60 0.15, 0.06
6	SMPTE 170M	0.3127, 0.3290	0.630, 0.340 0.310, 0.595 0.155, 0.070
7	SMPTE 240M (1987)	0.3127, 0.3291	0.630, 0.340 0.310, 0.595 0.155, 0.070
8–255	reserved		

Table 10.8 Transfer Characteristics.

Value	Transfer Characteristic
0	(forbidden)
1	Recommendation ITU-R BT.709 $V = 1.099\,L_c^{0.45} - 0.099$ for $1 \geq L_c \geq 0.018$ $V = 4.500\,L_c$ for $0.018 > L_c \geq 0$
2	Unspecified video
3	Reserved
4	Recommendation ITU-R BT.470–2 System M gamma = 2.2
5	Recommendation ITU-R BT.470–2 System B, G gamma = 2.8
6	SMPTE 170M $V = 1.099\,L_c^{0.45} - 0.099$ for $1 \geq L_c \geq 0.018$ $V = 4.500\,L_c$ for $0.018 > L_c \geq 0$
7	SMPTE 240M (1987) $V = 1.1115\,L_c^{0.45} - 0.1115$ for $L_c \geq 0.0228$ $V = 4.0\,L_c$ for $0.0228 > L_c$
8	Linear transfer characteristics $V = L_c$
9–255	Reserved

Table 10.9 Color Transforms. Values 4, 5, and 6 are essentially identical. Different numbers were published in the original standards owing to different rounding in the calculations.

Value	Matrix
0	(Forbidden)
1	Recommendation ITU-R BT.709 $E'_Y = +0.2125\,E'_R + 0.7154\,E'_G + 0.0721\,E'_B$ $E'_{PB} = -0.115\,E'_R - 0.386\,E'_G + 0.500\,E'_B$ $E'_{PR} = +0.500\,E'_R - 0.454\,E'_G - 0.046\,E'_B$
2	Unspecified Video
3	Reserved
4	US FCC $E'_Y = +0.30\,E'_R + 0.59\,E'_G + 0.11\,E'_B$ $E'_{PB} = -0.169\,E'_R - 0.331\,E'_G + 0.500\,E'_B$ $E'_{PR} = +0.500\,E'_R - 0.421\,E'_G - 0.079\,E'_B$
5	Recommendation ITU-R BT.470–2 System B, G $E'_Y = +0.299\,E'_R + 0.587\,E'_G + 0.114\,E'_B$ $E'_{PB} = -0.169\,E'_R - 0.331\,E'_G + 0.500\,E'_B$ $E'_{PR} = +0.500\,E'_R - 0.419\,E'_G - 0.081\,E'_B$
6	SMPTE 170M $E'_Y = +0.299\,E'_R + 0.587\,E'_G + 0.114\,E'_B$ $E'_{PB} = -0.169\,E'_R - 0.331\,E'_G + 0.500\,E'_B$ $E'_{PR} = +0.500\,E'_R - 0.419\,E'_G - 0.081\,E'_B$
7	SMPTE 240M (1987) $E'_Y = +0.212\,E'_R + 0.701\,E'_G + 0.087\,E'_B$ $E_{PB} = -0.116\,E'_R - 0.384\,E'_G + 0.500\,E'_B$ $E'_{PR} = +0.500\,E'_R - 0.445\,E'_G - 0.055\,E'_B$
8–255	Reserved

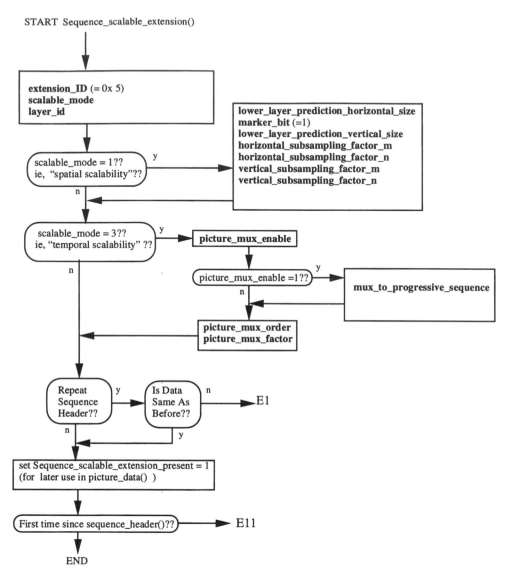

Fig. 10.6   Sequence Scalable Extension.

Table 10.10   scalable_mode

scalable_mode	Meaning
00	Data partitioning
01	Spatial scalability
10	SNR scalability
11	Temporal scalability

10.2.7   Group_of_pictures_ header( ) (Fig. 10.7)

group-start-code   Start-Code Prefix Psc followed by hexadecimal B8.

time_code   25-bit integer containing the SMPTE time code, as shown in Table 10.11. The time refers to the first of the ensuing pictures to be displayed after the group of pictures header that has a temporal_reference of zero.

closed_gop   1 indicates that B-pictures following the first I-picture have been encoded using only backward prediction or intra coding. Thus, they can be used following a random access.

broken_link   1 indicates that the first consecutive B-pictures (if any) immediately following the next I-frame following the group of picture header may not

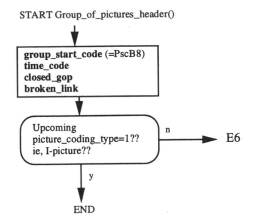

Fig. 10.7 Group of Pictures Header.

Table 10.11 SMPTE time codes are carried by the 25-bit parameter time_code. drop_frame_flag is used only for 29.97 frames/s.

time_code	Range	No. of bits
drop_frame_flag	0 to 1	1
time_code_hours	0 to 23	5
time_code_minutes	0 to 59	6
marker_bit	1	1
time_code_seconds	0 to 59	6
time_code_pictures	0 to 59	6

be correctly decoded because the previous reference is not available.

10.2.8 PICTURE_HEADER( ) (FIG. 10.8)

picture_start_code Start-Code Prefix Psc followed by hexadecimal 00.

temporal_reference 10-bit unsigned integer associated with frame. The temporal_reference is the same for both field pictures of a frame. The temporal_reference increments by one (modulo 1024) for each frame in display order. However, after a group of pictures header, the temporal_reference of the first frame to be displayed is zero. If low_delay = 1 and a *big picture* occurs: then the VBV buffer is reex-

amined several times (after the first examination) before removing the big picture. The temporal_reference of the picture immediately following the big picture equals the temporal_reference of the big picture incremented by $N + 1$ (modulo 1024), where N is the number of times that the VBV buffer is reexamined.* If the big picture is immediately followed by a group of pictures header, the temporal_reference following the group of pictures header equals N.

picture_coding_type 3-bit variable that indicates whether a picture is an I-, P-, or B-picture, as shown in Table 10.12.

vbv_delay A 16-bit unsigned integer indicating the amount of time the picture header† should reside in the VBV buffer before being extracted for decoding (see Section 8.3.3). The actual delay is given by vbv_delay $\times$ 300/27e6 seconds, where timing is locked to the 27MHz Systems Time Clock. If vbv_delay = 0xFFFF, then basically the VBV is not used.‡ If low_delay = 1 and a Big Picture occurs, the vbv_delay value may be wrong.

For constant bitrate (CBR) operation, vbv_delay may be calculated from the state of the VBV as follows:

$$\text{vbv\_delay}_n = 27e6 \times B_n^- /(300 \times R)$$

where B_n^- = VBV buffer fullness, measured in bits, immediately before removing picture n, but after removing any header(s), user data, etc. that immediately precedes the picture_data( ), and R = the actual bitrate. See also Eq. (8.9)

For variable bitrate (VBR) operation, the situation is much more complicated.[1] Basically, data enter the VBV at a rate that is constant for each picture, but which may vary from picture to picture (see Section 8.3.3). In this case, it is the responsibility of the encoder to ensure correct values for vbv_delay.

The following four variables are left over from MPEG-1 and are not used in MPEG-2.

full_pel_forward_vector (1 bit) Set to 0.
forward_f_code (3 bits) Set to "111."

---

*Thus with low_delay = 1, the temporal_reference of the two field pictures of a frame may differ if the first field is a Big Picture.

†Actually, the final byte of the Picture Start-Code.

‡Actually, the specification states that data intermittently enters the VBV buffer at a rate of $R_{\max} = 400 \times bit\_rate$ bits/s until it is full.

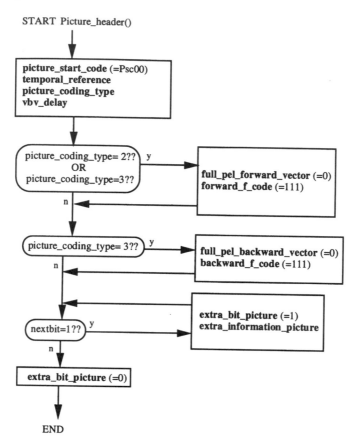

Fig. 10.8   Picture Header.

Table 10.12   picture_coding_type.

picture_coding_type	Coding method
000 and 100	Forbidden
001	Intra-coded (I)
010	Predictive-coded (P)
011	Bidirectionally-Predictive-coded (B)
101, 110 and 111	Reserved

full_pel_backward_vector (1 bit)   set to 0.

backward_f_code (3 bits)   set to "111."

extra_bit_picture   1 implies extra_information_ picture will follow, which should be discarded.

extra_information_picture (8 bits)   Reserved.*

10.2.9 Picture coding extension( ) (Fig. 10.9)

extension_start_code   Start-Code Prefix Psc followed by hexadecimal B5.

extension_ID   Hexadecimal 8.

f_code[s = 0,1][t = 0,1]   Four hexadecimal characters (taking values 1 to 9, or F) used in the decoding of motion vectors to set the parameter **r_size = f_code-1;** see Section 7.4. Unused values are set to 0xF. $s = 0,1$ for forward and backward motion vectors, respectively. $t = 0,1$ for horizontal and vertical components, respectively.

intra_dc_precision   2-bit integer used to define stepsize for inverse quantisation of the Intra DC

---

*Reserved values may be defined in the future.

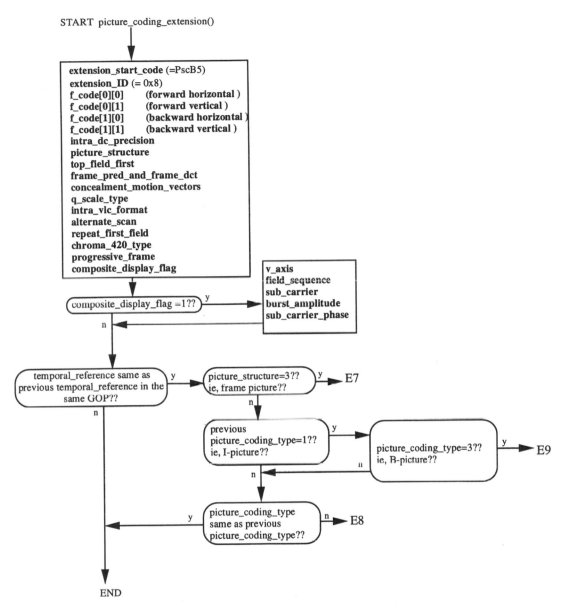

Fig. 10.9 Picture Coding Extension.

coefficients. See Section 8.4.6.1. intra_dc_precision $= 3 - \log_2$ stepsize.

picture_structure 2-bit integer defined in Table 10.13 that specifies the picture structure. With interfaced sources, a frame consists of a top field and a bottom field, either of which may be temporally first. Interlace may be coded using either frame-pictures or field-pictures. With progressive sources, only frame-pictures are used.

Table 10.13 picture_structure.

picture_structure	Meaning
00	reserved
01	Top Field Picture
10	Bottom Field Picture
11	Frame Picture

The following 10 variables are 1-bit flags:

top_field_first Set to 0 for field pictures. For frame pictures its meaning depends on the variables progressive_sequence and repeat_first_field.

If progressive_sequence = 0, this variable specifies which field of the decoded frame is output first. 1 indicates that the top field is output first. Otherwise, the bottom field is output first.

If progressive_sequence = 1, the 2-bit pair (top_field_first, repeat_first_field) specifies how many times (one, two or three) the reconstructed frame is output by the decoding process. Values (0,0), (0,1), (1,1) indicate 1, 2, and 3 times, respectively.

frame_pred_and_frame_dct* 1 indicates that only frame-DCT and frame prediction are used. Set to 0 for a field picture. Set to 1 if progressive_sequence = 1.

concealment_motion_vectors 1 indicates motion vectors are in Intra MBs. They may be used for concealment of transmission errors.

q_scale_type Used with quantizer_scale_code to determine quantizer_scale, which is used in the inverse quantisation process. See Section 8.4.6.1.

intra_vlc_format Used in the decoding of transform coefficients. See Section 8.4.6.2.

alternate_scan Used in the decoding of transform coefficients. 0 indicates use of the zigzag scan. 1 indicates use of the alternate scan. See Section 8.3.1.

repeat_first_field Set to 0 for field pictures or if progressive_sequence = 0 and progressive_frame = 0, in which case the variable has no effect.

If progressive_sequence = 0 and progressive_frame = 1 and repeat_first_field = 1, then the output of the decoder for this reconstructed frame consists of three fields, namely the first field, followed by the other field, followed again by the first field repeated. This is the 3:2 pulldown process. If repeat_first_field = 0, output consists of two fields as normal.

If progressive_sequence = 1, see top_field_first.

chroma_420_type If chroma_format is 4:2:0, set chroma_420_type = progressive_frame (for historical reasons).

progressive_frame 1 indicates that the two fields (of the frame) correspond to the same time instant, for example, film. In this case we must also have picture_structure = "frame" and if progressive_sequence = 1 we must also have frame_pred_and_frame_det = 1. See Table 10.14.

composite_display_flag 1 indicates the presence of the following five fields, which are of use when the original signal was analog composite video.

v_axis (1 bit) Used only when the original is PAL. 1 indicates a positive sign.

field_sequence (3 bits) Defines the field number in the eight field sequence used in PAL or the four field sequence used in NTSC.

Table 10.14 Legal combinations of **progressive_sequence, progressive frame, picture_structure, top_ field_ first,** and **repeat_ first_ field.** All other combinations are illegal.

progressive_ sequence	progressive_ frame	picture_ structure	top_field_ first	repeat_ first_field	Display order of Fields in a Frame
0	0	field	0	0	First coded field displayed first
0	0	frame	0	0	Bottom field first, 2 fields displayed
0	0	frame	1	0	Top field first, 2 fields displayed
0	1	frame	0	0	Bottom field first, 2 fields displayed
0	1	frame	0	1	Bottom field first, 3 fields displayed
0	1	frame	1	0	Top field first, 2 fields displayed
0	1	frame	1	1	Top field first, 3 fields displayed
1	1	frame	0	0	One progressive frame displayed (MP@ML)
1	1	frame	0	1	Two progressive frames displayed
1	1	frame	1	1	Three progressive frames displayed

---

*MPEG calls this frame_pred_frame_dct.

sub_carrier (1 bit)   0 indicates the subcarrier/line frequency relationship is correct.

burst_amplitude   7-bit integer defines the color subcarrier burst amplitude (in the pel domain) for PAL and NTSC.

sub_carrier_phase   8-bit integer defines the phase in degrees of the color subcarrier with respect to the field start. Phase = **sub_carrier_phase** × 360/256.

10.2.10   Quant matrix extension( ) (Fig. 10.10)

extension_ID   Hexadecimal 3.

Default matrix for Intra blocks (both luminance and chrominance)

8	16	19	22	26	27	29	34
16	16	22	24	27	29	31	37
19	22	26	27	29	34	34	38
22	22	26	27	29	34	37	40
22	26	27	29	32	35	40	48
26	27	29	32	35	40	48	58
26	27	29	34	38	46	56	69
27	29	35	38	46	56	69	83

Default matrix for Non Intra blocks (both luminance and chrominance)

16	16	16	16	16	16	16	16
16	16	16	16	16	16	16	16
16	16	16	16	16	16	16	16
16	16	16	16	16	16	16	16
16	16	16	16	16	16	16	16
16	16	16	16	16	16	16	16
16	16	16	16	16	16	16	16
16	16	16	16	16	16	16	16

load_intra_quantizer_matrix   1 indicates intra_quantizer_matrix follows.

intra_quantizer_matrix   Sixty-four 8-bit unsigned integers in the default zigzag scanning order. The first value = 8, but is not used for decoding.

load_non_intra_quantizer_matrix (1 bit)   1 indicates non_intra_quantizer_matrix follows.

non_intra_quantizer_matrix   Sixty-four 8-bit unsigned integers in the default zigzag scanning order.

load_chroma_intra_quantizer_matrix (1 bit)   1 indicates chroma_intra_quantizer_matrix follows. Set to 0 for 4:2:0.

chroma_intra_quantizer_matrix   Sixty-four 8-bit unsigned integers in the default zigzag scanning order. The first value = 8, but is not used for decoding.

load_chroma_non_intra_quantizer_matrix (1 bit)   1 indicates chroma_non_intra_quantizer_matrix follows. Set to 0 for 4:2:0.

chroma_non_intra_quantizer_matrix   Sixty-four 8-bit unsigned integers in the default zigzag scanning order.

10.2.11   Copyright_extension( ) (Fig. 10.11)

extension_ID   Hexadecimal 4.

copyright_flag (1 bit)   1 indicates that the video material up to the next copyright_extension or end_of_sequence code is copyrighted.

copyright_identifier   8-bit integer identifying a Registration Authority

original_or_copy (1 bit)   1 indicates the material is original, 0 indicates it is a copy.

reserved (7 bits)   Set to 0.

copyright_number_1   20 most significant bits of copyright_number.

copyright_number_2   22 middle bits of copyright_number.

copyright_number_3   22 least significant bits of copyright_number.

10.2.12   Picture display extension( ) (Fig. 10.12)

extension_ID   Hexadecimal 7.

This extension allows the display rectangle to be moved on a picture-by-picture basis. The number of variables following is 2 × number_of_frame_center_offsets (see Fig. 10.12), that is, one for each field to be displayed.

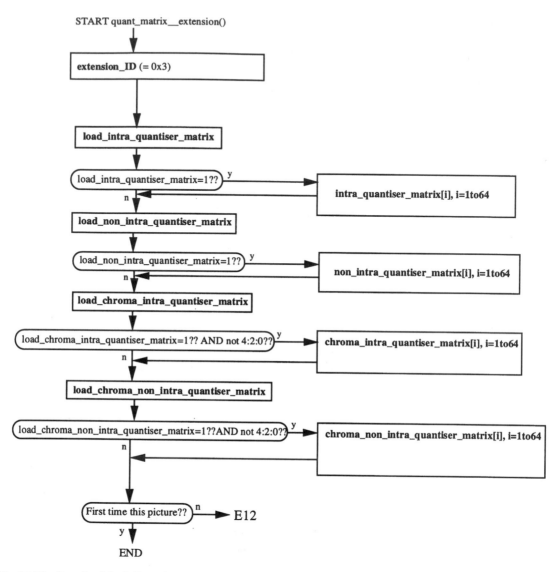

Fig. 10.10 Quantizer Matrix Extension allows updating quantizer matrices between two coded pictures without sending a new Sequence Header.

frame_center_horizontal_offset 16-bit signed integer in units of 1/16th pel. A positive value means the center of the decoded frame is to the right of the center of the display rectangle.

frame_center_vertical_offset 16-bit signed integer in units of 1/16th pel. A positive value means the center of the decoded frame is below the center of the display rectangle.

If a picture does not have a picture_display_ extension( ) then the most recent frame center offset is used.

10.2.13 Picture spatial scalable extension( ) (Fig. 10.13)

extension_ID Hexadecimal 9.

lower_layer_temporal_reference 10-bit unsigned integer that indicates the temporal reference of the lower layer frame to be used to provide the prediction for the current enhancement layer.

lower_layer_horizontal_offset, lower_layer_ vertical_offset 15-bit signed integers specifying the horizontal and vertical offsets, respectively, of the

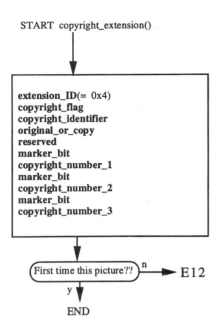

START copyright_extension()

extension_ID(= 0x4)
copyright_flag
copyright_identifier
original_or_copy
reserved
marker_bit
copyright_number_1
marker_bit
copyright_number_2
marker_bit
copyright_number_3

First time this picture?? — n — E12

y

END

Fig. 10.11   Copyright Extension.

upsampled lower layer picture relative to the enhancement layer picture

spatial_temporal_weight_code_table_index
2-bit integer used in combination with Macroblock parameters to determine the spatial-temporal weights w1 and w2 (see Chapter 9).

lower_layer_progressive_frame (1 bit)   1 indicates the lower layer frame is progressive.

lower_layer_deinterlaced_field_select (1 bit) Used in the spatial scalable upsampling process (see Chapter 9).

10.2.14 Picture temporal scalable extension( ) (Fig. 10.14)

extension_ID   Hexadecimal A.

reference_select_code   2-bit code that identifies reference frames or reference fields for prediction depending on the picture type as shown in Table 10.15.

forward_temporal_reference, backward_temporal_reference   10-bit unsigned integer values that indicate the temporal references of the lower layer frames to be used to provide the forward and backward predictions, respectively.

10.2.15 Picture_data( ) (Fig. 10.15)

Picture_data( ) consists of the slices of the picture. Each slice begins with a slice_header followed by the MBs contained in the slice. For all profiles defined so far, there must be no gaps of untransmitted MBs between slices.

slice_start_code   The slice_start_code consists of Psc followed by slice_vertical_position, which has a value in the range 0x01 through 0xAF inclusive.

slice_vertical_position   8-bit unsigned integer ranging from 1 to 175. Used to calculate MB_row, the slice vertical address in macroblocks.

In very large pictures, the slice vertical address (MB_row) calculation requires the **slice_vertical_position_extension** (3 bits)   (see Fig. 10.15). For the first row of macroblocks, slice_vertical_position = 1 and MB_row = 0.

priority_breakpoint   7-bit integer indicating where the video stream is partitioned. Used only with data partitioning. See Chapter 9.

quantizer_scale_code   5-bit unsigned integer in the range 1 to 31. Used with q_scale_type to determine quantizer_scale, which is used in the inverse quantisation process. See Section 8.4.6.1. The decoder uses this value until another quantizer_scale_code is received either in slice( ) or macroblock( ).

slice_extension_flag (1 bit)   1 indicates the presence of intra_slice, slice_picture_id_enable and slice_picture_id in the stream.

intra_slice(1 bit)   1 indicates all of the macroblocks in the slice are intra macroblocks. This flag assists certain trick display modes.

slice_picture_id_enable (1 bit)   1 indicates slice_picture_id may be nonzero. Otherwise, slice_picture_id = 0. slice_picture_id_enable must have the same value in all the slices of a picture.

slice_picture_id(6-bit integer)   If slice_picture_id_enable = 1, slice_picture_id may have any value. However, it must have the same value in all the slices of a picture. Intended to aid recovery from severe bursts of errors

extra_bit_slice (1-bit)   1 indicates extra_information_slice will follow, which should be discarded.

extra_information_slice (8 bits)   Reserved.

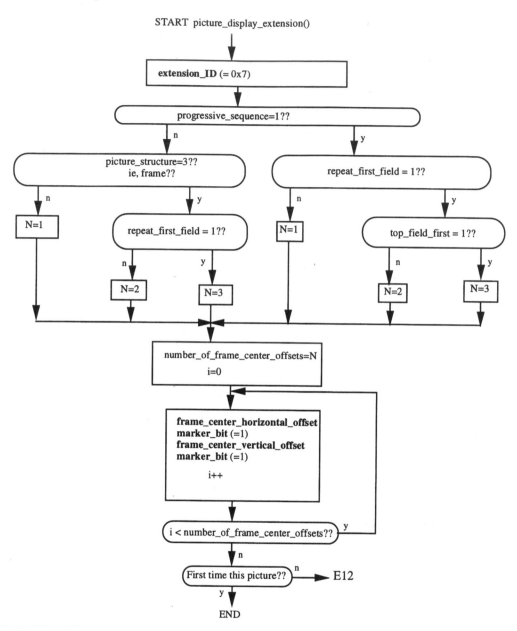

Fig. 10.12   Picture Display Extension.

10.2.16 Macroblock (Figs. 10.16 and 10.17)

macroblock_escape   Bit-string = 0000 0001 000. Used if the difference between macro-block_address and previous_macroblock_address exceeds 33.

macroblock_address_increment   VLC integer (see Table 8.3) used to calculate macroblock_address according to the formula

$$\begin{aligned} \text{macroblock\_address} = \ & \text{previous\_macroblock\_address} \\ & + \text{macroblock\_address\_increment} \\ & + N \times 33, \end{aligned}$$

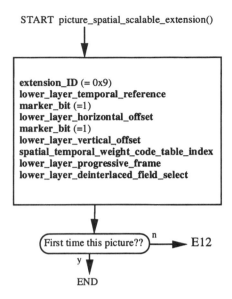

Fig. 10.13 Picture Spatial Scalable Extension.

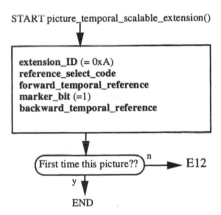

Fig. 10.14 Picture Temporal Scalable Extension.

where N is the number of macroblock_escape's received.

The macroblock_address defines the position of the current macroblock in the picture. The top-left macroblock has macroblock_address = zero, MB_row = 0 and MB_column = 0.

The previous_macroblock_address variable defines the position of the last nonskipped macroblock. However, at the start of a slice this variable is reset to:

$$previous\_macroblock\_address = (MB\_row \times MB\_width) - 1$$

The horizontal spatial position in macroblock units of a macroblock in the picture (MB_column) can be computed from the macroblock_address as follows:

$$MB\_column = macroblock\_address \% MB\_width$$

where MB_width is the number of macroblocks in one row of the picture.

macroblock_type VLC variable that specifies the content of the macroblock. The VLC table selection is according to picture_coding_type and scalable_mode as shown in Table 10.16. Temporal scalable enhancement layers use the same VLCs as the base layer.

macroblock_quant Derived from macroblock_type. 1 indicates that quantizer_scale_code is present in the stream.

macroblock_motion_forward, macroblock_motion_backward Derived from macroblock_type. Used later by the decoding process.

macroblock_pattern Derived from macroblock_type. 1 indicates that a coded_block_pattern is present in the stream and that one or more blocks are coded.

Table 10.15 Reference pictures to be used for prediction of current temporal scalability enhancement layer.

reference_select_code	Forward prediction reference for P-pictures	Forward prediction reference for B-pictures	Backward prediction reference for B-pictures
00	Most recent decoded enhancement picture(s)	Forbidden	Forbidden
01	Most recent lower layer frame in display order	Most recent decoded enhancement picture(s)	Most recent lower layer picture in display order
10	Next lower layer frame in display order	Most recent decoded enhancement picture(s)	Upcoming lower layer picture in display order
11	forbidden	Most recent lower layer picture in display order	Next lower layer picure in display order

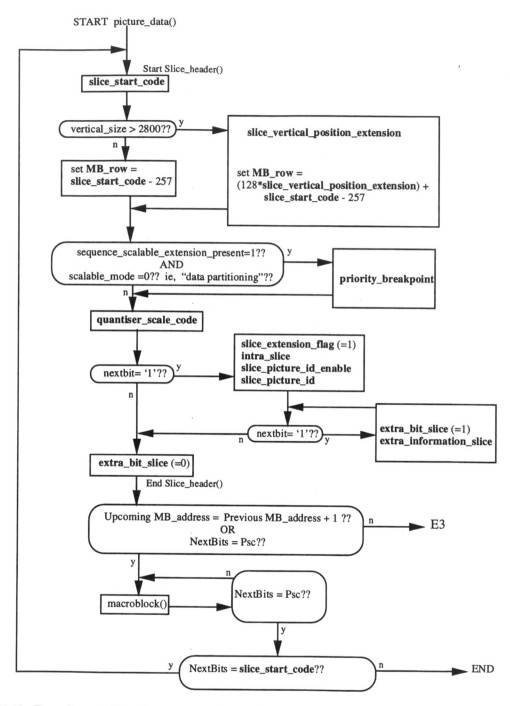

Fig. 10.15 Picture Data and Slices. slice_start_code values range from Psc01 to PscAF.

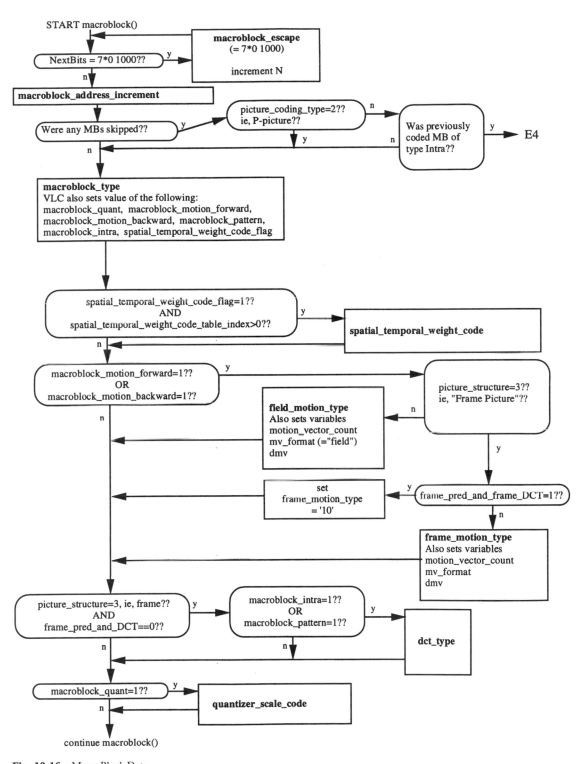

Fig. 10.16 MacroBlock Data.

249

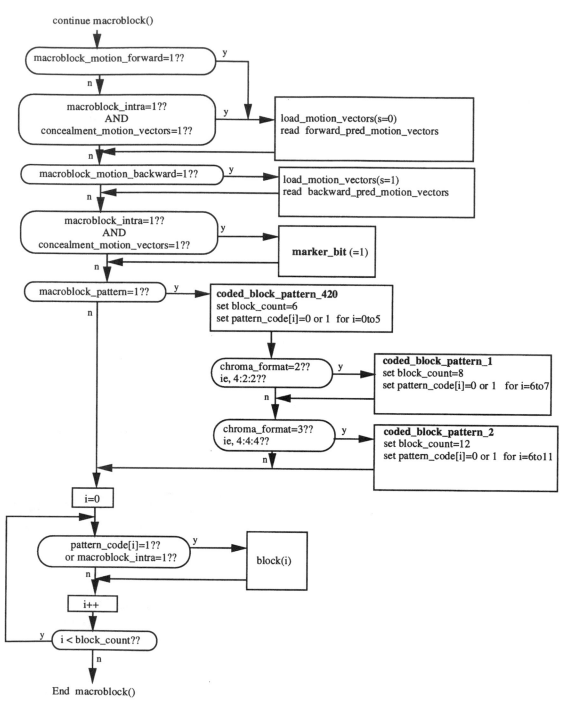

Fig. 10.17 MacroBlock Data continued.

Table 10.16a MacroBlock_type VLCs and the variables that depend on the value of macroblock_type for nonscalable I-Pictures. 1 indicates a true value for the variable. Otherwise, the variable has value 0 = false.

macroblock_ type	macroblock_ quant	macroblock motion_ forward	macroblock_ motion_ backward	macroblock_ pattern	macroblock_ intra	VLC
Intra					1	1
Intra, macroblock_quant					1	01

Table 10.16b MacroBlock_type VLCs and the variables that depend on the value of macroblock_type for nonscalable P-Pictures.

macroblock_ type	macroblock_ quant	macroblock_ motion_ forward	macroblock_ motion_ backward	macroblock_ pattern	macroblock_ intra	VLC
MC_Coded		1		1		1
No MC, Coded				1		01
MC Not Coded		1				001
Intra					1	00011
MC, Coded macroblock_quant	1	1		1		00010
No MC, Coded, macroblock_quant	1			1		00001
Intra, macroblock_quant	1				1	000001

Table 10.16c MacroBlock_type VLCs and the variables that depend on the value of macroblock_type for nonscalable B-Pictures.

macroblock_ type	macroblock_ quant	macroblock_ motion_ forward	macroblock_ motion_ backward	macroblock_ pattern	macroblock_ intra	VLC
MC Interpolated, Not Coded		1	1			10
MC Interpolated, Coded		1	1	1		11
MC Backward, Not Coded			1			010
MC Backward, Coded			1	1		011
MC Forward, Not Coded		1				0010
MC Forward, Coded		1		1		0011
Intra					1	0001 1
MC Interpolated, Coded, macroblock_quant	1	1	1	1		0001 0
MC Forward, Coded, macroblock_quant	1	1		1		0000 11
MC Backward, Coded macroblock_quant	1		1	1		0000 10
Intra macroblock_quant	1				1	0000 01

Table 10.16d MacroBlock_type VLCs and the variables that depend on the value of macroblock_type for spatial scalable I-Pictures.

macroblock_type	macroblock_quant	macroblock_motion_forward	macroblock_motion_backward	macroblock_pattern	macroblock_intra	w1	w2	VLC
Coded				1		1	1	1
Coded macroblock_quant	1			1		1	1	01
Intra					1	not used	not used	0011
Intra, macroblock_quant	1				1	not used	not used	0010
Not Coded						1	1	0001

Table 10.16e Macroblock_type VLCs and the variables that depend on the value of macroblock_type for spatial scalable P-Pictures. See Table 10.17 for spatiotemporal weights w1 and w2 that are unspecified here.

macroblock_type	macroblock_quant	macroblock_motion_forward	macroblock_motion_backward	macroblock_pattern	macroblock_intra	spatial_temporal_weight_code_flag	w1	w2	VLC
MC, Coded		1		1			0	0	10
MC, Coded		1		1		1			011
No MC, Coded				1			0	0	0000 100
No MC, Coded				1		1			0001 11
MC, Not Coded		1					0	0	0010
Intra					1		0	0	0000 111
MC, Not Coded		1				1			0011
MC, Coded macroblock_quant	1	1		1			0	0	010
No MC, Coded, macroblock_quant	1			1			0	0	0001 00
Intra macroblock_quant	1				1		0	0	0000 110
MC, Coded, macroblock_quant	1	1		1		1			11
No MC, Coded macroblock_quant	1			1		1			0001 01
No MC, Not Coded						1			0001 10
Coded				1			1	1	0000 101
Coded, macroblock_quant	1			1			1	1	0000 010
Not Coded							1	1	0000 011

Table 10.16f Macroblock_type VLCs and the variables that depend on the value of macroblock_type for spatial scalable B-Pictures. See Table 10.17 for spatiotemporal weights w1 and w2 that are unspecified here.

macroblock_type	macroblock_quant	macroblock_motion_forward	macroblock_motion_backward	macroblock_pattern	macroblock_intra	spatial_temporal_weight_code_flag	w1	w2	VLC
MC, Interpolated, Not Coded		1	1				0	0	10
MC Interpolated, Coded		1	1	1			0	0	11
MC Backward, Not Coded			1				0	0	010
MC Backward, Coded			1	1			0	0	011
MC Forward, Not Coded		1					0	0	0010
MC Forward, Coded		1		1			0	0	0011
MC Backward, Not Coded			1			1			0001 10
MC Backward, Coded			1	1		1			0001 11
MC Forward, Not Coded		1				1			0001 00
MC Forward, Coded		1		1		1			0001 01
Intra					1		0	0	0000 110
MC Interpolated, Coded, macroblock_quant	1	1	1	1			0	0	0000 111
MC Forward, Coded macroblock_quant	1	1		1			0	0	0000 100
MC Backward, Coded macroblock_quant	1		1	1			0	0	0000 101
Intra macroblock_quant	1				1		0	0	0000 0100
MC Forward, Coded, macroblock_quant	1	1		1		1			0000 0101
MC Backward, Coded, macroblock_quant	1		1	1		1			0000 0110 0
Not Coded							1	1	0000 0111 0
Coded, macroblock_quant	1			1			1	1	0000 0110 1
Coded				1			1	1	0000 0111 1

Table 10.16g Macroblock_type VLCs and the variables that depend on the value of macroblock_type for SNR scalable I-, P-, and B-Pictures.

macroblock_type	macroblock_quant	macroblock_motion_forward	macroblock_motion_backward	macroblock_pattern	macroblock_intra	VLC
Coded				1		1
Coded, macroblock_quant	1			1		01
Not Coded						001

macroblock_intra Derived from macroblock_type. Used later by the decoding process.

spatial_temporal_weight_code_flag Derived from macroblock_type. 1 indicates that Table 10.17 is to be used for the values of w1 and w2.

spatial_temporal_weight_code (2 bits) Used in Table 10.17 with spatial-temporal_weight_code_table_index to indicate the spatial scalability spatial_temporal weights w1 and w2 that were not specified in Table 10.16. (See also Chapter 9).

frame_motion_type (2 bits) Indicates the macroblock prediction type for frame pictures, as shown in Table 10.18. If not present, a default value of "10" is used.

field_motion_type (2 bits) Indicates the macroblock prediction type for field pictures, as shown in Table 10.19. If not present, a default value of "01" is used.

motion_vector_count Number of forward MVs, if present. Also, the number of backward motion vectors, if present.*

mv_format Specifies either field or frame MV.

dmv 1 indicates the presence of **dmvector** (see below).

dct_type(1 bit) Indicates in frame pictures whether the macroblock is frame DCT coded or field DCT coded. 1 means field DCT coded, 0 means frame DCT coded. If frame_pred_and_frame_dct = 1, then dct_type = 0 is assumed.

quantizer_scale_code 5-bit unsigned integer in the range 1 to 31. Used with q_scale_type to determine quantizer_scale, which is used in the inverse quantisation process. See Section 8.4.6.1. The decoder uses this value until another quantizer_scale_code is received either in slice( ) or macroblock( ).

coded_block_pattern_420 Variable length coded variable used to derive pattern_code[i = 0 to 5] and block_count (see Fig. 8.12). pattern_code[i] = 1 if block i is coded, =0 otherwise.

coded_block_pattern_1 (2 bits) Fixed length coded variable used to derive pattern_code[i = 6 to 7] and block_count.

coded_block_pattern_2 (6 bits) Fixed length coded variable used to derive pattern_code [i = 6 to 11] and block_count.

block_count Total number of blocks in the macroblock, as shown in Table 10.5.

Table 10.17 Spatial scalability spatiotemporal weights w1 and w2 that were not specified in Table 10.16 by macroblock_type. See Chapt. 9.

spatial_temporal_weight_code_table_index	spatial_temporal_weight_code	w1	w2
00	Not used	0.5	Not used
01	00	0	1
01	01	0	0.5
01	10	0.5	1
01	11	0.5	0.5
10	00	1	0
10	01	0.5	0
10	10	1	0.5
10	11	0.5	0.5
11	00	1	0
11	01	1	0.5
11	10	0.5	1
11	11	0.5	0.5

Table 10.18 **Frame-motion-type** specifies the macroblock prediction mode for frame-pictures, plus values for the variables **motion_vector_count, mv_format,** and **dmv**

frame_motion_type	Prediction type	motion_vector_count	mv_format	dmv
01	Field-Prediction	2	Field	0
10 (default)	Frame-Prediction	1	frame	0
11	Dual-Prime	1	Field	1

Table 10.19 **Field-motion-type** specifies the macroblock prediction mode for field-pictures, plus values for the variables **motion_vector_count, mv_format,** and **dmv**

field_motion_type	Prediction type	motion_vector_count	mv_format	dmv
01	Field-Prediction	1	Field	0
10	16 × 8 MC	2	Field	0
11	Dual-Prime	1	Field	1

See Chapter 9 for definition and use of w1 and w2.

---

*However, if w1 = 1 then no MV is sent for the top field. Similarly, if w2 = 1 then no MV is sent for the bottom field.

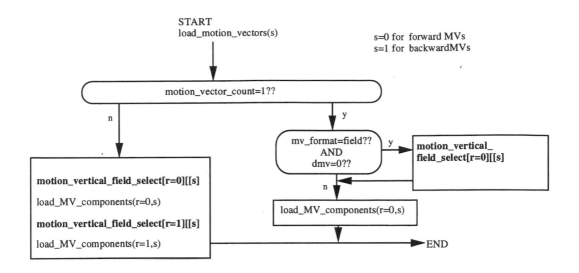

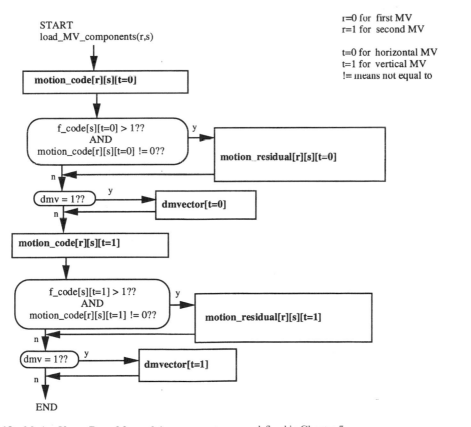

Fig. 10.18 Motion Vector Data. Many of these parameters were defined in Chapter 7.

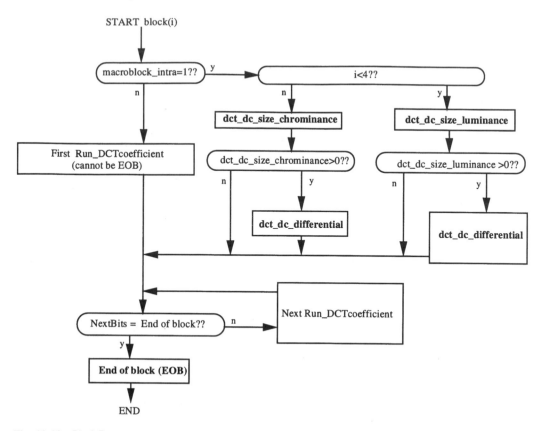

Fig. 10.19   Block Data.

In the following sections, index r = 0 indicates the first MV, r = 1 the second MV. Index s = 0 indicates the forward MV, s = 1 the backward MV. Index t = 0 indicates the horizontal MV component, t = 1 the vertical MV component.

10.2.17   load_motion_vectors(s) (Fig. 10.18)

motion_vector_count, mv_format, dmv   Derived from field _motion_type or frame_motion_type (see above). mv_format specifies field_motion vector or frame_motion vector. dmv is only used for dual-prime.

motion_vertical_field_select[r][s]   Array of 1-bit flags. Specifies which reference field shall be used for the prediction. Zero indicates the top field, one indicates the bottom field.

10.2.18   load_MV_components (r, s) (Fig. 10.18)

motion_code[r][s][t]   VLC variables used in decoding motion vectors (see Section 7.4).

r_size[s][t]   Derived from f_code[s][t]. Indicates the number of bits for the variables motion_residual[r][s][t].

r_size[s][t]   is given by

$$r\_size[s][t] = f\_code[s][t] - 1$$

motion_residual[r][s][t]   Integer of size rsize[s][t] bits. Used in motion vector decoding. (see Section 7.4).

dmvector[t]   VLC variable used in decoding motion vectors for dual-prime prediction (see Section 7.4).

10.2.19 Block( ) (Fig. 10.19)

dct_dc_size_luminance, dct_dc_size_chrominance VLC variables (see Section 8.4.6.2) indicating the size in bits of the Intra DC coefficient variable dct_dc_differential.

first_dc_coefficient VLC (see Section 8.4.6.2) NonIntra DC coefficient.

subsequent_dct_coefficients VLC (see Section 8.4.6.2) runlength_coefficient pairs.

end_of_block VLC variable (see Section 8.4.6.2) that ends every coded block.

REFERENCES

1. A. R. REIBMAN and B. G. HASKELL, "Constraints on Variable Bit-Rate Video for ATM Networks," *IEEE Trans. Circuits Syst. Video Technol.* 2 (4): 361–372 (December 1992).

2. "Coding of Moving Pictures and Associated Audio for Digital Storage Media, at up to about 1.5 Mbit/s," ISO/IEC 11172–2: Video (November 1991).

3. "Generic Coding of Moving Pictures and Associated Audio Information: Video," ISO/IEC 13818–2: Draft International Standard (November 1994)

4. T. SAVATIER, "Difference between MPEG-1 and MPEG-2 Video," ISO/IEC JTCI/SC29/WG11 MPEG94/37 (March 1994).

5. A. PURI, "Video Coding Using the MPEG-2 Compression Standard," Proc. SPIE Visual Commun. and Image Processing, SPIE Vol 1199, 1701–1713 (November 1993).

11

Requirements and Profiles

One of the main philosophies in design of the MPEG-2 Video[1] standard was that it be "generic." This philosophy was the outcome of the experience with MPEG-1, which although designed specifically for Digital Storage Media applications at about 1.5 Mbits/s, was usable for a large range of applications, at bitrates of 10 times or higher than what it was designed for. Of course, the MPEG-1 standard was designed only for coding of noninterlaced low-resolution video and was not as efficient on interlaced video, but it was usable. Before we proceed to discuss the requirements[2,17,18] of MPEG-2 video and its organization into profiles, it is relevant to study what led to its design by discussion of the MPEG-1 case.

The MPEG-1 video standard[16] was designed primarily for the following application:

- Video on Digital Storage Media—CD-ROMs, video CDs

The list of requirements for the MPEG-1 standard is as follows:

- Coding of SIF noninterlaced video at about 1.5 Mbits/s with good quality
- Random access to a frame in limited time (0.5 s or less)
- Trick modes such as fast forward/fast reverse
- Tradeoff of video quality with coding/decoding delay

- Bitstream editability
- A practical decoder.

As mentioned earlier, although the development of MPEG-1 Video primarily targeted coding of noninterlaced video of SIF resolution coded at 1.2 Mbits/s (later generalized to about 1.5 Mbits/s) while permitting random access and trick modes, the restrictions on bitrates and resolutions were somewhat arbitrary.

Since the syntax of the MPEG-1 Video standard already supported very large size pictures and high bitrates, it became necessary to define a minimum set of universal parameters, which could be decodable by any decoder. Those parameters are called "constrained parameters" and the bitstreams satisfying those parameters are called "constrained bitstreams." Constrained parameters for the MPEG-1 video standard are shown in Table 11.1.

11.1 MPEG-2 REQUIREMENTS

Earlier we mentioned that MPEG-2 is a generic standard. By generic, we mean that the coding is not targeted for a specific application but that the standard includes many algorithms/tools that can be used for a variety of applications under different operating conditions. Some explanation is in

258

Table 11.1 MPEG-1 Constrained Parameter Upper Bounds

Picture Parameter	Value
Horizontal size	720 pels/line
Vertical size	576 lines/frame
Number of MacroBlocks per picture	396
Number of MacroBlocks per second	$396 \times 25 = 330 \times 30$
Picture rate	30 frames/s
Bitrate	1.86 Mbits/s
Decoder buffer	376,832 bits

order to cut through the mumbo-jumbo. By different operating conditions, we mean different bitrates, different types of channel or storage media, normal or stringent delay needs (as in conversational applications), etc. So what we are saying is that the MPEG-2 video standard is quite flexible and can be used for a very large number of digital video applications. So, how do we even generate a set of requirements and develop a means of organizing a standard such as MPEG-2? The answer is, it can be done, but it has to be done very very carefully!

Presumably, having gone through the previous chapters in this book, you may already be familiar with some of the intended applications of MPEG-2, but then again, there may be some new ones lurking around, so here is a semiformal representative list of applications:

- TV—Broadcast TV, Satellite TV, Cable TV
- HDTV—Broadcast HDTV, Cable HDTV, Electronic Cinema
- Video on Digital Storage Media—CD-ROM, High-Density CDs, DVD
- Computer Video—Video e-mail, multimedia information systems
- Video on Demand—Movies on demand, live events on demand
- Videocommunication—Multipoint video, multiquality video services
- Networked Video—Video on ATM, video on Ethernet and LANs
- Layered Video—Windowed video on workstations, video database
- Layered HDTV—Compatible TV/HDTV, higher progressive HDTV
- Professional Video—Nonlinear editing, studio post production

- 3D Video—Compatible TV/stereoscopic TV, video games, surgery

If the list of potential applications seems overwhelming, it really is. The MPEG standards have had participation from a wide variety of industries: consumer electronics, microelectronics, communications, broadcasters, computers, and service providers. Representatives of these industries, universities, and other international standard bodies participate, ensuring that the standard being developed takes into account the general requirements of each industry, while maintaining an overall generic nature. In turn, the MPEG standards have reshaped the field of digital multimedia from a fledgling industry, full of wild, unsubstantiated claims, lack of interoperability, and lack of integrated products and services, to a dynamic integrated industry with standards promoting interoperability, vertically integrated products and services, and productive competition. Since there was a substantial participation from many industries, the MPEG-2 video requirements were many and had to be carefully analyzed to satisfy as many as possible as well as prevent functional duplication of similar requirements of different applications.

With that as the background, we are now ready to list the important requirements[2,17,18] for MPEG-2 video.

- Coding of CCIR-601 4:2:0 interlaced video at 4 to 9 Mbits/s with good quality
- Random access/channel-hopping in limited time (0.5 s or less)
- Trick modes such as fast forward/fast reverse
- Special modes with low coding/decoding delay
- Variety of picture resolutions and formats including 4:2:2 and 4:4:4 chroma formats
- Compatibility with MPEG-1 video coding
- Bitstream scalability
- Resilience to errors typical in storage or transport
- Bitstream editability

Originally, the main requirement for MPEG-2 was coding of video at under 10 Mbits/s with good quality. It was anticipated that, in the 3 to 5 Mbits/s range its quality would be comparable to

NTSC/PAL/SECAM and in the 8 to 10 Mbits/s range, its quality would surpass that provided by other standards (in the 15 to 30 Mbits/s range). The bitrates beyond 10 Mbits/s were originally reserved for HDTV, and an MPEG-3 work item was planned for those bitrates. However, in July 1992, the MPEG-2 and the MPEG-3 work items were merged as it was realized that at bitrates of interest for consumer HDTV, MPEG-2 could provide the high quality needed. Thus the MPEG-2 standard's scope was enlarged to cover a wider range of picture formats and bitrates.

Coding/Decoding Delay is an important requirement within the context of conversational applications. A low-delay coding mode (with a delay less than 150 ms), while maintaining good coding efficiency, is one of the very important requirements.

Scalability is a property that allows decoding of appropriate subsets of a bitstream to generate complete pictures of size and quality dependent on the proportion of total bitstream decoded. MPEG-2 requirements include the ability to perform Spatial, Temporal, and SNR (quality) scalability.

Compatibility is a specific form of spatial scalability that requires a new standard to interoperate with the established base of equipment designed for the old standard. Forward compatibility means that the decoders designed for the new standard can decode existing bitstreams corresponding to the old standard, while backward compatibility means that the decoders designed for the old standard can decode a subset of the bitstreams of the new standard.

Error Resilience is an important requirement as errors do occur during transmission or on data stored on any storage medium. The characteristics of the various media typically determine the type and length of errors that may occur and hence determine the strategy for error control. Since MPEG-2 is generic, the medium-specific strategies for error control are outside of the scope of the standard; however, the video coding algorithm should be robust to propagation of errors, and facilitate containment and concealment of errors.

The requirements analysis procedure led to guidelines for the development of mandatory features of the MPEG-2 video standard, including a single integrated syntax to support these features.

We now discuss the development process of MPEG-2 video in detail.

11.2   THE DEVELOPMENT PROCESS

The development[2,18] of the MPEG-2 standard, very much like the development of the MPEG-1 standard, was partitioned into two phases, the competitive phase and the collaborative phase. The purpose of the competitive phase was to seek the best video coding schemes and test them under standardized conditions to evaluate video coding performance, implementation complexity, random access, coding/decoding delay, and resilience to errors. The purpose of the collaborative phase was to iteratively refine the performance of video coding algorithms/tools identified to be promising, to find the best tool when multiple tools existed addressing the same problem, as well as to test any promising new tools.

For the competitive phase the MPEG video subgroup, which had then nearly completed the MPEG-1 work, and the MPEG implementation subgroup finalized the test conditions, test sequences, and the contents of the Proposal Package Description (PPD) to be submitted by each proposer. The PPD also included an evaluation of requirements claimed to be met but not tested, including a framework to evaluate the complexity of a proposal that was provided by the implementation subgroup. The MPEG tests subgroup undertook the brunt of the task of organizing video subjective quality tests and analyzing the results of these tests. The video coding performance was evaluated by requiring each proposer to encode standardized video test sequences by software simulations of their encoding system and produce encoded bitstreams at fixed bitrates of 4 and 9 Mbits/s. Each proposer was then asked to decode these bitstreams to generate decoded results which were to be provided on a digital video tape. The video data on the digital tapes of proposers were extracted to generate master digital tapes which were then used for subjective testing of video quality under standardized conditions by expert as well as nonexpert assessors. The results of the subjective tests were used to rank each coding scheme in terms of video coding performance. Besides the video coding performance, the complexity of video coding schemes was analyzed by an implementation experts evaluation panel, who rank ordered proposed video coding schemes in terms of their implementation complexity. Yet another expert evaluation panel evaluated other functionalities, such as random access delay, coding/decoding delays, resilience to errors, and so forth. A weighted combined score was assigned to each proposal based on its performance in the subjective tests, complexity evaluation, and functionality evaluation.

The collaborative phase was begun by analyzing the top performing proposals from the competitive phase for commonalities and attempting to merge these proposals to begin convergence. In MPEG-2, it turned out that there were substantial commonalities in these proposals, as many proposers had capitalized on the excellent framework of MPEG-1, and they further improved performance substantially by incorporating adaptation for interlace. Since MPEG-1 dealt with coding of noninterlaced video, a good way of dealing with interlace seemed to be the primary difference in terms of the coding efficiency requirement in the early stages of MPEG-2. However, soon several other requirements for MPEG-2 became obvious, as well as the fact that the decoder complexity had to be considered before deciding to further investigate coding algorithms/tools. The MPEG-2 standard was developed via a series of Test Models; the next version of a Test Model was arrived at by evaluating by simulations the performance of the previous Test Model and any "core" experiments testing specific algorithms/tools. If a core experiment was successful, the item being tested was incorporated into the next Test Model. The requirements work was ongoing in parallel with the development of Test Models; likewise, the implementation complexity analysis was also performed in parallel. Thus, for approval of a new Test Model, not only the requirements subgroup but also the implementation subgroup was consulted to ensure that algorithms/tools being experimented upon did address real applications for which the requirements subgroup had collected requirements and whose implementation complex-

ity was not prohibitive. Figure 11.1 illustrates the development process followed.

There were a total of five Test Models, TM1 through TM5, that iteratively refined not only the coding efficiency performance but also other functionalities, primarily low-delay coding and scalable coding. Various forms of scalability were incorporated into the standard as long as they solved some unique problem for which there were distinct requirements.

In the meantime, the requirements group was working in parallel to determine the best way to organize the MPEG-2 video standard. After some effort it was recognized that the best approach would be the one that offered a compromise between the two extremes, that is, a single standard in which every decoder would be capable of decoding bitstreams generated for any application, versus a standard in which no decoder would be capable of decoding bitstreams generated for any but its own application. The chosen solution maintains a hierarchical structure between groups of applications (Fig. 11.2), striking a balance between complete interoperability between applications, using expensive decoders, and no interoperability between applications, using very inexpensive decoders.

The hierarchical organization of the MPEG-2 video standard implied that the syntax had to be somewhat flexible to allow subsets of the syntax to be defined to correspond to each of the levels in the hierarchy. This type of organization of MPEG-2 was dubbed as the concept of "profiles."

11.3 BASIC PROFILES HIERARCHY

As discussed earlier, the MPEG-2 video standard specifies rules that allow a coded representation of video using algorithms/tools that it supports, as well as a description of the exact operations in decoding a coded representation. The former is called the bitstream syntax and the latter the decoding semantics. Since many algorithms/tools are supported, the MPEG-2 video standard is generic; that is, it can be used for a variety of applications. However, it may not be economically viable to design decoders that can support all algorithms/tools, and the solution of independent

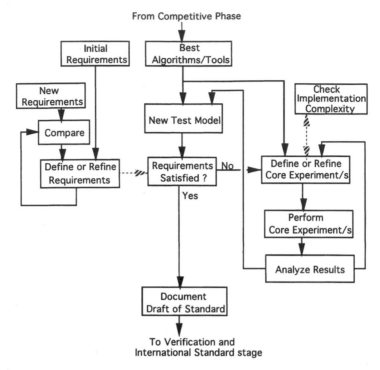

Fig. 11.1   Diagram of Flow of the MPEG-2 video development process.

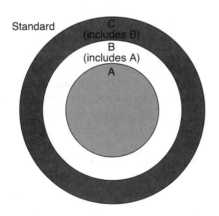

Fig. 11.2   A standard with hierarchical defined subsets.

decoders for each application without any interworking is not a good one either. Thus, as a compromise, subsets of the standard are defined to address specific classes of applications with similar functional requirements. This is done via the concept of *profiles*.

A *profile* is thus a defined subset of entire bitstream syntax of MPEG-2 video. The bitstream

syntax included in a profile corresponds to the algorithms/tools it supports. Further, since a profile may cover a large range of processing complexity, say from in videoconferencing to HDTV, profiles are partitioned into levels. Each *level* of a profile represents constraints on values of parameters in the bitstream. A level can also be thought of simply as specifying a range of allowable values for the parameters in a bitstream. Therefore, a profile and level combination implicitly specifies the decoding complexity of a bitstream.

The concept of profiles and levels is key to defining interoperability between different MPEG-2 applications. The defined profiles at defined levels specify points of conformance for implementers of MPEG-2 video decoders and bitstreams to adhere to. *Conformance* points facilitate interchange of bitstreams between different applications of the standard.

A *conformant decoder* to a particular profile at a particular level is expected to properly interpret all allowed values of all syntax elements for that profile at that level and perform all decoding opera-

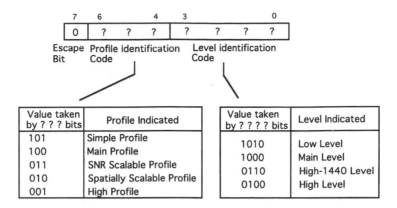

Fig. 11.3 Profile and Level indication when Escape bit is "0."

tions following the decoding semantics for the syntax supported by that profile. A *conformant bitstream* to a particular profile at a particular level does not contain any disallowed syntax elements for that profile, and the value taken by syntax elements do not exceed values allowed for that level.

The next obvious question is: How does a decoder know which bitstreams it can decode? The answer is contained in the *profile_and_level_indication* syntax element carried by the *sequence_extension* layer and contains the profile and level conformance information for the bitstream. This syntax element is 8 bits; the most significant bit is called the *escape_bit* and when its value is zero it indicates that profiles and levels are structured in a specified hierarchical manner. In this case, the following 3 bits indicate the profile and the next 4 bits indicate the level.

For instance, there are five profiles organized in an "onion" layer structure[1,2,20]—Simple profile, Main profile, SNR profile, Spatial profile, and High profile. Among the profiles listed, the SNR profile, Spatial profile, and High profile include scalability tools and thus allow scalable coding, whereas the Simple profile and Main profile allow only nonscalable (single-layer) coding. Further, there are four possible levels in each profile—Low, Main, High-1440, and High. The values of syntax elements for the Main level approximately correspond to normal TV resolution; the values for the Low level correspond to the SIF resolution for which MPEG-1 was optimized, and the values for High and High-1440 levels to HDTV resolution.

The meaning of different bits of *profile_and_level_indication* and the allowed values when the *escape_bit* is "0" is shown by Fig. 11.3.

When the *escape_bit* is "0," profiles are labeled with identification numbers opposite of the actual hierarchical relationship between profiles; that is, a profile with a numerically larger identification value is actually a subset of a profile with a numerically smaller identification value. As an example, the Simple profile is identified by value 5 and the Main profile is identified by value 3, although in terms of hierarchical relationship, the Simple profile is a subset of the Main profile. Likewise, in terms of the hierarchical relationship of levels, a level with a numerically larger identification value is a subset of a level with numerically smaller identification value. It is worth noting that not all levels are defined in all profiles. Only those for which immediate real-world applications exist are currently allowed.

The *escape_bit* when "1" means that there is no hierarchy implied by *profile_and_level indication;* that is, profiles are not necessarily subsets of others and likewise, some levels are not necessarily subsets of others.

At the beginning of this chapter we mentioned that all legal MPEG-1 video decoders must be able to decode all bitstreams compliant with MPEG-1 video-constrained parameters. For such bitstreams, the *constrained_parameters_ flag* is "1," indicating that the bitstream is conformant to constrained parameters. The profiles concept of MPEG-2 video ensures full interoperability with constrained parameter MPEG-1 video bitstreams; that is, all Simple, Main, SNR, Spatial, and High profile conformant decoders shall be able to decode all constrained parameters conformant bitstreams of MPEG-1 video. Moreover, decoders compliant to these profiles shall also be able to decode a special type of MPEG-1 bitstream containing D-pictures (consisting of DC coefficients obtained in DCT coding of each block of pixels) only.

Now that we have some idea of profiles and levels supported by MPEG-2 video, and before we discuss the detailed structure of profile and level organization including various types of constraints and interoperability, it may be useful to understand what type of applications are expected to be addressed by each of these profiles.

11.3.1 Main Profile and Simple Profile

The Main profile is intended for applications requiring good to high quality nonscalable video at bitrates commensurate with the picture resolution and format used. It is primarily useful for video coding for broadcast TV, broadcast HDTV, and high-quality Digital Storage Media applications. Typically in Main profile applications, video encoding delay is not very critical; however, decoding delay is expected to be short.

The Simple profile also requires good to high video quality at similar bitrates, picture resolutions, and formats as the Main profile, but with the restriction that encoding delay should also be short, in addition to the decoding delay. The Simple profile is primarily intended for conversational applications and other real-time video communication applications where coding delay and decoder cost are very critical, but good quality video is needed. These applications are not addressed by the MPEG-1 or by other video coding standards with the quality desired. This is so because MPEG-1 is not optimized for coding of interlaced video of standard TV resolution and format or other higher resolutions, which is needed by the aforementioned applications.

As a brief reminder, nonscalable MPEG-2 video coding allows use of I-pictures, P-pictures, and B-pictures and the group-of-pictures structure,

similar to MPEG-1; however, it also includes tools to deal with interlace, such as frame motion compensation, frame/field DCT, and efficient scanning of DCT coefficients. Thus, MPEG-2 is quite efficient in dealing with interlaced video and shows significant improvement over MPEG-1. To achieve low-delay video coding with high efficiency, MPEG-2 includes dual-prime motion compensation that can reach a performance comparable to that of B-pictures, without incurring the delay associated with B-pictures.

We now list anticipated target applications of these nonscalable profiles:

- Consumer broadcast TV and broadcast HDTV
- Cable TV and Satellite TV
- Real-time Video Conferencing and Interactive TV (Simple profile only)
- Digital Video Recorders—video CD players, video DAT recorders
- Entertainment Systems—electronic cinema, games
- Computer Video—multimedia, video e-mail, archival, Internet video

The list of applications of the Main profile video is rather large and growing fast. It is anticipated that the MPEG-2 Main profile video will eventually replace analog video in nearly all current consumer and professional applications as well as spawn a new generation of applications. The request for the Main profile came from various sectors of industry—broadcasters, video programming providers, consumer electronics manufacturers, the computer industry, and the communications industry. In addition, the standards bodies such as ITU-T, ITU-R, SMPTE, and others either directly or indirectly participated or encouraged completion of work for this Profile. The Simple profile originally had support from mainly two sectors, Cable TV providers and the ITU-T. At that time, for Cable TV application, the cost of the decoder was considered quite crucial and it was expected that the Simple profile, owing to the lack of B-pictures, would result in cheaper decoders; this point is not as relevant now. The ITU-T's primary interest in the Simple profile has been its potential for use in low-delay conversational services. The ITU-T has actively participated in MPEG, and in fact the MPEG-2 video standard, ISO 13818–2, and ITU-T's H.262 are essentially the same standard and use shared text.

Now that we understand the applications of these profiles, the MPEG-2 video coding algorithms/tools to resolve the functional needs of these applications, as well as the segments of industry interested in these profiles, we can list a set of formal requirements as follows:

1. Good to high video quality
2. Interlaced and progressive video of 4:2:0 formats
3. Channel hopping/random access within 0.5
4. Low encoding/decoding delay of 0.25 (Simple profile)
5. Variety of resolutions and picture rates
6. Flexibility in bitrates
7. Some level of editability of compressed bitstreams
8. Cost-effective implementation.

For the Main profile, all four levels, Low, Main, High-1440, and High, are allowed. However, for the Simple profile, only the Main level is allowed. A Simple profile decoder is expected to decode both the Simple profile at Main level bitstreams and the Main profile at Low level bitstreams. It is worth noting that a Simple profile at a Main level decoder is required to be able to decode the Main profile at low-level bitstreams that may contain B-pictures, albeit of smaller size, although the Simple profile itself does not support B-pictures. This is done to ensure that all profiles are capable of decoding MPEG-1 bitstreams and to enhance interoperability with the Main profile.

11.3.2 The SNR Profile

The SNR scalable profile is intended for video applications requiring two simultaneous layers of video of different qualities but the same picture resolution and formats in each layer. The video layers are efficiently coded with bitrates commensurate with the quality of each layer. Syntactically, the SNR profile is a superset of the Main profile with the main difference being that two layers of video at different qualities are allowed. This profile

can be used whenever an application requires graceful degradation of quality in the presence of errors. SNR profile uses an SNR scalability tool, in which the base layer of full spatial resolution is coded by itself and the enhancement layer, which is of same resolution as the base layer, is decoded and combined with the decoded base layer to produce improved quality video. The base layer is normally assigned higher transmission priority than the enhancement layer so that in case of errors, the corrupted enhancement layer can be temporarily discarded and only the base layer may be decoded and displayed.

We now list some target applications of this profile:

• Broadcast TV at two qualities
• Two-quality service for conversational applications
• Two-layer video on LANs and other networks Subject to congestion

The SNR profile, like other profiles that include scalability, addresses niche applications and thus the list of its applications is somewhat limited. Nevertheless, the SNR profile is useful as decoders corresponding to this profile are not very complex, and a high degree of error resilience is possible, which may be critical in some applications. The request for the SNR profile came from a few European organizations who are investigating the performance of this profile for broadcast TV application.

Now that we understand the applications of the SNR profile and the MPEG-2 video coding algorithms/tools to solve the functional needs of these applications, we state a few requirements for this profile (in addition to the requirements of the Main profile) as follows:

1. Two layers with different qualities
2. Interoperability with the Main profile
3. Low cost implementation of a two-layer decoder
4. Flexibility in bitrates for each layer.

For the SNR profile, only two levels, Low and Main, are allowed. The SNR profile at each level supports two layers of video of a 4:2:0 format. The profile/level of the simplest base layer decoder is the Main profile at the same level. As an example, the SNR profile at the Main level decoder shall decode SNR profile bitstreams at the Low level, Main profile bitstreams at the Main level, and Main profile bitstreams at the Low level, besides being able to decode SNR profile bitstreams at the Main level. Furthermore, owing to the hierarchical structure of profiles, since the Simple profile is a syntactic subset of the Main profile, it is implicit that an SNR profile at the Main Level decoder shall decode the Simple profile at Main level bitstreams.

11.3.3 The Spatial Profile

The Spatial scalable profile is intended for video applications requiring up to three simultaneous layers of video. A combination of SNR and Spatial scalability is allowed in any order to form the two enhancement layers, and these two layers cannot use the same type of scalability. It is also possible to have two layers only, in which case only one type of scalability will be used. If all three layers are present, any two consecutive layers can be coded to yield video of different quality by SNR scalability, while the other layer can yield different picture resolution (and even different format, if needed) by Spatial scalability. The three layers are coded efficiently at bitrates commensurate with the resolutions or qualities needed. Syntactically, the Spatial profile is a superset of the SNR profile with the main difference that in addition to syntax for SNR scalability it also supports syntax for Spatial scalability. This profile can be used whenever an application requires graceful degradation or good quality coding of up to three layers. Thus the list of applications for this profile includes some of the applications of the SNR profile as well as a few new applications.

We now list some potential applications of this profile:

- Two-quality broadcast HDTV
- Broadcast compatible TV with HDTV, coded as two resolution layers
- Two-quality conversational applications
- Two-quality compatible TV with HDTV, coded as three layers

- Layered video on LANs and other networks subject to congestion
- Two- or three-layer video for video database applications.

The Spatial profile addresses applications employing Spatial scalability, SNR scalability, as well as hybrids of Spatial and SNR scalability. Since the Spatial profile supports three layers, Spatial profile decoders can be relatively more complex than that for the SNR profile. The request for the Spatial profile mainly came from several European organizations who are investigating the possibility of using this profile for either two-layer compatible TV/HDTV or a three-layer application using either two-quality compatible TV with HDTV or as compatible TV with two-quality HDTV.

Now that we understand the applications of the Spatial profile and MPEG-2 video coding algorithms/tools to resolve the functional needs of these applications, we state a few additional requirements for this profile (besides the requirements for the Main profile) as follows:

1. Up to three layers by a combination of Spatial and SNR scalability
2. Interoperability with the Main profile and the SNR profile
3. Flexibility in bitrates for each layer.

In the Spatial profile, only one level, High-1440, is allowed. Up to three layers of video of 4:2:0 format can be coded a as combination of SNR and Spatial scalability coding. The three layers are referred to as base layer, enhancement layer 1, and enhancement layer 2. The profile/level of the simplest base layer decoder is the Main profile at the same level or the Main profile at a previous level, depending on the scalability used in enhancement layer 1 or enhancement layer 2, if it exists. The reference to levels is made with respect to the highest layer, enhancement layer 1, if only two layers are used, or enhancement layer 2 if three layers are used. In Fig. 11.4 we show the two-layer and three-layer combinations of scalability allowed in this profile.

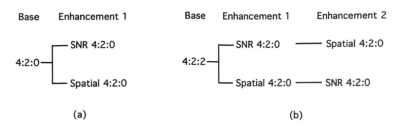

Fig. 11.4 Layer combinations for the Spatial profile.

11.3.4 The High Profile

The High profile is intended for video applications requiring up to three simultaneous layers of video, where each layer of video could have one of the two possible chroma formats (some combinations are disallowed). As in the Spatial profile, a combination of SNR and Spatial scalability is allowed in any order to form the two scalable layers, and these two layers cannot use the same type of scalability. If all three layers are present, any two consecutive layers can be coded to yield video of different quality by SNR scalability, while the other layer can yield different picture resolution (and even different format, if needed) by Spatial scalability. The three layers are coded efficiently at bitrates commensurate with resolutions or qualities needed.

Syntactically, the High profile is a superset of the Spatial profile with the main difference being that in addition to syntax for Spatial scalability it also supports syntax for the 4:2:2 format. It is, however, possible that in some cases only two layers are used although up to three layers are allowed. This profile can be used whenever an application requires graceful degradation or good quality coding with 4:2:0 or 4:2:2 format video as three layers. Thus the list of applications for this profile includes some applications of the Spatial profile as well as a few new applications.

We now list a few potential applications of this profile:

- Studio applications using compatible coding of chroma formats
- Contribution/distribution quality compatibly coded video

- Compatible TV with two-quality HDTV, coded as three layers
- Two-quality Compatible TV with HDTV, coded as three layers.

The High profile addresses applications employing Spatial scalability, SNR scalability, as well as hybrids of Spatial and SNR scalability. Moreover, some applications require support of higher resolution chroma formats, so these scalabilities can also be used with, or to achieve, different chroma format and layer combinations. The High profile was conceived as the ultimate profile, supporting a wide range of applications simultaneously requiring scalability and multiple chroma formats. However, many applications requiring only scalability can be handled by either the SNR or Spatial profile, and with a new profile in the works to separately deal with multiple chroma formats (discussed later in this chapter), it is doubtful that this profile will find widespread use. This view is further supported by the fact that the complexity of the High profile seems to be a major problem preventing cost-effective implementation in the near future.

Now that we understand the applications of the High profile and MPEG-2 video coding algorithms/tools to resolve the functional needs of these applications, we state a few additional requirements for this profile (besides the requirements of Main profile), as follows:

1. Up to three layers by a combination of Spatial and SNR scalability
2. 4:2:2 (and 4:2:0) chroma format
3. Partial interoperability with the Main profile and SNR profile
4. Flexibility in bitrates for each layer.

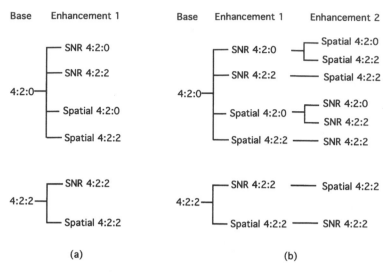

Fig. 11.5   Layer combinations for the High profile.

In the High profile, three levels, Main, High-1440, and High, are allowed. Up to three layers of video of 4:2:0, 4:2:2, or some combinations of 4:2:0 and 4:2:2 formats can be coded with SNR or Spatial scalability coding. As in the case of the Spatial profile, the three layers are referred to as base layer, enhancement layer 1, and enhancement layer 2. The profile/level of the simplest base layer decoder is the High profile at the same level or the High profile at the next lower level, depending on the scalability used in enhancement layer 1 or enhancement layer 2, if it exists. Notice that something unusual has just been said; the simplest decoder for the base layer in the High profile is in fact a High profile decoder, not a Main profile decoder. One reason is that the High profile allows the 4:2:2 format as the base layer, while the Main profile does not allow the 4:2:2 format at all; thus such bitstreams would not be decodable by the Main profile. There are other reasons as well; as just one example, the High profile allows 8-, 9-, 10-, or 11-bit precision for DC coefficients of Intra-coded macroblocks, while the Main profile allows only 8-, 9-, or 10-bit DC precision. However, we are probably getting ahead of ourselves, as syntax constraints of profiles will be dealt with in the next section. For now, in Fig. 11.5, we show the two-layer and three-layer combinations of scalability allowed in this profile.

11.4 BASIC PROFILES INTEROPERABILITY

Although we have clarified the meaning of each profile, given example applications, stated which levels are allowed for which profiles, and even hinted on what picture resolutions correspond to which levels, the profile/level structure may still be rather murky. The discussion thus far was basically intended to give enough insight, and getting that out of the way, we will try to give you an integrated view of what is going on, as well as provide specifics when necessary. As a major step in that direction we present Fig. 11.6, which provides a snapshot of the structure of the currently finalized MPEG-2 video profiles and levels that follow a hierarchical organization.

As mentioned previously, five profiles, each with a maximum of four levels each, are possible. In reality, a total of 11 combinations of profiles and levels out of a potential 20 combinations are defined. The Main profile allows all four levels, the Simple and the Spatial profiles allow one level

Level \ Profile	SIMPLE	MAIN	SNR	SPATIAL	HIGH
HIGH		1920 pels/line 1152 lines/frame 60 frames/s 62.7 Msamples/s 80 Mbit/s			1920 pels/line 1152 lines/frame 60 frames/s 62.7 Msamples/s @ 83.5 Msamples/s * 100 Mbit/s All 3 layers 80 Mbit/s Base+Middle 25 Mbit/s Base layer
HIGH-1440		1440 pels/line 1152 lines/frame 60 frames/s 47.0 Msamples/s 60 Mbit/s		1440 pels/line 1152 lines/frame 60 frames/s 47.0 Msamples/s 60 Mbit/s All 3 layers 40 Mbit/s Base+Middle 15 Mbit/s Base layer	1440 pels/line 1152 lines/frame 60 frames/s 47.0 Msamples/s @ 62.7 Msamples/s * 80 Mbit/s All 3 layers 60 Mbit/s Base+Middle 20 Mbit/s Base layer
MAIN	720 pels/line 576 lines/frame 30 frames/s 10.4 Msample/s 15 Mbit/s	720 pels/line 576 lines/frame 30 frames/s 10.4 Msamples/s 15 Mbit/s	720 pels/line 576 lines/frame 30 frames/s 10.4 Msample/s 15 Mbit/s Both layers 10 Mbit/s Base layer		720 pels/line 576 lines/frame 30 frames/s 11.06 Msamples/s @ 14.75 Msamples/s * 20 Mbit/s All 3 layers 15 Mbit/s Base+Middle 4 Mbit/s Base layer
LOW		352 pels/line 288 lines/frame 30 frames/s 3.04 Msamples/s 4 Mbit/s	352 pels/line 288 lines/frame 30 frames/s 3.04 Msamples/s 4 Mbit/s Both layers 3 Mbit/s Base layer		
	SIMPLE nonscalable 4:2:0	**MAIN** nonscalable 4:2:0 with B- pictures	**SNR** scalable 4:2:0	**SPATIAL** scalable 4:2:0	**HIGH** nonscalable 4:2:2 scalable 4:2:0/4:2:2 * refers to 4:2:0 @ refers to 4:2:2

Profile ⟶

Fig. 11.6 MPEG-2 profile and level structure with parameter upper bounds.

each, the SNR profile allows two levels, and the High profile allows three levels. Thus the profile and level combinations left blank are currently not defined. For the allowed profile and level combinations the upper bounds on picture size parameters are provided. For example, the Main profile at the Main level (MP@ML) supports up to 720 pels/line and up to 576 lines/frame with a frame rate of up to 30 frames/s, for a total sampling rate of up to 10.4 Msamples/s. The sampling rate of 10.4 Mbits/s is in fact smaller than what would be obtained by the product of pels/line, lines/frame, and frames/s, as either a combination of 480 lines/frame at 30 frames/s or a combination of 576 lines/frame at 25 frames/s is expected, but the combination of 576 lines/frame and 30 frames/s

is not expected. This is consistent with the two versions of CCIR-601 video format standards in use.

11.4.1 Syntax Constraints

Because of the hierarchical organization of the Simple, Main, SNR, Spatial, and High profiles, the syntax for a profile[1] includes all syntax elements of the immediately lower profile in the hierarchy. As an example, the syntax for the Spatial profile includes all the syntax elements of SNR profile. Although one or more levels may be allowed in a profile, the syntax elements supported by a profile are the same regardless of the level. In Table 11.2 we illustrate some constraints on syntax elements between various profiles.

Table 11.2   Syntax Constraints Comparison for Profiles

Syntax Element	Allowed Value for Profile				
	Simple	Main	SNR	Spatial	High
chroma_format	4:2:0				4:2:2, 4:2:0
frame_rate_extension_n	0				
frame_rate_extension_d	0				
aspect_ratio_information	1.0, 4/3, 16/9				
picture_coding_type	I, P	I, P, B			
repeat_first_field	Constrained		Unconstrained		
sequence_scalable_extension( )	No		Yes		
scalable_mode	No		SNR	SNR, Spatial	
picture_scalable_extension( )	No			Yes	
intra_dc_precision	8, 9, 10				8, 9, 10, 11
Slice_structure	Restricted				

11.4.2   Semantics and Syntax Constraints Explanation

In addition to syntax constraints there are some semantic constraints on the bitstreams as well. For example, there is a constraint on the maximum number of bits that a coded Macroblock (MB) can produce. When employing CCIR-601 4:2:0, 4:2:2, and 4:4:4 video formats, the corresponding maximum number of bits allowed for an MB are 4608, 6144, and 9216. However, a maximum of two MBs in each horizontal row of MB may exceed this limit. As a slight detail, counting of bits of a MB starts from the first bit of *macroblock_address_increment* (or *macroblock_escape*, if any) and continues until the last bit of "End of Block" symbol of the last coded block (or the last bit of *coded_block_pattern* if there are no coded blocks).

Earlier in Table 9.2 we showed that the syntax element *repeat_first_field* is constrained for the Main and Simple profiles and unconstrained for the SNR, Spatial, and High profiles; its allowed values are as follows:

- When the sequence is progressive (*progressive_sequence* = 1), for frame rates of 23.976, 24, 25, 29.97, 30, and 50 frames/s *repeat_first_field* = 0, while for frame rates of 59.94 and 60 frames/s, *repeat_first_field* = 0 or 1.
- When the sequence is not progressive (*progressive_sequence* = 0), for frame rates of 23.976 and 24 frames/s, *repeat_first_field* = 0, while for

frame rates of 25, 29.97, 30, 50, 59.94, and 60 frames/s, *repeat_first_field* = 0 or 1.

Further, additional constraints are imposed on the Main profile at the Main level and the Simple profile at the Main level as follows:

- If vertical_size is greater than 480 lines and if *picture_coding_type* is B-picture, *repeat_first_field* shall be 0. Also, if *vertical_size* is greater than 480, the frame rate shall be 25 frames/s.

We have already pointed this out earlier, but once again as a reminder, decoders for the Simple profile at the Main level shall be capable of decoding the Main profile at Low level bitstreams, including B-pictures.

One last point worth noting is that the High profile has different constraints on sample rate and maximum bit rate as compared to other profiles, as can be noted from Fig. 11.6.

11.4.3   Parameter Constraints

In Table 11.3 we illustrate some of the constraints on parameters[1] for various levels of a profile.

11.4.4   Scalability Layers

We now provide a clear specification of the scalability combinations allowed for each of the scalable profiles. Table 11.4 shows upper bounds for scalable layers for each level of each scalable profile.

Table 11.3 Constraints on Parameters for Each Level

Syntax Element and Meaning	Allowed Value for Level			
	Low	Main	High-1440	High
f_code[0][0] or f_code[1][0] forward or backward horizontal motion vector range	[1:7] −512 to 511.5	[1:8] −1024 to 1023.5	[1:9] −2048 to 2047.5	[1:9] −2048 to 2047.5
Frame-Picture f_code[0][1] or f_code[1][1] forward or backward vertical motion vector range	[1:4] −64 to 63.5	[1:5] −128 to 127.5	[1:5] −128 to 127.5	[1:5] −128 to 127.5
Field-Pictures f_code[0][1] or f_code[1][1] forward or backward vertical motion vector range	[1:3] −32 to 31.5	[1:4] −64 to 63.5	[1:4] −64 to 63.5	[1:4] −64 to 63.5
frame_rate_code frame rate values	[1:5] 23.976, 24, 25, 29.97, 30		[1:8] 23.976, 24, 25, 29.97, 30, 50, 59.94, 60	
Sample rate and Bitrates, etc.		See Fig. 11.6		

Table 11.4 Scalable Layers in SNR, Spatial, and High Profiles

Level and Type of Layers	Number of Layers for Profile		
	SNR	Spatial	High
High			
All layers (base and enhancement)			3
Spatial enhancement layers			1
SNR enhancement layers			1
High-1440			
All layers (base and enhancement)		3	3
Spatial enhancement layers		1	1
SNR enhancement layers		1	1
Main			
All layers (base and enhancement)	2		3
Spatial enhancement layers	0		1
SNR enhancement layers	1		1
Low			
All layers (base and enhancement)	2		
Spatial enhancement layers	0		
SNR enhancement layers	1		

11.5 BASIC PROFILES VERIFICATION TESTS

The purpose of verification tests is to confirm if the MPEG-2 standard has achieved its picture quality objectives at the bitrates and other condi-tions under which it is expected to be used in vari-ous applications. It is worth noting that the prima-ry purpose of these tests is verification of the pic-ture quality obtained, and not the verification of syntax or decoding semantics. For that a separate major effort was undertaken by many organiza-tions within MPEG to cross-verify each other's bit-stream syntax and decoding semantics. The Bit-stream Exchange ad hoc group was primarily responsible for undertaking this huge effort to veri-fy the syntax and semantics of all the tools includ-ed in the MPEG-2 video standard, even those that are not explicitly supported by profiles thus far.

Since MPEG-2 video coding, to achieve the high compression needed, exploits various forms of spatial, temporal, and perceptual redundancies, and since the decoded video is intended to be viewed by human viewers, visual subjective quality assessment tests[3] are needed to verify if video qual-ity objectives are met by MPEG-2 video coding or not. Various international standards organizations such as EBU, ITU-R, SMPTE, and so forth, with the expertise in this area have collaborated with MPEG in planning these tests.

The test method used is called DSCQS (Double Stimulus Continuous Quality Scale) and is based on CCIR Rec. 500–5 for normal TV resolution video and on CCIR Rec. 710–1 for HDTV video. This method consists of showing pairs of video

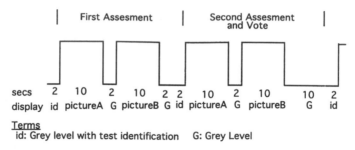

Terms
id: Grey level with test identification   G: Grey Level

Fig. 11.7   A complete presentation cycle for the DSCQS test.

sequences to observers such that one of the sequences in each pair is a coded video sequence to be tested, and the other sequence is a reference (in this case, uncoded) video sequence. During the tests, the observers do not know if the video sequence they are viewing is coded or not. The observers are shown each pair of video sequences two times, and during the second viewing of the pair, are asked to grade the quality of the two sequences on a continuous scale from 1 to 5 (where discrete values represent the following: 1; excellent; 2; good; 3; fair; 4; poor; and 5; bad). Each of the two sequences forming a pair is 10 s long with a 2-s gap during which a gray level is shown. One complete presentation cycle for the DSCQS test is summarized in Fig. 11.7.

Each presentation (of about 60 s) is repeated for a number of different video sequences being tested. A session is an assemblance of such presentations that lasts for a maximum of 30 min. Although a session may include 30 such presentations, typically, the first five presentations may be for training purposes only, with video sequences not actually being tested. The subjective quality assessors who participated in MPEG-2 verification tests are a combination of experts and nonexperts. The video sequences used in the tests are selected to represent a range of source material typically encountered in TV programming, such as fast paced sports, natural scenes with lots of details, scenes with flesh tones, scenes with camera zoom, graphics, and so forth.

Following the CCIR Rec. 500–5, the assessors participating in the tests view scenes at a distance of four and six times the picture height. The statistics of results are tabulated separately for each viewing distance and separately for each video coding bitrate. Many organizations have participated in MPEG-2 verification tests; for some tests there have been more participants than others owing to a higher degree of interest in certain profiles. Although not all levels of all profiles have been fully tested, a representative set of tests have been conducted on profile/level combinations for which immediate applications are anticipated. Tests are conducted on nonscalable and scalable profiles; some of the scalability tests are performed using picture resolutions and bitrates that are a scaled version of what actual applications would use. In all cases, it has, however, been ensured that although reduced resolution video sequences at lower bitrates may have been used in scaled tests, the picture formats have been kept identical to what may actually be encountered in real applications, where scalability may be used for each resolution layer, that is, progressive or interlaced as applicable.

As mentioned earlier, the CCIR continuous quality scale allows assessors to mark the absolute quality of each video sequence on a continuous five point scale, representing the range of qualities, from excellent to bad. After the tests, these markings on a graphical scale are converted to a range of discrete numbers from 0 to 100 and averaged for all viewers but separately processed for each viewing distance for the bitrate tested. We now list the results of MPEG-2 verification tests. A note of caution in comparing results across the various tests: for instance, comparing the performance of a lower layer in a Scalable profile that may be compliant with the Main profile at the Main level with results obtained in separate Main Profile verification tests. There are two difficulties in such comparisons. First, owing to many fewer participants in the scalability tests the statistics differ significantly from the Main profile tests, and second, scalability applications may make different tradeoffs in bitrates of each layer to generate the quality of each layer as may be necessary in the layered application of interest, rather than optimize quality of one layer only.

Based on the results in Table 11.5 and results of other tests; experts[4,5,6,7,10] have been able to conclude that MPEG-2 Main profile video coding does indeed often achieve good picture quality in the coding of TV resolution video at 4 Mbits/s, and in many cases, the picture quality exceeds that provided by NTSC/PAL analog TV systems currently used in our homes. Further, in coding of same resolution pictures at 9 Mbits/s, the picture quality is high and definitely exceeds the currently available picture quality to consumers.

The testing of the SNR profile was conducted to verify the feasibility of two-quality layered service for telecommunications or broadcast applications. The lower layer is assigned 4 Mbits/s, and the enhancement layer uses an additional 5 Mbits/s to refine the lower layer for a total bitrate of 9 Mbits/s. For the purpose of comparison, results for Main profile coding at 9 Mbits/s as a single layer are also included for reference. Since the Main profile results were obtained as part of the SNR profile test, they are more consistent with the rest of the statistics and can be used for comparisons.

From the results of the SNR profile tests[8,10,12] shown in Table 11.6 it can be observed that the Main profile results at 9 Mbits/s significantly outperform the results produced at 9 Mbits/s by combined base and enhancement layers in SNR scalable coding. However, two-layer SNR scalable coding, although somewhat less efficient than Main profile coding, does provide two layers of the

Table 11.6 SNR Profile at Main Level Verification Results

| Bitrates, Mbits/s: | 525 line/60 Hz | | 625 line/50 Hz | |
Profile and Layer	View 4H	View 6H	View 4H	View 6H
4: SNR profile base	21.3	14.4	18.0	13.8
9: SNR Profile base + enh.	12.3	9.0	8.5	6.8
9: Main profile	6.4	5.7	4.9	4.4

same resolution for applications requiring either a higher error resilience or two quality layers.

For the Spatial profile at the High-1440 level, verification tests have been conducted[9,10] under scaled down test conditions in terms of picture sizes and bitrates. The actual application envisaged is two-layer digital HDTV compatible with normal resolution digital TV. The interest in this application has been primarily in Europe and a limited number of tests have been conducted. The tests use coding of interlaced SIF resolution for the lower layer at 2.25 Mbits/s and interlaced CCIR-601 resolution for the enhancement layer with Spatial scalability at 6.75 Mbits/s, for a total of 9 Mbits/s. In two-layer HDTV/compatible TV application the actual bitrates could be 9 Mbits/s for the base layer and 27 Mbits/s for the enhancement layer, for a total of 36 Mbits/s. Based on the tests it has been concluded that for a combined base and enhancement layer bitrate of 9 Mbits/s, high quality can be obtained, suggesting that two-layer HDTV at 36 Mbits/s is feasible. Also, the quality of the base layer at 2.25 Mbits/s was judged to be satisfactory in the context of a two-layer HDTV service. As a reminder, various issues relevant to scalable HDTV were addressed in Chapter 19. Although basics of HDTV and results of nonscalable HDTV coding will be presented in Chapter 14, here, within the context of subjective tests, for nonscalable HDTV a few comments are necessary.

Since the FCC recommendation[19] for an HDTV standard in the United States uses MPEG-2 Main profile video coding, MPEG has also verified the picture quality at bitrates and resolutions suitable for HDTV. Thus, the Main profile is also tested at the High level (which includes the High-1440 level). The bitrates used are 18, 30, and 45

Table 11.5 Main Profile at Main Level Verification Results

| Bitrate, Mbit/s | 525 line/60 Hz | | 625 line/50 Hz | |
	View 4H	View 6H	View 4H	View 6H
4	19.1	13.5	15.0	13.6
6	10.3	8.4	7.5	7.2
9	5.8	5.0	3.8	4.0

Mbits/s; the bitrate of 18 Mbits/s is at present more suitable for terrestrial broadcast HDTV, whereas the two higher bitrates are better suited for HDTV on Cable, Fiber, or other means. The results of the HDTV verification tests indicate that acceptable consumer grade HDTV of acceptable quality is indeed possible at 18 Mbits/s. These results are consistent with findings of the FCC Advanced Television Systems Committee (ATSC) which evaluated the various candidate proposals for HDTV standardization. Furthermore, the MPEG-2 Main profile at the High level tests also suggest that significant improvement in picture quality is possible only by increasing the bitrate to 45 Mbits/s. By further encoding optimizations, it is envisaged that 18 to 30 Mbits/s range is sufficient for very good quality HDTV for consumer applications.

11.6 NEW PROFILES

Although the MPEG-2 video coding standard is complete, it is recognized that the process of defining profiles should remain open. Thus, looking into the near-term future, some tools have been included in the video standard with the anticipation that new profiles that employ these tools may soon be necessary. There are at least two new application areas for which new profiles are judged to be necessary. The first application area is intended for high-quality professional video and the second area is that of multiviewpoint video (two views means stereoscopic video).

11.6.1 Profile

The 4:2:2 profile[14] is intended for professional video applications where ease of editing of compressed video and multigeneration coding/decoding of video are important requirements. These applications are not addressed by other MPEG-2 video profiles. Quite often, in studio applications, where this profile may be most applicable, contribution quality video of very high quality and CCIR-601 4:2:2 video format is needed owing to ease of chroma keying and other special effects. Because of the requirement of ease of editing,

more frequent Intra pictures are necessary which also result in high coding bitrates. Further, in a typical TV broadcast chain, video goes through multiple generations of encode/decode as local TV stations add local programming information and commercials to video before it gets distributed to consumers for reception at home. With analog TV today, multiple generations of encode/decode can result in significant picture quality losses.

We now list anticipated target applications of this profile:

- Studio postproduction of high-quality video sequences
- Disk storage for nonlinear editing
- Creation of visual effects
- Archiving of video
- Videotape recording for professional use
- Efficient transmission for storage and distribution

At the request of the industry and several standards organization such as the ITU-R and the SMPTE, and in considering the market needs for professional applications at reasonable cost while requiring interoperability, MPEG, with the participation of requesting organizations, is developing a specification for the 4:2:2 profile. This profile contains nonscalable video coding similar to the Main profile except for one big difference; it supports both CCIR-601 4:2:2 and 4:2:0 video formats. It is worth noting here that the High profile also supports both these video formats as well as scalability; however, a High profile decoder may be too complex (and expensive) owing to the large number of video coding tools it supports. As pointed out earlier, the Main profile on the other hand only supports the CCIR-601 4:2:0 video format.

Now that we understand the applications of this profile, MPEG-2 video coding algorithms/tools to resolve the functional needs of these applications, as well as segments of the industry interested in this profile, we can list a set of formal requirements as follows:

1. High to very high video quality
2. High quality after eight generations of encoding/decoding
3. Both 4:2:2 and 4:2:0 video formats

4. Compatibility with the Main profile at the Main level

5. Editability of compressed bitstreams (accurate to one frame sometimes)

6. Flexibility in bitrates

7. Low encoding/decoding delay

8. 512-line (60 Hz) video and 608-line (50 Hz) video capable

9. Cost-effective implementation

The following application-specific experiments were conducted for testing the performance of this profile:

- Intrastudio environment
  —I coding only or I–B coding structure at 30 and 50 Mbits/s
  —Up to eight generations
  —No picture shift between generations
  —Two picture shifts between eight generations
  —Shift in GOP structure
- Interstudio environment
  —$M = 3$ with $N = 12$ coding structure at 30 Mbits/s
  —Up to four generations
  —No picture shift between generations
  Shift in GOP structure between generations.

The complete specification of the 4:2:2 profile is currently in progress.

11.6.2 Multiview Profile

The Multiview profile[15] (MVP) is intended for video applications where many views of a scene may be needed. The simplest example is that of stereoscopic video, which contains two monoscopic video channels, pretty much in the same manner that stereophonic audio contains two channels of monophonic audio. In the case of stereoscopic video, the two channels represent two slightly different views, called the left and right views as they are intended for the left and right eyes of a human visual system, and help generate the effect of depth in scenes. To ensure that each eye of the human visual system sees only one view, typically stereoscopic video requires use of special viewing glasses and modified TV or projection display. There are some display methods that do not require special viewing

glasses and allow several people to view simultaneously; however, they require many simultaneous views of video, typically eight or more. There is yet another type of application in which multiple views are used but not simultaneously, that is, only one view may be selected at a time for display.

We now list anticipated target applications of this profile:

- Education and Training—multimedia presentations, flight simulation
- Entertainment—3D movies, 3D video catalogs, virtual reality, games
- Medical—surgical planning, surgery
- Industrial—animated CAD, engineering simulations
- Remote Sensing—teleoperations, navigation, undersea exploration
- Video Communication—virtual meetings, telepresence
- Scientific Visualization—geographical, metereological.

At the request of industry and standards organizations such as the ITU-R, and anticipating the market needs for emerging and future applications requiring low-cost solutions as well as interoperability, MPEG, with the participation of the requesting organizations, is developing a specification for the Multiview profile. One important requirement[11] is that rather efficient coding of multiple views should be possible; the other equally important requirement[11] is that the coding method should support interoperability with monoscopic decoders and displays. Of course, the issue of decoder complexity is also rather important since for a multiview decoder, the complexity can get rather high owing to the number of views that need to be decoded. Since the scalability tools of the MPEG-2 Video standard allow correlations to be exploited between two or more layers of video and allow coding of one layer of video independently, they are suitable tools for exploiting correlations that exist between different views of a scene. More specifically, Temporal scalability has been found to offer a good solution for coding of stereoscopic video. In Chapter 15, we will discuss, among other issues, the details of coding and display aspects of stereoscopic video, as well as generalized multiview video.

Now that we understand the applications of this profile and MPEG-2 video coding algorithms/tools to resolve the functional needs of these applications, as well as segments of industry interested in this profile, we can list a set of formal requirements as follows:

1. Efficient coding of two views, better than simulcast for more views
2. Compatibility with the Main Profile
3. Syntax extension for coding of more than two views
4. Support for various resolutions and number of views
5. Multiview descriptors to specify imaging setup
6. Random access and trick modes
7. Editability of bitstreams
8. Cost-effective implementation
9. Extensibility

In the following we list the experimental conditions for testing the performance of this profile:

- Stereoscopic Video Application
  —Two views, one view coded independently for compatibility
  —Bitrate combinations of independently coded and dependently coded views are 6 Mbits/s and 3 Mbits/s in one case, and 4 Mbits/s and 2 Mbits/s in another case.
  —Prediction distance $M = 3$ (two B-pictures) and Intra distance $N = 12$ for 625 lines with 25 frames/s sequences and $N = 15$ for 525 lines with 30 frames/s sequences
  —Frame picture structure for both views
  —Independent view coded as the base layer by the Main profile at the Main level; the dependent view coded as enhancement layer by using Temporal scalability.
- Multiviewpoint Video Application
  —Although some proposals have been made on experiments for multiviewpoint applications, experiments have not been finalized yet. Currently, it appears that the main work on this application will be performed within the framework of MPEG-4, the next phase MPEG standard currently in progress.

The complete specification of the Multiview profile is currently in progress.[11,15,21,22] This profile is expected to be interoperable with the Main profile, allowing coding of two layers using the Temporal scalability tool.

11.7 NEW PROFILES INTEROPERABILITY

Among the 4:2:2 profile and the Multiview profile in progress, the specification of the 4:2:2 profile, although not complete, is in a fairly advanced stage and is discussed here. Unlike the hierarchical relationship between the Simple, Main, SNR, Spatial, and High profiles, the 4:2:2 profile does not have a hierarchical relationship with other profiles.

11.7.1 4:2:2 Profile Indication

The 4:2:2 profile is defined only at the Main level and is identified by an 8-bit code whose most significant bit, also called the escape_bit, is "1." It should be noted that levels of a profile that does not fit into a hierarchical structure of profiles do not necessarily have a relationship with similarly named levels of the profiles that fit into a hierarchical structure. The complete code identifying the 4:2:2 profile at the Main level is "10000101."

11.7.2 4:2:2 Profile Syntax Constraints

Earlier, for profiles that fit into a hierarchical structure, Table 11.2 showed the constraints on syntax. In Table 11.7 we now list the syntax constraints for the 4:2:2 profile.

The *repeat_first_field* is constrained as in the case of the Main and the Simple profiles with the main difference as follows:

- If *vertical_size* is greater than 512 lines and if *picture_coding_type* is B-picture, *repeat_first_field* shall be 0. Also, if *vertical_size* is greater than 512, the frame rate shall be 25 frames/s.

Table 11.7　Syntax Constraints Comparison for Profiles

	Allowed value
Syntax Element	4:2:2 Profile
chroma_format	4:2:2, 4:2:0
frame_rate_extension_n	0
frame_rate_extension_d	0
aspect_ratio_information	1.0, 4/3, 16/9
picture_coding_type	I, P, B
repeat_first_field	Constrained
sequence_scalable_extension()	No
scalable_mode	—
picture_scalable_extension()	No
intra_dc_precision	8,9,10,11
Slice_structure	Restricted

Table 11.8　Constraints on Parameters for the 4:2:2 Profile at Main Level

Parameter	Upper bound value
Horizontal size	720 pels/line
Vertical size	608 lines/frame
Frame rate	30 frames/s
Sampling density	11.0592 Msamples/s
Bitrate	50 Mbits/s

11.7.3　4:2:2 Profile Parameter Constraints

Many of the parameter constraints for the 4:2:2 profile are the same as the constraints on the profiles that fit into a hierarchical structure. In Table 11.8 we now show the parameter constraints of the 4:2:2 profile that are different than other profiles.

11.8　DECODERS AND BITSTREAM COMPATIBILITY

For profiles and levels that obey a hierarchical structure it is recommended that each layer bitstream should contain *profile_and_level_indication* of the "simplest" decoder capable of decoding that layer of the bitstream. Although we have discussed the relationship between various profiles in a hierarchical structure and thus implied which bit-

stream is decodable by which decoder, we have not shown a complete picture showing interoperability between decoders and bitstreams. Now that we have also discussed an example of a profile (the 4:2:2 profile) that does not fit into the hierarchical structure, it is time to present an integrated picture illustrating the forward compatibility between the profiles and levels that a decoder is designed for and the profile and level indication in a bitstream for all profiles. This is shown in Table 11.9. A little explanation regarding the abbreviations used is in order. The Simple, Main, SNR, Spatial, and High profiles are correspondingly abbreviated as SP, MP, SNR, Spat, and HP, whereas the High, High-1440, Main, and Low levels are correspondingly abbreviated as HL, H-14, ML, and LL. The entry "y" in the table means that the decoder indicated by the column where this entry is located would be able to decode the bitstream indicated by the corresponding row.

11.9　SUMMARY

In this chapter, we first learned about the MPEG-2 video requirements, the process of developing the MPEG-2 video standard, and the organization of MPEG-2 video into profiles. Next, the details of profile and level structure were discussed, followed by a discussion of how the various profiles were verified by MPEG. Then the work in progress in MPEG on defining new profiles for upcoming applications was discussed. We now summarize the highlights of what we discussed in this chapter as follows:

- Many general requirements of existing and anticipated applications were satisfied and a generic standard was successfully designed.
- MPEG-2 video development consisted of competitive phase tests and collaborative phase simulations via Test Model experiments.
- The Test Model work was iterative and considered new requirements as well as the implementation complexity of the Test Model.
- Profiles and levels allow subsets of syntax and semantics to be organized to trade off interoper-

Table 11.9   Forward Compatibility Between Decoders and Bitstreams

Decoder for a Profile and Level

Bistream Profile and Level Indication	HP @ HL	HP @ H-14	HP @ ML	Spat @ H-14	SNR @ ML	SNR @ LL	MP @ HL	MP @ H-14	MP @ ML	MP @ LL	SP @ ML	4:2:2 @ ML
HP@HL	y											
HP@H-14	y	y										
HP@ML	y	y	y									
Spat@H-14	y	y		y								
SNR@ML	y	y	y	y	y							
SNR@LL	y	y	y	y	y	y						
MP@HL	y						y					
MP@H-14	y	y		y			y	y				
MP@ML	y	y	y	y	y		y	y	y			
MP@LL	y	y	y	y	y		y	y	y			y
SP@ML	y	y	y	y	y	y	y	y	y	y	y	y
MPEG-1	y	y	y	y	y	y	y	y	y	y	y	y
422@ML												y

ability with implementation cost. Current profiles are hierarchically structured based on the syntax they support. Each profile is subdivided into one or more levels.

- There are a total of five profiles: Simple, Main, SNR, Spatial, and High profile. The Simple and the Main profiles are nonscalable, the SNR profile supports SNR scalability, and both the Spatial and the High profiles support SNR scalability as well as Spatial scalability. A total of 11 profile and level combinations are currently defined.

- There is considerable interoperability between profiles. In addition, decoders conforming to any of the hierarchically structured profiles are expected to decode MPEG-1 constrained parameter bitstreams.

- Some constraints are imposed on syntax, semantics, and parameter values to ensure practical decoders without losing flexibility.

- Verification Tests were conducted on profiles at some representative levels to confirm the quality for intended applications.

- Work on two new profiles is in progress; they are the 4:2:2 profile and the Multiview profile, and they expect to be finalized later in 1996. The interim specification of the 4:2:2 profile is included.

REFERENCES

1. "Generic Coding of Moving Pictures and Associated Audio Information: Video," ISO/IEC 13818-2: Draft International Standard (November 1994).

2. S. OKUBO, K. McCANN, and A. LIPPMAN, "MPEG-2 Requirements, Profiles and Performance Verification," in *Signal Processing of HDTV*, V, Elsevier Science, Amsterdam, pp. 65–73 (1994).

3. T. HIDAKA, K. OZAWA, and D. NASSE, "Assessment of the MPEG-2 Performance," in *Signal Processing of HDTV*, V, Elsevier Science, Amsterdam, pp. 97\N104, (1994).

4. M. DELAHOY, "Subjective Viewing Testing Conducted in Australia," ISO/IEC JTC1/SC29/WG11 MPEG94/047 (February 1994).

5. K. OZAWA, "Results of MP@ML Verification Tests in YTSC," ISO/IEC JTC1/SC29/WG11 MPEG94/028 (March 1994).

6. J. URANO and H. YAMAUCHI, "Results of MPEG-2 MP@ML Verification Test at NTV," ISO/IEC JTC1/SC29/WG11 MPEG94/029 (March 1994).

7. D. NASSE, "Revision of Draft Doc N0673 on Results of MP@ML Video Quality Verification Tests," ISO/IEC JTC1/SC29/WG11 MPEG94/250 (July 1994).

8. T. Hidaka, "Comparison of Each Organization's Results for SNR@ML Verification Test," ISO/IEC JTC1/SC29/WG11 MPEG94/222 (July 1994).

9. D. Nasse, "Tests on the Video Quality of SSP@H-14," ISO/IEC JTC1/SC29/WG11 MPEG94/350 (November 1994).

10. T. Alpert, "Summarized Results of MPEG-2 MP@ML SNRP@ML, SSP@H-14 Video Quality Verification Tests," ISO/IEC JTC1/SC29/WG11 MPEG95/024 (March 1995).

11. Requirements Subgroup, "Status Report on the Study of Multi-view Profile," ISO/IEC JTC1/SC29/WG11 N0906 (March 1995).

12. D. Nasse, "Revision of Draft Doc. N0907," ISO/IEC JTC1/SC29/WG11 MPEG95/205 (July 1995).

13. A. Tabatabai, "Report of the Adhoc Group on 4:2:2 Profile Testing," ISO/IEC JTC1/SC29/WG11 MPEG95/185 (July 1995).

14. Requirements and Video Subgroups, "Draft Amendment to ISO/IEC 1318-2-4:2:2 Profile," ISO/IEC JTC1/SC29/WG11 N0907 (July 1995).

15. Video Subgroup, "Proposed Draft Amendment No. 3 to 13818-2 (Multi-view Profile)," ISO/IEC/JTC1/SC29/WG11 N1088 (November 1995).

16. "Coding of Moving Pictures and Associated Audio for Digital Storage Media at up to about 1.5 Mbit/s," ISO/IEC 11172-2: Video, (November 1991).

17. S. Okubo, "Requirements for High Quality Video Coding Standards," *Signal Proc. Image Commun.*, pp. 141–151 (1992).

18. L. Chiariglione, "The Development of an Integrated Audiovisual Coding Standard:MPEG," Proc. of the IEEE, Vol. 83, No. 2, pp. 151–157 (February 1995).

19. United States Advanced Television Systems Committee, "Digital Television Standard for HDTV Transmission," ATSC Standard, (April 1995).

20. R. L. Schmidt, A. Puri and B. G. Haskell, "Performance Evaluation of Nonscalable MPEG-2 Video Coding," Proc. SPIE Visual Commun. Image Proc., (Oct. 1994).

21. A. Puri and B. G. Haskell, "A revised proposal for Multi-view Coding and Multi-view Profile," ISO/IEC JTSI/SC29/WG11 MPEG 95/249 (July 1995).

22. T. Homma, "Report of Adhoc Group on MPEG-2 Requirements-Applications for Multiview Pictures," ISO/IEC JTCI/SC29/WGII MPEG 95/193 (July 1995).

12

DIGITAL VIDEO NETWORKS

Broadband access networks to support digital multimedia services present two interrelated challenges: (1) design of the best network architecture that is cost effective and provides easy migration from the current situation, and (2) creating a uniform control and software infrastructure for end-to-end support and modular growth of multimedia services. The current situation is rather complicated because of regulation and participation of different industry segments (see Fig. 12.1). Direct Broadcast Satellite (DBS) and Terrestrial Broadcast both transmit directly to the consumer's home. Cable Television (CATV) offers a large analog bandwidth over coaxial (coax) cable but over a shared channel (referred to as logical bus); Telephone Companies (Telco) offer a reliable, low bandwidth but with an individual connection (referred to as a star topology).

For delivery of digital multimedia information, the facilities may employ constant bitrate (CBR) or variable bitrate (VBR) transmission. The elementary streams will usually be packetized using the MPEG-2 Systems multiplexing standard. Further processing may also be carried out for particular transmission networks. For example, Asynchronous Transfer Mode (ATM) networks further packetize the digital data into 53 byte *Cells* having 47 or 48 byte payloads. Satellite and other wireless transmission channels typically require the addition of extra bits for Forward Error Correction (FEC).

CATV and Telco networks are both wired connections. Both are being upgraded using optical fiber technology to provide enhanced transmission capacity—both analog and digital. Three dominant wired architectures have emerged over time without a clear consensus on their comparative advantages and disadvantages. The first one, called the hybrid fiber-coax (HFC), can deliver 60 to 150 channels of analog (AM-VSB*) video while at the same time providing digital services using time division and frequency division multiplexing (TDM and FDM) all within a 500- to 1000-MHz bandwidth.

Service providers with more faith in switched digital services prefer point-to-point fiber-to-the-curb (FTTC) architectures, typically served by a 1 Gbits/s curbside switch. As an example, one such switch could provide switched 50 Mbits/s to each of about two dozen homes. Providing analog video is problematic on FTTC and usually needs an analog overlay on top of this architecture. The overlay could be configured along the HFC architecture.

In Europe and Japan, with less aggressive bandwidth expectation and less emphasis on analog video, point-to-multipoint passive optical networks (PONs) are used. PONs use passive splitters to

---

*Amplitude Modulation—Vestigial Side Band.

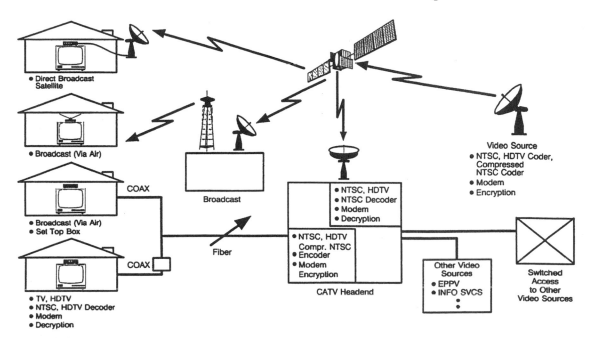

Fig. 12.1 Delivery of broadband video is currently over several media: (a) Direct Broadcast Satellite, (b) Terrestrial Broadcast, and (c) Wired systems (CATV and Telco). Each of these media are in different stages of becoming digital. This does not include delivery using stored media such as video-cassettes and CD-ROMS.

route a fiber feeder to a group of closely located homes, thereby reducing cost.

Each of the three architectures can provide on-demand digital video services. The difference is in the digital modulation schemes used and how the bandwidth is shared. The FTTC architecture uses baseband digital modulation and point-to-point distribution to individual customers from a time-multiplexed fiber feeder. On the other hand, the HFC architecture utilizes advanced digital modulation technology, for example, QAM (Quadrature Amplitude Modulation) or VSB (Vestigal Sideband) and a coax bus fed by a frequency multiplexed fiber feeder for distribution to customers. Between these two wired alternatives, there is a wide choice in other parameters, such as packetization, protocols, signaling and control, and so forth. Some of these will be discussed in this chapter.

There are nonwired alternatives as well. Terrestrial broadcasting, which has served analog television for many decades, may become digital as well. Although digitization is led by HDTV, digital

SDTV* and other data-broadcasting may follow soon. In addition, high-power satellites are giving rise to direct broadcasting of digital video using small dish antennas and inexpensive electronics.

Figure 12.2 provides an end-to-end functional block diagram for delivering server content to the consumer terminals independent of the specific medium. Content is stored by service providers in multimedia servers (see Chapter 13). The access network includes all active and passive devices, as well as all cabling and service interface technologies. The access node provides appropriate format conversions to meet internal network interfaces for all the services. It is typically located within the Telco Central Office or CATV headend, which is usually within several tens of miles of the consumer home. The service node is usually a switch that connects a consumer to the server(s) of any service provider. Service management provides the gateway(s) through which a consumer selects information providers, manages service options, and navigates among multiple choices. Integrated

*Standard Definition TV, for example, NTSC in North America and Japan, PAL and SECAM in Europe and Asia.

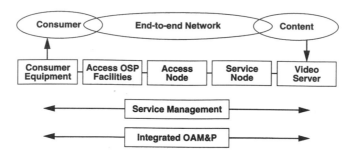

Fig. 12.2   To provide a full spectrum of normal and broad-band services, the platform must contain consumer equipment, an access network, and servers with content.

OAM&P (Operations, Administration, Maintenance, and Provisioning) refers to all the operations capabilities designed to operate a full-service network with high efficiency and minimal human intervention.

In this chapter, we describe each of the media alternatives from a full-service end-to-end network perspective, cover their strengths and weaknesses, and describe a possible evolution path into the future.

12.1   CURRENT METHODS OF ACCESS[1-5]

Telcos provide telephone service over a network constructed of cabled sets of wire pairs, varying in count between six and 3600 wire pairs per cable (see Fig. 12.3). This supports a 4-kHz analog bandwidth for voice. Individual wire pairs are bundled together at various points in the network, to aggregate larger and larger groups of wire pairs, eventually connecting to the telephone central office. An important point is that each customer premise equipment (CPE) has a separate connection to the Telco central office and that the bandwidth is not shared. Therefore, the Telco architecture is said to have a star topology.

Cable TV service providers transmit analog video signals from the headend (HE) to the consumer's home (see Fig. 12.4) over fiber coax networks. Lasers at the headend transmit the full RF spectrum (from about 50 MHz to typically either 330 MHz or 550 MHz) to "fiber nodes" serving 500 to 2000 homes. From the fiber node the signal is distributed electrically to homes via an amplified tree and branch coaxial cable network. Multiple-home taps are spaced along the coax distribution bus, and from each tap individual homes are connected via coax "drops." All consumers receive the same 40+ channels of composite VSB video, each with a bandwidth of 6 MHz. Trap filters are often placed in series with the drop termination to control access to premium services.

Long amplifier cascades for the older "all-coax networks" have been replaced by direct fiber links,

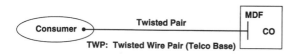

Fig. 12.3   Telco Architecture is narrowband, with twisted copper pairs connecting consumer equipment to the Main Distribution Frame (MDF) in the Central Office (CO).

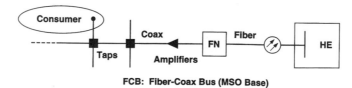

Fig. 12.4   Current CATV delivery system. Signals travel from the Head End (HE) to a fiber node in the CATV plant, followed by a tree-branch network of coaxial cables (coax) to the consumer homes.

resulting in a maximum of three to ten amplifiers between a consumer and the HE. This improves both reliability and signal quality. CATV systems do not employ sophisticated operations support systems, and their networks provide little or no power redundancy. They are also cheaper on a per consumer basis than Telco networks.

12.2 NEW BROADBAND WIRED ACCESS SCENARIOS[6-12]

Both Telcos and CATV operators can provide incremental upgrades for the short term. The Asymmetrical Digital Subscriber Line (ADSL) offers 1.5 to 6.0 Mbits/s downstream transport, depending on the length and the quality of the loop (see Fig. 12.5). However, at this time, it is too expensive, and moreover it does not provide an

integrated means of both digital and AM-VSB transmission of video. While it has the advantage of serving a sparsely distributed subscriber base with investment added largely as subscribers are added, it also requires central office (CO) modems and extra switching functionality in the CO to accomplish a channel for "broadcast" services. These aspects limit the overall popularity of the ADSL technology.

CATV systems can construct incrementally a Bidirectional Fiber Coax Bus (BFCB) by (1) increasing downstream bandwidth to 750 MHz and (2) providing a very narrowband upstream channel for limited interactive control, as shown in Fig. 12.6. These steps are equivalent in cost to that required to build a hybrid fiber coax network, with substantial improvement in service capability.

Three dominant architectures are gaining momentum among service providers. The first of these, called Hybrid Fiber Coax (HFC), is being considered by all the CATV operators (and some Telcos) as an extension of the current CATV fiber-coax topology (Fig. 12.7). The second architecture, called the Fiber to the Curb (FTTC), is being studied by the Telcos (Fig. 12.8). The third architecture, popular in Europe and Japan, is called the Passive Optical Network (PON) as shown in Fig. 12.9. All three architectures and their numerous variations have relative strengths and weaknesses, and the choice between them is influenced to a

Fig. 12.5 Asymmetrical Digital Subscriber Line (ADSL) provides a substantial increase in bitrate compared with the current telephone plant. However, it does not provide for evolution to the higher bidirectional bitrates that may be required in the future.

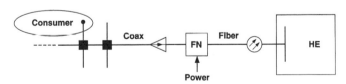

Fig. 12.6 A Bidirectional Fiber Coax Bus (BFCB) provides an in-between step from the current CATV architecture to the Hybrid Fiber Coax (HFC).

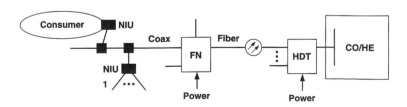

Fig. 12.7 HFC is a strong candidate for evolution of the current CATV architecture and has also been chosen by several Telcos. Each leg of the coax may service over hundred consumer homes.

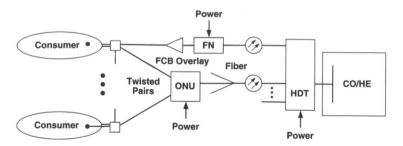

Fig. 12.8   The fiber-to-the-curb (FTTC) system provides fiber almost to the customer premise (HE/CO to ONU). Consumer homes connect to the ONU by either twisted pair (as shown) or by a coax. Provision of analog broadcast video requires a Fiber-Coax Bus (FCB) overlay.

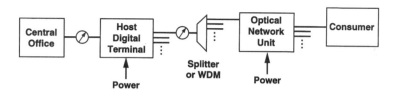

Fig. 12.9   PON systems are basically point-to-multipoint in that they use passive splitters to share a fiber feeder among many closely spaced homes.

large extent by the relative weight given to these factors by service providers.

The HFC architecture provides significant upstream bandwidth in the 5 to 40 MHz band. From the headend in the "downstream" direction, linear photonic transmission equipment is used to transport telephony, various data services, switched digital video, broadcast digital video, and broadcast analog AM-VSB cable TV signals to a "fiber node" (FN) which typically serves up to 500 homes. From the fiber node, transmission is bidirectional over a coaxial cable tree and branch topology to Network Interface Units (NIUs), which serve a very small group of homes. Broadband video (analog and digital) signals are carried into the home on coaxial cable for distribution to cable set-top boxes (STBs). Digital signals are encoded and decoded only once at the HDT (or elsewhere) using standard modulation techniques (e.g., 64-QAM) without any form of further transcodings. This improves the quality of the signal displayed on the consumer TV.

In the FTTC architecture, all traffic except analog video is carried digitally over point-to-point fiber facilities from the central office to the Optical Network Units (ONUs) which may serve a few to

several dozen homes. However, two practical considerations force a significant variation from this ideal. To power the ONUs and to transport broadcast analog TV, practical FTTC systems require an overlay carrying electrical power and analog TV channels.

A single point-to-point fiber might serve a single ONU, and through it a single home (Fiber-to-the-Home or FTTH), giving unlimited bidirectional digital bandwidth to the home. However, to reduce costs, both the fiber and ONU must be shared. Such sharing converts the point-to-point physical fiber star architecture into a point-to-multipoint architecture.

Hybrid twisted-pair/coax connects the ONU to the home. Twisted pair cable is used to carry the digital video signal in the 5 to 40 MHz band from the ONU to a distribution terminal no more than 500 feet away, where an RF combiner puts this signal on the coaxial drop carrying the analog TV signal. It is this 500 foot limitation on the range of the digital video signal on twisted pairs that limits the number of homes served by a single ONU.

The PONs currently under development provide narrowband services to multiple ONUs in a point-to-multipoint configuration, as shown in Fig.

12.9. Each fiber leaving the CO or host digital terminal serves multiple (8 to 32) ONUs. Various types of ONUs are used, depending on whether the fiber is terminated at a home, business, curbside pedestal, or multiunit building. Each ONU receives a baseband broadcast optical signal at approximately 50 Mbits/s, from which it selects the appropriate information. Upstream transmission from each ONU must be timed, in accordance with provisioned time-slot allocations, to avoid colliding at the HDT with transmission from other ONUs. Upstream and downstream transmissions are separated into alternating frames, using time-compression multiplexing, or can be full duplexed at two different wavelengths, for example, 1.30 and 1.55 micron.

12.3 WIRED ACCESS COMPARED

All the above access architectures, HFC, FTTC, and PON, provide full service capability with differing difficulty. The current generation of analog services can be better handled by HFC systems, since both FTTC and PON require special additions for providing analog video. In addition, the three systems can support: (1) data services, for example, 1.5 Mbits/s data that require wideband upstream access and (2) a goal of $1000 per home.

The future generation of services that need to be supported require all-digital transport and a broadband capability even in the upstream direction. Broadband bidirectional access could be provided by either HFC, FTTC, or PONs and many of their variations.

Investment-wise, the Bidirectional Fiber Coax Bus (Fig. 12.6) represents a relatively modest upgrade of existing fiber coax networks by providing:

- Digital downstream capability for video on-demand channels frequently shared with today's analog cable TV
- Very limited upstream bandwidth for control information of Pay-Per-View and Digital Video services.

To satisfy the full-service objective, the HFC architecture will have to be used by enabling more upstream bandwidth and making the bidirectional bandwidth both more reliable and available. This may require fiber to be driven to smaller nodes than the existing coax plant, and the drops may have to be reengineered.

Both Telcos and CATV companies have a significant investment in their current plant and must evolve this to meet the needs of the future services. CATV systems can either evolve through the intermediate step of a bidirectional fiber coax bus or go directly to full-service HFC. For Telcos, the ADSL technology provides an increment in bandwidth, but does not provide an evolutionary path toward full-service capability. Telcos must therefore begin constructing broadband access either via HFC or FTTC systems. The deciding factor may be provision of analog broadcast video. From this broadband base, both Telcos and CATV companies can evolve to an all-digital network as the consumer environment evolves and technology makes costs fall.

The HFC architecture, relative to FTTC and PON, has a lower first cost of implementation. The major reason for this is that, in the case of FTTC or PON, the analog cable TV must be carried with special arrangements such as a broadcast fiber coax overlay in the case of FTTC. For the CATV companies, the HFC architecture has a perfect "fit" to the embedded plant.

For the Telco, FTTC is a more natural fit to their operations plans and craft skills. In the currently deployed configurations, FTTC or PON get fiber closer to the home than does HFC. Some service providers want to get fiber physically closer to the home, while others focus on a predominantly passive plant and see little difference in the two architectures.

Current economics permits HFC systems to put the last point of network intelligence very close to the customer, even all the way to the home, while FTTC or PON systems are economical only with ONUs (the last point of network intelligence for these systems) which serve a few dozen homes. Points of network intelligence closer to the customer facilitate more complete automated fault isolation and more automated service provisioning. HFC systems provide bandwidth closer to the customer because all of the bandwidth carried on the medium is available to every home in HFC systems, while in FTTC or PON the broad bandwidth of the fiber is terminated at the ONU. In FTTC systems, the hybrid twisted-pair/coax drop limits the bandwidth available to or from any home to only a small fraction of that on the fiber bus serving the ONU.

HFC systems being deployed by both Telcos and the CATV providers usually retain the set-top-box signaling and control technologies of the cable TV industry, while FTTC or PON systems typically plan to use the ATM protocol for the upstream signaling. This consideration allows FTTC or PON systems easier migration to packet-based services through ATM connectivity. However, ATM-capable modems are emerging from several vendors that will provide equivalent capability to HFC systems as well.

Evolving technology might make these three architectures ultimately converge. Over time, high-powered laser transmitters and linear optical fiber amplifiers will make video distribution directly to FTTC ONUs, or to HFC fiber nodes both serving approximately the same number of homes. Sometime thereafter, steadily declining electronics costs will permit further convergence of the architectures. Eventually the long held dream of "fiber directly to the home" will become possible. But far

more important than this long-term system evolution view is the view of how the broadband access systems being deployed today will be evolvable to provide service parity with the newer systems that eventually emerge. All the three architectures, HFC, FTTC and PON, have this *in situ* evolutionary potential.

12.4 TERRESTRIAL BROADCASTING[12]

Current advanced television (ATV) research for terrestrial broadcasting in the VHF/UHF bands is converging toward a fully digital implementation. In the United States, a Grand Alliance* has been formed by proponents of digital HDTV[†] systems. The planned system is a digital simulcast with the current analog NTSC transmission (see Fig. 12.10). The goal of the Grand Alliance is to introduce an ATV terrestrial transmission standard for the United States. Similar digital initiatives are beginning to emerge in Europe and Japan.

In a digital ATV system, an MPEG-2 compressed HDTV signal is transmitted at approximately 15 Mbits/s. This data rate is sufficient to provide a satisfactory distribution[‡] quality HDTV service. The aggregate data rate for an ATV sys-

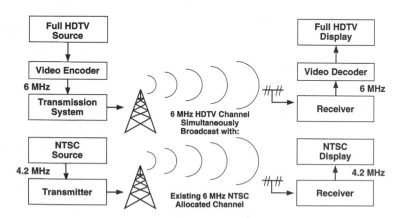

Fig. 12.10   Terrestial Simulcast of digital HDTV and analog NTSC with similar content. See Chapter 14.

---

*AT&T, David Sarnoff Research Center, General Instrument, M.I.T. North American Philips, Thomson Consumer Electronics, and Zenith Electronics.

†See Chapter 14 for a description of HDTV.

‡TV studios typically produce *contribution* quality TV, which is then encoded and transmitted as *distribution* quality TV.

tem, which includes compressed video, compressed audio, conditional access, and an auxiliary data channel, is around 18 to 20 Mbits/s. In North America, this data rate must be squeezed into a channel of 6 MHz bandwidth for terrestrial broadcasting. For some other countries, the same data rate only needs to fit into a 7 or 8 MHz channel. Other considerations are: (1) the system must be able to sustain multipath distortion; (2) the system should be robust to interference to and from existing analog TV services; (3) the system should provide fast carrier recovery and system synchronization for rapid channel changing; and (4) the system should provide easy transcoding to and from other transmission media, such as satellite, cable, fiber, mobile reception, and computer networks.

Transmission of 18 to 20 Mbits/s in a 6 MHz terrestrial band could be achieved by a 16-QAM scheme where in-phase (I) and quadrature (Q) components each carries data of 2 bits/symbol. The symbol rate of each component is approximately equal to the usable bandwidth. Another alternative is a four-level VSB modulation scheme, where each symbol again carries 2 bits and the symbol rate is about twice the usable bandwidth. Thus, the bit rate is about the same as for 16-QAM. The noise performances of 4-VSB and 16-QAM are similar.

Another choice is between single-carrier modulation (SCM) and multicarrier modulation (MCM), such as orthogonal frequency division multiplexing (OFDM) currently under investigation in many countries in Europe and elsewhere. In SCM, the information bits are used to modulate a carrier that occupies the entire bandwidth of the RF channel. The SCM-QAM technique has long been used in satellite communications, line-of-sight digital microwave radio, telephone modems, and so forth.

In MCM, QAM symbols are used to modulate multiple low data-rate carriers, which are transmitted in small bands at different frequencies within the channel. Since each QAM modulated carrier has a $\sin(x)/x$ frequency spectrum, the carrier spacing is selected so that each carrier is located at the zero crossing points of the other carriers to achieve frequency domain orthogonality. Although the carrier spectra overlap, they can be orthogonally demodulated without intercarrier interference. MCM can be implemented via the digital Fast Fourier Transform (FFT), where an inverse FFT is used as the modulator and an FFT as the demodulator. One important feature of MCM is that, by inserting a guard interval of small duration between the MCM symbols, or FFT blocks, and using "cyclic extension," the intersymbol interference (ISI) can be eliminated. However, the amplitude/phase distortion within a MCM symbol still exists. Another advantage is that the MCM signal spectrum is more rectangular in shape, which means a higher spectrum efficiency than the SCM system. There are many projects in Europe to implement an MCM system for digital SDTV/ HDTV broadcasting. However, more experience is necessary before a detailed comparison can be made.

MCM is more robust to time domain impulse interference, since the interference will be averaged over the entire FFT block. However, MCM is vulnerable to single frequency tone interference. SCM needs to reserve part of the spectrum for pulse shaping (frequency domain), while MCM needs to insert guard intervals (time domain). For channels with multipath distortion, SCM may need a training sequence to assist adaptive equalizer convergence and synchronization. MCM usually sends our reference carriers to obtain channel state information for frequency domain equalization and channel decoding.

The Grand Alliance HDTV system uses a concatenuated forward error correction (FEC) code with the inner code implemented using trellis coded modulation (TCM) and the outer code implemented using a Reed–Solomon (R–S) code. The inner code combines coding and modulation into one step to achieve high coding gains without affecting the bandwidth of the signal. In a TCM encoder, each symbol of n bits is mapped into a constellation of $(n + 1)$ bits, using a set partitioning rule. To achieve a spectrum efficiency of 4 bits/s/Hz, n is set to 4. The resulting TCM constellation is 32-QAM. The 4-VSB modulation, having 2 bits/symbol, can be trellis coded into 8-VSB, having 3 bits/symbol, including one error protection bit, for transmission. This scheme also achieves a spectrum efficiency of 4 bits/s/Hz, since the symbol rate of VSB system is twice that of the QAM system. It should be noted that TCM coding only increases the constellation size and uses this additional redundancy to trellis-code the signal; there is no bandwidth expansion. A TCM code can be decoded with the Viterbi algorithm that exploits the soft decision nature of the received signal. The coding gain for a two-dimensional TCM code over a Gaussian channel is around 3 dB for a BER of 10^{-15}.

The output of the inner code delivers a symbol error protection rate, of approximately 10^{-3} to the outer code. In a R–S encoder, information data is partitioned into blocks of N information bytes (N is typically between 100 and 200). The output of the R–S encoder has a block size of $N + 2t$, where $2t$ additional bytes are generated for each block, which can correct t-byte transmission errors. Since more data are "generated" by a R–S encoder, additional bandwidth is required to transmit the coded data. Because of the bandwidth constraint in broadcasting, the redundancy must be limited to about 10 percent.

The motivation for using TCM as the inner code and R–S as the outer code is that the TCM code has good noise-error performance at low signal-to-noise ratios (SNR), and a BER* versus SNR curve with a relatively slow roll-off. On the other hand, an R–S code has a very fast BER roll-off, but requires additional bandwidth. In a concatenated system, the bandwidth efficient TCM code is implemented to achieve high coding gain and to correct short bursts of interferences, such as NTSC synchronization pulses from an interfering analog TV channel in a nearby city[†]. The required output BER is around 10^{-3} to 10^{-4}. The R–S code is used to handle the burst errors generated by the inner TCM code and to provide a BER of 10^{-9} or lower. The drawback of the concatenated coding system is that it has a brick-wall or threshold type of performance degradation. The signal dropout is within 1 dB of SNR change around the threshold. Some argue that such a sharp "cliff" is inappropriate for broadcasting and is one of the serious shortcoming of digital broadcasting.

The concatenated coding can be applied in both SCM and MCM systems. With today's implementation of this channel coding technique, system carrier-to-noise ratio (C/N) for a BER $\leq 10^{-9}$ needs to be about 15 to 16 dB for the ATV system. An interleave and deinterleave is also typically used to exploit the error correction ability of FEC codes. Since most errors in the raw bitstream occur in bursts, they often exceed the error correction ability of the FEC code. Interleaving can be used to decorrelate these bursts into isolated errors that can be handled by the FEC code.

For SCM systems, multipath distortion results in intersymbol interference (ISI) and may cause loss of data. For MCM systems, if a guard interval is implemented and the multipath spread is less than the guard interval, there will be no intersymbol interference. However, there are still in-band fadings, which could cause severe amplitude/phase distortion for high-order QAM signals. In both SCM and MCM some measures must be taken to combat the multipath distortion. One way to reduce the multipath distortion is to use a high gain directional antenna. For a SCM system, a decision feedback equalizer is also used to minimize the multipath distortion. A training sequence can be embedded in the transmitted signal to assist fast convergence and system synchronization. However, any adaptive equalizer can also cause noise enhancement, which would decrease the system noise performance under a multipath scenario.

For a MCM system, the use of a guard interval can eliminate the ISI, but it also reduces data throughput. To minimize the loss of throughput, the size of the FFT must be increased. However, the size of the FFT is limited by digital signal processing (DSP) hardware capability and receiver phase noise. To compensate for in-band fading, a frequency domain equalizer can be used in combination with soft decision trellis decoding using channel state information. One of the distinct advantages of MCM over SCM with an adaptive equalizer is that MCM is not sensitive to variations in delay as long as the multipath falls within the

---

*Bit Error Rate

[†]Interference from a nearby city using the same channel is called *co-channel interference*. See Chapter 14.

guard interval and the interleave can effectively decorrelate the faded signal. Adaptive equalization performs better on short-delay multipaths than on long-delay multipaths. Therefore, MCM is potentially a better candidate for cell based networks, where there are long delay manmade multipaths. For a SCM system, a directional antenna must be used at some sites where strong multipath distortions are encountered.

The composite analog TV signal has a nonflat frequency spectrum, with three distinct carriers for luminance, chrominance, and audio. The luminance carrier has the highest power, of which the main contributor is the synchronization pulses. For an SCM system, an adaptive equalizer can be used to reduce the impact of co-channel analog TV interference. A few decibels gain of carrier-to-interference ratio (C/I) can be achieved when adaptive equalization is used. An adaptive equalizer can also exploit the cyclostationary nature of the co-channel ATV interference. Other approaches include the use of a comb filter to create spectrum notches to reduce interference at the expense of a 3 dB noise enhancement, and using spectrum gaps to avoid interference with the penalty of reducing throughput. An MCM system is vulnerable to co-channel analog TV interference because of its nonflat spectrum. MCM carriers colocated with interfering analog luminance, chrominance, and audio carriers may suffer from strong interference. To avoid this, some MCM carriers can be turned off to create spectrum "holes" with the penalty of reducing system throughput. This can take up a substantial fraction of the 6-MHz bandwidth. A better solution is to rely on an error correcting code to recover lost data. An interleaver and channel estimator combined with a soft decoding algorithm and error erasure could be used to combat co-channel analog TV interference. The performance of this approach has yet to be demonstrated.

For a SCM system, the peak-to-average power ratio depends on the channel filter roll-off. A faster roll-off, which has a higher spectrum efficiency, will result in a higher peak-to-average ratio. The typical peak-to-average ratio is about 7 dB. Some ATV systems take advantage of the nonsymmetric property of the analog VSB TV receiver input filter to transmit more power or to implement a pilot tone without increasing the co-channel interference into the analog channel of a nearby city. For an MCM system, the amplitude distribution is closer to Gaussian, having a relatively high peak-to-average ratio. Tests show that a SCM system is marginally better, by 1 to 2 dB, than a flat spectrum MCM system. However, if spectrum shaping is used, opening holes in the spectrum where analog carriers are normally located, a few decibels gain may be achieved by MCM systems.

12.5 Direct Broadcast Satellites[14]

A Direct Broadcast Satellite (DBS) system provides approximately 150 channels of direct-to-the-home TV in a compressed form that can be received by an 18-inch diameter dish antenna, decoded, and made ready for display to the consumer television set (see Fig. 12.11). In a little over a year since its introduction, more than 1 million direct broadcast satellite receivers have been installed in United States homes. Equally impressive, the entire receiver electronics system, including the antenna, low-noise front end, and set-top box are sold at less than $800 per system. The quality of video is generally comparable to that of broadcast or cable video.

The information flow from the video source to the satellite and back to the consumer's home is shown in Fig. 12.12. Video signals on earth from a variety of sources are compressed with the MPEG-2 algorithm, multiplexed perhaps using statistical multiplexing (see Chapter 8), and then sent using

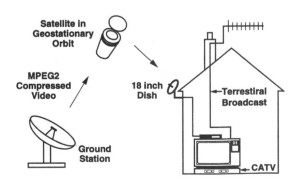

Fig. 12.11   Direct Broadcast Satellite System.

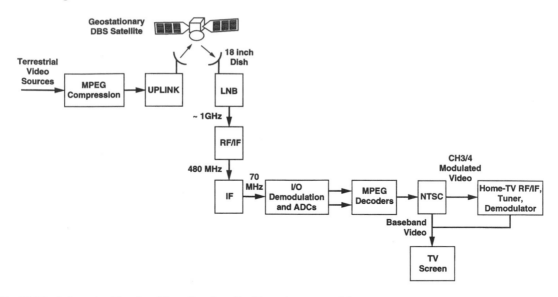

Fig. 12.12   Information Flow from Direct Broadcast Satellite to the consumer's home.

an existing uplink to a geostationary statellite. The signal representing this bitstream is amplified by transponders within the satellite and beamed directly to the small dish antenna at each consumer's home. The modulation technique that is used often is Quadrature Phase-Shift Keying (QPSK), which provides 2 bits per symbol. The modulator creates In-phase (I) and Quadrature (Q) components of bitstreams which are then added before up-conversion to a higher frequency for transmission. Forward Error Correction (FEC) is required for noise tolerance and correction of transmission errors. Convolution (trellis) encoding provides protection against short impulsive error bursts. Reed–Solomon block codes help detect and correct longer duration errors.

The first of the DBS satellites currently in orbit was launched in December, 1993. It carries 16 transponders. Each transponder has a 120W output,* about 10 to 20 times the output of a typical communications satellite designed for a much larger receiver antenna. The transponder and antenna gain on the satellite yield a downlink radiated signal of more than 50 dBW, depending on receiver location. The receiver's parabolic antenna provides a gain of approximately 30 dB at 12 GHz. The

antenna must be installed with a clear view of the southern horizon and pointed at the satellite, which is geosynchronously stationed at 101° W longitude and aimed at the continental United States.

In spite of the high power and high antenna gain of the transponders and the satellite, high performance is still required of the front-end receiver circuitry. The gain of the receiver antenna partially compensates for the 200 dB path loss incurred at 12 GHz over the 37,000 km synchronous orbit distance. The DBS receiving antenna uses a Low-Noise Block (LNB) mounted at the focal point of the dish to provide gain for the approximately −100 dBm received signal. The RF and IF stages of the receiver circuitry are designed by balancing the allocation of gain, noise, and cost among various blocks, while providing protection against overload, which can easily occur at signal levels higher than −80 dBm. The low-noise block is more than just an amplifier. It also down-converts the 12 GHz untuned signal to a range of 950 to 1450 MHz. The next stage is to further down-convert the signal to a single channel at around 70 MHz and apply synchronous demodulation to get both the In-Phase (I) and Quadrature (Q) digital

*An alternative configuration allows eight transponders to operate at 240 W.

components. Such a tuned signal is then ready for data recovery.

The processing performed in the set-top box does the QPSK demodulation to get a symbol rate of approximately 21 MHz, which produces a raw bitrate of about 42 Mbits/s. The output data from the demodulator is then fed to an FEC decoder, which incorporates a Viterbi decoder concatenated with a Reed–Solomon decoder. The final error-corrected payload is either 23 Mbits/s or 30 Mbits/s, depending on configuration and FEC settings.

The output of the FEC decoder is fed to an MPEG Systems demultiplexer, followed by the audio/video decompression devices. The resulting $Y\ Cb\ Cr$ video components are then usually combined to produce a composite NTSC signal. This is then either passed directly to the baseband inputs of the TV or up-converted to channel 3 or 4 and passed to the RF antenna inputs.

The current Direct Broadcast Systems distribute many of the same programs seen on the cable television systems. However, there is no local programming. In addition, DBS is able to show special interest and special event programming. An attractive user-friendly interface is created on a channel that is reserved for on-screen menus and selection guides. The growth of the DBS system in the United States has been much higher than expected. The early experience in Europe and Japan also shows significant promise. With the cost of the receiver electronics coming down, we expect strong demand and growth of direct broadcast satellite services.

REFERENCES

1. R. C. MENENDEZ, D. L. WARING, and D. S. WILSON, "High-Bit-Rate Copper and Fiber-to-the Curb Video Upgrade Strategies," *Proc. IEEE International Conference on Communications ICC '93* (May 23–26, 1993).

2. LIPSON et al. "High Fidelity Lightwave Transmission of Multiple AM-VSB NTSC Signals," *IEEE Trans. Micro. Theory Techniques*, 38:483–493 (1990).

3. OLSHANKSKY et al., "Subcarrier Multiplexed Broad-Band Service Network: A Flexible Platform for Broad-Band Subscriber Services," *J. Lightwave Technol.*, 11:60–69 (1993).

4. R. JONES and R. B. SHARPE, "A Cost Effective Digital Broadband Passive Optical Network," *Proc. NFOEC*, Nashville, TN (1991).

5. X. P. MAO et al., "Brillouin Scattering in Lightwave AM-VSB, Transmission Systems," in *Tech. Dig. Opt. Fiber Comm.*, San Jose, CA (February, 1992).

6. S. D. DUKES, "Photonics for Cable Television System Design," *Commun. Eng. Design*, pp. 34–48 (May 1992).

7. L. ELLIS, "The 5–30 MHz Return Band: A Bottleneck Waiting to Happen?," *Commun. Eng. Design*, pp. 40–43 (December 1992).

8. W. SHUMATE, "Economic Considerations for Fiber to the Subscriber," *Proc. 1993 European Conf. in Optical Communication*, 1:120–123 (1993).

9. J. A. CHIDDIX, "Fiber Optic Technology for CATV Supertrunk Applications," in *NCTA '85 Technical Papers*, 1985.

10. J. A. CHIDDIX, "Optical Fiber Super-Trunking, The Time Has Come: A Performance Report on a Real-World System," in *NCTA '86 Technical Papers, 1986 and IEEE Journal on Selected Areas in Communications*, SAC-4 (5) (August 1986).

11. P. A. ROGAN, R. B. STELLE, III and L. D. WILLIAMSON, "A Technical Analysis of a Hybrid Fiber/Coaxial Cable Television System," in *NCTA '88 Technical Papers*, 1988.

12. D. L. WARING and W. Y. CHEN, "Applicability of ADSL to Support Video Dial Tone in the Copper Loop," *IEEE Commun. Mag.* 32 (5):102–109 (May 1994).

13. W. F. SCHRUBER, "Advanced TV Systems for Terrestial Broadcasting: Some Problems and Some Proposed Solutions," *Proc. IEEE*, pp. 958–978 (June 1995).

14. J. WOOD, *Satellite Communications DBS Systems*, Focal Press (1992).

13

INTERACTIVE TELEVISION

Interactive television (ITV) is a new form of digital consumer multimedia service that can give viewers much greater control over the content of the programs than is possible with conventional analog television.[1-3] Advances in audio/video compression, multimedia database systems, ongoing deployment of broadband networks, inexpensive home terminals, and user-friendly interfaces will provide the infrastructure for offering a wide variety of interactive video services to consumers at home via standard television. Such services include video on demand, video telephony, multimedia information retrieval, distance-education, home shopping, and multiplayer, multilocation video games, and so forth. This chapter deals with a platform over which many of these services can be implemented.

The most important aspect of interactive television is the ability of the viewer to exercise both coarse and fine-grain control over the contents of the programming being viewed. By comparison, current television allows the viewer to select only among a number of channels that broadcast a predetermined presentation to a large audience. This presentation follows a predefined sequence from start to finish and cannot be customized in any manner by an individual viewer. In interactive television, viewers can directly control the content displayed on their TV screens. With a properly designed user interface, a viewer can choose not

only a program, but also other related information that is referenced in the program. The selected multimedia material could contain still images, moving video, and audio. Creating a customized entertaining experience with audio/video whose quality is comparable to state-of-the-art conventional TV programming is the challenge of interactive TV. Computer video games, for example, have always been interactive, but in most cases the pictures that were controlled were cartoon-like. The quality of these pictures has been steadily rising as the result of the growing programming sophistication, inexpensive hardware, and the desire of the viewers to see exactly what they want in a time frame that they can allocate among other competing demands on their time.

ITV programs generally consist of a collection of media elements together with software that controls the program flow in response to consumer inputs and directs how the media elements create the multimedia presentation. Media elements, such as audio and video clips, still images, graphics, text, etc., are the primitives of the presentation and are created as part of the process of producing the ITV application.

Each viewer of an ITV application is totally independent. This requires that a separate instance of an application be executed by the ITV system for each viewer, although the stored or synthesized media elements are shared. In addition, communi-

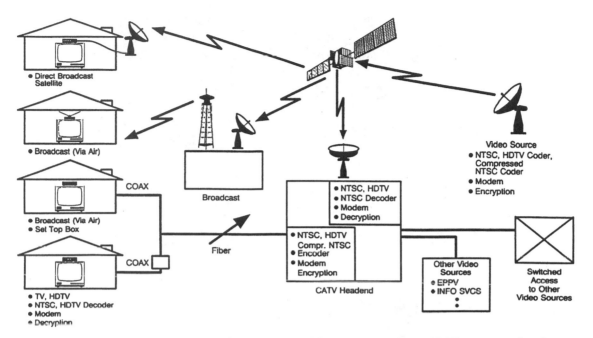

Fig. 13.1 This block diagram shows a variety of services such as video-telephony, games, and ITV. The system consists of servers, a distribution network, and the set-top box (customer-premise equipment, CPE).

cations bandwidth must be allocated for each viewer, a situation that requires a logical star topology similar to the telephone network rather than the shared bandwidth topology of the conventional broadcast or cable TV (see Chapter 12).

Economically providing the appropriate computing, storage, and communication facilities required for each viewer of an ITV system is the key enabler for such services. An ITV system deals with data types (e.g., audio and video clips) that are typically large (even after compression) and must be processed within real-time constraints. The data throughput or bandwidth required for each viewer can be 25 to 100 times that required for standard voice telephony.

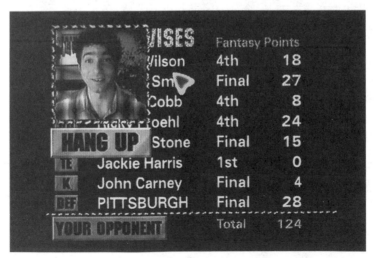

Fig. 13.2 An experimental fantasy sports league application. A videophone call is in progress in the window at the upper left corner. The rectangular area labeled "Hang Up" is a button to be clicked by the user (via the remote control) to terminate the call.

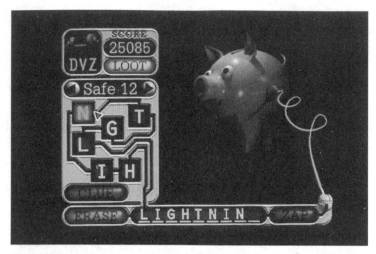

Fig. 13.3   A screen from an experimental children's educational application. The areas labeled "Zap, Erase, Clue, Loot" and the boxed letters *N, L, G, T, I,* and *H* are all buttons. The button *N* is shown highlighted after being clicked.

Fig. 13.4   A screen from an experimental home shopping application. As the user clicks the buttons labeled with manufacturer names, video presentations about various juice extractors are played in the window to the right. Clicking the button labeled Consumer Service will initiate a telephone call (audio or video) to a customer sales representative who can answer questions about the products.

The four main components of interactive television systems are: (1) a home terminal, commonly known as the set-top box (STB) or customer premises equipment (CPE), (2) an access network, (3) a network-based server, and (4) a powerful user-friendly interface. Figure 13.1 shows a typical configuration of an ITV system. Figures 13.2 to 13.4 show typical video screens from an experimental ITV system.[5] In this chapter, we discuss the CPE, network-based servers, and user interface. The access network has already been discussed in Chapter 12.

13.1   CUSTOMER PREMISES EQUIPMENT

The cost of the CPE is a dominant factor in the overall cost of the ITV system. The less expensive the CPE, the more likely it is that people will try the service and become repeat users. The CPE (or STB) typically takes the form of a box sitting on top of the TV set.[4] This CPE connects to both the television and to an external communication network (e.g., telephone, CATV, or direct broadcast satellite) via a subscriber "drop" or "loop" from an outside pole, underground conduit, or direct broadcast antenna. Each TV set in a home typically requires its own STB.

For standard, noninteractive, analog TV programs delivered via a CATV plant, a STB includes analog frequency translation, program descrambling circuitry, and a simple wireless remote control. Such a STB has a very low cost and is often owned and supplied by the network operator.

When interactive services are offered, the complexity and cost of the STB increases. Since the digital services are new "add-ons," the new STB must provide both the analog functions outlined above as well as the digital functions of audio and video decompression, demodulation to recover the digital feeds, decryption, an upstream modem for communicating consumer control requests back to the program source, plus a user-friendly interface. This type of STB, for use on a CATV plant, often costs several times the basic analog STB. However, this cost is expected to go down in the future.

The basic interactive STB just described is capable of supporting a variety of interactive video services. Among them are the popular ones such as movies-on-demand and home shopping. To support these services, a source of interactive service programming (called a server) sends an individual, digitally compressed bitstream to the STB of the consumer currently using a particular service provided from that server. As noted previously, each feed will be distinct because each consumer may be viewing a different portion or a different presentation of the program.

The bitstream representing the audio/video program sent to the basic interactive STB needs to be decompressed and fed to the TV set. All of the processing required to create the real-time interactive program is usually performed at the program server. This might include the composition of multiple audio and video elements to create the final presentation. This partitioning of functionality assures that the cost of the STB is minimized. Sharing of network-based functionality to reduce

CPE, and thereby the total system cost, is common in the network-based services industry, including telephony. However, this strategy of reducing CPE costs by sharing functionality in the network can restrict the types of interactive applications that can be effectively supported. Highly interactive applications, such as rapid reaction video games (called "twitch" games), require extremely low latency from the time a consumer presses a button until the effect is seen on the screen.

A consumer's remote control input is transmitted from the STB back to the server in the network. This input can change the flow of the program being executed by the server on behalf of that consumer. The audio and video bitstream produced by the server is compressed, often in real-time,* and modulated for transmission to the consumer's STB. The STB then simply demodulates the received signal to recover the transmitted bitstream and decompresses it to create the presentation on the TV set.

The principal latency in this architecture does not arise from the transmission time of either the control input from the STB to the server or from the transmission of the video from the server to the STB. The latency arises because compression schemes, like MPEG, require one or more frame delays in the encoding process plus additional frame delays in the decoder. In interactive graphics, 100 ms, or roughly 3 NTSC frame times, is typically regarded as the upper bound on latency for maintaining the consumer's feeling of instant control response. To reduce the decompression latency, the STB could support the local execution of highly interactive graphical applications. In this case, the STB has direct access to the locally synthesized graphics without the delay imposed by compression and decompression for transmission from the server. However, local execution of applications requires additional capabilities in the STB, with a corresponding increase in cost. At a minimum, it requires a processor with memory for program and data storage. Additional capabilities could include sound generators, video layering, special effects, control ports, and so forth. The fea-

---

*In many instances the server may store bitstreams compressed previously in non-real-time. In some other instances, an uncompressed bitstream may be sent to the STB.

tures that can be added to an intelligent high-end STB are essentially unlimited, and such a STB could resemble a high-end multimedia personal computer with a modem, but without the display (since a TV is used as the display).

In addition to the reduction in interactive latency, STBs with significant local processing power can assume more of a client–server relationship with the server as opposed to the source–sink model of the basic STB and server. The client–server relationship complicates the development of applications and introduces potential security and authentication problems. However, it can define open interfaces between the different software modules (in the server and the client) and therefore offers flexibility in developing applications by multiple vendors. As the cost of digital hardware continues to decline, the price difference between the basic and high-end STBs will shrink significantly. In the long run, all STBs will have substantial capability, and the functionality of the basic STB will be incorporated directly into television sets. With further increases in local processor power, it will be possible for most functions of a STB, including decompression, modems, graphics, and so forth to be implemented entirely in software.

In the telephone network, the network–CPE interface is well defined and open, enabling consumers to purchase or lease telephone sets with features to match their needs. On the other hand, most CATV system operators view the STB as part of a network, and lease STBs to their customers as part of their service package. STBs used in CATV systems differ in their frequency plans, scrambling and address control systems, signaling, and modem capabilities. This approach has limited the services available to the consumers to only those supplied by the CATV operator. It has led to unfortunate situations where some features of expensive TVs, such as picture-in-picture, cannot be used on a CATV system, and consumers cannot use a VCR to record one program while watching another. If an interface between a CATV network and the STB were well defined and open, consumer electronics firms could manufacture STBs with a variety of feature sets to meet different consumer needs, or build the basic

STB functionality directly into the TV set to lower the cost.

13.2   INTERACTIVE TELEVISION SERVERS

ITV servers are a collection of computing, storage, and communications equipment that implement interactive video services. A service may require more than one server to be implemented, or a server may implement more than one service. The overall architecture of a server is shown in Fig. 13.5. All subsystems of the server communicate with each other via a local high-bandwidth interconnect and a switch. This architecture can be used to scale the capabilities of the server incrementally, and it provides isolation between the various subsystems.

A storage subsystem stores the media (audio/video/data) samples. It may use different combinations of tapes, disks, and semiconductor storage technologies to meet the capacity and latency requirements of the various applications. As discussed later, these storage technologies are typically combined into a hierarchy to optimize performance for a given cost.

Application processors interact with the storage subsystem to deliver the output of various media to the subscribers' CPE or to compose media for eventual presentation. Media composition subsys-

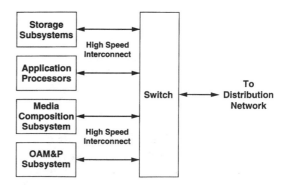

Fig. 13.5   Architecture of ITV server includes several subsystems optimized for specific functions and a high-speed network connecting them. OAM&P refers to operations, administrated maintenance, and provisioning.

tems use media samples from the storage subsystem, along with control commands from the application processors, to create a single composite compressed audio and video bitstream as required by each STB.

As mentioned previously, when driving basic low-cost STBs, servers emulate compressed digital program sources. Here the servers generate compressed audio and video streams in standard formats that require only decompression in the STB. Consumers' remote control button-press signals are carried back to the server for interpretation by the service application. With larger processing power, STBs will be able to combine and manipulate media samples. In such cases, the server need only send primitive media elements to the STB to be processed under the control of the locally executing services application.

A special class of servers called the gateway server (see Fig. 13.1) provides the navigation and other functions required to support the consumer's interface for the selection of services. These servers present "menus" of available services (e.g., movies-on-demand, home shopping, and broadcast television), accept the consumer's selection, and then hand off control to the server providing the selected service. The visual interface and navigation scheme of a gateway server are completely programmable. A high-end interactive media server can present a very rich full-motion audio/video interface, whereas a low-end system can provide simple text-based menus. Gateway servers allow customization by individual consumers of the on-screen appearance and operation of the system (see Fig. 1.5). Consumers might simply designate their favorite services to be shown first or they might engage "intelligent agents" to notify them of programs or events of potential interest. This could also provide different views of the system for each individual in a household. For example, when the TV is turned on with a child's remote control, the system could show only programs or services suitable for children.

Two key components of the server technology are (1) the logical organization of the multimedia samples in a file system or database and (2) techniques by which media components can be "continuously" (real-time) recorded or played back from the server. We review these two aspects below.

13.2.1 Multimedia Data Storage

Multimedia storage systems are characterized by many new capabilities as compared to standard ASCII storage systems (either files or databases); for example, massive storage volumes, large objects, rich object types, more flexible relationships among objects, temporal composition of media objects, synchronization of various media for effective presentation, etc. Among the choices for logical organization of the multimedia data are:

- Multimedia file systems
- Extended relational database management systems (RDBMS), which support multimedia as binary objects and in some cases support the concepts of inheritance and classes
- Object-oriented databases.

13.2.1.1 Multimedia file systems

File systems provide support for storing data in most computer systems. Interfaces between file systems and other parts of the software system of a computer are well known, understood, and continue to evolve. This is true of both single standalone computers as well as networked computers with a shared-network file system. In cases where additional functionality is desired (e.g., concurrent transactions, flexible queries, indexing, report generation) database are used.

There are many disadvantages to file systems when they are used as DBMSs:

- They do not provide physical data independence. Application developers must be intimately familiar with the physical layout of data.
- They do not provide backup or recovery mechanisms beyond those provided by the underlying operating system, which are often inadequate.
- They are not supported by a query language such as the Structured Query Language (SQL). Consequently, application programs must manipulate the file data using lower-level programming languages.
- They are not very portable, since file systems vary among different operating systems.

However, the additional functionality provided by a database system comes at the expense of substantial overhead. File systems are therefore used when the above functionality is not required or when the overhead of a database system cannot be tolerated. Typically, movies-on-demand systems, where records are almost always accessed in the same sequence, may be implemented via a file system. To satisfy the stringent requirements of continuous media, many modifications are necessary. Much work has been done in this area and a number of implementations of file systems exist that can support multimedia.

13.2.1.2 Relational Database Management Systems (RDBMS) for Multimedia

Relational databases are a collection of tables (called relations) containing data supplied by the user. Data are extracted by composing queries that "cut and paste" selected tables. Rules for cutting and pasting are based on relational algebra and are embodied as a query language (e.g., SQL). Standard database features include functionality such as concurrency control, recovery, and indexing. Relational databases have traditionally been designed to operate mostly with small fields (large multimedia objects will lose efficiency in a DBMS). Also the shared data is usually of a single type, namely ASCII characters.

The key limitations of relational databases are in two areas: the relational data model and the relational computational model. Relational databases have been extended by incorporating an additional data type, commonly known as a *binary large object* (BLOB), for binary free-form text images or other binary data types. Relational database tables include location information for the BLOBs, which may be stored outside the database on separate servers, thereby extending the database to access these BLOBs. Extended relational databases provide a gradual migration path to more object-oriented databases. Relational databases also have the strength of rigorous management for maintaining the integrity of the database, an important feature that has been lacking in early object-oriented databases. A number of database vendors, with a large invested base in relational database products, offer a variety of extensions to handle multimedia data, usually with a relational data model and pointers to continuous media data (i.e., audio and video).

13.2.1.3 Object-Oriented Database Management Systems (OODBMS) for Multimedia

Object-oriented database management systems incorporate a collection of user-defined object types (called *classes*). For each class, appropriate operations and data structures are defined either by the system or by the user. Most object databases are modeled around the class paradigm of the object-oriented languages (such as C++ or small talk). In the context of multimedia, OODBMS provide the following useful characteristics:

- The natural ability to express relationships between objects in different classes (e.g., relationships between component audio and video objects to create a "composite" multimedia presentation).
- Unlike in a relational database, there is no mismatch between the application's and the database's view of the world, since both have the same data model.

Different components (e.g., a sequence of frames representing a scene in a movie) of an MPEG-2 bitstream can be stored as different objects. This will allow the original MPEG-2 stream to be customized by selecting a subset of the components. An object-oriented database supports both *encapsulation* (the ability to deal with software entities as units) and *inheritance* (the ability to create new classes derived in part from existing object classes). These concepts constitute the fundamental tenets of the object-oriented paradigm.

The class definition in the object model has special applicability for multimedia data. Text, audio, still images, and video can each be of a different object class. Stated differently, the class definitions associate specialized processing to typed data, that is, the ways in which the various primitive media types are accessed and used. OODBMS also provide advantages in terms of the speed and a wider range of capabilities for developing and maintaining complex multimedia applications. OODBMS capabilities, such as extensibility, hierarchical structures, and message passing, are important for multimedia applications. *Message passing*, for example, allows objects to interact using each other's methods, which can be customized by the programmer.

Complex "composite" objects can be constructed that collect various "primitive" media objects such that the desired presentation and interaction are captured in the composite object. OODBMS are *extensible* (i.e., users can define new operations as needed) and allow incremental changes to the database applications. These changes would be more difficult to express in a relational language (e.g., SQL).

Summing up, the object-oriented paradigm offers several advantages for multimedia systems:

- Databases that understand the types of multimedia objects and will ensure that appropriate operations are applied to each class.
- Easy to manipulate component multimedia objects that allow the making of a composite multimedia presentation
- Facilities for the user to customize object classes and processing methods for each class
- Better sharing of component media objects and therefore the need to store only one copy for each media object.

The OODMS must be capable of indexing, grouping, and storing multimedia objects in distributed hierarchical storage systems, and accessing these objects on a keyed basis. A multimedia object (e.g., a video presentation) may be a component of a multiple hypermedia* document such as memos, presentations, video sales brochures, and so forth. The design of the object management system should be capable of indexing objects in such a manner that there is no need to store multiple copies.

13.2.2 Recording and Retrieval for Multimedia[5-9]

Digital audio/video is a sequence of quanta (audio samples or video frames). Retrieval and presentation of such quanta require continuity in time; that is, a multimedia database server must ensure that the recording and presentation follows the real-time data rate. During presentation, for example, the server must retrieve data from the disk at a rate that prevents the output device (such as a speaker or video display) from overflowing or running out of samples.

To reduce the initiation latency and buffering requirement, continuous presentation requires either (1) a sequence of periodic tasks with deadlines, or (2) retrieval of data from disks into buffers in rounds. In the first approach, tasks include retrieval of data from disks, and deadlines correspond to the latest time (e.g., MPEG Systems Time Stamps) when data must be retrieved from disks into buffers to guarantee continuous presentation (see Fig. 13.6). In the latter approach, in every round sufficient data for each media stream is retrieved to ensure continuous presentation. The server has to supply the buffers with enough data to ensure continuity of the playback processes. Efficient operation is then equivalent to preventing buffer overflow/underflow while minimizing buffer size and the initiation latency. Since disk data

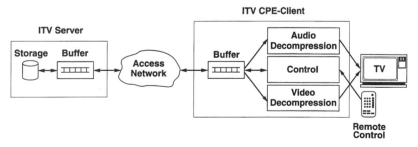

Fig. 13.6 Since the data flow is bursty owing to varying compression ratio and jitter in the storage system, buffers are required in the CPE and the server to ensure smooth data flow to the presentation hardware (i.e., TV).

---

*Hypermedia documents contain pointers to other documents or to other places in the current document.

transfer rates are significantly higher than a single stream's data rate, a small buffer should suffice for conventional file and operating systems support for several media streams.

The real challenge is in supporting a large number of streams simultaneously. ITV servers must process retrieval requests from several CPEs simultaneously. Even when multiple CPEs access the same file (such as a popular movie), different streams might access different parts of the file at the same time. Since disk I/O rates significantly exceed those of single streams, a larger number of streams can be supported by multiplexing a disk head among several streams. This requires careful disk scheduling so that each buffer is adequately served. In addition, new streams should not be admitted unless they can be properly served along with the existing ones.

13.2.2.1 Disk Scheduling

Database systems employ disk scheduling algorithms, such as "first come, first serve" and "shortest seek time first," to reduce seek time and rotational latency, to achieve high throughput, and to enforce fairness for each stream. However, continuous media constraints usually require a modification of traditional disk-scheduling algorithms so they can be used for multimedia servers.

One commonly used disk scheduling algorithm, called the *Scan-EDF* (Earliest Deadline First), services the requests with earliest deadlines first. However, when several requests have the same deadline, their respective blocks are accessed with a shortest-seek-time algorithm. Scan-EDF becomes more effective as the number of requests having the same deadline increase. This happens more frequently as the ITV server services a larger number of CPEs.

If data are retrieved from disks into buffers in rounds, a commonly used disk scheduling algorithm is *C-LOOK* (Circular LOOK), which steps the disk head from one end of the disk to the other, retrieving data as the head reaches one of the requested blocks. When the head reaches the last block that needs to be retrieved within a round, the disk head is moved back to the beginning of the disk without retrieving any data, after which the process repeats.

13.2.2.2 Buffer Management

Most ITV servers process multimedia stream requests from the CPE in rounds. During a round, the amount of data retrieved by the buffer could be made equal to the amount consumed by the CPE. This means that, on a round-by-round basis, data production never lags display, and there is never a net decrease in the delay of buffered data. Therefore, such algorithms are referred to as buffer-conserving. A non-buffer-conserving algorithm would allow retrieval to fall behind display in one round and compensate for it over the next several rounds. In this scenario, it becomes more difficult to guarantee that no starvation of the display will ever take place.

For continuous display, a sufficient number of blocks must be retrieved for each CPE-client during a round so that starvation can be prevented for the round's entire duration. To determine this number, the server must know the maximum duration of a round. As round length depends on the number of blocks retrieved for each stream, unnecessarily long reads must be avoided. To minimize the round length, the number of blocks retrieved for each CPE during each round should be proportional to the CPE's display rate.

13.2.2.3 Admission Control

To satisfy the CPE's real-time requirements, an ITV server must control admission of new CPEs so that existing CPEs can be serviced adequately. Strict enforcement of all real-time deadlines for each CPE may not be necessary since some applications can tolerate occasional missed deadlines. A few lost video frames or a glitch in the audio can occasionally be tolerated, especially if such tolerance reduces the overall cost of service. An ITV server might accommodate additional streams through an admission control algorithm that exploits the statistical variation in media access times, compression ratios, and other time-varying factors.

In admitting new CPEs, ITV servers may offer three categories of service:

- *Deterministic.* Real time requirements are guaranteed. The admission control algorithm must consider worst-case bounds in admitting new CPEs.

- *Statistical.* Deadlines are met only with a certain probability. For example, a CPE may subscribe to a service that guarantees meeting deadlines 95% of the time. To provide such guarantees, admission control algorithms must consider the system's statistical behavior while admitting new CPEs.
- *Background.* No guarantees are given for meeting deadlines. The ITV server schedules such accesses only when there is time left after servicing all guaranteed and statistical streams.

In servicing CPEs during a round, CPEs with deterministic service must be handled before any statistical ones, and all statistical-service CPEs must be serviced before any background CPEs. Missed deadlines must also be distributed fairly so that the same CPE is not dropped each time.

13.2.2.4 Physical Storage of Multimedia

A multimedia server must divide video and audio objects into data packets on a disk. Each data packet can occupy several physical disk blocks. Techniques for improving performance include optimal placement of data packets on the disk, using multiple disks, adding tertiary* storage to gain additional capacity, and building storage hierarchies.

Packets of a multimedia object can be stored contiguously or scattered across the storage device. Contiguous storage is simple to implement, but subject to editor fragmentation. It also requires copying overheads during insertions and deletions to maintain contiguity. Contiguous layouts are useful in mostly-read servers such as video-on-demand, but not for read-write servers. For multimedia, the choice between contiguous and scattered files relates primarily to intrafile seeks. Accessing a contiguous file requires only one seek to position the disk head at the start of the data, whereas with a scattered file, a seek may be required for each packet read. Intrafile seeks can be minimized in scattered layouts by choosing a sufficiently large packet size and reading one packet in each round. It improves disk throughput and reduces the overhead for maintaining indexes that a file system requires.

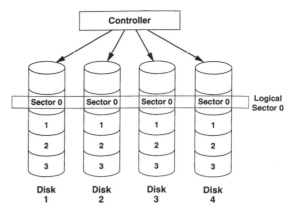

Fig. 13.7 Data is striped across several disks and is accessed in parallel.

To increase the number of possible concurrent access of a stored multimedia object, it may be scattered across multiple disks. This scattering can be achieved by using two techniques: data striping and data interleaving. RAID (redundant array of inexpensive disks) technology (see Fig. 13.7) uses an array of disks and stripes the data across each disk. Physical sector "n" of each disk in the array is accessed in parallel as one large logical sector "n." In this configuration, the disks in the set are spindle synchronized and operate in lock-step mode. Thus, the transfer rate is effectively increased by the number of disk drives employed. Striping does not improve the seek time and rotational latency, however.

In data interleaving, the packets of multimedia objects are interleaved across the disk array with successive file packets stored on different disks. A simple interleave pattern stores the packets cyclically across an array of N disks. In data interleaving, the disks in the array operate independently without any synchronization. Data retrieval can be performed in one of two ways: (1) One packet is retrieved from each disk in the array for each stream in every round or (2) data are extracted from one of the disks for a given CPE in each round. Data retrieval for the CPE cycles through all disks in N successive rounds. In each round, the retrieval load can be balanced by staggering the streams. Thus, all streams still have the same round

---

*The first two storage media are buffer RAM and main disk.

length, but each stream considers the round to begin at a different time.

A combination of striping and interleaving can be used to scatter a file in many disks from a server cluster. This technique is often used in constructing a scalable video that can serve a large number of CPEs accessing the same copy of the file.

The reliability of the server will become an important issue as the number of users who rely on ITV services increases. A major failure point in the server is the storage subsystem. With the increase in the number of disks, the reliability of the server decreases. Even if the mean time between failures (MTBF) of a single disk is between 20 and 100 years, as the number of disks increases, the MTBF of the disk subsystem decreases to weeks. This is because the MTBF of a set of disks is inversely proportional to the number of disks, assuming that the MTBF of each disk is independent and exponentially distributed. Further, owing to disk striping, the number of files affected per failure increases.

To provide continuous, reliable service to CPEs, the disk subsystem should employee redundancy mechanisms such as data replication and parity. However, these common techniques need to be extended for ITV servers so that in the presence of a failure all the admitted CPEs can be still serviced without interruption.

13.2.2.5 Updating ITV Servers

Copying of multimedia objects is expensive because their size is large. Copying may be reduced by considering object files to be immutable. Editing is then enabled by manipulating tables of pointers to the media component objects. Obviously, small insertions will not find pointers as valuable as large ones. Also, for deletion, if the multimedia object is an integral number of disk blocks, then the file map could simply be modified.

It is possible to perform insertion and deletions in a time that is proportional to the size of the data inserted/deleted rather than to the whole file. A common scheme is where packets must be filled to a certain minimum level to support continuous retrieval. The insertion/deletion will then consist of some number of full packets plus a remaining partially filled packet. All the packets are then inserted/deleted by modifying the file map, and then data are distributed among adjacent packets of the file to meet the required fill level.

13.2.2.6 VCR-like Functions

An ITV server must support VCR-functions such as pause/resume, fast-forward, and fast-rewind. The pause/resume function is a challenge for buffer management because it interferes with the sharing of streams among different viewers. The fast-forward and fast-rewind functions are implemented by playing back at the normal rate while skipping packets of data. This is easy to do by skipping from one I-frame to another; otherwise interframe dependencies complicate the situation.

The simplest method of fast-forward is to display only I-frames. However, this reduces flexibility. Another way is to categorize each frame as either relevant or irrelevant to fast-forward. During normal operation, both types of frames are retrieved, but during fast-forward, only the fast-forward frames are retrieved and transmitted to the CPE-buffer. It is assumed that the decoding of fast-forward blocks by CPE is straightforward. Scalable compression schemes are readily adapted to this sort of use, although the drawback here is that it poses additional overhead for splitting and recombining blocks.

13.3 USER INTERFACES[10-12]

User interface designs for ITV are more involved than for standard TV owing to the richness of the types of possible interactions with the consumer. As mentioned earlier, they need to be extremely easy to use, as well as intuitive. They will most likely vary among different applications. There will be a small number of ITV generic program types, such as movies-on-demand, shopping, and so forth, and certain "look-and-feel" guidelines will emerge for user interfaces and for ITV programming in general. Typically, multimedia interfaces are designed using a mixture of four kinds of modules:

1. Media editors
2. Multimedia object locators and browsers (navigation)
3. Authoring tools for creating program logic
4. Hypermedia object creation.

The last item is currently less relevant to ITV and is therefore not discussed here. However, it may become more widely available in the future. A number of media editors already exist to create "primitive" media objects such as text, still images, and video and audio clips. A media editor is then used for editing primitive multimedia objects to construct a composite object that is suitable as a multimedia presentation. The basic functions provided by the editor are delete, cut, copy, paste, move, and merge. The real challenge is to present these basic functions in the most natural manner that is intuitive to the user. For example, the move operation is usually represented by highlighting text and dragging and dropping it to the new location. However, the same metaphor does not work for all multimedia objects. The exact implementation of the metaphor and the efficient implementation for that metaphor will distinguish such editors from each other.

Navigation refers to the sequence in which the objects are browsed, searched, and used. Navigation can be either direct or predefined. If direct, then the user determines the next sequence of actions. If predefined, for example, searching for the next program start, the user needs to know what to expect with successive navigation actions. The navigation can also be in a *browse* mode, in which the user wants to get general information about a particular topic. Browsing is common in systems with graphical data. ITV systems normally provide direct navigation as well as browse options.

The browse mode is typically used to search for a specific attribute or a subobject. For example, in an application using dialog boxes, the user may have traversed several dialog boxes to reach a point where the user is ready to play a sound object. A browse mode may be provided to search the database for all sound objects appropriate for that dialog box. An important aspect of any multimedia system is to maintain a clear perspective of the objects being displayed and the relationship between these objects. The relationship can be associative (if they are a part of a set) or hierarchical. Navigation is undoubtedly the first generic application for ITV. Various flavors have been cre-

ated, for example, mapping of ITV programs to channel numbers, and interfaces that show "picture-in-picture." Navigation is rightly the first application focus, acknowledging the real challenge ahead in helping viewers find their way in the (hoped for) proliferation of programming in the interactive world.

The interactive behavior of ITV determines how the consumer interacts with the services available. Some of the design elements are:

- Data entry dialog boxes
- A source-specific sequence of operations depicted by highlighting or enabling specific menu items
- Context-sensitive operation of buttons
- Active icons that perform ad hoc tasks (e.g., intermediate save of work).

The look and feel of a service depends on a combination of the metaphor (e.g., VCR interface) being used to simulate real-life interfaces, the windows guidelines, the ease of use, and the aesthetic appeal. An elegant service has a consistent look and feel. The interface will be more friendly if, for example, the same buttons are found in the same location in each dialog box and the functions of an icon toolbar are consistent. User interfaces for multimedia applications require careful design to ensure that they are close adaptations of the metaphors to which consumers are accustomed in other domains. Two metaphors relevant to ITV are the Aural and VCR interfaces described below.

A common approach for speech recognition (also called the Aural metaphor) user interfaces has been to graft the speech recognition interface into existing graphical user interfaces. This is a mix of conceptually mismatched media which makes the interface cumbersome. A true aural user interface (AUI) would accept speech as direct input and provide a speech response to the user's actions. The real challenge in designing AUI systems is to creatively substitute voice and ear for the keyboard and display, and to be able to mix and match them. Aural cues can be used to represent icons, menus, and windows of a GUI.* Besides computer technology, AUI

---

*Graphical User Interface

designs involve human perception, cognitive science, and psychoacoustic theory. The human response to visual and aural inputs is markedly different. Rather than duplicate the GUI functions in an AUI, this requires new thinking to adapt the AUI to the human response for aural feedback. AUI systems need to learn so that they can perform most routine functions without constant user feedback, but still be able to inform the user of the functions performed on the user's behalf with very succinct aural messages. This is an area of active research.

A common user interface for services such as video capture, switching channels, and stored video playback is to emulate the camera, TV, and VCR on the screen. A general TV/VCR and camera metaphor interface will have all functions one would find in a typical video camera when it is in a video capture mode. It will have live buttons for selecting channels, increasing sound volume, and changing channels. The video is shown typically as a graduated bar, and the graduations are based on the full length of the video clip. The user can mark the start and end points to play the clip or to cut and paste the clip into a new video scene. The screen duplicates the cut and paste operation visually. Figure 1.5 shows an interface with the TV/VCR metaphor. A number of video editors emulate the functionality of typical film-editing equipment. The editing interface shows a visual representation of multiple film strips that can be viewed at user-selected frame rates. The user may view every nth frame for an entire scene or may wish to narrow down to every frame to pinpoint the exact point for splicing another video clip. The visual display software allows index markers on the screen for splicing another video clip down to the frame level. The screen display also shows the soundtracks and allows splicing soundtracks down to the frame level.

A good indexing system provides a very accurate counter for time as well as for individual frames; it can detect special events such as a change in scenes, start and end of newly spliced frames, start and end of a new soundtrack, etc. Also, the user can place permanent index markings to identify specific points of interest in the sequence. A number of indexing algorithms have been published in the literature.[13]

Indexing is useful only if the audio/video are stored. Since audio and video may be stored on separate servers, synchronization must be achieved before playback. Indexing may be needed to achieve the equivalent of a sound mixer and synthesizer found, for example, in digital pianos. Exact indexing to the frame level is desirable for video synchronization to work correctly. The indexing information must be stored on a permanent basis, for example, as SMPTE time codes or MPEG Systems time stamps. This information can be maintained in either the soundtrack or the video clip itself and must be extracted if needed for playback synchronization. Alternatively, the composite object created as a result of authoring can maintain the index information on behalf of all objects.

13.4   CONTENT CREATION FOR ITV[14-19]

A good authoring system must access predefined media and program elements, sequence these elements in time, specify their placement spatially, initiate actions in parallel, and indicate specific events (such as consumer inputs or external triggers) and the actions (some may be algorithmic functions) that may result from them. Some of these algorithmic functions may be performed by both the CPE and the server when the application is executed. Existing commercial tools for creating interactive multimedia applications run the gamut from low-level programming languages to packages that support the relatively casual creation of interactive presentations. Two very broad categories are (1) visual tools for use by nonprogrammers and (2) programming languages. The latter might include specialized "scripting" languages as well as more traditional general-purpose programming languages, perhaps extended by object-types or libraries that add ITV functionality. Usually the media elements used by these tools are created with a variety of other tools designed for particular media types, such as for still images, video and audio segments, and animation.

The design of an authoring system requires careful analysis of the following issues:

1. Hypermedia design details
2. User interfaces

3. Embedding/linking multimedia objects to the main presentation
4. Multimedia objects—storage and retrieval
5. Synchronization of components of a composite multimedia stream.

Hypermedia applications present a unique set of design issues not commonly encountered in other applications. A good user interface design is crucial to the success of a hypermedia application. The user interface presents a window to the user for controlling storage and retrieval, connecting objects to the presentation, and defining index marks for combining different multimedia streams. While the objects may be captured independently, they have to be played back together for authoring. The authoring system must allow coordination of many streams to produce a final presentation. A variety of authoring systems exist depending on their role and flexibility required. Some of these are described below.

13.4.1 Dedicated Authoring Systems

Dedicated authoring systems are designed for single users and single streams. The authoring is typically performed using desktop computer systems on multimedia objects captured by a video camera or on objects stored in a multimedia library. Dedicated multimedia authoring systems are not usually used for sophisticated applications. Therefore, they need to be engineered to provide user interfaces that are extremely intuitive and follow real-world metaphors (e.g., the VCR metaphor). Also flexibility and functionality has to be carefully limited to prevent overly complex authoring.

13.4.2 Timeline-Based Authoring

In a time-line-based authoring system, objects to be merged are placed along a timeline. The author specifies objects and their position in the timeline. For presentation in a proper time sequence each object is played at the prespecified point in the time scale.

In early timeline systems, once the multimedia object was captured in a timeline, it was fixed in location and could not be manipulated easily. Editing a particular component caused all objects in the timeline to be reassigned because the positions of objects were not fixed in time, only in sequence. Copying portions of the timeline became difficult as well. Newer systems define timing relations directly between objects. This makes inserting and deleting objects much easier because the start and end of each object is more clearly defined.

13.4.3 Structured Multimedia Authoring

Structured multimedia authoring is based on structured object-level construction of presentations. A presentation may be composed of a number of video clips with associated sound tracks, separate music tracks, and other separate soundtracks. Explicit representation of the structure allows modular authoring of the individual component objects. The timing constraints are also derived from the constituent data items. A good structured system also allows the user to define an object hierarchy and to specify the relative location of each object within that hierarchy. Some objects may have to undergo temporal adjustment, which allows them to be expanded or shrunk to better fit the available time slot. The navigation design of the authoring system should allow the author to view the overall structure, while at the same time being able to examine specific object segments more closely. The benefit of an object hierarchy is that removing the main object also removes all subobjects that are meaningful only as overlays on the main object. The design of the views must show all relevant information for a specific view. For example, views may be needed to show the object hierarchy plus individual component members of that hierarchy, and to depict the timing relation between the members of the hierarchy.

13.4.4 Programmable Authoring Systems

Early structured systems did not allow authors to express automatic functions for handling certain routine tasks in the authoring process. Programmable authoring systems provide such additional capability in the following areas:

- Functions based on audio and image processing and analysis
- Embedded program interpreters that use audio and image-processing functions.

Thus the main improvement in building user programmability into the authoring tool is not only to perform the analysis, but also to manipulate the stream based on the analysis results. This programmability allows the performing of a variety of tasks through the program interpreter rather than manually. A good example of how this is used is the case of locating video "silences" which typically occur before that start of a new segment (cut). The program can be used to help the user to get to the next segment by clicking a button rather than playing the video. Advances in authoring systems will continue as users acquire a better understanding of stored audio and video and the functionalities needed by users.

The development and deployment of interactive television on a large scale presents numerous technical challenges in a variety of areas. As progress continues to be made in all these areas, trial deployments of the interactive services are necessary to ascertain consumer interest in this new form of communication. We are certainly rather naive today about the future of these services. However, it seems certain that the variety of trials now being conducted will find a subset of this technology that is popular with consumers and a solid business proposition for service providers. This will provide the definition of services and the economic drivers necessary for these services to prosper.

REFERENCES

1. D. E. BLAHUT, T. E. NICHOLS, W. M. SCHELL, G. A. STORY, and E. S. SZURKOWSKI, "Interactive Television," *Proc. of IEEE*, pp. 1071–1086 (July 1995).

2. L. CRUTCHER and J. GRINHAM, "The Networked Jukebox," *IEEE Trans. Circuits Systems*, pp. 105–120 (April 1994).

3. G. H. ARLEN, *Networked Multimedia: A Review of Interactive Services and the Outlook for Cable-Delivered Services*, Arlen Communications (April 1993).

4. E. G. BARTLETT, *Cable Television Technology and Operations*, McGraw-Hill, New York (1990).

5. D. ANDERSON, Y. OSAWA, and R. GOVINDAM, "A File System for Continuous Media," *ACM Trans. Computer Systems* 10(4):311–337 (November 1992).

6. A. L. NARASIMHA REDDY and J. C. WYLLIE, "I/O Issues in a Multimedia System," *Computer*, 27(3): 69–74 (March 1994).

7. P. LOUGHER and D. SHEPHERD, "The Design of a Storage Server for Continuous Media," *Computer J.* 36(1):32–42 (February 1993).

8. P. VENKAT RANGAN and H. M. VIN, "Efficient Storage Techniques for Digital Continuous Multimedia," *IEEE Trans. Knowledge Data Eng.* 5(4):564–573 (August 1993).

9. B. OZDEN, R. ROSTOGI, and A. SILBERSCHATZ, "A Framework for the Storage and Retrieval of Continuous Media Data," *Proc. IEEE Conf. on Multimedia Computing and Systems* (May 95).

10. J. KUEGEL, C. KESLEIN, and J. RUTLEDGE, "Toolkits for Multimedia Interface Design," *Exhibition 92*, pp. 275–285.

11. R. HAYAKAWA, H. SAKAGAMI, and J. BEKIMOTO, "Audio and Video Extensions to Graphical User Interface Toolkits," *3rd Int'l Workshop on Operating System Support for Audio/Video*, No. 92.

12. M. BLATHER, R. DENEBURG (Eds.), *Multimedia Interface Design*, Addison-Wesley, Reading, MA (1991).

13. A. HAMPAPUR, R. JAIN, and T. E. WEYWOUTH, "Indexing in Video Databases," *SPIE*, 2420: 293–305 (1995).

14. J. CONKLIN, "Hypertext: An Introduction and Survey," *IEEE Computer* (September 1987).

15. L. HARDMAN, G. VAN ROSSUM, and O. BULTERMANS, "Structured Multimedia Authoring," *ACM Multimedia*, pp. 283–289 (August 1993).

16. MACRO MEDIA Direction Overview Manual, Macromedia Inc. (1991).

17. Script X Language Guide, Kaleida (1993).

18. MACRO MEDIA Authorwave Working Model, Macromedia, Inc. (1993).

19. APPLE COMPUTER, *Inside MacIntosh: Quick Time*, Addison-Wesley, Reading, MA (1993).

14

HIGH-DEFINITION TELEVISION (HDTV)

Analog television was invented and standardized over fifty years ago, mainly for over-the-air broadcast of entertainment, news, and sports. In North America, the basic design was formalized in 1942, revised in 1946, and shortly after, limited service was offered. Color was added in 1954 with a backward compatible format. European broadcasting is equally old; monochrome television dates back to the 1950s and color was introduced in Germany in 1965 and in France in 1966. Since then, only backward compatible changes have been introduced, such as multichannel sound, closed-captioning, and ghost cancellation. The standard has thus evolved[+] in a backward compatible fashion, thereby protecting substantial investment made by the consumers, equipment providers, broadcasters, cable/ satellite service providers, and program producers.

Television is now finding applications in telecommunications (e.g., video conferencing), computing (e.g., desktop video), medicine (e.g., telemedicine), and education (e.g., distance learning). To enable many of these applications, television is being distributed by a variety of means not imagined at the time of standardization, such as cable (fiber or coaxial), video tapes or discs (analog or digital), and satellite (e.g., direct broadcast). It has also been recognized that the television signal packaged for terrestrial broadcasting, while cleverly done in its time, is wasteful of the broadcast spectrum and can benefit from modern signal pro-

cessing. In addition, the average size of television sets has grown steadily, and the average viewer is getting closer to the display.

These newer applications, delivery media, and viewing habits are exposing various limitations of the current television system. At the same time a new capability is being enabled owing to the technology of compression, inexpensive and powerful integrated circuits, and large displays. Thus, we have an unstoppable combination of technological push with commercial pull.

High-Definition Television (HDTV) systems now being designed are more than adequate to fill these needs in a cost-effective manner. HDTV will be an entirely new high-resolution and wide-screen television system, producing much better quality pictures and sound. It will be one of the first systems that will not be backward compatible.

Worldwide efforts to achieve practical HDTV systems have been in progress for over 20 years. While we do not provide a detailed historical perspective on these efforts, a summary of the progress made towards practical HDTV is as follows:

- HDTV in Japan
  - —1970's NHK started systems work; infrastructure by other Japanese companies
  - —1980's led to MUSE, 1125/60 2:1 interlaced, fits 27 MHz satellite transponder
  - —Satellite HDTV service operating; NHK's format international exchange standard

—Narrow-MUSE proposed to FCC for USA

—Future of MUSE not promising, switch to all digital system expected

- HDTV in Europe
  —In 1986, broadcasters and manufacturers started HD-MAC effort
  —HD-MAC intended backward compatible with PAL and D-MAC; satellite (DBS)
  —HD-MAC obsolete; new system with digital compression/modulation sought
- HDTV in North America
  —In 1987, broadcasters requested FCC which resulted in formation of Advisory Committee (ACATS)
  —ACATS proposed hardware testing of 23 systems, 5 survived around test time
  —All 4 digital systems in running after tests, formed Grand Alliance in 1993

This chapter summarizes the main characteristics of HDTV and its expected evolution.

14.1   CHARACTERISTICS OF HDTV

HDTV systems are currently evolving from a number of proposals made by different organizations in different countries. While there is no complete agreement, a number of aspects are common to most of the proposals. HDTV is characterized by increased spatial and temporal resolution; larger image aspect ratio (i.e., wider image); multichannel CD-quality surround sound; reduced artifacts compared to the current composite analog television; bandwidth compression and channel encoding to make better utilization of the terrestrial spectrum; and finally, digital to provide better interoperability with the evolving telecommunications and computing infrastructure.

The primary driving force behind HDTV is a much better quality picture and sound to be received in the consumer home. This is done by increasing the spatial resolution by more than a factor of 2 in both the horizontal and vertical dimensions. Thus, we get a picture that has about 1000 scan lines and more than 1000 pels per scan line.

In addition, it is widely accepted (but not fully agreed to) that HDTV should eventually use only progressive (noninterlace) scanning at about 60 frames/s to allow better fast-action sports and far better interoperability with computers, while also eliminating the artifacts associated with interlace. The result of this is that the number of active pels in an HDTV signal increases by about a factor of 5, with a corresponding increase in the analog bandwidth.

From consumers' experience in cinema, a wider image aspect ratio (IAR) is also preferred. Most HDTV systems specify the image aspect ratio to be 16:9. The quality of audio is also improved by means of multichannel, CD-quality, surround sound in which each channel is independently transmitted.

Since HDTV will be digital, different components of the information can be simply multiplexed in time instead of frequency multiplexed on different carriers as in the case of analog TV. For example, each audio channel is independently compressed, and these compressed bits are multiplexed with compressed bits from video, as well as bits for closed-captioning, teletext, encryption, addressing, program identification, and other data services in a layer fashion.

Unfortunately, increasing the spatial and temporal resolution of the HDTV signal and adding multichannel sound also increases its analog bandwidth. Such an analog signal cannot be accommodated in a single channel of the currently allocated broadcast spectrum.

Moreover, even if bandwidth were available, such an analog signal would suffer interference both to and from the existing TV transmissions. In fact, much of the available broadcast spectrum can be characterized as a fragile transmission channel. Many of the 6-MHz TV channels are kept unused because of interference considerations, and are designated as *taboo* channels.

Therefore, all the current HDTV proposals employ digital compression, which reduces the bitrate from approximately 1 Gbit/s to about 20 Mbits/s. The 20 Mbits/s can be accommodated in a 6-MHz channel either in a terrestrial broadcast spectrum or a cable television channel. This digital signal is incompatible with the current television system and therefore can be decoded only by a special decoder.

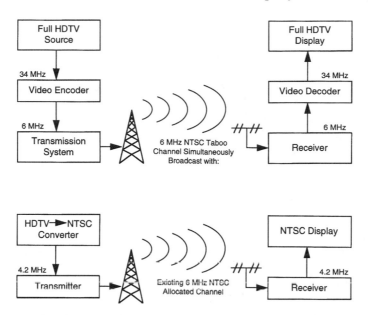

Fig. 14.1   Simulcasting of digital HDTV with NTSC analog TV. Under current channel assignments, many channels remain vacant (i.e., taboo) to avoid co-channel and adjacent channel interference. Since digital signals have lower transmission power, they can use the taboo channels in the new channel assignment.

In the case of terrestrial broadcast it is also necessary to reduce interference to and from other broadcast signals, including the analog television channels. Therefore, digital HDTV modulates the compressed audio and video bits onto a signal that has much lower power than the analog TV signal and has no high-power carriers, while requiring the same analog bandwidth as the current analog TV. In addition, most HDTV proposals simulcast* the HDTV signal along with the current analog TV having the same content (see Fig. 14.1).

A low-power, modulated digital HDTV signal is transmitted in a taboo channel that is currently unused because of co-channel and adjacent channel interference considerations. Simultaneously, in an existing analog channel, the same content (down-converted to the current television standard) will be transmitted and received by today's analog television sets. By proper shaping of the modulated HDTV signal spectrum, it is possible to avoid its interference into distant co-channel as well as near-by adjacent channel analog television signals. At the same time, distant co-channel and nearby adjacent channel analog television signals must not

degrade the HDTV pictures significantly because of interference.

This design allows previously unused terrestrial spectrum to be used for HDTV. Of course, it could also be used for other services such as digital Standard Definition Television (SDTV), computer communication, and so forth. Consumers with current television sets will receive television as they do now. However, consumers willing to spend an additional amount will be able to receive HDTV in currently unused channels. As the HDTV service takes hold in the marketplace, spectrally inefficient analog television can be retired to release additional spectrum for additional HDTV, SDTV, or other services. In the United States, the FCC plans to retire NTSC after 15 years from the start of the HDTV broadcasts.

Another desirable feature of HDTV systems is easy interoperability with the evolving telecommunications and computing infrastructure. Such interoperability will allow easy introduction of different services and applications, and at the same time reduce costs because of commonality of components and platforms across different applica-

---

*The same program is sent on two channels.

tions. There are several characteristics that improve interoperability. Progressive scanning, square pels, and digital representation all improve interoperability with computers. In addition, they facilitate signal processing required to convert HDTV signals to other formats either for editing, storage, special effects, or down-conversions to current television. The use of a standard compression algorithm (e.g., MPEG[2,3]) transforms raw, high-definition audio and video into a coded bitstream, which is nothing more than a set of computer instructions and data that can be executed by the receiver to create sound and pictures.

Further improvement in interoperability at the transport layer is achieved by encapsulating audio and video bitstreams into fixed-size packets. This facilitates the use of forward error correction (necessary to overcome transmission errors) and provides synchronization and extensibility by the use of packet identifiers, headers, and descriptors. In addition, it allows easy conversion of HDTV packets into other constant size packets adopted by the telecommunications industry, for example, the Asynchronous Transfer Mode (ATM) standard for data transport.

Based on our discussion thus far the requirements for HDTV can be summarized as follows:

- Higher spatial resolution
  —Increase spatial resolution by at least a factor of 2 in horizontal and vertical direction
- Higher temporal resolution
  —Increase temporal resolution by use of progressive 60 Hz temporal rate
- Higher Aspect Ratio
  —Increase aspect ratio to 16:9 from 4:3 for a wider image
- Multichannel CD quality surround sound
  —At least 4 to 6 channel surround sound system
- Reduced Artifacts as compared to Analog TV
  —Remove composite format artifacts, interlace artifacts etc
- Bandwidth compression and channel coding to make better use of Terrestrial spectrum
  —Digital processing for efficient spectrum usage
- Interoperability with evolving telecommunication and computing infrastructure
  —Digital compression and processing for ease of interworking

14.2 PROBLEMS OF CURRENT ANALOG TV

Current analog TV was designed during the days when both the compression, modulation, and signal processing algorithms and the circuitry to process them were not adequately developed. Therefore, as the current TV began to get deployed for different applications, its artifacts as well as inefficiency of spectral utilization became bigger problems. Digital HDTV attempts to alleviate many of these problems. *Interline flicker, line crawl,* and *vertical aliasing* are largely a result of interlaced scanning in current television. These effects are more pronounced in pictures containing slowly moving high-resolution graphics and text with sharp edges. Most pictures used for entertainment, news, and sports did not historically contain such sharp detail, and these effects were tolerably small. The computer industry has largely abandoned interlace scanning to avoid these problems. Progressive scanning and transmission of HDTV can eliminate these effects almost entirely.

Large area flicker as seen from a close distance on large screens is the result of our ability to detect temporal brightness variations, even at frequencies beyond 60 Hz, in the peripheral areas of our vision. The cost of overcoming this problem in terms of bandwidth (i.e., by increasing the frame rate) is too large and is therefore not addressed in any current HDTV proposal. Computer displays employ much higher frame rates (up to 72 Hz) to overcome large area flicker. However, the brightness of a computer display is usually much higher than that of home TV, and this accentuates the effect, thereby requiring the larger frame rate.

Static raster visibility is more apparent on larger displays since viewers can visually resolve individual scan lines. The increased spatial resolution in HDTV will reduce this effect substantially.

Cross-color and cross-luminance are a result of the color modulation used to create the analog composite signal. HDTV will use component signals. Moreover, in digital HDTV, each color component will be compressed separately, with the compressed bits multiplexed in time. Thus, these two effects will be eliminated entirely.

With the present analog television, a number of

other artifacts also arise, owing to transmission. For example, *ghosts*, which are due to multiple receptions of the same signal delayed with respect to each other, cannot be entirely removed from the analog signal. In digital HDTV, ghost cancellation can be done more successfully using adaptive equalizers, generally leaving no impact on the transmitted bitstream.

Another transmission problem is the interference from the other TV signals. As an example, the NTSC signal, containing high-energy sync pulses with sharp transitions, has poor spectrum utilization because of a high degree of *co-channel* and *adjacent channel* interference. Co-channel interference constrains the minimum geographic distance between transmitters on the same frequency band. Adjacent channel interference precludes using spectrally adjacent channels to serve the same area unless the transmitters are located together, thus ensuring near equivalent, noninterfering signal strengths at all receiving locations. However, this is usually not possible. Both of these interferences are reduced by keeping adjacent channels in the spectrum vacant (i.e., taboo channels) at any one place, while using these vacant channels at other surrounding places. As a result, in the United States, only about 20 channels are usable in each location out of a total of 68. In the United Kingdom, only about 4 out of 44 channels are usable.

The problems of current analog TV can be summarized as follows:

- Interline Flicker, Line Crawl and Vertical Aliasing
  - More predominant with high resolution moving graphics and text
  - Can be eliminated by 60 Hz progressive format
- Large Area Flicker
  - Seen on closeup of screen; due to brightness variations at freq < 60 Hz
  - Can be resolved by going to 72 Hz, but bandwidth expensive
- Static Raster Visibility
  - Visible on large displays, higher spatial resolution can reduce this
- Cross-color and Cross-luminance
  - Caused by color modulation in composite; component signals can eliminate it

- Ghosts
  - Multiple receptions of same signal, ghost cancellation easier for digital system
- Co-channel and Adjacent Channel Interference
  - Co-channel limits distance between transmitters on same frequency band
  - Adjacent channel precludes spectrally adjacent channels to serve same area
  - Currently reduced by keeping spectrally adjacent channels vacant or taboo

14.3 GRAND ALLIANCE HDTV SYSTEM

The FCC considered various options for introducing HDTV service in the United States and finalized the following requirements:

- HDTV service should not interfere with current broadcast TV
  - Digital HDTV and analog TV will be simulcast so as not to make current TV receivers immediately obsolete
  - HDTV and TV services are expected to coexist for many years so HDTV should not cause interference to analog TV reception
- HDTV service must fit in a 6 MHz channel
  - Only one 6 MHz channel will be available per broadcaster, so HDTV while providing fairly good picture quality must fit in bandwidth of analog channel
  - More bandwidth may be available later if analog TV is retired
- Terrestrial solution for HDTV to be usable on other Media
  - Interoperability between Terrestrial HDTV and HDTV on Cable
- Other aspects of HDTV to be considered
  - Channel switching in reasonable amount of time
  - Interworking of contribution and distribution quality HDTV
  - Recording of HDTV signal for professional and consumer use

Without going into a detailed historical perspective on U.S. HDTV efforts we present a snapshot of events leading to formation of a joint effort on part of various proponents for U.S. HDTV.

- FCC: Advisory Committee on Advanced TV (ACATS) in 1987
- To recommend toFCC a standard for terrestrial broadcast HDTV
- From 1987 to 1991 many proposals to ACATS, 5 survived
- Tests Sept 1991–Oct 1992; 4 digital, 1 analog proposal
- Results Feb 1993, major advantages of digital, drop analog
- May 1993 all 4 digital proponents joined in Grand Alliance (GA)
- GA participants: AT&T,[8] Sarnoff,[9] GI,[6,7] MIT,[7] Philips,[9] Thomson,[9] Zenith[8]

The Grand Alliance HDTV (GA-HDTV) system is a proposed[10–11,13–16] *transmission* standard considered by the FCC for North America. It is expected that a *production* standard that is optimized for operation within a studio, but at the same time interoperable with the transmission standard, will emerge once the transmission standard is approved.* Efforts to this end are underway within the FCC Advanced Television Systems Committee (ATSC). Similarly using the transmitted signal, it is expected that different manufacturers of consumer television receivers will create optimum displays consistent with picture quality obtainable for a given cost.

14.3.1 Overall GA-HDTV System

The GA-HDTV supports signals in multiple formats as follows:

1280 pels × 720 lines at frame rates of 24, 30, 60 Hz
progressive

and

1920 pels × 1080 lines at frame rates of 24 Hz, 30 Hz
progressive, and 30 Hz interlaced.

Thus, HDTV may start with available, inexpensive interlaced equipment, but migrate to progressive scanning at some later date. In addition, higher frame rate film sources are supported directly.

Frame rates may also be divided by 1.001 in systems that require compatibility with NTSC. With a 16:9 image aspect ratio these formats create square pels for both 720 and 1080 line rasters. Such flexibility in scanning allows different industries, program producers, application developers, and users to optimize among resolution, frame rate, compression, and interlace artifacts.

The GA-HDTV system incorporates a subset of the MPEG-2 video compression standard[3] described in Chapter 8. In addition, it includes several enhancements in the encoder built for initial FCC testing. For example, the GA-HDTV system can process film originated material at 24 or 30 frames/s by adaptively adjusting the compression algorithm at the encoder (see Fig. 14.2). Also, compression efficiency is improved by appropriate prefiltering/subsampling to handle signals of different formats and of varying quality (e.g., noisy signals). Among the compression techniques used are: DCT of motion-compensated signals with field/frame prediction, flexible refreshing (both I-frame as well as pseudo-random Intra-blocks), bidirectional prediction (B-frames), and forward analysis/perceptual selection for rate control and best picture quality.

The GA-HDTV system uses five-channel audio with surround sound, which compresses into a 384 kbits/s bit stream. Audio, video, and auxiliary data are packaged into fixed-length packets in conformity with the MPEG-2 Systems transport layer. MPEG-2 transport packets are 188 bytes long, consisting of up to 184 bytes of payload and 4 bytes of header. These packets easily interoperate with ATM since they have approximately a 4:1 ratio with ATM cells, which are 48 bytes long plus 5 bytes of header.

The modulation scheme chosen for over-the-air is 8-VSB (Vestigial Side Band), which gives 32.28 Mbits/s of raw data in a 6-MHz bandwidth. After error correction, 19.3 Mbits/s are available. Coaxial cable can carry more than this in a 6-MHz channel, since there is no co-channel and very little adjacent channel interference. Cable standards are under discussion. Thus, the GA-HDTV system attempts to

---

* The production standard needs to be interoperable with, but is not constrained by the transmission requirements. Production requirements are derived from storage, computer processing, multiple edits, very high quality and special effects.

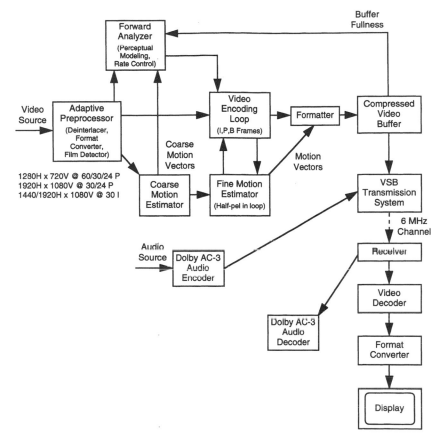

Fig. 14.2 Grand Alliance system overview. The block diagram shows major function blocks for the compression encoder–decoder for audio and video as well as the transmission system.

make the best compromise to suit different applications and provide for maximum interoperability.

We now summarize the aforementioned high level characteristics of the GA-HDTV system

- Video Compression
    —SMPTE video format standards, 16:9 image aspect ratio, square pels
    —MPEG-2 Video standard at Main Profile at High Level, bitstream packetized
- Audio Compression
    —Audio sampling at 48 kHz
    —AC-3 multichannel, bitstream packetized
- Transport
    —Audio, Video PES packets along with ancillary data presented to mux
    —Output of mux is fixed length 188 byte MPEG-2 Transport stream packets
- Modulation

—MPEG-2 Transport stream packets go to modulator for channel coding
—VSB transmission system, two modes 8-VSB and 16-VSB

The Grand Alliance would have wished to include a 1920 × 1080 progressive scan format at 60 Hz. However, the 6-MHz terrestrial broadcast channel does not allow for the required bitrate. As the NTSC audience declines and the interference into NTSC becomes less of a concern, there will be an opportunity to increase the data rate to support, in a compatible manner, 1920 × 1080 progressive scan at a 60-Hz frame rate.

The GA-HDTV system may allow for temporal scalability as well, so that an enhanced bitstream to handle 1920 × 1080 progressive scan at a 60-Hz frame rate can be created in the future. Figure 14.3 shows one possible scenario. A 30 frames/s subset

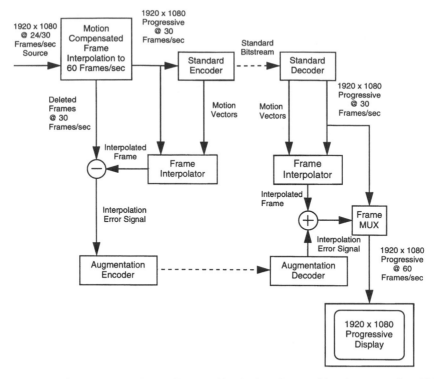

Fig. 14.3   An arrangement for using temporal scalability to enable a backward compatible enhancement from 30 Hz to 60 Hz progressive HDTV. Standard GA-HDTV cannot currently accommodate 60-Hz, 1080-line progressive HDTV. If transmission capacity is available in the future, this diagram shows a possible system for enhancing 24- or 30-Hz film for display at 60 Hz.

would be coded in the standard way while the remaining frames would be predicted by an interpolator. The difference between the predicted interpolation and the actual frame would be coded using the MPEG-2 algorithm[2] and sent to the decoder. Appropriately computed motion vectors could be used for both motion compensation and interpolation. This scheme would have the advantage of providing improved temporal performance for existing film material when properly applied.

14.3.2   GA-HDTV Video Issues and Parameters

We now list the main issues in GA-HDTV video and the choice of parameters in MPEG-2 video coding that can address these issues. The list of issues is summarized as follows:

- Raw rate of 1 Gbit/s or more to be compressed to about 20 Mbit/s, ie, compression by a factor of 50 or more

- Picture formats with 16:9 image aspect ratio and square pels
  —1280×720 at 60, 30 and 24 Hz progressive
  —1920×1080 at frame rates of 30 Hz interlaced and 30 and 24 Hz progressive
- Compression scheme with following properties for High Efficiency:
  —Source Adaptive Preprocessing
  —Temporal Redundancy Reduction
  —Spatial Redundancy Reduction
  —Perceptual Irrelevancy Exploitation
  —Efficient Entropy Coding
  —Frequent Refresh

Next in Table 14.1 we summarize[11,13,16] specific MPEG-2 Video coding parameters suitable for GA-HDTV.

14.3.3   GA-HDTV Audio Issues and Parameters

We now list the main issues in GA-HDTV audio, and the choice of parameters in Dolby's

Table 14.1 GA-HDTV Video Parameters Summary

Video Parameter	Value for Format 1	Value for Format 2
Active pels	1280 (hor) × 720 (vert)	1920 (hor) × 1080 (vert)
Total Samples	1600 (hor) × 787.5 (vert)	2200 (hor) × 1125 (vert)
Frame Rate	60 Hz progressive/ 30 Hz progressive/ 24 Hz progressive	30 Hz interlaced/ 30 Hz progressive/ 24 Hz progressive
Chrominance Sampling	4:2:0	
Image Aspect Ratio	16:9	
Data Rate	Selected Fixed Rate (10 − 45 Mbit/s)/Variable	
Colorimetry	SMPTE 240M	
Picture Coding Types	Intra (I-), Predictive (P-), Bidirectionally Predictive (B-) pictures	
Video Refresh	I-picture/progressive refresh	
Picture Structure	Frame	Frame/Field (interlace only)
Coefficient Scan	ZigZag	Zigzag/Alternate
DCT coding Modes	Frame	Frame/Field (interlace only)
Motion Comp Modes	Frame	Field/Dualprime (interlace only)
Motion Vector Range	Horizontal: Unlimited by Syntax Vertical: −128, + 127.5	
Motion Vector Precision	1/2 pixel precision	
DC Coefficient Precision	8 bits/9 bits/10 bits	
Rate Control	Modified TM5 with Forward Analyzer	
Film Mode Processing	Automated 3:2 pull down detection and coding	
Max VBV Buffer size	8 Mbits	
Inter/Intra Quantization	Download scene dependent matrices	
VLC Coding	Coefficient VLC Intra and Inter 2D VLC tables	
Error Concealment	Motion compensated frame holding (slice level)	

AC-3 audio coding address these issues. The list of issues is summarized as follows.

- Main audio service can range from monophonic, through stereo upto 6 channel surround service (left, center, right, left surround, right surround, and subwoofer). Sixth channel conveys only low frequency (subwoofer) info and is called 0.1 channel; total of 5.1

- Additional service can be provided, for example:
  —Service for hearing or visually impaired
  —Multiple simultaneous languages

- Audio sampling rate is 48 kHz, with 6 channels and 18 bits per sample. Raw data rate is 5 Mbit/s and needs to be compressed to 384 kbit/s, a compression factor of 13

- From a multichannel service, if mono or stereo outputs are needed, downmix is done at the decoder
  —Downmix may be done in frequency domain to reduce decoder complexity
  —Program originator indicates in bitstream which downmix coefficients for program

Next, in Table 14.2 we summarize specific Dolby AC-3 audio coding parameters suitable for GA-HDTV

14.3.4 GA-HDTV Transport Issues and Parameters

We now list the main issues in GA-HDTV Transport and the choice of parameters in MPEG-2 Transport system. The list of issues is summarized as follows.

- A Transport packet is composed of 4 bytes of header followed by 184 bytes of payload
  —Header identifies content of packet and the nature of the data. It also provides info about packet synchronization, error handling and conditional access
  —For conditional access, header indicates if the payload data is scrambled
  —Sometimes additional header information in adaptation header, a variable length field in

Table 14.2 GA-HDTV Audio Parameters Summary

Audio Parameter	Value
Number of Channels	5.1
Audio Bandwidth	10–20 kHz
Sampling Frequency	48 kHz
Dynamic Range	100 dB
Compressed Data Rate	384 kbit/s

Table 14.3 GA-HDTV Transport Parameters Summary

Transport Parameter	Value
Multiplex Technique	MPEG-2 Systems Layer
Packet Size	188 bytes
Packet Header	4 bytes including sync
Services	
Conditional Access	Payload scrambled on service basis
Error Handling	4-bit continuity counter
Prioritization	1 bit/packet
System Multiplex	Multiple program capability described in PSI Stream

payload of Transport Packet. This is used for audio-video synchronization, random access into compressed bitstream, and local program insertion

- Transport stream provides interoperability with ATM
  —ATM cell contains 5 byte header and 48 byte payload; a transport packet (188 bytes) can fit into four ATM cells ($4 \times 48 = 192$)

Next, in Table 14.3 we summarize[11,13] specific MPEG-2 Transport parameters suitable for HDTV.

14.3.5 GA-HDTV Transmission Issues and Parameters

We now list the main issues[11,13] in GA-HDTV Transmission and the choice of parameters in the modulation scheme chosen.

- Vestigial Sideband (VSB) transmission system with 2 modes
  —8-VSB for terrestrial broadcasting
  —16-VSB for high data rate cable transmission
- Both Modes use Reed-Solomon coding, segment sync, a pilot, and a training signal. Terrestrial mode adds trellis coding

- Symbol rate and Payload Data Rate
  —Terrestrial mode uses 10.76 Msymbols/s at 3 bits/symbol; payload 19.3 Mbit/s
  —Cable mode uses 10.76 Msymbols/s but 4 bits/symbol; payload 38.6 Mbit/s
- Data transmitted according to a data frame
  —It contains, a first data field sync segment, 312 data segments, a second data field sync segment, 312 data segments. Each segment is 4 symbols for sync followed by 832 symbols of data
- Terrestrial Mode Segments
  —One segment corresponds to 1 MPEG-2 Transport packet. Number of bits plus FEC per symbol is 832 symbols $\times$ bits/symbol = 2496. MPEG-2 transport packet is 188 bytes, RS encoding adds 20 bytes, further trellis coding adds 1 bit for every 2 input bits increasing ratio by 3/2, for total of 312 bytes
- Symbols modulate a carrier using suppressed-carrier modulation
  —Before transmission most of lower sideband is removed, resulting in flat spectrum except at band edges. At receiver, a small pilot is used to achieve carrier lock is added 310 kHz above lower band edge

Next, in Table 14.4 we summarize[11,13] specific Transmission parameters suitable for GA-HDTV.

14.4 PERFORMANCE OF MPEG-2 VIDEO ON HDTV PICTURES

In the previous section we have covered the GA-HDTV system in detail. Since MPEG-2 video is a main component of this system we now evaluate the performance of various encoding options and tools on HDTV pictures. Simulations on several sequences are performed. Although our simulations[5,12] primarily report results on Peak-Signal to RMS-Noise Ratio (PSNR), we are aware that a subjective measure of performance is usually needed to truly evaluate the video coding quality. Also, the simulation results presented here are for illustration purposes and do not represent the results of an optimized MPEG-2 video encoder. Generally, they use the various macroblock mode decisions and rate control of the MPEG-2 Test Mode.[1]

In our simulations we employ three interlaced

Table 14.4 GA-HDTV Transmission Parameters Summary

Transmission Parameter	Terrestrial Mode	High Data Rate Cable Mode
Channel Bandwidth	6 MHz	6 MHz
Excess Bandwidth	11.5%	11.5%
Symbol Rate	10.76 Msymbols/s	10.76 Msymbols/s
Bits per Symbol	3	4
Trellis FEC	2/3 rate	None
Reed-Solomon FEC	(208, 188) T = 10	(208, 188) T = 10
Segment Length	836 Symbols	836 Symbols
Segment Sync	4 symbols/segment	4 symbols/segment
Frame Sync	1 per 313 segments	1 per 313 segments
Payload Data Rate	19.3 Mbit/s	38.6 Mbit/s
NTSC cochannel Rejection	NTSC Rejection Filter inReceiver	N/A
Pilot power contribution	0.3 dB	0.3 dB
C/N threshold	14.9 dB	26.3 dB

HDTV sequences of 1920 × 1024 active resolution, derived from a 1920 × 1035 resolution SMPTE 240M source. For each of these sequences, namely Marching in, Leaves, and Basketball, 60 frames were employed.

14.4.1 Group-of-Pictures Length $N = 9$ versus $N = 15$ Comparison for HDTV

We now show PSNR results[12] of experiments of frame-picture coding for P-picture prediction distances $M = 1$ and $M = 3$ and group-of-pictures of size $N = 9$ and 15 on HDTV sequences. We perform simulations with and without dual-prime motion compensation, but do not use concealment motion vectors. Adaptive motion compensation and adaptive DCT coding modes are allowed. Table 14.5 indicates that the use of $N = 9$ instead of $N = 15$ degrades the PSNR performance only by a small amount, up to 0.2 dB. Thus, $N = 9$ can

be used for faster switching between TV channels without noticeable degradation in performance.

Table 14.6 shows that Dual-prime achieves a higher PSNR without the extra delay of B-pictures. However, as shown in Table 14.3 B-pictures achieve a higher PSNR than Dual-prime.

14.4.2 No B-Pictures versus with B-Picture Comparison for HDTV

With the adaptive motion compensation and adaptive DCT coding conditions for frame-pictures we conduct simulations on interlaced HDTV video to estimate performance tradeoffs with ($M = 3$ case) and without B-pictures ($M = 1$ case). The detailed PSNR statistics are provided for each picture types only for reference. Table 14.4 illustrates that $M = 3$ case outperforms $M = 1$ without dual prime by as much as 0.25 to 0.6 dB with the possibility of 0.3 to 0.6 additional dB improvement

Table 14.5 Luminance PSNR [dB] for interlaced HDTV: $M = 1$ versus 3 and $N = 9$ versus 15 comparison, bitrate = 18 Mbits/s

Sequence	$M = 1$ without Dual-prime MC		$M = 1$ with Dual-prime MC		$M = 3$	
	$N = 9$	$N = 15$	$N = 9$	$N = 15$	$N = 9$	$N = 15$
Marching in	25.40	25.51 (+0.11)	25.96	26.12 (+0.16)	25.70	25.75 (+0.05)
Leaves	24.05	24.17 (+0.12)	24.91	25.14 (+0.23)	24.70	24.75 (+0.05)
Basketball	28.60	28.78 (+0.18)	29.36	29.60 (+0.24)	28.90	29.07 (+0.17)

Table 14.6   Coding statistics for interlaced HDTV $M = 1$ with and without Dual prime, $N = 15$, bitrate = 18 Mbits/s. (a) "Marching in," (b) "Leaves," (c) "Basketball"

"Marching In"	$M = 1$ Without Dual-Prime MC			$M = 1$ with Dual-Prime MC		
Statistic	I-Picture	P-Picture	Avg. Picture	I-Picture	P-Picture	Avg. Picture
Bits	1,866,871	509,462	599,956	1,973,228	501,979	600,062
Avg. Qnt step	31.32	31.14	31.16	29.50	28.40	28.48
SNR, Y	26.89	25.41	25.51	27.10	26.05	26.12

"Leaves"	$M = 1$ Without Dual-Prime MC			$M = 1$ with Dual-Prime MC		
Statistic	I-Picture	P-Picture	Avg. Picture	I-Picture	P-Picture	Avg. Picture
Bits	1,888,359	508,093	600,111	1,964,643	502,588	600,059
Avg. Qnt step	43.36	43.36	43.36	41.00	39.50	39.60
SNR, Y	26.14	24.03	24.17	26.40	25.05	25.14

"Basketball"	$M = 1$ Without Dual-Prime MC			$M = 1$ with Dual-Prime MC		
Statistic	I-Picture	P-Picture	Avg. Picture	I-Picture	P-Picture	Avg. Picture
Bits	1,662,531	523,859	599,771	1,739,009	518,647	600,005
Avg. Qnt step	28.72	26.04	26.22	27.48	22.94	23.24
SNR, Y	29.97	28.69	28.78	30.14	29.56	29.60

Table 14.7   Coding statistics for interlaced HDTV sequence $M = 3, N = 15$, bitrate = 18 Mbits/s. (a) "Marching in," (b) "Leaves," (c) "Basketball"

Statistic	I-Picture	P-Picture	B-Picture	Avg. Picture
Avg. Bits	2,060,618	762,721	377,477	599,830
Avg. Qnt. step	28.46	29.16	40.92	36.82
SNR, Y	27.27	25.82	25.56	25.75

Statistic	I-Picture	P-Picture	B-Picture	Avg. Picture
Avg. Bits	2,140,284	930,871	297,573	599,360
Avg. Qnt. step	37.30	38.68	52.16	47.42
SNR, Y	26.89	24.72	24.54	24.75

Statistic	I-Picture	P-Picture	B-Picture	Avg. Picture
Avg. Bits	1,941,327	713,456	405,251	596,209
Avg. Qnt. Step	22.44	22.62	33.02	29.42
SNR, Y	30.78	29.35	28.76	29.07

Table 14.8   Luminance PSNR [dB] for interlaced HDTV: with B- versus without B-picture comparison, $N = 15$, bitrate = 18 Mbits/s

Sequence	P-Pictures Only			Average of All Pictures		
	$M = 1$, No Dual-Prime	$M = 1$ and Dual-Prime	$M = 3$	$M = 1$, No Dual-Prime	$M = 1$ and Dual-Prime	$M = 3$
Marching in	25.51	26.05 (+0.54)	25.82 (+0.31)	25.51	26.12 (+0.61)	25.75 (+0.24)
Leaves	24.03	25.05 (+1.02)	24.72 (+0.69)	24.17	25.14 (+0.97)	24.75 (+0.58)
Basketball	28.69	29.56 (+0.87)	29.35 (+0.66)	28.78	29.60 (+0.82)	29.07 (+0.29)

when dual-prime is included in adaptive motion compensation.

14.4.3 MPEG-1 Quantization versus MPEG-2 Nonlinear Quantization Scale Evaluation for HDTV

Using an interlaced HDTV video source we investigate[12] the effect of choosing linear or nonlinear quantization scale tables. In all the previous experiments, the linear quantization table was chosen. As expected, Tables 14.9 and 14.10 show that based on PSNR statistics it is not possible to draw any conclusion regarding which table is the best. The improvements are strictly subjective. The non-

linear table has an additional benefit for rate control in difficult sequences.

14.4.4 Evaluation of HDTV Coding Performance at Various Bitrates

Tables 14.11 to 14.13 show the PSNR results[12] of coding interlaced HDTV source material at a variety of bitrates and under various motion com-

Table 14.9   Luminance PSNR [dB] for interlaced HDTV sequences for quantizer comparison: $M = 3$, $N = 15$, bitrate = 18 Mbits/s

Sequence	Linear Quant Scale	Nonlinear Quant Scale
Marching in	25.75	25.72
Leaves	24.75	24.71
Basketball	29.07	29.03

Table 14.10   Luminance PSNR [dB] for interlaced HDTV sequences for quantizer comparison: $M = 3$, $N = 15$, 45 Mbits/s coding

Sequence	Linear Quant Scale	Nonlinear Quant Scale
Marching in	28.23	28.24
Leaves	28.95	28.96
Basketball	32.29	32.33

Table 14.11   Luminance PSNR [dB] for interlaced HDTV sequences: $M = 1$ without Dual-prime, $N = 15$, bitrates = 18, 30, and 45 Mbits/s

Sequence	18 Mbits/s	30 Mbits/s	45 Mbits/s
Marching in	25.51	26.95 (+1.44)	28.09 (+2.58)
Leaves	24.17	26.58 (+2.41)	28.70 (+4.53)
Basketball	28.78	30.58 (+1.80)	32.08 (+3.30)

Table 14.12   Luminance PSNR [dB] for interlaced HDTV sequences: $M = 1$ with Dual-prime, $N = 15$, bitrates = 18, 30, and 45 Mbits/s

Sequence	18 Mbits/s	30 Mbits/s	45 Mbits/s
Marching in	26.12	27.47 (+1.35)	28.62 (+2.50)
Leaves	25.14	27.61 (+2.47)	29.42 (+4.28)
Basketball	29.60	31.26 (+1.66)	32.66 (+3.06)

Table 14.13   Luminance PSNR [dB] for interlaced HDTV sequences: $M = 3$, $N = 15$, bitrates = 18, 30, and 45 Mbits/s

Sequence	18 Mbits/s	30 Mbits/s	45 Mbits/s
Marching in	25.75	27.10 (+1.35)	28.23 (+2.48)
Leaves	24.75	26.99 (+2.24)	28.95 (+4.20)
Basketball	29.07	30.84 (+1.77)	32.29 (+3.22)

pensation conditions. The coder uses adaptive frame/field DCT, zigzag scan, concealment motion vectors, and the MPEG-1 VLC table.

14.5 EVOLUTION OF HDTV

HDTV faces significant challenges to its widespread deployment. One line of thinking indicates that the current resolution of television is sufficient for most applications and that the television should evolve first by being digital and then by being more interactive. This would help consumers get the content they want with modern computer processing in a form (resolution, duration, etc.) that they choose. For example, better use of bandwidth could be made by filling it with a larger number of lower resolution programs and then interactively choosing between them. This is quite feasible since with 20 Mbits/s, which can be accommodated in the 6-MHz bandwidth, one can fit 4 to 10 compressed digital 525-line NTSC programs. Thus, a 60-channel cable television system would be converted into a 240- to 600-channel digital NTSC system. Viewers could then select between them with ease, with the help of a good user interface. However, good interoperability with progressive scan computer displays is foregone.

Many people also believe that the precious radio spectrum should not be used to transmit television to fixed locations. They argue that the terrestrial spectrum should be reserved exclusively for mobile services, which are in great demand as a result of cost reductions of wireless terminals owing to low power, compact electronics, and better batteries. Larger and larger fractions of modern industrialized countries are installing fiberoptic cables having enormous traffic capability. Television could reach many people through such fixed wired connections. However, the cost of "wiring" any country is substantial, and therefore it may take quite a while before television could be distributed entirely by land-based cables.

The high cost of HDTV sets may be another inhibiting factor, although with growing demand, both the electronics and displays should become cheaper over time. Previous experience from color television and VCRs in terms of market pene-

tration, combined with cost projections from the electronics and display industries, give much hope for a low-cost HDTV by the year 2000. While the costs look more manageable than ever before, broadcasters are still unenthusiastic about the business value of HDTV. However, they are also concerned that the other media (tape, DBS, cable) might deliver HDTV to the public before them or that the FCC might assign the taboo channel spectrum to other services if the broadcasters do not use it.

HDTV may also grow by other means. In the United States, the National Information Infrastructure (NII) is being created. Digital HDTV, with its flexible data transport structure, is an immediate opportunity to be deployed as an important part of the interconnected web of information superhighways that make up the NII. GA-HDTV's interoperability with the computing and telecommunication platforms may allow many uses of HDTV for medicine, information access, and education. Thus, it is clear that many of the technical impediments to HDTV are being removed. Customers and the marketplace will decide which way it will evolve.

REFERENCES

1. Test Model Editing Committee, "MPEG-2 Video Test Model 5," ISO/IEC JTC1/SC29/WG11 Doc. N0400 (April 1993).

2. "Generic Coding of Moving Pictures and Associated Audio Information: Systems ISO/IEC 13818–1: Draft International Standard (November 1994).

3. "Generic Coding of Moving Pictures and Associated Audio Information: Video, ISO/IEC 13818–2: Draft International Standard (November 1994).

4. A.N. NETRAVALI and B.G. HASKELL, *Digital Pictures: Representation, Compression, and Standards*, Second Edition, Plenum Press, 1995.

5. A. PURI and B.G. HASKELL, "Digital HDTV Coding with Motion Compensated Interpolation," *Proc. of the Fourth Int. Workshop on HDTV and Beyond,* (September 1991); *Signal Processing of HDTV,* III, Elsevier Science, 1992.

6. GI CORP., "Digicipher™ HDTV System Description," (August 1991).

7. MIT, "Channel Compatible Digicipher HDTV System," Amer. Television Alliance (April 1992).

8. Zenith Electronics Corp. "Digital Spectrum Compatible HDTV System," (November 1991).

9. Advanced Television Research Consortium, "Advanced Digital Television System Description," Jan 1992.

10. R. Hopkins, "Progress on HDTV Broadcasting Standards in the United States, Signal Processing: Image Commun., vol. 5, pp. 355–378, 1993.

11. R. Hopkins, "Digital Terrestrial HDTV for North America: The Grand Alliance HDTV System," *IEEE Transac. on Consumer Electronics,* vol. 40, No. 3, August 1994.

12. R.L. Schmidt, A. Puri, and B.G. Haskell, "Performance Evaluation of Nonscalable MPEG-2 Video Coding," *Proc. SPIE Visual Commun and Image Proc.,* (October 1994).

13. United States Advanced Television Systems Committee, "Digital Television Standard for HDTV Transmission," ATSC Standard Doc A/53 (April 1995).

14. The Grand Alliance, "The U.S. HDTV Standard," *IEEE Spectrum magazine,* pp. 36–45 (April 1995).

15. Y. Ninomiya, "High Definition Television Systems," *Proceedings of the IEEE,* vol. 83, No. 7, pp. 1086–1093 (July 1995).

16. E. Petajan, "The HDTV Grand Alliance System," *Proceedings of the IEEE,* vol. 83, No. 7, pp. 1094–1105 (July 1995).

15

Three-Dimensional TV

In this chapter, we describe an exciting new application area called three-dimensional (3D) TV.[1,16] By 3DTV we mean a broad range of applications where one or more of 3D video, 3D animations, and 3D graphics may be used to present a sophisticated multimedia experience. The principal distinguishing feature of 3DTV as compared to normal TV is that it adds a sense of depth to the scenes being viewed, resulting in increased realism. As you may already know, the analog broadcast TV system in use came into existence in 1941 and since then has had only a few major changes thus far other than addition of color in 1953 and stereophonic sound in 1984, both in a compatible manner. As noted elsewhere in this book, over the last several years considerable progress has been made toward establishing standards for compression of digital video. The completion of all major technical work for MPEG-1 in 1990 and the completion of all primary parts of MPEG-2 by the end of 1994 were major steps in this direction. If you have gone through Chapter 14 on HDTV you may have noted that efforts for HDTV in the United States started in 1987 and converged in 1993 by the formation of the Grand Alliance whose recommendations for the United States digital HDTV and digital TV include the MPEG-2 Video standards among other items. Digital HDTV is one natural direction of evolution for normal analog TV; digital 3DTV is another. The two forms of

TV may appear to be different, but are very related to each other and have a potential synergy capable of enabling each other. You are probably already aware by now that the MPEG committee with its foresight decided to include several tools in the MPEG-2 Video coding standard, anticipating the future direction of evolution of applications. 3DTV is one such application that can be enabled by tools already included in the MPEG-2 standard. During our discussion on Profiles in Chapter 11, we mentioned ongoing efforts within MPEG to define a profile supporting the coding of stereoscopic (two views) video and generalized multiviewpoint video. One way of achieving 3DTV is via use of stereoscopic video; another is via the use of more than two views in multiviewpoint video.

The distinction between computers and TV is fast blurring. Home TV sets are getting increasingly smarter with the help of a new generation of set-top boxes which contain a processor–memory arrangement similar to that in your computer, albeit not as powerful. As noted elsewhere in this book, set-top boxes are basically the control units that enable reception of digital satellite broadcast TV or digital cable TV and incorporate much more capabilities than current cable TV control boxes. Furthermore, it is currently possible to view TV on your computer or workstation by simply adding a TV tuner card. The situation with 3DTV is also quite similar. The technology of 3DTV has significant commonalities with the technology used for Virtual Reality (VR) applications on computers and with video gaming systems. As a consequence, what form[10,16] 3DTV will take is not completely clear. For example, it may be a new type of digital TV set capable of receiving normal as well as special 3DTV broadcasts, it may be a future digital HDTV set capable of also receiving 3DTV broadcasts, it may be simply a plug-in card in a computer (used with a virtual reality headset for viewing in 3D), or it may not be broadcast at all, but used mainly in specialized multimedia presentation systems, video games, and virtual reality. Our favorite scenario is that of an HDTV set also capable of delivering 3DTV and being able to interface to a computer.

In this chapter, we will discuss much more than how MPEG-2 can compress 3DTV video signals. We will in fact discuss everything that is needed to make practical, usable[1,6,7,11,12,14] 3DTV systems. Our discussion will involve basics of 3D vision, techniques for viewing in 3D, stereoscopic imaging and displays, comfortable 3D viewing issues, and, of course, compression of 3DTV signals. As an additional item, we also discuss multiviewpoint video because of its close relationship with 3DTV and emerging future applications. Before going any further, it is perhaps important to provide a more complete list of example applications of 3D:

- Education and Training: flight simulators
- 3D Movies and Entertainment: concerts and live events
- Medical visualizations and surgery: laparoscopic surgery
- Video games and sci-fi theme parks: virtual reality games
- Virtual Travel and Virtual Shopping: home shopping
- Industrial and Remote Tele-operations: undersea and nuclear sites
- Geographical and Meteorological Visualizations: weather patterns

At this point you may be wondering why 3DTV is needed for accomplishing the various listed applications since many of them can be done with normal TV. A very brief answer is that the noted applications can be done much better in 3D owing to the extra sense of realism provided by it. A slightly longer list of reasons is provided next:

- Better depth and distances perception
- Better sense of relative location of objects in a scene
- Greater effective image quality and perceived resolution
- Visibility of luster, surface sheen, and scintillation
- Realism owing to a wider and variable field of view
- Better visual noise rejection
- Lower perceived workload in interactive manipulation.

15.1 BASICS OF 3D

In this section we start with a slightly historical perspective about advancements in technology over the last 150 years or so to illustrate that viewing in 3D is not a new desire, as well as to learn from past failures and build on recent successes. Toward this end, we present a review of the operation of the human visual system, a discussion of visual depth cues, and an analysis of typical impairments in 3D vision. We also include a discussion on the basic principles of optics used in the development of techniques for 3D imaging and display.

15.1.1 Historical Perspective

Our overview of the history of devices and applications involving pictures starts around the early 18th century, when photography was invented and became quite popular because of its ability to capture and accurately reproduce images of faraway places. Soon after that, in 1833, stereoscopy was invented. Stereoscopes allowed viewing of images that appeared to capture the three-dimensional (3D) nature of objects and became an instant rage. Fast forwarding over half a century, by 1895 the first real film was projected for a private audience in Paris. By 1923, the first film with a soundtrack recorded on it was shown in New York. In the early 1950s a new format for movies called Cinerama was developed which used three synchronized projectors and special curved screens to provide a complete 180 degree horizontal field of view, with a total of six times the visual image and six-track stereophonic sound. This process was expensive and enjoyed limited popularity and was abandoned in the early 1960s. During the 1950s, 3D movies became popular as they were able to add the sensation of depth to the movie viewing experience. However, the poor quality of the 3D effect, excessive 3D effect, and funny red-and-blue glasses contributed to a gimmickiness that resulted in its demise. By 1960, Sensorama, a novel personal movie experience that was developed, used 3D pictures, a wide field-of-vision, motion, color, stereo sound, aromas, wind, and vibrations to realize a multisensory experience. It was then ahead of its time and did not get the necessary support needed from Hollywood to ensure its success. During the 1980s video games became a huge craze owing to their interactive nature, only to disappear for several years until they were reincarnated on high-tech game platforms and on personal computers. In the 1990s the technology of video games has evolved toward higher realism by increased interactivity, sophisticated sound, and complex 3D graphics. Theme Parks, over the last 10 years or so, have combined the best of their experience with thrilling rides with the borrowed experience from Hollywood moviemaking to create superb new entertainment experiences involving multiviewpoint movies on huge screens, 3D special effects, realistic animations, and Virtual Reality experiences, with ever increasing interactivity for audience participation.

15.1.2 The Human Visual System

The anatomy of the human eye was discussed in Chapter 5. Here we review some points of interest and briefly discuss how the human visual system, which is inherently binocular, works. The human visual system consists of a pair of eyes, a left eye and a right eye, the back of each of which, called the retina, is connected via optic nerves to a region called the lateral geniculate body, and via pathways to the visual cortex. Light bouncing off an external

object enters the eye and is focused by the cornea and lens to the retina. The amount of refraction of light depends on the cornea as well as the overall thickness of lens controlled by muscles in the eye. Involuntary eye movements maintain visibility of the image on the retina. The retina consists of light-sensitive photoreceptors called rods and cones, and the light absorbed by these receptors initiates electrochemical reactions leading to generation of electrical pulses that pass through the optic nerve. By the way, the image produced at each of the two retinas is upside down. Significant processing of visual information takes place beyond the retina. The pathway beyond the retina mentioned earlier can be best explained by Fig. 15.1.

As can be observed, fibers of each optic nerve split at the optic chiasma and cross over to the opposite sides. Thus, fibers from the right retina go over to the left lateral geniculate body and vice versa, at the underside of brain. Some degree of

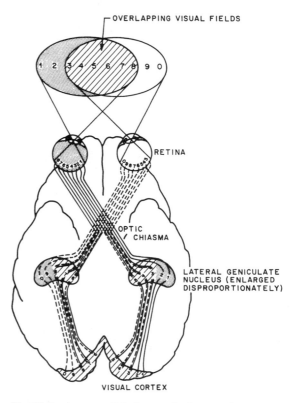

Fig. 15.1 Anatomy of the human visual system, from eyes to visual cortex.

information processing occurs here. Further, via pathways, visual information from each of the two lateral geniculate bodies arrives at visual areas 1 and 2 in the visual cortex. Cortical processing provides the remaining functions for visual perception. Overall, the end result of the processing is a single perceived upright image, although, as hinted earlier, an upside-down, left-eye image is processed by the right half of brain and another upside-down, right-eye image is processed by the left half of the brain.

15.1.3 Visual Depth Cues

Humans learn from the world around us by experience. The outcome of that experience is a huge database of visual information that the human brain accumulates. This lets us perform everyday tasks, many of which involve judging relative sizes of objects, shapes, distances, colors, and much more. The ability to accurately perceive depth in the 3D world is an important attribute of the human visual system. There are many cues[6,7] that assist us in this task, and these can be classified in at least two different ways: monocular versus binocular, and physiological versus psychological. For now, we provide a brief introduction using the latter approach.

15.1.3.1 Physiological Cues

Accommodation is the change in focal length of the lens by the muscles in the eyes.

Convergence is the rotation of the eyes to focus on objects as they get closer.

Binocular Disparity is the dissimilarity in views due to the relative location of each eye.

Motion Parallax is the difference in views caused by moving the scene, objects in the scene, or the viewer.

15.1.3.2 Psychological Cues

Image Size is useful but not sufficient to determine size or depth of objects.

Linear Perspective is the inverse change in size of the object on the retina with distance.

Aerial Perspective means that distant objects appear hazier and bluish.

Shading implies that objects farther from the source of light are darker.

Shadowing of an object on others provides clues about position and size.

Occlusion of objects by other objects provides clues about their relative location.

Color, if bright, implies closer objects and when dark implies distant objects.

Texture Gradient provides clues regarding distance and relative location.

15.1.3.3 Other Cues

Experience lets us estimate relative size, shape, distance, and color of objects.

Overall, the many cues listed are generally additive and all contribute to the process of 3D vision. However, some cues are more powerful than others and this may lead to contradictory depth information and needs to be minimized.

15.1.4 More on Binocular Cues

As mentioned earlier, an alternate classification of visual depth cues is possible based on whether they are binocular or monocular. Because of the special relevance of binocular cues to our discussion, we need to understand them a little better and thus provide further explanation.

Accommodation consists of adjustment of the focal length of the lens in the human eye by the surrounding muscles; the lens, in effect, changes shape, becoming thicker or flatter as needed. In contrast, the lens in a camera does not change its shape; it only moves in or out relative to the iris.

Convergence consists of bringing an object into focus by rotating each eyeball, and thus adjusting the optic axis of each eye to get the two axes to meet at the same point. The angle between the optic axis of the two eyes is called the *convergence angle*.

In normal viewing, accommodation and convergence usually work together. For example, if an object suddenly moves from a distance toward us, our eyes quickly change the focus and the orientation in an attempt to track the movement of the object.

Binocular Disparity is the dissimilarity or the difference in views observed by each human eye owing to their relative offset. The spacing between the pupils of adult human eyes is typically 62 to 70 mm, and for young children its value is a little lower, 50 to 58 mm. In viewing a real-world scene consisting of different objects at different depths, light reflecting from each object enters each eye and hits the retina at slightly different positions. This process is called retinal disparity and is one of the contributing factors to 3D vision; other factors are due to visual processing that occurs elsewhere on the way to brain. By the way, binocular disparity in itself is a sufficient cue to provide perception of depth in a scene, although it is not absolutely essential for depth perception.

15.1.5   Impairments to 3D Vision

It is estimated that at least 10% (and as high as 20%) of the human population has some type of problems with 3D vision. The near and far sightedness problems, are, of course, correctible; however there are other problems that cause difficulties in 3D vision.

15.1.5.1   Night and Snow Blindness

As mentioned earlier, the retina of the human eye contains two kinds of photoreceptors called *rods* and *cones*. Rods are responsible for low light monochrome vision such as night-time vision, while cones are responsible for normal daylight color vision. In low light situations, people with poorly functioning rods have difficulty with their vision, and this condition is referred to as *night blindness*. The opposite condition is referred to as *snow blindness*, which occurs in people with poorly functioning cones, when exposed to bright daylight. People with night blindness or snow blindness have difficulties with 3D vision.

15.1.5.2   Color Blindness

The most common form of color blindness involves inability to distinguish between reds and greens, as persons afflicted with this condition can distinguish only two color bands—yellow and blue. In a rarer form of color blindness, there is inability to distinguish any colors except different shades of gray. People with color blindness have difficulty with their 3D vision as well.

15.1.5.3   Amblyopia

This is also referred to as lazy eye and is caused by slowness in the focussing ability of one eye as compared to the other eye. When this happens, the brain tires of waiting for a response from the slow eye and simply uses the signal from the other eye. Eventually, one gets used to seeing with only one eye, losing the ability to resolve in three dimensions, depending more on other cues such as experience and reference.

15.1.5.4   Astigmatism

This is the condition in which the cornea in the eye is not exactly spherical and, to make matters worse, the shape of the cornea in each eye can be different. As with near or far sightedness, astigmatism can be corrected to a significant degree by special eyeglasses or contact lenses, although some distortions in visualization of small objects still remain. This also contributes to problems in 3D vision.

15.1.5.5   Atmospheric Conditions

The reduction of contrast due to atmospheric changes can result in reduction of visual cues essential to 3D viewing.

15.1.6   Properties of Light

We now briefly review a few facts about light that we will exploit later when discussing the basic principles of techniques for viewing in 3D. Light is a form of electromagnetic radiation visible to the human eye. By *propagation* of light we mean that light travels from one point to another. Furthermore, when light travels in a single medium, it travels in a straight line only and its direction of travel is called the *direction of propagation*. When a light ray travels, it pulsates, and has amplitude and frequency like other waveforms. Light is sometimes modelled as a wave, and other times as a particle, leading to what is known as the *wave-particle duality* of light. Depending on the application, the more appropriate model can be chosen. The model of light most relevant to our discussion is the wave model. Light waves extend outward in all directions, perpendicular to the direction of propagation. Thus, as a ray of light travels toward our eye the electromagnetic field is varying in all directions perpendicular to it. As we will see later, it is relatively straightforward to modify light rays selectively to retain electromagnetic fields of certain orientations only; this fact is exploited to permit us to view in 3D.

15.2 TECHNIQUES FOR 3D VIEWING

In this section we discuss the basic principles used in the various types[1,6,16] of 3D displays. 3D displays can be classified into three classes[6]: Real 3D, True 3D, and Not-True 3D. Real 3D displays recreate a pattern of light rays in actual 3D space. True 3D displays recreate two sets of images, one intended for the left eye and the other for the right eye. Not-True 3D displays present the same image to both eyes. We are primarily interested in Real 3D and True 3D displays. Using a somewhat different approach, 3D displays can be classified into two classes—autostereoscopic and stereoscopic. Autostereoscopic displays are the ones that do not require the use of special viewing glasses or any other optical means for viewing in 3D, and conversely, stereoscopic displays require special viewing glasses or optical means for viewing in 3D. Yet another way of classifying 3D displays is based on the technology[6,7] or some other primary characteristic, such as holographic, volumetric, lenticular, parallax based, or stereoscopic; we follow this approach because of its simplicity.

15.2.1 Holographic

A hologram represents a real 3D nature of a scene as it captures a scene's original pattern of light rays in real x–y–z space. A hologram is thus able to capture a sense of depth[6] in a scene so that the scene looks three-dimensional and on moving the viewpoint, one can look around objects and uncover details that may have been covered by other objects. A holographic image when projected seems to occupy real 3D space right in front of us.

A hologram is not directly a record of the scene itself, but rather interference patterns at microscopic scale, containing information about the scene. When a hologram is illuminated properly, only then can the information contained in a hologram become visible and appear as a reconstructed scene. A hologram is created by combining reflected monochromatic light with nonreflected coherent light from the same source. Monochromatic light is light of a single color or frequency and is quite different from ordinary sources of light such as sunlight or a flashlight. An example of a common and inexpensive source of a monochromatic light is the helium–neon laser, which produces a deep red light. In holographic imaging[32], two monochromatic light paths are combined, one after reflection from objects in a scene and the other, a reference, such that differences in phase, called interference patterns, are recorded onto photographic emulsion and developed. When the same source of light is applied to this photo, the image of the object used for reflection appears. Holograms have the interesting property that any small portion of a holographic image contains information regarding the entire image, albeit at lower resolution and reduced viewing angle. The principle of regenerating a holographic image from a hologram can be best explained with the help of Fig. 15.2.

Figure 15.2 also shows the principle employed in the earliest holographic movies. These movies used 360 degree multiplexed holograms on a cylinder, which was rotated before the viewer to give moving action that lasted for a few seconds. In another case, holograms have been recorded in a sequence of frames on film that was illuminated by flashes of laser light for viewing, with one viewer able to watch at a time. The concept of holographic TV could be developed in the same way as experimental movies, that is, coherent light could be combined with the reflected light from a scene illuminated by coherent light and either recorded for storage or directly converted by a special transducer into an electrical signal for broadcasting to holographic TV receivers. At the receiver, the scene could be painted line-by-line on the raster or stored as an entire frame prior to display. However, there are many problems with the scenario just described.

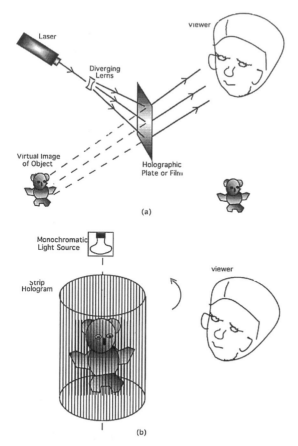

Fig. 15.2 Principle of holographic displays.

First, holographic imaging, in many cases, requires a dark room without dust or other particles or any vibrations. Second, it requires a flexible and erasable material used for recording and reconstructing holographic images. The next problem is caused by the enormous amount of bandwidth needed, for example, for a screen of small size (4 inches × 4 inches), a rate of about 100,000 times as compared to that of normal TV. For larger size screens, the bandwidth required is truly astronomical. With some restrictions, such as transmission of only strip holograms, the bandwidth requirements can be somewhat reduced; however, some of the 3D affect would also be lost. Although many experiments have been conducted, holographic TV displays are not expected to be practical for quite some time to come.

The general characteristics of holographic display systems are as follows:

- Real 3D, as it reconstructs the original light pattern of a scene in x–y–z space.
- Does not require special viewing devices.
- Provides look-around capability with full motion parallax, such that the viewed scene changes in accordance with the observer's viewpoint.
- Recreates the optical distance of objects in the real scene.
- Not yet practical owing to strict imaging conditions, the need for erasable recording, small display size, and excessive bandwidth requirements.
- Color display is difficult to achieve.
- Not yet real-time.

15.2.2 Volumetric

Volumetric multiplanar[2,6] systems display images composed a large number of planes. Such systems generate hologram-like images that occupy real space or imaginary 3D space. Each displayed point in three dimensional space is really a cube referred to as a *voxel* (for volume-pixel).

An example of a volumetric multiplanar display is the *varifocal mirror,* which consists of a vibrating circular mirror along with a monitor. The monitor is connected to a woofer such that the woofer can be synchronized to the monitor. A flexible, circular mirror is attached to the front of the woofer, and the monitor is pointed toward the mirror. With the vibrations from the woofer, the mirror changes focal length and the different points being displayed on the monitor seem to appear at different physical points in space, giving the appearance of different depths to different objects in the scene being displayed. The principle of a varifocal mirror based display is shown in Fig. 15.3.

Varifocal mirror based systems are primarily limited by the size of the mirror, since this mirror has to vibrate. Another type of volumetric display uses lasers to draw cross-sections of the image on a helical surface. Yet other examples of volumetric displays are *rotating mirrors* and *rotating LEDs.*

The general characteristics of volumetric multiplanar display systems are as follows:

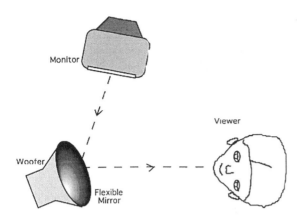

Fig. 15.3   Principle of volumetric-multiplanar varifocal display.

- Real 3D as it reconstructs the original light pattern of a scene in x–y–z space.
- Provides look-around capability with full motion parallax such that the viewed scene changes in accordance with the observer's viewpoint.
- Autostereoscopic: does not require special viewing devices.
- Recreates the optical distance of objects in a real scene.
- Some systems are quite practical, while others are complex.
- Generally lack color and produce wireframe images only.
- Real-time.

15.2.3 Parallax-Based

Parallax-based displays[1,2,6] are based on the principle of parallax stereograms, which consists of a plate of vertical slits, called the parallax barrier. An image consisting of alternating vertical stripes of left- and right-eye perspectives is placed behind the parallax barrier in a manner such that the slits force each eye to view the perspective intended for it, resulting in the 3D effect. An improvement on the parallax stereogram is the parallax panoramagram, which uses fewer slits and parallax barrier farther away from image, resulting in more perceived depth and a wider viewing angle.

A practical example of a parallax-based system for 3D display uses a parallax illumination method.

In this technique an Liquid Crystal Display (LCD), which is a two-dimensional array of individually addressable pixels, is illuminated from behind. To generate the 3D effect, the LCD displays left and right views on alternate columns of pixels, with the left view forming the odd columns and the right view the even columns. Thus if the horizontal resolution of the LCD display is the same as that of the original horizontal resolution of each of the two views, only half the horizontal resolution for each view can be retained. A special illumination plate located behind the LCD generates half as many thin very bright uniformly spaced lines at the horizontal resolution of the LCD display. There is a fixed relationship of distance between LCD, illumination plate, and the distance of the viewer from the LCD display that decides the number of viewing zones. The principle of the parallax-based display and an example of a practical display is shown in Fig. 15.4.

Parallax-based systems suffer from a few drawbacks. For one, the parallax barrier blocks some amount of light, resulting in the relatively dark

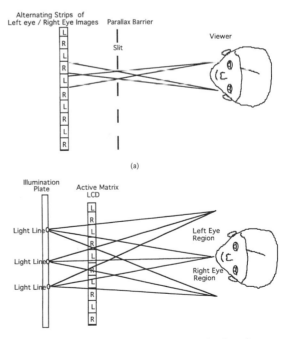

Fig. 15.4   Principle and an example of parallax-based displays.

appearance. Second, a certain amount of distortion may be caused when the viewing angle is not very precise owing to some refraction of light at the slits. Parallax-based systems allow a limited look-around capability. However, by increasing the number of views this can be increased. But if more views are used, the slits have to be even narrower, resulting in an even darker appearance of images.

The general characteristics of parallax-based display systems are as follows:

- True 3D rather than real 3D, as it does not recreate the optical light pattern in $x–y–z$ space. With many views, it can be thought of as in between true 3D and real 3D.
- Limited look-around capability with the degree of motion parallax directly dependent on the number of views employed; more views result in better motion parallax.
- Autostereoscopic; does not require special viewing devices.
- Does not recreate the optical distance of objects in a real scene.
- Some systems are quite practical.
- Color display possible. Because of some blocking of light by the barrier, images tend to appear darker.
- Real-time.

15.2.4 Lenticular

Lenticular lenses have been used in the past in novelty picture postcards and baseball cards to view them in 3D. A lenticular lens is a transparent plate with a large number of closely spaced narrow cylindrical lenses on its surface. If several multiplexed images are placed behind the lens sheet, a lenticular lens can generate a hologram-like image with look-around capability with movement of the viewer's head position. The same basic principle is utilized to build lenticular displays. In a lenticular display, views captured by several parallel video cameras representing different viewpoints are spatially multiplexed and projected from the rear onto the diffusion screen and appear as a set of vertical stripes of images. The diffusion screen is placed at the focal plane behind the lenticular sheet, where the lenticular sheet consists of columns of narrow cylindrical lenses covering it in its entirety. This dis-

play arrangement[1,9] results in a number of viewing zones in front of the lenticular screen where the scene can be viewed in 3D. When viewers move their head position, the feeling of depth persists owing to the look-around capability, which is caused by the availability of several views, such that another pair of views is seen by the visual system of the viewer. Large lenticular displays can be viewed by several people simultaneously and have a certain number of viewing zones in which 3D visualization is possible. The number of viewing zones is directly related to number of views available. We show the principle of operation of a lenticular display system in Fig. 15.5.

Although lenticular display is one of the most promising methods of 3D viewing without glasses, there are several partially unresolved problems. One is that the lenticular lens and the striped image must be placed very accurately relative to each other. If there are errors in positioning, the result can range from distortions in 3D images to loss of 3D effect. Another problem is that such systems require fairly high horizontal resolution since there is loss in horizontal resolution due to partitioning of all available horizontal resolution to display several perspectives. Currently, HDTV projectors allowing four views of one-quarter horizontal resolution are being used for 3D viewing with lenticular displays; however, for better 3D effect, more views are necessary. Research is in progress to resolve the aforementioned problems by tracking the position of the head of the viewer and displaying only the appropriate pair of views corresponding to the perspectives expected; how-

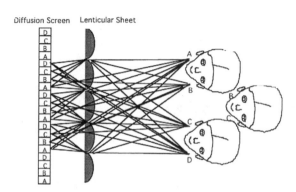

Fig. 15.5 Principle of lenticular displays.

ever, even that has limitations as it currently works for one viewer only.

The general characteristics of lenticular display systems are as follows.

- True 3D rather than real 3D as it does not recreate optical light pattern in x–y–z space. With many views, it can be thought of in between true 3D and real 3D.

- Some look-around capability with degree of motion parallax directly dependent on number of views employed; more views result in better motion parallax.

- Autostereoscopic; does not require special viewing devices; however, viewed scene is viewer position dependent.

- Does not recreate the optical distance of objects in the real scene.

- Some systems are quite practical. If more views are used, accurate alignment of lenticular lens and striped image may be difficult.

- Color display easily possible.

- Real-time.

15.2.5 Stereoscopic

In stereoscopic viewing,[1,6,7,16] two sets of images are used, one intended for viewing by the left eye and the other for the right eye of a human visual system. Each right-eye and left eye view is a normal two-dimensional view, as may be achieved by imaging using two separate video cameras or a single video camera with two separate lenses. Further, the two views are required to be imaged under specific constraints as one is intended, for each eye of the human visual system; the human brain fuses these two slightly different views, resulting in sensation of depth in the scene being viewed. The principal characteristic of stereoscopic systems is that they require viewers to wear some sort of specialized glasses such as red-and-blue glasses or polarized glasses, or other optical means, such as head-mounted displays (HMDs) which consist of a pair of small LCD screen and associated optics, or simply a hood containing an arrangement of mirrors. We now briefly discuss the principles used in various approaches for stereoscopic 3DTV viewing and then list the characteristics of stereoscopic 3D systems.

15.2.5.1 Color Separation

The *color separation* method used in red-and-blue (or red-and-green) glasses is also called the *anaglyph* method and consists of generating two sets of images corresponding to the perspectives intended for the left eye and the right eye. The left images are tinted blue and right images are tinted red; a pair of glasses with a red filter on the right-eye lens and a blue filter on the left-eye lens is used and masks off the opposite-eye color. The choice of colors to use is guided by the fact that colors should be far apart in frequency on the color spectrum and easily separated by inexpensive filters. One problem with a color-separated stereoscopic display is that actual colors in the scene are difficult to reproduce and the resulting images appear heavily tinted.

In the context of color separation, *Chroma Stereopsis* deserves mention as it also uses color separation but provides a significantly enhanced depth effect. This method typically uses a new type of glasses called *superchromatic prisms*. These lenses are made by joining a pair of prisms back to back; one of the prisms is a high-dispersion prism that produces a large dispersion of color but distorts the angle of travel of light rays, and the other is a low-dispersion prism that keeps the colors intact while correcting for the angular distortion. Superchromatic viewing glasses can be made in RTTF, red-to-the-foreground or BTTF, blue-to-the-foreground formats. Chroma stereopsis can produce a very striking 3D effect but suffers from the fact that some colors can cause depth inversion, appearing in different order from the way they are in a scene.

15.2.5.2 Passive Separation

When light is filtered in a way such that the electric field of a specific direction is allowed, it is said to have been *polarized*.[13] For this purpose, a polarizing filter, a lens with extremely thin molecular level stripes, is used to polarize light and only the light with its electromagnetic field aligned with that of the stripes passes through. The light that passes through is now polarized, with its field oscillating in one direction instead of many. For 3D viewing, light with two perpendicular polarization directions is necessary, each intended for one eye. This is pos-

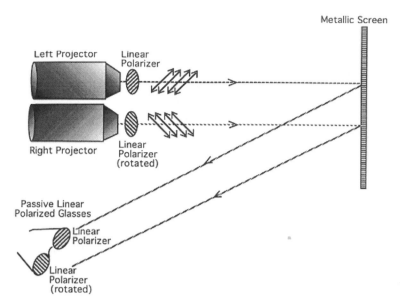

Fig. 15.6   Stereoscopic display using passive separation.

sible by using two polarizing filters that polarize light in two very different directions, say horizontally and vertically. Normally, the two directions used correspond to two diagonals. For 3D viewing, in the path of light to each eye, not one but two filters are necessary, one at the source of light to actually polarize the light in one direction, and the other on the eye, to allow light of only the same direction to pass through, blocking the light polarized in the opposite direction. The principle of passive separation of light is shown in Fig. 15.6.

With polarized light, 3D viewing becomes possible since each of the two eyes of the human visual system receives different images and the human brain does the rest by fusing these images to produce the impression of 3D very similarly to the way we see the natural world around us.

Linear polarization is the technique for polarization of light discussed so far, and suffers from a big drawback in that tilting of the head by the viewer by a small angle causes misalignment between the direction of polarized light from the lens of a projector and corresponding matched filter in linear polarized eyeglasses worn by the viewer. Depending on the tilt angle of head of viewer, this can cause the image intended for one eye to be seen by the other eye, resulting in loss of the 3D effect.

Circular polarization is another technique for polarization of light and it offers an alternative to the aforementioned problem introduced by linear polarization. In circular polarization, two different polarizations are used, clockwise and counterclockwise, generated by use of right and left circular polarizing filters. The polarized light is also referred to as left-circular polarized and right-circular polarized. Circular polarization can be a bit more expensive owing to the higher cost of circular polarizing filters as compared to linear polarizing filters, but can generate a fairly high-quality 3D effect, largely free from problems caused by a small tilting of the head of the viewer. Circular polarizers are basically a combination of linear polarizers followed by quarter-wave plates (also called quarter-wave retarders) oriented at an angle of 45 degrees to the incident light.

15.2.5.3 Active Separation

This technique is used in active LCD glasses which are synchronized with display such as a TV monitor or computer monitor to view in 3D. The LCD glasses act as an alternating electronic shutter such that when the shutter closes on one eye, preventing any light from reaching it, it opens on the other eye, allowing full light to pass through. If the left- and the right-view frames are alternatingly multiplexed in time, LCD glasses can be used such that the shutter on only one eye is open at a time, letting through the frame corresponding to the appropriate view. Since these glasses contain the electronics needed for shuttering, they are called active glasses as compared to passive glasses, which contain no electronics. Another reason for calling them active has to do with alternating separation of left and right views for the appropriate eyes. These glasses are commonly referred to as active shuttered glasses, active LCD glasses, or LCD shuttered glasses. The principle of active separation of light is shown in Fig. 15.7.

The LCD shuttered glasses may be wired or wireless and are controlled by a controller synchronized with a display monitor. Wireless glasses are preferable for obvious reasons when more than viewer may want to watch a single monitor and could be used for viewing 3DTV at home in the future.

A bit of explanation is in order regarding how electronic shuttering may be accomplished. Consider that each lens of LCD shuttered glasses consists of three layers, a linear polarizer in the front, another linear polarizer in the back, and LCD cells between them. When the lens is "off," light polarized through the front polarizer passes through LCD cells but is blocked by the rear polarizer. When the lens is "on," LCD cells are used to rotate the polarized light by 90 degrees so that it can pass through the rear polarizer, reaching the eye. At any given point in time, only one lens is kept "on" whereas the other is kept "off," accomplishing the shuttering.

15.2.5.4 Active and Passive Separation

The method of active and passive separation, as the name implies, employs both active and passive separation and is thus a hybrid of the previous two methods. In this method, an alternating polarizing plate, say using circular polarization, is placed in front of a monitor on which stereoscopic 3DTV is to be viewed and the viewer uses circular polarizing glasses. The arrangement used is shown in Fig. 15.8.

The alternating polarizing plate, if it uses circular polarization, switches at a frame rate of the monitor between left-circular and right-circular polarization states and is used in conjunction with passive circular polarized glasses worn by the viewer to deliver a different view to each eye.

We have now discussed the basic principles involved in several types of methods for stereoscopic display. Using these principles, a variety of stereoscopic display systems become feasible. We describe various example display systems in more detail in the next section; for now, we simply list the general characteristics of stereoscopic display systems as follows:

- Not real 3D but true 3D as it uses only two views with horizontal disparity, discarding much depth information in the scene.
- Does not provide look-around capability and thus normally does not provide motion parallax.
- Stereoscopic; requires special viewing devices.
- Does not recreate optical distance of objects in a real scene. Instead, it provides only binocular disparity, leading to differences in focus and fixation.

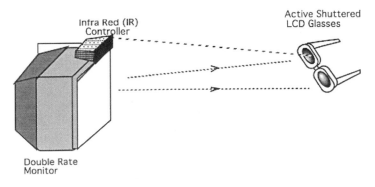

Fig. 15.7 Stereoscopic display with active separation.

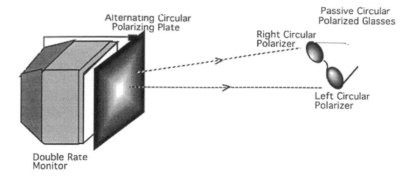

Fig. 15.8 Stereoscopic display with active and passive separation.

- Quite practical.
- Color display is easily possible.
- Real-time.

15.3 STEREOSCOPIC IMAGING AND DISPLAY

We are now ready to tackle the first main issue in realization of practical 3DTV, a good display system. In the previous section, we introduced the basic concepts behind various techniques for 3D viewing including that of stereoscopic 3D viewing, currently the most practical technology for 3DTV. In this section we first briefly discuss the camera and display setup for stereoscopic viewing,[1,6,7,14,16] followed by examples of stereoscopic 3DTV display systems employing principles introduced in the previous section.

Consider a display screen of width s at a viewing distance v from the viewer. Next, consider stereoscopic imaging in which two cameras separated by approximately the distance between human eyes are used to image objects in a scene. Assuming Charge Coupled Device (CCD) cameras, each camera consists of an imaging sensor, a CCD chip of width w, and a lens of focal length f.

A general question that arises is how should the geometry of camera be arranged? The answer to that question is that the display geometry determines the camera geometry; in other words, the geometric arrangement of the eye to display should be matched by the geometric arrangement of the lens to the CCD imaging chip. This implies that the following ratios should be maintained:

$$f/w = v/s$$

This allows the camera field-of-view to be matched by the display-field-of-view, meaning that the display field-of-view is neither magnified nor reduced to prevent any perspective distortions. Figure 15.9 shows a simplified view of the camera and display geometries.

We discuss the issue of camera and display geometries including complete terminology a little later; for now, we get back to our discussion of examples of stereoscopic display systems.

15.3.1 Time-Sequential Display

In the time-sequential display,[1,6,10,11,16] frames of left view and right view are alternately multi-

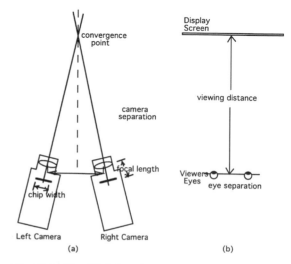

Fig. 15.9 Simplified view of a stereoscopic camera and display geometries.

plexed to produce a single stream of frames at twice the rate. By the way, time-sequential displays are also known as time-multiplexed displays for obvious reasons. Anyway, in such systems this single multiplexed stream of frames is sent to a display which may be a TV monitor, computer monitor, or a video projector. Further, some sort of switching mechanism is employed to ensure that alternate frames from this single stream are viewed by the left eye and the right eye. We describe two such systems; the first uses active glasses and the second uses polarizing plate and passive glasses. Although the time-sequential system can produce a fairly good quality 3D effect, it suffers from the obvious limitations of requiring double rate monitors or projectors which can be expensive; and that the monitor or projectors employ phosphors with faster decay rate, otherwise crosstalk between left and right views results.

15.3.1.1 With Active Glasses

Currently, a simpler form of the time-sequential system with active glasses is used in commercial 3D video game systems in conjunction with home TV sets, which in the United States are capable of displaying 30 interlaced frames/s (or 60 fields/s). The video game system retains only half of the vertical resolution of frames by multiplexing left-view and right-view fields alternately and using active glasses that operate at 60 fields/s rate, offering both the left eye and the right eye of the viewer appropriate views at 30 fields/s only. Note that in normal TV viewing, this rate would cause flicker, however, owing to the nature of video game material which basically consists of moving graphics, flicker is not as noticeable, and besides, the human brain adjusts

to it. Other ways[12,33] of time multiplexing the left and the right views have also been proposed.

For stereoscopic 3DTV, a higher quality time-sequential system[1,6,7,11,16] using alternately multiplexed frames of the left view and right view is more appropriate, requiring TV sets (or computer workstations) and active shuttered glasses to operate at twice the frame rate. Figure 15.10 shows an example of a practical system.

A pair of video cameras separated by approximately the distance between the two human eyes, forming a stereoscopic video camera, images the scene and feeds the left-view frames and the right-view frames to a Frame Multiplexer, which outputs a multiplexed sequence consisting of corresponding pairs of left- and right-view frames. The multiplexed frames are input to a double rate monitor to which is connected an Infrared (IR) controller capable of controlling wireless shuttered LCD glasses. When a left-view frame appears on the monitor, the shuttered LCD glasses worn by the viewer allow this frame to be presented to the left eye only while blocking it from the right eye. When the next frame, a right-view frame, appears, the LCD shutter on the left eye closes, while that of the right eye opens, allowing the frame to be input to the right eye. This process continues resulting in left- and right-eye viewing frames that were intended for each one of them only; human brain fuses the continuous sequence of left- and right-eye frames to provide a stable 3D effect.

15.3.1.2 With Polarizing Plate and Passive Glasses

Although this technique of time-sequential display also uses alternately multiplexed frames of

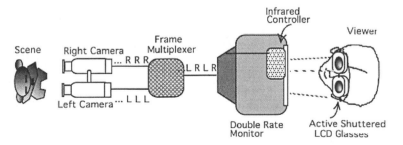

Fig. 15.10 Time-sequential display with active shuttered glasses.

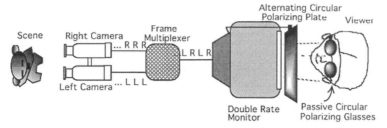

Fig. 15.11   Time-sequential display with polarizing glasses.

left-view and right-view, in this approach, an alternating polarizing plate is placed in front of the screen of the monitor while the viewer wears passive polarized glasses. An example of such a system is shown in Fig. 15.11.

A pair of video cameras separated by approximately the distance between human eyes, forming a stereoscopic video camera, images the scene and feeds the left-view frames and the right-view frames to the Frame Multiplexer, which outputs a multiplexed sequence consisting of corresponding pairs of left- and right-view frames. The multiplexed frames are input to a double rate monitor, in front of which is placed an alternating polarizing plate. Here, we assume that the polarization of the plate to be circular and alternately change directions in synchronization with the frame rate of the monitor. Thus, the polarization of the plate alternates between left-circular polarization and right-circular polarization at a rate that allows left-view frames to be viewed by the left eye and right-view frames to be viewed by the right eye of a human viewer whose brain fuses the two, resulting in perception of 3D.

15.3.2   Time-Simultaneous Display

In time-simultaneous display,[1,6,7,16] left-view and right-view frames are simultaneously sent to display systems which may be a pair of TV monitors, a pair of computer monitors, or a pair of video projectors. Time-simultaneous display is also known as time-parallel display for obvious reasons. In any case, in such systems, the left and the right views are intended to be viewed simultaneously. We describe three such systems; the first uses a mirrored hood; the second uses polarizing plates, mirror, and passive glasses; and the third uses passive glasses. Although time-simultaneous display produces a very good quality 3D effect, it suffers from the obvious limitation that it requires a matched pair of monitors or a matched pair of projectors which may make the system bulkier and expensive.

15.3.2.1   With Mirrored Hood

This is one of the simplest techniques for time-simultaneous display and uses two monitors and a hood with an adjustable optical arrangement of mirrors positioned in a manner that permits left-view frames displayed on the left monitor to be seen only by the left eye and right-view frames simultaneously displayed on the right monitor to be seen only by the right eye. In Fig. 15.12 we show an example of such an arrangement.

A pair of video cameras separated by approximately the distance between human eyes, forming a stereoscopic video camera, images the scene and feeds the left-view frames and the right-view frames to two separate monitors positioned side by side. The viewer wears a hood which includes an optical arrangement of four mirrors positioned such that each eye looks at a different monitor, as shown. Consequently, the left eye sees the left view and the right eye the right view, resulting in fusion of the two views by the human brain to a produce sensation of 3D. Further drawbacks of this scheme are that the hood is quite constraining for viewing for a longer period of time, and only one viewer can view in 3D at one time.

15.3.2.2   With Polarizing Plate, Mirror, and Passive Glasses

In this technique for time-simultaneous display, an arrangement of two monitors each with a different polarized plate in front of its screen is used and the viewer wears passive polarizing

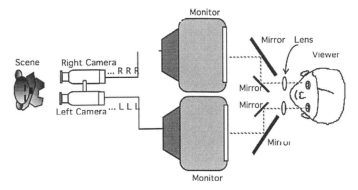

Fig. 15.12 Time-simultaneous display with mirrored hood.

glasses with filters matched with polarized plates used in front of monitors to view in 3D. This technique can be best explained with the help of Fig. 15.13.

A pair of video cameras, with one camera viewing the scene directly and the other located perpendicular to it, both of them viewing the scene through a half-silvered mirror oriented at 45 degrees, is employed as shown. This camera pair forms a stereoscopic video camera that images the scene and feeds the left-view frames and the right-view frames to two separate monitors also located perpendicular to each other. A different polarized plate is placed in front of the screen of each monitor; for example, in a linear polarizing system, a linear polarized plate in front of the left monitor and a similar plate with its direction of polariza-tion perpendicular to the plate in front of the left monitor is placed in front of the right monitor. A half-silvered mirror is placed between the two monitors oriented at 45 degrees, similar to that used in front of cameras. The viewer wears passive glasses with linear polarized lenses in each eye matching the polarized plates on each monitor, thus allowing only left-view frames to be seen by the left eye and right-view frames to be seen by the right eye, resulting in the brain fusing the two views, producing a sensation of 3D.

15.3.2.3 *With Passive Glasses*

In this technique for time-simultaneous display, an arrangement of two projectors each with a different polarized lens and with the viewer wearing passive polarizing glasses with filters matched to

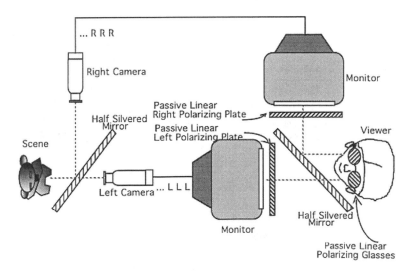

Fig. 15.13 Time-simultaneous display with polarizing plate and polarized glasses.

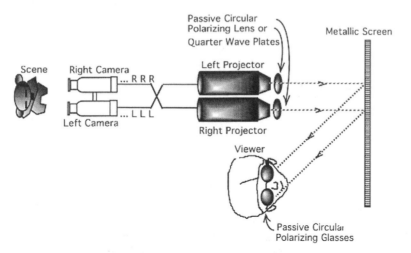

Fig. 15.14   Time-simultaneous display using polarizing glasses.

the polarized lens of each projector is used. This type of system is shown in Fig. 15.14.

A pair of video cameras separated by approximately the distance between the human eyes, forming a stereoscopic video camera, images the scene and feeds the left-view frames and the right-view frames to two separate projectors positioned such that their images can be shifted to coincide (other than the difference of views due to the differing position of cameras). The two projectors, each through their polarizing lenses, project a different view of the scene onto a metallic screen that has the special property of being polarization preserving, that is, it does not depolarize the light. If a nonmetallic screen such as the normal glass-beaded screen was used instead, the light incident on the screen would get depolarized and would be oscillating in all directions, making 3D viewing nearly impossible. Anyhow, the viewer wears passive polarizing glasses with filters on each eye matched to the polarizing lenses to view in 3D. Earlier, a system using two projectors and passive linear polarizes was described; here we discuss details of an example system with circular polarizes.

The projectors used can be CRT based or LCD based. If LCD projectors are used, light emitted by them is already polarized; for 3D viewing only those projectors are preferable that polarize all three components—red, green and blue—equally. Then, if circular polarization of light is to be used,

quarter-wave retarders, arranged at 90 degrees orientation with respect to each other, are sufficient in front of each projectors lens, such that the resulting light from the left projector is now right-circular polarized and the light from the right projector is now left-circular polarized. Although this may seem the reverse of what is needed, in fact, this light when it bounces off the metallic screen reverses its direction of polarization such that the light emitted from the left projector is now left-circular polarized and the light from the right projector is right-circular polarized. Now glasses with left-circular filters on the left eye and right-circular filters on the right eye can be used to view in 3D. This simultaneous display system with circular polarization produces a very high quality 3D effect and is quite suitable for viewing by either a few people or a large audience.

15.4   STEREOSCOPIC GEOMETRY AND HUMAN FACTORS

Earlier, we informally (and perhaps not very precisely) mentioned some terms needed to explain basics of stereoscopic imaging and display. We also briefly introduced the geometry of stereoscopic imaging and display. In this section we discuss these two issues a bit more formally as well as visual distortions, range of parameters to minimize distortions, and human factor issues in stereoscopic viewing.

15.4.1 Basic Terminology

In stereoscopic imaging,[1,4,6,7,16] two cameras separated by approximately the distance between the human eyes and image a scene from two slightly different perspectives. In stereoscopic viewing, which is the inverse of stereoscopic imaging, if these two sets of images representing the two views are presented correctly, the human brain interprets the geometry of mapped objects in the scene to create a 3D view at the intersection point of rays leading from each eye. We now present a brief explanation of a few commonly employed terms in stereoscopy.

Disparity is the dissimilarity in views observed by the left and the right eyes, forming a human visual system.

The *Convergence Point* is the point where rays leading from the left and right eyes forming a human visual system meet. This is also called the object point.

The *Convergence Plane* is the plane of no disparity between left and right views.

Positive Parallax, Negative Parallax, and Zero Parallax respectively mean that the convergence point appears to lie behind the display screen, in front of the display screen, and on the display screen. The maximum possible value of positive parallax is the separation between human eyes, whereas negative parallax can be arbitrarily small.

Homologous Points are points where rays from the left and right eyes forming a human visual system intersect the screen.

Horizontal Parallax is the horizontal distance between two homologous points on the display screen.

Vertical Parallax is the vertical distance between two homologous points on the display screen.

The *Epipolar Plane* is the plane containing eyes of a viewer and the object point.

Figure 15.15 shows an illustration further clarifying the meaning of the various terms just discussed.

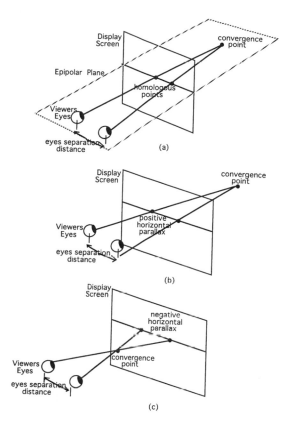

Fig. 15.15 Terminology in stereoscopic imaging and display.

15.4.2 Camera and Display Geometry

Before we discuss the set-up geometry[4,6,7] of cameras and stereoscopic display, we need to formalize terminology in imaging and display geometries by a more precise definition of terms that we have discussed earlier in this chapter, define some new terms, and assign notations to these terms.

The *Interocular Distance, e,* is the distance between the right and the left eyes.

The *Viewing Distance, v,* is the distance between the viewer and the screen.

The *Convergence Distance, c,* is the distance between the midpoint of a line passing through camera lenses and the convergence point. This is also called the pick-up distance.

Camera Separation, d, is the distance between centers of camera lenses. This is also called the baseline distance.

CCD Sensor Width, w, is the width of the CCD chip used as the imaging sensor for imaging in the CCD camera.

CCD Sensor Shift, h, is the distance by which the CCD sensor is shifted laterally.

Focal Length, f, is the focus distance or the relative distance of the camera lens from the imaging sensor.

In stereoscopic imaging, two types of camera configurations can be used to achieve the same convergence distance. The first is called the *angled* configuration, and the second the *parallel* configuration. In the angled configuration, the two cameras are rotated inward by a small angle with respect to each other so that convergence can be achieved at a distant point. The advantage of this method is its simplicity but a disadvantage is introduction of distortions in the imaging geometry. On the other hand, parallel configuration involves a lateral shift between image plane and optics. This

is done by shifting the CCD imaging sensor of each camera away from the center by a small distance. The advantage of this approach is that it does not introduce any distortions in imaging geometry but a disadvantage is that such adjustment requires optics with increased field-of-view as well as specialized mechanical adjustment which may not be possible in available cameras. The two camera configurations are illustrated in Fig. 15.16.

We now consider the mapping between imaging and display geometries.[5,6,7] The complete geometry of a stereoscopic video system can be quite complicated because of many transformations, such as from 3D object space to 2D CCD coordinates for left and right cameras to 2D screen coordinates for left and right object points and finally to 3D image space. Since in stereoscopic imaging, vertical parallax is expected to be almost zero, the problem of determining exact object location can be simplified to 2D if we place coordinates in the epipolar plane. More precisely, we place the origin of our 2D coordinate system on the midpoint between the left and the right eyes. The top view of such a coordinate system with the display geometry is shown in Fig. 15.17.

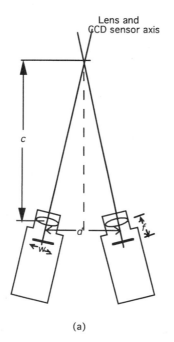

(a)

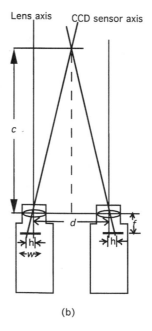

(b)

Fig. 15.16   Stereoscopic camera configurations and imaging geometry.

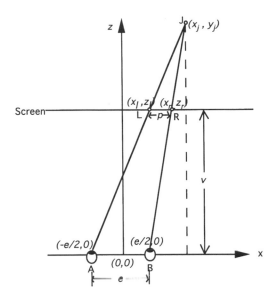

Fig. 15.17 Stereoscopic display geometry.

In Fig. 15.17, the left and right eyes of a viewer are shown as *A* and *B* whereas the rays leading from the viewer's eyes intersect the screen at points *L* and *R* and converge at point *I*. In *x–z* 2D coordinate space, the corresponding coordinates of points *A*, *B*, *L*, *R*, and *J* are $(-e/2,0)$, $(e/2,0)$, (x_l, z_l), (x_r, z_r), and (x_j, z_j). The interocular distance, *e*, horizontal parallax, *p*, and the distance of the viewer from the screen, *v*, is also shown. Although the details of transformations from camera (object) space to display (image) space are outside the scope of this chapter, the results can be stated as follows.

$$x_j = (x_l + x_r)e/2(e - p)$$

$$z_j = ve/(e - p)$$

The parallax, *p*, can be derived by simply rewriting the previous equation as

$$p = (z_j - v)e/z_j$$

We had stated earlier that the maximum value of parallax is the distance between the pair of eyes of the viewer; this can be verified from the parallax equation and happens when object distance (z_j) is very large compared to viewing distance (*v*). Further, for comfortable stereoscopic viewing, the needed optimization of the parallax can be performed by mathematical minimization of the parallax over all visible points in the scene.

15.4.3 Visual Distortions

We now discuss potential visual distortions[3,4,5,6,7,8] resulting from peculiarities of stereoscopic geometry and how they can be minimized.

15.4.3.1 Depth Plane Curvature

As discussed earlier, in stereoscopic imaging it is possible to achieve the same convergence distance by using either an angled camera configuration or a parallel camera configuration. One problem resulting from the use of the angled camera configuration is that of depth plane curvature. This results in objects at the corner of the image appearing further away from the viewer as compared to objects at the center of the image, even if they are located at the same depth. This could lead to the wrong perception of relative distance between objects. The depth plane distortion can be avoided with a parallel camera configuration for which depth plane is linear.

15.4.3.2 Keystone Distortion

This is another distortion that results when the angled camera configuration is used. This is also referred to as trapezoidal distortion. This causes vertical parallax owing to differences in planes of location of imaging sensors in the two cameras. In viewing images produced by angled cameras imaging an equally spaced grid, one notices that for one of the cameras the image of the grid appears larger on one side of the screen and shorter on the opposite side of the screen. The image produced by the other camera has the same properties but is the reverse, longer on the same side where the first camera's image is shorter and shorter on the side where first camera's image is longer. This results in vertical parallax whose value is largest at the edges of the screen. Vertical parallax also increases with increased camera distance, decreased convergence distance, and decreased focal length. Besides vertical parallax, keystone distortion also produces horizontal parallax at the edges of image. It is worth noting that the use of a parallel camera configuration can eliminate this type of distortion completely.

15.4.3.3 Shear Distortion

One known property of stereoscopic images is that they seem to follow the viewer around when the viewer changes viewing position. A sideways movement of the viewer results in sideways shear of stereoscopic image on the surface of the screen. Shear distortion can lead to a problem in judging relative distances between objects. Another problem that may be caused by shear distortion is a false perception of motion in a scene.

15.4.3.4 Depth Nonlinearity

The nonlinearity in depth between actual objects and as seen on a stereoscopic display may provide an incorrect estimate about speed of movement of an object. For example, a moving vehicle equipped with a stereoscopic camera imaging a stationary object appears to suddenly accelerate when it gets closer to the object despite no actual increase in speed. This effect can be minimized if a linear relationship can be preserved between the object depth and image depth. This can be done by configuring stereoscopic camera and display system such that an object at infinity appears at infinity on the displayed image.

15.4.3.5 Radial Distortion

This type of distortion is caused by use of spherical lens, which in effect has nonuniform focal length at various radial distances from the center of the lens. This type of distortion can cause vertical parallax. Typically, focal length is highest at the center and lower at the edges of the lens. The net effect of this type of distortion is that a uniform grid appears to be distorted with distortion severe at the corners of image. The mismatch in distortions between the two cameras, especially at the corners of image, results in vertical parallax. Also, such distortion appears to be worse for lenses with short focal length. This distortion can be reduced by use of aspherical lenses.

15.4.4 Parameter Ranges

In design of comfortable stereoscopic imaging and display systems, the following parameters should be used as a guideline to limit the differences between left and right images:

- Horizontal misalignment—no divergence beyond parallel
- Vertical misalignment of less than 10 minutes of arc
- Rotational misalignment of less than 0.5 degree
- Picture size or focal difference of less than 1%
- Luminance difference of less than 15% for white and 1% for black level
- Color difference of less than 10%
- Adjustment for eye separation in the range of 50 to 75 mm.

15.4.5 Human Factors Issues

While the stereoscopic 3DTV is distinctly superior to normal TV, there are several challenges[1,3,4,8] in design of practical display systems for comfortable viewing. Even if all the necessary precautions are taken in imaging a scene appropriately to minimize distortions and displaying it correctly, there are unique human factors issues that need to be addressed. We study two such issues, one is *visual fatigue* and the other is the *puppet theater effect*.

15.4.5.1 Visual Fatigue

Visual fatigue is also called *eye fatigue* or mental fatigue and refers to the sensation some viewers experience on prolonged viewing of stereoscopic 3DTV. A primary cause of visual fatigue is believed to be inconsistency between accommodation and convergence (or vergence) in stereoscopic 3DTV displays. To be more precise, in viewing stereoscopic 3DTV, the accommodation (focus) of eyes is at a fixed distance where the display screen is located, whereas vergence (fixation) has to continuously vary as the viewer looks at various objects at different depths in the scene and tracks their movement. This mismatch of accommodation and vergence is unique to stereoscopic viewing and for obvious reasons is also referred to as mismatch of focus and fixation. Studies to measure the amount of mismatch that can be tolerated by viewers reveal a wide variation in viewers' responses, leading to the conclusion that for comfortable viewing by many viewers, the range of depth should be minimized.

In general, the relationship between accommodation and vergence is quite complicated and is not fully understood. However, the eye accommodation by itself is considered to reflect visual fatigue fairly well in viewers using computer monitors and can be measured in terms of *critical fusion frequency (CFF)*, the ability to see flicker-free images. Generally, a drop of 5% in CFF is considered quite critical and reflective of visual fatigue. Studies indicate that for nonexpert viewers, there is in fact a drop of approximately 5% in CFF after 30 minutes of viewing stereoscopic 3DTV as compared to no drop when viewing normal TV. However, after a 30-minute rest, the CFF of these viewers recovered to near normal. For expert viewers the drop in CFF was much lower, indicating that with experience, visual fatigue becomes less of a problem.

Overall, to minimize visual fatigue and facilitate fusion, the presence of excessive parallax or significant continuous variations in parallax should be avoided, as it can increase mismatch between accommodation and vergence. Horizontal parallax should thus be kept at values sufficient for 3D viewing, meaning that objects should not be too close to the camera and camera separation should be approximately the same as eye separation. Further, vertical parallax should kept to the absolute minimum possible value to prevent eyestrain.

15.4.5.2 The Puppet Theater Effect

The puppet theater effect means that objects reproduced on 3DTV screen appear nearer and much smaller than the size they appear on normal TV. Experiments indicate that this effect is stronger when camera separation increases and the focal length of the camera decreases. This effect can thus be reduced by constraining the imaging arrangement. Although this effect is still not completely understood, it appears to be caused by mismatch between camera geometry and display geometry. For example, if 3DTV viewing is done on a much smaller size screen than assumed while imaging, the result is that infinitely distant objects appear at finite distance, giving the impression of whole display resembling a puppet theater. An illustration of this effect is shown in Fig. 15.18.

The net result of the puppet theater effect is that distances in the scene get shrunk and the objects appear smaller when viewed on a smaller screen. A possible way for compensating for this effect is by means of a global shift between the two views before display. The amount of shift is not dependent on the contents of the scene but rather on the size of the display screen and needs to adjusted only once as long as the same imaging

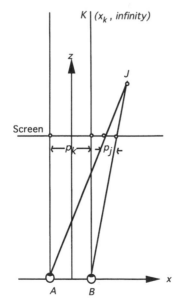
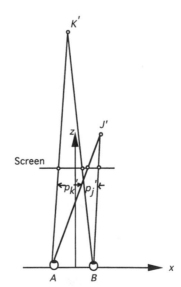

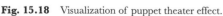

Fig. 15.18 Visualization of puppet theater effect.

geometry is used. However, such a shift can cause problems at image boundaries if the screen is not wide enough to entirely cover the two views after shifting, and in this case, the image may have to be cut by the amount of shift at the boundaries. In imaging new scenes, cameras can be calibrated, assuming a display screen of standard size at a standard viewing distance. Existing recorded stereoscopic scenes need to be visually calibrated on different size screens to determine the correct shift which ensures that disparity of objects located at infinity equals the eye separation.

15.4.6   Commercial Products

At the present time there are a number of inexpensive ways[34–36] of capturing as well as displaying stereoscopic images and video. It is expected that with ongoing research in 3DTV area, the ease of use as well as the display quality of various systems will improve further.

15.5   VIDEO COMPRESSION FOR STEREOSCOPIC 3DTV

We are now ready to tackle the second main issue in realization of practical 3DTV, the issue of efficient representation.[15–18,21–22] We are already aware of the significance of compression in efficient transmission or storage of video and multimedia; the need for video compression becomes even greater for 3DTV. For example, in stereoscopic 3DTV, the raw bandwidth of the video signal is twice that of normal video, since stereoscopic 3DTV contains two channels, each carrying a different viewpoint, one intended for the left eye, and the other intended for the right eye of a human visual system. Thus compression of 3DTV video

signals is absolutely essential; moreover, we are interested in digital compression techniques only, for the same main reasons of ease of manipulation, processing flexibility, and robustness to errors, as for TV and HDTV. Our complete stereoscopic 3DTV system consists of cameras for capturing a scene from two viewpoints; digitizing the corresponding video signals in an analog-to-digital converter; compressing them using an encoder; transmitting or storing a compressed representation; decompressing it using a decoder; and forwarding the two video signals to a display processor, digital-to-analog converter, and stereoscopic 3D display for viewers. In Fig. 15.19 we show a simplified representation of such a system by not explicitly including analog-to-digital and digital-to-analog converters.

When the system just described is used for broadcast transmission, the compressed digital video signals would need to be modulated prior to actual transmission over-the-air or on cable, and demodulated at the receiver prior to decompression. The techniques for modulation and demodulation are like the ones mentioned in Chapter 14 on HDTV, and since for the most part the same techniques apply, they will not be repeated here. Our discussion will mainly focus on compression of stereoscopic video signals, and more particularly on how MPEG compression can be applied for efficient compression.

15.5.1   Coding Approaches

There are primarily three types of approaches that can be used for coding of stereoscopic 3DTV video and are explained as follows:

Simulcast Stereoscopic Coding: This approach involves independent coding of left and right views. Thus, any of the two views may be decoded

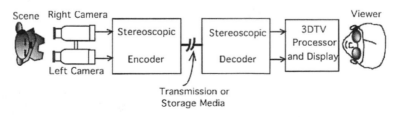

Fig. 15.19   Stereoscopic system for transmission or storage.

and shown on normal TV as the compatible signal from 3DTV.

Compatible Stereoscopic Coding: In this approach, one view, say the left view, is coded independently; the other view, in this case the right view, is then coded with respect to the independently coded (left view). Thus, only the independently coded view may be decoded and shown on normal TV as the compatible signal from 3DTV.

Joint Stereoscopic Coding: In this approach, both the left and the right views are coded together. Since both views are coded jointly, normally neither the left view nor the right view may be decoded by itself as the compatible signal. However, there may be exceptions where by imposing some restrictions, a subset of 3DTV signal may be decoded and shown on normal TV as a pseudo-compatible signal.

As you may have already guessed, in the general sense of compression efficiency, the simulcast stereoscopic coding approach is likely to be the least efficient, followed by compatible stereoscopic coding and the joint stereoscopic coding approaches. This is so because simulcast stereoscopic coding independently codes the two views whereas compatible stereoscopic coding independently encodes only one view, exploiting the correlations between the two views for coding the other view dependently. Likewise, since joint stereoscopic coding does not involve any independent coding at all, in principle, it may allow an even higher compression efficiency, albeit at the cost of lack of compatibility.

For practical reasons, compatible stereoscopic coding offers a good solution as it provides a reasonable compromise between the conflicting requirements of higher compression and compatibility with normal TV. This also allows for the possibility of gradual introduction of stereoscopic 3DTV services without making millions of existing TV receivers obsolete. In a rather oversimplified scenario, consumers can keep receiving normal TV while compatible 3DTV broadcast service is being introduced, and when they are ready to replace their TV in the future can simply buy a 3DTV set. However, this scenario makes best sense when over-the-air broadcast TV becomes digital, since then compatible stereoscopic coding can be used to provide a compatible signal that can be decoded by home TV receivers. In the United States, Direct Broadcast Satellite (DBS)-based TV is already digital, and cable TV is expected to migrate to digital soon, so it is likely that over-the-air broadcast TV will also be digital in some form, such as digital TV or digital HDTV. Perhaps the solution to this dilemma lies in somewhat flexible HDTV sets including decoding and display mechanisms not only for digital HDTV but also for digital TV and digital 3DTV. If this sounds far fetched, consider that, to enable the scenario just described, currently nearly all parts necessary to enable 3DTV algorithms, both hardware and software, either already exist or are relatively feasible in the near future.

15.5.2 Compatible Stereoscopic Coding

There are several techniques for compatible stereoscopic video coding and generally are an outgrowth of whatever methods exist for normal video coding. Most of the practical methods are based on the motion-compensated DCT coding structure, and although some variations have been reported, the compression performance is believed to be similar. Considering that MPEG-based video compression is already used in DBS-based TV, about to appear in digital cable TV, and already accepted for digital HDTV, to promote interoperability, it makes sense to use MPEG-based compression for 3DTV as long as its performance is good. In fact, based on real experiments we suggest that MPEG-2-based video coding performs[15,17,21–23,25] as good if not better than any other compatible stereoscopic coding approach practical today.

15.5.2.1 Disparity Between Views

As discussed earlier, since the left and the right views forming stereoscopic video are usually generated under the constraints imposed by the human visual system, they are highly correlated. Normally, since the two views are expected to be input to the human visual system, they are generated with a camera separation fairly similar to the separation between a pair of human eyes, with exceptions to create special effects only such as

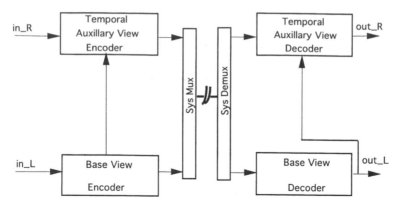

Fig. 15.20   A high-level structure of 3DTV codec derived from Temporal scalability.

hyperstereo. Thus the two views of a stereoscopic scene are only slightly different and are easily fused by the human brain to generate the perception of depth. As noted earlier in this chapter, the difference in perspective between the views is referred to as *disparity*. Stereoscopic video compression attempts to exploit the correlations between the two views by compensating for disparity. Disparity compensation can potentially be more complex than motion compensation, although it can be based on principles similar to those for motion compensation. As an example of disparity values, considering CCIR 601 resolution video and cameras in convergent configuration, disparity values can be as much as up to 50 pixels or so in the horizontal direction and up to 5 pixels in the vertical direction between a consecutive pair of frames. However, constraints of camera set-up for stress-less viewing of 3DTV and other restrictions can be used to reduce the complexity of disparity compensation; for example, for a parallel camera configuration, disparity becomes purely horizontal, since in this case epipolar lines become identical to scan lines in image frames.

15.5.2.2   *Codec Structure from Temporal Scalability*

As mentioned in Chapter 9 on Scalability, the scalability tools of MPEG-2 video coding are intended for applications where the same video is needed at different resolutions or formats simultaneously. Among the scalability tools in MPEG-2, Temporal scalability can code two video signals of

full resolution with very high coding efficiency and is particularly relevant to our discussion. In fact, the high-level two-layer stereoscopic codec structure[22,23] of Fig. 15.20 is directly derived from the high-level two-layer Temporal scalability codec structure. We are now ready to discuss the operation of various elements of this block diagram.

With stereoscopic video, the left view is input to a Base View Encoder, a nonscalable MPEG-2 video encoder, and the right view input to a Temporal Auxillary View Encoder, an MPEG-2 Temporal interlayer encoder that uses temporal prediction from the decoded lower layer, in this case, the decoded left view. Thus, while the left view is coded standalone and provides the compatible signal for normal TV viewing, the right view is encoded with respect to the left view using MPEG-2 Temporal scalability and the two resulting bitstreams are packetized by the system multiplexer, Sys Mux, for transmission or storage. The systems demultiplexer, Sys Demux, performs the inverse operation, separating packets representing each layer and feeding unpacketized bitstreams to the Base View Decoder and Temporal Auxillary View Decoder, which are the nonscalable MPEG-2 video decoder and MPEG-2 Temporal interlayer decoder respectively. Although we have chosen to code the left view independently and the right view dependently, if needed, the structure could easily be reversed such that the right view would be the independently coded signal and the left view the dependently coded signal. For the purposes of illustration, we will continue with the use of the left view as the

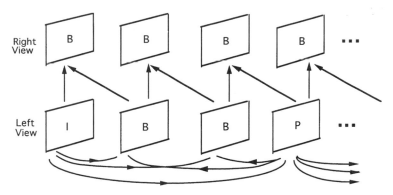

Fig. 15.21 Disparity-compensated prediction only.

independently coded signal and the right view as the dependently coded signal.

15.5.2.3 Disparity-Compensated Prediction Only

A little earlier, in our discussion about disparity we noted the similarities, at least, in principle, between disparity compensation and motion compensation. Also, in Chapter 9 on Scalability, we explained the concept of interlayer prediction when using Temporal scalability. There we mentioned that the time of occurrence of enhancement-layer pictures may either be in between base-layer pictures or may sort of coincide with base-layer pictures. However, in coding of stereoscopic video, the left-view pictures and the right-view pictures are assumed to be independently generated; this has the implication that for each right-view picture there is a corresponding left-view picture, coincident in time. Of course, each of the corresponding left-view and right-view pictures, although taken at the same instance in time, are different depending on the amount of disparity in the scene.

Since nonscalable MPEG-2 coding is quite efficient because of bidirectional motion-compensated prediction in B-pictures, and also since Temporal scalability is highly efficient mainly owing to use of bidirectional prediction in B-pictures, it is conceivable that high coding efficiency may also be obtained by coding the right view as B-pictures using only bidirectional prediction. The bidirectional prediction employed here uses two disparity-based predictions[22,23]; the resulting prediction configuration is shown in Fig. 15.21.

As mentioned earlier, the left view may be coded with nonscalable MPEG-2 coding. More specifically, it can employ I-pictures, P-pictures, and B-pictures, arranged in an $M = 3$ structure of basic MPEG-2 video compression. As a reminder, I-pictures do not use any prediction, P-pictures use forward prediction, and B-pictures use bidirectional prediction employing two different kinds of prediction—forward and backward prediction, which result in three different modes—forward, backward, and interpolated prediction. Further, in using frame-picture coding, the $M = 3$ structure means that there are two B-frame pictures between a pair of reference frames; each of the reference frames can be an I-frame picture or a B-frame picture.

Also, as mentioned earlier, the right view can be coded with MPEG-2 Temporal interlayer coding with predictions with respect to decoded left-view frames. More specifically, it employs B-pictures, as they use bidirectional prediction, consisting of two different kinds of prediction, both referencing decoded left-view frames. The two reference frames used for prediction are the left-view frame coincidental with the right-view frame to be predicted, and the left-view frame temporally next in display order to the right-view frame being predicted. As before, the two predictions result in three different modes called forward, backward, and interpolated prediction. In this context, owing to interlayer predictions, the terminology of forward versus backward gets confusing. Here, we use forward prediction to signify prediction from the coincidental frame in the left view and backward

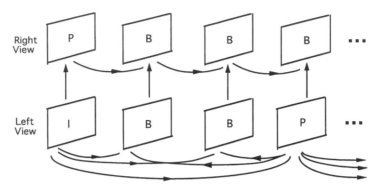

Fig. 15.22 Disparity and motion-compensated prediction.

prediction to mean prediction from the left-view frame, next in display order to the coincidental frame. Since prediction of right view frames is done by two left-view frames, this type of prediction attempts to predict disparity between the views and is called disparity-compensated prediction, requiring the need to calculate disparity vectors at the encoder and including them in the right-view coded bitstream, using which the right-view decoder builds disparity-compensated bidirectional prediction.

It is worth noting that the specific prediction configuration just described can have variations also supported by the MPEG-2 Temporal scalability syntax which offers significantly higher flexibility. For example, if we need to include an occasional right-view frame as a P-frame picture, it is possible to do so as explained in the prediction configuration described next.

15.5.2.4 Motion- and Disparity-Compensated Prediction

This prediction configuration[22–24,30] also uses bidirectional prediction for the same reasons of high prediction efficiency of nonscalable MPEG-2 video coding with B-pictures as well as that of coding using Temporal scalability with B-pictures. Thus it is again conceivable that efficient coding of the right view may be possible using B-pictures owing to bidirectional prediction. However, the bidirectional prediction here involves a disparity-based prediction and a motion-based prediction; the resulting prediction configuration is shown in Fig. 15.22.

The coding of the left view may use the same nonscalable MPEG-2 coding as described for the disparity-compensated prediction configuration and will not be discussed again here.

The right view may use MPEG-2 Temporal interlayer coding with predictions with respect to the decoded left-view frame and a decoded right-view frame. More specifically, it employs B-pictures, since they use bidirectional prediction, consisting of two different kinds of prediction, one referencing a decoded left-view frame and the other referencing a decoded right-view frame. The two reference frames used for prediction are the left-view frame coincidental with the right-view frame to be predicted, and the previous decoded right-view frame in display order. As before, the two predictions result in three different modes called forward, backward, and interpolated prediction. Again, owing to interlayer prediction, the terminology of forward versus backward gets confusing. Here, we use forward prediction to signify prediction from the previous decoded right-view frame in display order and backward prediction to mean prediction from the left-view frame, coincidental with the right-view frame being predicted. Since one prediction is done with respect to the left-view, this is a disparity-compensated prediction, whereas in the second prediction, a right view frame is predicted from a previous decoded right-view frame; this prediction is simply a motion-compensated prediction. Both disparity vectors and motion vectors needed for bidirectional prediction of the right view are calculated at the encoder and included in the right-view coded bistream, using which, the right-view decoder computes disparity- and motion-compensated bidirectional prediction.

Perhaps a comment about the first frame of the right view is in order, since it is coded as a P-frame picture and thus uses only the coincidental left-view frame for prediction. In all generality, P-frames can occur whenever desired in coding of the right view and besides the prediction just mentioned can also use either the left-view frame next to the temporally coincidental frame in display order or the previous decoded frame of the right view for prediction. Again, it is worth noting that the specific prediction configuration just described can have other variations also supported by the MPEG-2 Temporal scalability syntax.

15.5.2.5 *Stereoscopic Codec Structure*

Now that we have a somewhat clearer understanding of which prediction configurations to use, it is time to revisit the high-level codec structure of Fig. 15.20 and derive a more precise structure[22,23] of a compatible stereoscopic codec. We have already indicated that the Base-View Encoder and the Base-View Decoder are simply a nonscalable MPEG-2 encoder and decoder respectively. For clarity, it may be a better idea to rename them as a Motion-Compensated DCT Encoder and Motion-Compensated DCT Decoder to show the coding method used. Also, we mentioned that the Temporal Auxillary View Encoder performs the function

of Temporal interlayer encoder, using prediction with respect to decoded base view frames. Based on our discussion of prediction configurations, in the generalized case, both disparity- and motion-compensated prediction may be employed. Considering this and the renaming convention used for the base view, the Temporal Auxillary View Encoder and Decoder are renamed as Disparity- and Motion-Compensated DCT Encoder and Decoder, respectively. Further, disparity-compensated encoding requires a Disparity Estimator and a Disparity Compensator, pretty much the same way that motion-compensated encoding requires a Motion Estimator and a Motion Compensator. On the other hand, disparity-compensated decoding requires only a Disparity Compensator. The resulting codec structure with renamed encoders and decoders as well as more details is shown in Fig. 15.23.

The motion-compensated encoder included in the Motion- and Disparity-Compensated DCT Encoder is referred to as the Modified Motion-Compensated DCT Encoder owing to the fact that it employs modified B-pictures which can use motion- and disparity-compensated predictions. Disparity vectors are generated by the encoder and are used for disparity compensation both at the encoder and the decoder. These vectors, along with motion vectors, coded DCT coefficients, and overhead indicating specific motion and coding modes, form the bitstream produced by the Disparity- and Motion-Compensated DCT Encoder

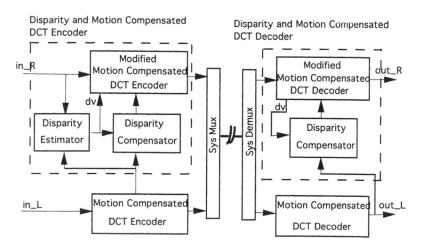

Fig. 15.23 A high-level stereoscopic codec with some details.

which is sent to the corresponding decoder for decoding. Each of the aforementioned encoders and decoders produces disparity- and motion-compensated predictions using decoded frames and thus remain synchronized in their feedback loops, in the absence of channel errors. The operation of Sys Mux and Sys Demux is exactly the same as described earlier, that is, Sys Mux packetizes and multiplexes the encoded left-view and right-view bitstreams, while Sys Demux performs the inverse operation.

Although we have not discussed the details of the internal structures of each encoder and decoder here, it is assumed that by this time you are fairly familiar with interframe block motion-compensated DCT coding which forms the core of MPEG-1 and MPEG-2 standards. As a very brief review, the encoding process consists of block motion estimation and compensation, DCT of blocks of motion-compensated prediction errors, its quantization, and VLC encoding, whereas the decoding process simply consists of VLC decoding, inverse quantization, inverse DCT, and motion-compensated prediction. With frame-pictures, MPEG-2 motion estimation uses 16×16 frame-blocks or 16×8 field-blocks with half-pel accuracy with macroblock-based selection of the best motion mode while MPEG-2 coding first involves DCT of either 8×8 frame- or field-blocks of motion compensated prediction with macroblock-based selection of the best DCT coding mode. It is worth pointing out that disparity estimation and compensation can be performed exactly similar to motion estimation and compensation, that is, with half-pixel accuracy on 16×16 frame-blocks or 16×8 field-blocks, whichever give better results on a macroblock basis. A primary difference is that disparity estimation and compensation is performed between coincidental frames of the left and right view, and that for block matching motion estimation uses a much larger range for horizontal search as compared to vertical search owing to the geometry of stereoscopic imaging and display discussed earlier.

15.5.2.6 Example Results

We illustrate the performance[22,23,25] of MPEG-2 video coding on four stereoscopic 3DTV scenes

available in a European CCIR-601 4:2:2 format of 720 pixels/line and 576 lines/s at a rate of 25 interlaced frames/s. These test sequences are called Train, Manege, Tunnel, and Aqua and each sequence is available as two views, a left view and a right view. For coding, sequences are first reduced to the 4:2:0 format by vertical subsampling of chrominance signals vertically by a factor of 2. Each sequence is then coded using the two prediction configurations discussed earlier, one using disparity-compensated prediction only, and the other using both motion- and disparity-compensated predictions. The left view is coded independently by using nonscalable MPEG-2 coding using frame pictures with prediction distance $M = 3$ (two B-pictures) between a pair of reference pictures and Intra picture distance $N = 12$. The right view is coded dependently with respect to decoded left-view frames using modified B-pictures with the coding structure dependent on the specific prediction configuration employed.

In stereoscopic 3DTV coding, it is not necessary to code both views with the same quality; in most cases, the quality of one of the views can be significantly lower compared to the other view without incurring significant visual distortions. This fact is supported by several practical experiments that claim that quality of one view can be lower by 3 dB or more as compared to the other view. Further support for unequal distribution of quality between the two views is provided by the leading-eye theory, according to which, the overall subjective impression in stereoscopic 3D viewing depends mainly on the quality perceived by the leading eye of viewers. For a given limited bitrate, significant benefits can accrue from unequal distribution of bits between the two views, and thus this approach is adopted frequently. Since in our discussion, the left view is coded independently, as it is intended to be the compatible signal for normal TV, it is coded with high quality, say at 6 Mbits/s, whereas the right view, which is coded dependently and thus requires a lower bitrate, and moreover is also intended to have lower quality, is coded at a significantly lower bitrate, say 2 Mbits/s.

Table 15.1 shows SNR results of independent coding of the left view using MPEG-2 nonscalable coding with frame-pictures, $M = 3$ and $N = 12$.

Table 15.1   SNR for Left View Coded at 6 Mbits/s

Sequence	Left View
Train	37.70
Manege	31.91
Tunnel	37.55
Aqua	37.45

Table 15.2   SNR for Right View Coded at 2 Mbits/s

Sequence	Simulcast	Disp. Comp.	Disp. and Motion Comp.
Train	31.75	33.58 (+1.83)	34.56 (+2.81)
Manege	26.64	26.36 (−0.28)	28.84 (+2.20)
Tunnel	33.12	29.95 (−3.17)	34.36 (+1.24)
Aqua	30.59	27.78 (−2.81)	31.19 (+0.50)

Next in Table 15.2, SNR results of dependent coding of the right view using disparity compensation (Disp. Comp.) and disparity and motion compensation (Disp. and Mot. Comp.) are compared with Simulcast coding. All results are at 2 Mbits/s.

The results of Table 15.2 indicate that with chosen bits partitioning between left and right views, disparity and motion-compensated coding works significantly better by anywhere from 0.5 to about 3 dB when compared to simulcast. This table also shows that simulcast in some cases significantly outperforms disparity-compensated coding by as much as about 3 dB, whereas in other cases disparity only compensated coding either performs the same as simulcast or outperforms it. Thus, it can be concluded that disparity- and motion-compensated coding provides the best overall performance whereas disparity-compensated coding falls short of expectations. As a reminder, both these techniques use B-pictures which use bidirectional prediction, resulting in three basic prediction methods.

The results discussed here, which are based on MPEG-2 Temporal scalability, are among the best results reported using any block-based compatible stereoscopic video coding scheme. Other compatible approaches in the past either used disparity-compensated coding or disparity- and motion-compensated coding, but results have not been as good. One possible reason for good performance of the MPEG-2 Temporal scalability approach is that it uses an interpolated prediction mode that uses averaged prediction and when either disparity or motion compensation by themselves do not work as well, this interpolation mode still performs well. The frequency of selection of various prediction modes is compared in Table 15.3.

As Table 15.3 shows, in disparity-compensated coding the percentage of times the interpolation mode is used is about 65% on the average, indicating that, quite often, each of the disparity estimates cannot provide good prediction and thus the interpolation mode is needed. Also, each of the two disparity estimates are selected equally frequently. This table also shows that in disparity- and motion-compensated coding, the interpolation mode is selected about 40% on average; motion-compensated prediction is as selected at least as often if not more, and disparity-compensated prediction is selected even less frequently than in the previous case. Overall, when compared to simulcast coding, which uses only motion-compensated coding, the higher performance of disparity- and

Table 15.3   Percentage of Right View Coded Blocks in Different Prediction Modes

Sequence	Prediction Mode in Right-View Disp. Comp coded			Prediction Mode in Right-View Disp. and Mot. Comp coded		
	% age Disp. 1	% age Disp. 2	% age Both	% age Mot.	% age Disp.	% age Both
Train	18.5	8.8	71.6	24.1	15.2	59.0
Manege	20.6	14.9	55.5	58.6	2.4	34.1
Tunnel	16.4	9.8	68.1	50.4	1.7	30.9
Aqua	11.9	13.5	69.4	46.6	5.8	42.5

motion-compensated coding can be attributed to use of the interpolated mode which uses both motion- and disparity-compensated prediction.

15.5.3 Improvement to Compatible Stereoscopic Coding

The example results just discussed lead to a couple of inevitable questions. First, why is the performance of disparity-compensated prediction not as good as expected, and second, what if anything can be done to improve its performance?

A potential answer to the first question is that, since disparity estimates are calculated on large blocks (16 × 16 or 16 × 8), perhaps they are not very accurate and thus result in large disparity prediction errors. This also suggests an answer to the second question, that is, perhaps choosing smaller size blocks may be a way of improving efficiency of disparity compensation prediction. However, the situation here is similar to motion compensation, in which prediction errors can be reduced by going to smaller size blocks; however, the overall coding efficiency rarely increases by a significant amount since with smaller size blocks the motion vector overhead also increases. Thus smaller blocks for disparity estimation are also likely to result in higher disparity vector overhead, cancelling any significant benefits that may accrue from use of smaller blocks. As an alternative, perhaps a non-block-based disparity compensation technique may help; however, they have also met with limited success thus far.

Yet another potential answer to the first question is provided by stereoscopic imaging conditions and geometry itself. Earlier during our discussion on stereoscopic geometry and human factors, we indicated an acceptable range of differences in parameters between the left and the right views, for comfortable stereoscopic vision. It is possible that mismatch in parameters is an additional cause of lack of high efficiency of disparity-compensated prediction and that the mismatch may have to be even smaller than that acceptable for comfortable stereoscopic viewing. As before, the answer to the first question also suggests an answer to the second question. Given the left and right view sequences, perhaps mismatch of some imaging parameters[23,27] can be digitally corrected to improve efficiency of disparity-compensated prediction. Among the simple parameters of interest are luminance and chrominance mismatch between the two views. Further, we are interested in correcting for the mismatch in a way that is compatible with MPEG-2-based coding of stereoscopic video discussed thus far.

15.5.3.1 Gain Compensation

To correct for mismatch[22,31] in luminance and mismatch[22,31] in chrominance between the two views of a stereoscopic scene, the luminance and the chrominance of one of the views can be compensated with respect to the other view. Since we have assumed the left view to be coded independently it can be used as a reference for computation of luminance and chrominance mismatch and the right view can be compensated for this mismatch. This type of compensation can be applied globally, on an entire image basis, or locally, for each block of pixels. Both local and global compensation have some advantages and disadvantages and may use separate methods for estimating of differences. For example, global methods may use methods based on matching of histograms, whereas local methods may use minimizing of the absolute value of the difference between a reference block of the left image and a matching block of the right image. Moreover, this compensation can be applied as preprocessing, during coding, or as postprocessing. Since our interest is in increasing compression efficiency when using disparity-compensated coding, mismatch correction during preprocessing or during coding can be used; however, only the preprocessing approach is entirely compatible with the MPEG-2 Temporal scalability approach for coding of stereoscopic video discussed thus far.

The technique of luminance and chrominance gain-mismatch compensation can be briefly outlined as follows. Let $S_l(x, y)$ represent the luminance or chrominance intensity value of a pixel located at (x, y) in the left view and $S_r(x, y)$ represent the corresponding luminance or chrominance intensity value of a pixel located at (x, y) in the right view. Then, assuming a linear model using a contrast gain a and offset of b, $S_l'(x, y)$, a corrected lumi-

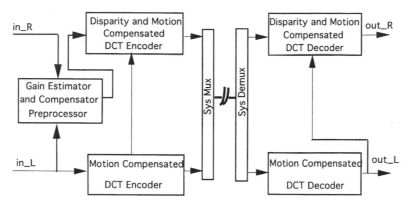

Fig. 15.24 Stereoscopic codec structure with gain compensation.

nance or chrominance can be written as follows:

$$S_r'(x,y) = a \times S_r(x,y) + b$$

Now, using real image intensities in a given stereoscopic scene, it becomes possible to mathematically compute the values of a and b for that scene by minimizing the difference between $S_r'(x, y)$ and $S_l(x, y)$, either globally on a picture basis or locally on a block basis. Once values of a and b are known, they are substituted into the very same equation to compute mismatch-corrected values of the right view of that scene. Of course, three such parameter computations are necessary, one for luminance and one for each of the two chrominance signals.

Figure 15.24 shows a high-level structure of a compatible codec for the stereoscopic video discussed earlier along with gain mismatch compensation just described. The left view is coded independently by a Motion-Compensated DCT Encoder as before; however, the right view is compared in the Gain Estimator and Compensator Preprocessor, with respect to the left view, to estimate its luminance and chrominance gain and produce a modified right-view signal after compensating for its gain. This modified right-view signal is then fed to the Disparity- and Motion-Compensated Encoder, which also uses the decoded left-view signal for prediction. The details of operation of the two Encoders, Sys Mux and Sys Demux, and the corresponding decoders is exactly same as that described for Figs. 15.20 and 15.23.

15.6 MULTIVIEWPOINT VIDEO

Multiviewpoint video is simply what is obtained when the scene is imaged from various viewpoints. In fact the stereoscopic video can be thought of as multiviewpoint video with the restriction that the number of views are limited to two and that the geometry of imaging these views is guided by the geometry of human binocular vision. Thus, generalized multiviewpoint video is composed of more than one view, with or without restrictions[20,26] on how imaging is accomplished. Quite related to the issue of multiviewpoint imaging is also the issue of multi-viewpoint display.

Since multiviewpoint video in fact includes 3D video, besides some new applications, the applications of multiviewpoint video also include applications of 3D video. The new applications of multiviewpoint video for consumers are still evolving although in some niche applications such as in theme parks, it has already been in use for more than 15 years. In any case, we next highlight the important applications of multiviewpoint video.

15.6.1 Applications of Multiview

- Interactive television
- Interactive education and training
- Specialized movies and in movie making
- Medical surgery and surgical planning
- New generation of video games
- Sophisticated virtual travel and virtual shopping
- Industrial and remote teleoperations.

As for the case of 3D video, it is probably useful to state some of the perceived benefits of multiviewpoint video.

15.6.2 Why Multiview?

- Several perspectives to satisfy curiosity of what is behind objects
- Realism owing to wider field-of-vision
- Higher level of immersion owing to the surround-effect
- Better estimation of distances and relative locations of objects
- Greater effective quality of the experience.

As for the case of 3D video, we will discuss imaging and display aspects of multiviewpoint video, the multiviewpoint system, coding approaches, and how MPEG-2 video compression can be applied for multiviewpoint coding. Admittedly, although a significant amount of work has been done on compression of stereoscopic video, the issue of compression of multiviewpoint video is relatively more complex owing to lack of constraints on organization of views, and also more new. The MPEG committee has jointly treated the issue of stereoscopic video and multiviewpoint video via the Multiview Profile[24,26,30] discussed in Chapter 11 on Requirements and Profiles.

15.6.3 Multiview Imaging and Display

Generalized multiview systems,[20,26] owing to potential complexities associated with multiviewpoint display, can be used in a manner such that while several perspectives are imaged and available, either only one or a pair of perspectives is shown simultaneously. When the pair of perspectives is generated under conditions of stereoscopic imaging, stereoscopic video results. However, the case of display on one view or a pair of views is only a subset of the generalized case of multiviewpoint imaging and multiviewpoint display and will not be explicitly covered here. We now consider several example arrangements of multiview imaging and display configurations to better understand the relationship between the different views and

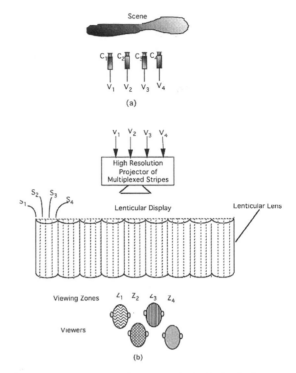

Fig. 15.25 Multiviewpoint capture for lenticular display.

what if anything can be exploited for efficient compression of generalized multiviewpoint video.

Figure 15.25 shows an arrangement of four cameras and the corresponding display arrangement. The cameras, C_1, C_2, C_3, and C_4, are arranged in a side-by-side configuration, imaging a scene; the resulting video signals are V_1, V_2, V_3, and V_4 and represent the four different viewpoints. In general, the number of views is assumed to be two or more, although four views are shown. If the cameras are located parallel to each other separated by approximately interocular distance and pointed in the same direction toward the scene then they can be use to image a scene that can be displayed on a lenticular display as mentioned in the section on techniques for 3D viewing earlier in the chapter. Each of the corresponding vertical stripes of pixels of each view are multiplexed to form a single spatially multiplexed image which is projected on the diffusion screen in the lenticular display whose lenticular lens is aligned with multiplexed columns of multiview images.

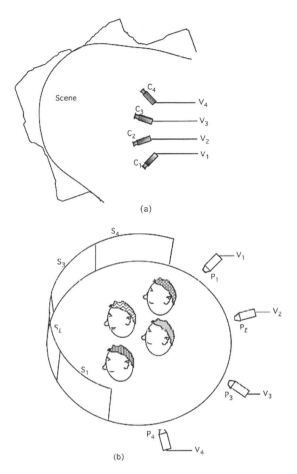

Fig. 15.26 Multiviewpoint capture for panoramic display.

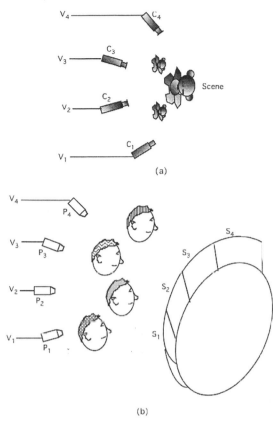

Fig. 15.27 Multiviewpoint capture and surrounded display.

Another multiview imaging and display configuration is shown in Fig. 15.26. In this configuration, the four cameras, C_1, C_2, C_3, and C_4, are arranged in a configuration, imaging a distant scene panoramically; the resulting video signals are V_1, V_2, V_3, and V_4 and represent the four different panoramic viewpoints. In this configuration, the scene surrounds the cameras entirely, and although only four cameras are shown, any number of cameras may be used. The display of such multiviewpoint scenes can also be panoramic, as shown, and uses projectors P_1, P_2, P_3, and P_4 to display separate viewpoints on corresponding screens S_1, S_2, S_3, and S_4. The viewers are positioned in the center of the theater surrounded by screens filling their field-of-vision with the panoramic multiviewpoint scene.

Yet another multiview imaging and display configuration is shown in Fig. 15.27. In this configuration, the four cameras, C_1, C_2, C_3, and C_4, are arranged in a configuration surrounding a close-by scene; the resulting video signals are V_1, V_2, V_3, and V_4 and represent the four different surround viewpoints. In this configuration, the cameras surround the scene, and although only four cameras are shown, any number of cameras may be used. The display of such multiviewpoint scenes can be done by projecting the scene using projectors P_1, P_2, P_3, and P_4 to display separate viewpoints on corresponding screens S_1, S_2, S_3, and S_4, and the viewers surround the screens and because of the relative motion between viewers and the screens (moving viewers, moving screens, or some other arrangement) are able to see the scene from multiple viewpoints. An alternative may be to simply use the panoramic display of the previous example.

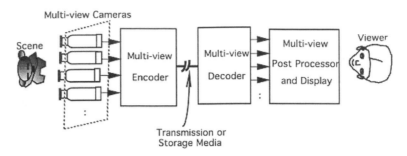

Fig. 15.28   Multiview transmission or storage system.

Although we have discussed several potential configurations in multiviewpoint imaging and display, in reality the number of possibilities are truly endless.

15.6.4   Multiview Video System

We now focus on systems aspects of a generalized multiview system. Figure 15.28 shows a system for multiview video transmission or storage. It shows four cameras forming a multiview camera system imaging a scene; these views of the scene are input to a Multiview Encoder that compresses these views, exploiting redundancy between the views. The resulting compressed bitstream is either transmitted or stored and awaits decoding by the Multiview Decoder, which regenerates the four views and sends them to the Multiview Postprocessor and Display.

15.6.5   Multiview Coding Approaches

Multiview coding, like stereoscopic 3DTV coding, may employ the same three basic approaches which are explained as follows.

Simulcast Multiview Coding: This approach involves independent coding of the various views forming multiview video. Thus, any of the views may be decoded and shown as the compatible signal. Further, if overall coding complexity becomes high, some of the views may be coded at lower resolution.

Compatible Multiview Coding: In this approach, one view is coded independently; the other views can be coded with respect to this view or with

respect to each other. Thus, only the independently coded view may be decoded and shown as the compatible signal. Further, to reduce coding complexity, one of the two approaches, the *Resolution Constrained* approach or the *Viewpoints Constrained* approach, may be used.

Joint Multiview Coding: In this approach, the various views are coded together. Normally, none of the views may be decoded by themselves as all the views are jointly coded. To reduce coding complexity, the views may be coded at lower resolution, the same as that for simulcast or compatible coding.

Overall, very much like the case of compatible stereoscopic coding, in the general sense of compression efficiency, the simulcast approach is likely to be the least efficient, followed by the compatible coding and joint coding approaches. Since our main interest is in multiview applications that can interwork with existing video systems, we are primarily interested in the compatible coding approach. Even more specifically we are interested in using MPEG-2 video coding methods for compression of multiview video signals.

15.6.6   Compatible Resolution Constrained Multiview Coding

In this compatible coding approach,[26] the spatial resolution of each view may be constrained to reduce the coding problem of more than two views to that of coding of two views and thus the same approaches as that used for coding of stereoscopic video can be applied for coding of the two views. This is shown in Fig. 15.29, where four input views are spatially decimated to generate four views of

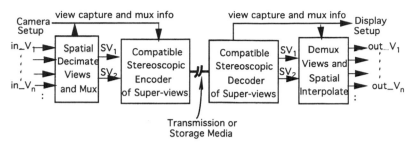

Fig. 15.29 Compatible resolution constrained multiview coding.

half-resolution each such that a pair of half-resolution views forms a superview and the coding problem of four views is reduced to that of coding of a pair of superviews. A Compatible Stereoscopic Encoder which uses Temporal scalability such as that of Fig. 15.23 can then be used such that one of the superviews is coded independently and the other superview is coded with the disparity- and motion-compensated prediction configuration of Fig. 15.22. The transmitted or stored bitstreams representing the compressed superviews are then decoded by a corresponding Compatible Stereoscopic Decoder, which generates the decoded superviews. Each of the superviews is then demultiplexed to individual views and then spatially interpolated (if required) for display. In fact the need for spatial decimation before encoding is dependent on the resolution that can be handled by the Compatible Stereoscopic Encoder and Decoder, and in some cases may not be necessary.

In Fig. 15.30 we show examples of how four or eight reduced resolution views may be multiplexed to form a pair of superviews. To obtain reduced spatial resolution, spatial decimation consisting of filtering and subsampling explained in Chapter 5 may be employed. Likewise to upsample to full resolution, the spatial interpolation operation also described in Chapter 5 may be employed.

15.6.7 Compatible Viewpoints Constrained Multiview Coding

In this compatible coding approach, the spatial resolution of each view is kept full; however the number of views to be coded is constrained. It is expected that typically either three or four views of

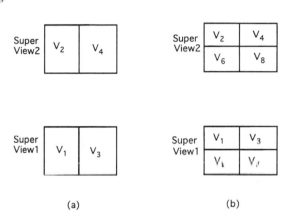

Fig. 15.30 Multiplexing of multiviews to form superviews.

full resolution would be coded and the remaining missing views would be interpolated at the receiver. The three or four views can be coded by extending the Compatible Stereoscopic Encoder and Decoders, still using Temporal scalability. Assuming the case of coding of three views, one such codec configuration can be derived by simply adding a third layer to a stereoscopic codec such that this third layer uses Disparity- and Motion-Compensated Coding with respect to the second layer, which also uses Disparity- and Motion-Compensated Coding with respect to the first layer, which is independently coded with Motion-Compensated Coding. The first layer provides also the compatible signal and can be decoded and displayed by itself.

Yet another variation to the compatible codec structure just described is shown in Fig. 15.31. In this configuration, three views are coded in a manner such that there is a center view and two side views, with the center view coded independently, and each of the two side views coded with dispari-

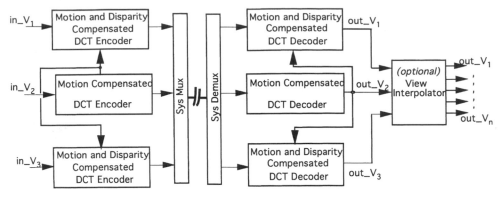

Fig. 15.31   Compatible viewpoints constrained multiview coding.

ty- and motion-compensated coding with respect to the center view. The Motion-Compensated DCT Encoder and Decoder can be used to encode and decode the center view, and the Disparity- and Motion-Compensated Encoders and Decoders to encode the two side views. For the coding to be MPEG-2 compliant, the Main Profile Encoder and Decoder can be used for coding of the center view, and MPEG-2 Temporal scalability based Interlayer Encoders and Decoders for side views. As mentioned earlier, if more views than that coded with MPEG-2 encoding are needed, they can be generated at the decoder by interpolation, although this task is not straightforward.

Besides the Multiview Profile in MPEG-2, the next MPEG standard (MPEG-4) is also planning to address[28,29] various issues in stereoscopic and multiviewpoint video.

15.7   SUMMARY

In this chapter we have addressed nearly all aspects of practical realization of 3DTV systems and shown how MPEG compression may play a potentially major role in enabling this exciting application area. More particularly, we have discussed the following:

- Applications involving 3D video, graphics, and animations as well as the reasons why 3D viewing may be preferable
- Basics of 3D vision including the human visual system, visual depth cues, details of binocular

depth cues, impairments to 3D vision, and properties of light

- Principles of techniques for viewing in 3D including holographic, volumetric, parallax based, lenticular, and stereoscopic
- Stereoscopic 3DTV display methods belonging to time-simultaneous and time-sequential classes
- Stereoscopic geometry and human factors issues
- Stereoscopic 3DTV compression with MPEG-2 video including prediction configurations, their performance, compatible stereoscopic codec, and improvements
- Multiviewpoint video including applications, imaging, and display; and multiviewpoint video system, coding methods, and application of MPEG-2 video coding.

REFERENCES

1. T. Motoki, H. Isono, and I. Yuyama, "Present Status of Three-Dimensional Television Research," *Proc. IEEE* 83(7):1009–1021 (July 1995).

2. J. Eichenlaub and A. Martens, "3D Without Glasses," *Information Display*, 8(3):9–12 (March 1992).

3. R. Kutka, "Reconstruction of Correct 3-D Perception on Screen Viewed at Different Distances," *IEEE Trans. Commun.* 42(1):29–33 (January 1994).

4. T. Mitsuhashi, "Subjective Image Position in Stereoscopic TV Systems—Considerations on Comfortable Stereoscopic Images," *Proc. SPIE,* 2179:259–266 (March 1994).

5. A. Woods, T. Docherty, and R. Koch, "Image Distortions in Stereoscopic Video Systems," *Proc.*

SPIE Stereosc. Displays Appl. IV, 1915:36–48 (February 1993).

6. J. O. MERRIT, "Stereoscopic Display Applications Issues—Part 1: Human Factor Issues," *Course Notes, IS&T/SPIE Symposium on Electronic Imaging Science and Technology* (February 1993).

7. D. F. McALLISTER, "Stereoscopic Display Applications Issues—Part 2: Introduction and Algorithms," *Course Notes, IS&T/SPIE Symposium on Electronic Imaging Science and Technology* (February 1993).

8. N. HIRUMA and T. FUKUDA, "Accommodation Response to Binocular Stereoscopic TV Images and Their Viewing Conditions," *SMPTE J.,* pp. 1137–1144 (December 1993).

9. H. ISONO et al., "Autostereoscopic 3-D Television," *Electron. Commun. Jpn.* 76(8):89–98 (January 1993).

10. M. STARKS, "Stereoscopic Video and the Quest for Virtual Reality," *Proc. SPIE Stereosc. Displays Appl. II,* pp. 327 342 (February 1991).

11. M. STARKS, "Stereoscopic Video and the Quest for Virtual Reality—Part II," *Proc. SPIE Stereosc. Displays Appl. III,* pp. 216–227 (February 1992).

12. L. LIPTON, "Stereoscopic Real-Time and Multiplexed Video System," *Proc. SPIE Stereosc. Displays Appl. IV,* pp. 6–11 (February 1993).

13. F. A. JENKINS and H. E. WHITE, *Fundamentals of Optics,* 4th edit., McGraw-Hill, New York (1976).

14. B. CHOQUET et al., "Development of European Stereoscopic Camera for Television Imagery," *Proc. SPIE Camera Scan. Image Acquis. Syst.* 1901:37–45 (February 1993).

15. M. ZIEGLER et al., "Digital Stereoscopic Television—State of the European Project DISTIMA," *4th European Workshop on Three-Dimensional TV,* pp. 247–255 (October 1993).

16. A. KOPERNIK, R. SAND, and B. CHOQUET, "The Future of Three-Dimensional TV," *International Workshop on HDTV'92, Signal Processing of HDTV, IV,* pp. 17–29 (1993).

17. R. HORST, "A Digital Codec for 3D-TV Transmission," *International Workshop on HDTV'92, Signal Processing of HDTV, IV,* pp. 489–495 (1993).

18. P. D. GUNATILAKE, M. W. SIEGEL, and A. G. JORDAN, "Compression of Stereo Video Streams," *International Workshop on HDTV'93, Signal Processing of HDTV, V,* pp. 173–185 (1994).

19. M. ZIEGLER, S. PANIS, and J. P. COSMAS, "Object Based Stereoscopic Image Coding," *International Workshop on HDTV'93, Signal Processing of HDTV, V,* pp. 187–193 (1994).

20. M. E. LUKACS, "Predictive Coding of Multi-Viewpoint Image Sets," *Proc. International Conf. on Accoustics, Speech and Signal Processing,* pp. 521–524 (April 1986).

21. B. TSENG and D. ANASTASSIOU "Scalability for Stereoscopic Video Coding," *International Workshop on HDTV'94, Signal Processing of HDTV, VI* (1995).

22. A. PURI, R. V. KOLLARITS, and B. G. HASKELL, "Stereoscopic Video Compression Using Temporal Scalability," *Proc. SPIE Visual Communications and Image Processing* (May 1995).

23. A. PURI, R. V. KOLLARITS, and B. G. HASKELL, "Compression of Stereoscopic Video Using MPEG-2," *SPIE Photonics East, Critical Review of Standards and Interfaces* (October 1995).

24. A. PURI and B. G. HASKELL, "Multi-View Profile Proposal for Discussion in MPEG-2 Multi-View Ad Hoc," ISO/IEC JTC1/SC29/WG11 Doc. MPEG95/091 (March 1995).

25. A. SHIGENAGA and T. HOMMA, "Experimental Results of Coding Stereo Sequences with Temporal Scalability," ISO/IEC JTC1/SC29/WG11 Doc. MPEG95/254 (July 1995).

26. A. PURI and B. G. HASKELL, "A Revised Proposal for Multi-View Coding and Multi-View Profile," ISO/IEC JTC1/SC29/WG11 Doc. MPEG95/249 (July 1995).

27. T. C. CHIANG and Y. ZHANG, "A Proposal on Coding Algorithms and Syntax for the Multi-View Profile," ISO/IEC JTC1/SC29/WG11 Doc. MPEG95/253, Tokyo (July 1995).

28. A. PURI, "3D/Stereo Video Applications and Their Initial Requirements for MPEG-4," ISO/IEC JTC1/SC29/WG11 Doc. MPEG94/093 (March 1994).

29. MPEG-4 REQUIREMENTS AD HOC GROUP, "MPEG-4 Functionalities," ISO/IEC JTC1/SC29/WG11 Doc. MPEG94/399 (November 1994).

30. VIDEO SUBGROUP, "Proposed Draft Amendment No. 3 to 13818–2 (Multi-View Profile)," ISO/IEC JTC1/SC29/WG11 Doc. N1088 (November 1995).

31. A. PURI, B. G. HASKELL, and R. V. KOLLARITS, "Gain Corrected Stereoscopic Coding Using SBASIC for MPEG-4 Multiple Concurrent Streams," ISO/IEC JTC1/SC29/WG11 Doc. MPEG95/0487 (November 1995).

32. J. E. KASPER and S. A. FELLER, "The Complete Book of Holograms," John Wiley and Sons, (1987).

33. L. LIPTON, "Compatibility issues and selection devices for stereoscopic television," Signal Processing: Image Commun., Vol. 4, pp. 15–20 (1991).

34. "Product Information", VRex Inc., N.Y. (1995).

35. "Product Information", 3DTV Corp. CA. (1995).

36. "Product Information", Stereographics Corp., CA (1995).

16

Processing Architecture and Implementation

by Dr. Horng-Dar Lin

MPEG video processing requires a significant amount of processing power. Depending on the MPEG-2 profile and level selected, the range of processing power also varies a lot. For example, the performance requirement for decoding applications employing the Main profile at Low level is relatively low and can be implemented in software on general microprocessors. To encode the Main profile at Main level video in real time, however, requires performance well into 10,000 million operations per second. Such high performance is often delivered through special video processing hardware.

Processing power is not the only factor that influences the design choice. Cost and flexibility play major roles in determining the success of the design as well. The emphasis often depends on target applications. For MPEG-2 video decoding in consumer products, minimum cost is critical. However, for postproduction editing and other MPEG-2 encoding applications, flexibility, quality, and performance certainly outrank cost in implementation considerations.

Like many systems, an MPEG processing system is a combination of software and hardware modules. The first design decision is to choose how much processing should be put in software and how much should be in hardware. This is more than merely a technical decision, as it also translates to required resources for hardware and soft-

ware development. The second design decision is what type of software and hardware should be put in place. For example, the implementation could be a special-purpose engine which runs hand-coded microcode or a generic processor that runs object codes compiled from high-level languages. Obviously the required development efforts vary significantly depending on the choice.

In this chapter we review implementation issues and also use several examples to illustrate different design balances. Different architectures are presented in the chapter. Readers should bear in mind that there is no one "optimal" solution for all kinds of systems. The choice depends on specific application requirements, the VLSI technology, and quite often other considerations beyond MPEG processing itself.

16.1 MPEG PROCESSING REQUIREMENTS

The MPEG-1 video standard was designed primarily for digital storage media such as CD-ROM. The standard is optimized for SIF noninterlaced video at 1.5 Mbits/s with features for random access and fast forward/reverse trick mode play. The MPEG-2 standard extends the MPEG-1 standard to support interlaced video, higher video resolutions, higher coded bit rates, and various scalabilities. Both

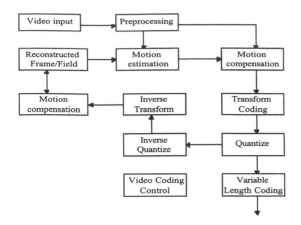

Fig. 16.1 MPEG encoding block diagram.

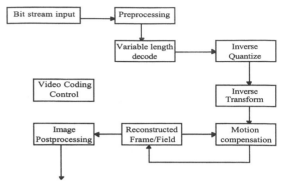

Fig. 16.2 Decoder signal flow.

MPEG standards share many common processing requirements. The video encode process usually consists of various processing tasks as illustrated in Fig. 16.1. The main modules include:

- Video preprocessing such as filtering and format conversions
- Motion estimation
- Motion compensation
- DCT and IDCT
- Quantization and inverse quantization
- Bitstream encoding
- Coding control.

The encoding itself includes a decoding loop. In addition, note that the encoding includes processing of different data types. At the video input, data are typically a stream of pels or scan lines. Filtering and preprocessing operates on pels directly. Motion estimation, motion compensation, transform coding, and quatization are all block-based processing. Variable length coding takes a block of quantized data and generates a bitstream. In designing the overall system, functionality as well as manipulation of data types must be carefully considered.

The decoder modules are similar to the encoder modules. Following the flow of signals as illustrated in Fig. 16.2, we see that the main modules include:

- Bitstream preprocessing
- Variable length decoding

- Inverse quantization
- IDCT
- Motion compensation
- Video postprocessing
- Decoding control.

Although the modules are similar, the nature of some modules could be quite different. For example, in the encoder design, the input is usually a constant rate video input with a constant or variable rate bitstream output. In the decoder design, the time it takes to process a single frame or field of image could vary significantly.

Although the MPEG standards specify the decoding procedure explicitly (and general encoding procedure implicity), the implementation still involves various design choices even at the algorithmic level. For example, at the encoding end, the choice of picture coding types, the motion estimation algorithm, the video preprocessing, and the buffer/delay control are open to various tuning and optimization options. For the decoder, the video quality also relies heavily on video postprocessing and decoding control for error concealment. These algorithm choices also need to be considered prior to the architecture design.

16.2 ARCHITECTURAL CHOICES AND EXAMPLES

The next step after determining the application requirements and video processing algorithms is to choose the right architecture for implementing the encoder or decoder modules. There are three base types of implementation:

- Generic processor
- Custom data path engine
- Application-specific processing engine.

A generic processor could be based on a Complex Instruction Set Computing (CISC), a Reduced Instruction Set Computing (RISC), or any other "all-purpose" architecture. In this style the implementation often involves software compiled from high-level codes or programmed in the assembly language. A custom data path engine is the one with special instruction sets designed for certain classes of applications. Typical examples of such custom data path engine are today's audio digital signal processors (DSPs). The instruction sets and architectures of these DSPs are tuned to make them extremely efficient in filtering and similar applications. However, these instruction sets also make DSPs harder to program and less suitable for generic data processing. Application-specific processing engines are special-purpose hardware dedicated for specific functions. For example, these could be transform engines for DCT and IDCT or motion estimator engines for performing the motion vector search.

Obviously the three choices represent different tradeoffs of flexibility and efficiency. For example, application-specific processing engines are not as flexible as generic processors, but they are much more efficient. For a complex system, mix and match is often the name of the game. For example, coding control maps naturally to a generic processor, while motion estimation for high-resolution video is often implemented with an application-specific engine for performance reasons.

Let us take a closer look at the three architectural options and the tradeoffs.

16.2.1 Generic Processor

The simplest form of implementation is based on compiled software codes. This approach is suitable for applications that do not demand high processing power or for those that have no real-time

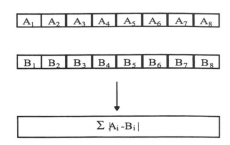

Fig. 16.3   Sample special-purpose instruction.

constraints. As opposed to implemention of hardware modules, the focus is on building software modules.

Many traditional generic processors have started to incorporate new instructions to assist multimedia processing. For example, in the UltraSparc™ design, the instruction set has been extended to include many video processing functions.[1] A single pel distance instruction compares two 64 bit registers each holding eight 8-bit components, calculates eight absolute differences between the corresponding components in parallel, and accumulates the absolute differences. The instruction is illustrated in Fig. 16.3. Note that commands like these are more for application-specific than generic computing. However, the same pel distance calculation done in generic form takes more than 20 subtract, absolute, and add operations to complete. This is a good indication that generic computing is not as effective as custom commands.

16.2.2 Custom Data Path Engines

Custom data path engines are special-purpose processors that are based on application-specific commands. These application-specific commands deliver higher performance for certain types of calculations. However, these commands are less convenient to learn and use effectively for general computing tasks. Most generic processors have standard compilers that apply advanced compiler technologies to optimize code performance. This is possible because the instruction set is simple and generic. Custom data path engines such as DSPs have complicated commands and are difficult to optimize automatically.

Designing of such custom data path engines is highly dependent on the target applications. The reason is obvious. The instruction set is limited. Within the limited set of commands there must be a careful balance between generic type commands and special commands. The architecture must also be carefully tuned to achieve maximum performance.

Video processing operations usually involve lots of parallelism. There are two ways to leverage this parallelism for high-performance computing engines. One is through parallel execution, and the other is through pipeline execution. In parallel execution independent tasks are assigned to separate processing units that operate in parallel. In pipeline execution the independent tasks are subdivided and then are processed as in an assembly line in an industrial plant; there are several tasks being processed simultaneously but all at different stages. In reality, the implementation could be combinations of parallel and pipeline executions on a single high-performance engine or on multiple computing engines. For example, one way to get high performance on a single computing engine is through the very-long-instruction-word (VLIW) architecture. In a VLIW architecture, the highly pipelined computing engine can execute complicated commands that yield high parallelism. For implementations with multiple computing engines, the single-instruction-multiple-data (SIMD) architecture and the multiple-instruction–multiple-data (MIMD) architecture are two popular choices. In the SIMD architecture the computing engines perform the same instruction on their local data synchronously. Such synchronous execution provides opportunities for hardware sharing and thus save implementation costs. However, the form of parallelism is quite rigid. In contrast, the MIMD architecture allows different computing engines to execute independently. This provides more flexibility at the expense of additional overhead for controlling these independent engines.

To date, there are many special-purpose engines that are SIMD and VLIW based. The parallelism within these engines is at the instruction or sub-instruction levels. The MIMD architecture is more popular with coarse-grain parallelism, although there are certain fine-grain implementations as well. For example, consider the Texas Instrument's MVP processor as illustrated in Fig. 16.4.[3] It contains multiple special-purpose DSP cores and memory banks connected through a crossbar switch. It also contains a master processor and other supporting units such as video and transfer control. The MIMD architecture is suitable for executing different coarse-grain tasks on different DSPs. For example, one DSP engine could be computing motion estimation while another is computing the DCT routine. Or all the engines could be computing DCT routines for different blocks independently. In contrast, a video signal processor by Philips[4] employs a MIMD architecture with a finer grain of parallelism. As illustrated in Fig. 16.5, the processor consists of arithmetic units, memory units, branch units, and output units interconnected with a crossbar switch. These

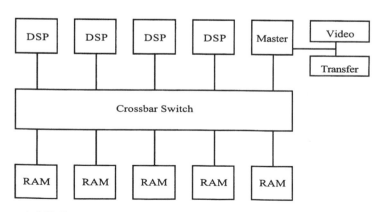

Figure 16.4   Coarse-grain MIMD architecture example.

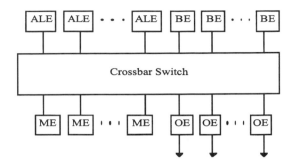

Figure 16.5 Fine-grain MIMD architecture example.

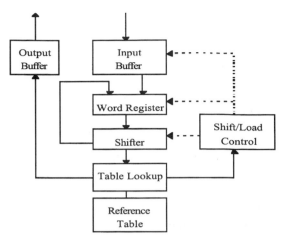

Figure 16.6 Example variable length decoder.

units are fully independent but they need to be carefully synchronized for performance reasons. From the programmer's view, the MIMD parallelism at this level corresponds to a resource scheduling problem. Good software programming tools are quite useful in improving code efficiency.

16.2.3 Application-Specific Processing Engine

Application-specific processing engines are data processing units optimized for specific functions. For example, such engines could be a DCT engine for efficient DCT and IDCT processing, a quantization processor for quantization and inverse quantization, a variable length encoder for encoding Huffman codes, a variable length decoder for decoding Huffman codes, and so forth. Because these engines are specifically tuned for a given function or functions, they are much more efficient in implementing the desired processing.

These application-specific processing engines differ significantly from generic microprocessors and DSPs. There are no instruction cache, instruction decode, or data paths that execute decoded instructions. Instead, the processing itself is captured directly in hardware. For example, consider building a variable length decoder. The functionality of the decoder is to recognize codewords embedded in a stream of bits and convert these codewords into proper data. Because the codeword boundaries are not aligned to fixed positions, the decoder needs to decide how many bits to discard for each decoded codeword. Figure 16.6 illustrates a conventional design. Conceptually the input bitstream is stored in an input buffer with bits aligned

at fixed word length boundaries. The initialization process includes loading data into the word register so that the word register contains at least the first codeword plus some trailing bits. The content of the word register is then compared against a lookup table without any bit shifting to determine the codeword size and the corresponding output value. The size of the codeword is fed back to the shift/load control which keeps a running count of the next leading bit position within the word register. The shifter relies on this running count to discard the decoded bits and to align the new leading bits for looking up the next codeword. If the number of valid bits within the word register drops below a certain threshold, the valid bits are shifted up and loaded in the word register together with new data bits from the input buffer. The running count is adjusted accordingly to reflect the new leading bit position.

An application-specific processing engine like the variable length decoder described above obviously lacks flexibility; it is quite hard to compute DCT on a variable length decoder and vice versa. The architecture style is also quite different; signal and data flow in prescribed directions and functional blocks work synchronously. The result is a highly efficient implementation that cannot be reused for other purposes.

The properties of these processing engines clearly display a range of tradeoffs between efficiency and flexibility. Generic processors are the most flexible; they are designed for general tasks. Custom data path engines are more specialized; they are like generic processors modified for certain sets of tasks. These engines achieve high performance for the tasks they are designed for at the expense of programming. The programmers need to understand the data path engines to write efficient programs. Application-specific processing engines have computing structures matched to the tasks they are designed for. These application-specific engines are compact and highly efficient. There is practically no need for software development since most of these engines have hardwired controllers and sequencers.

16.3   DESIGN ISSUES

Given the basic architecture choices, there are many ways to build a video processing IC for specific applications. Here is a quick revisit to possible applications:

- Consumer TV—broadcast, satellite, and cable TV
- Video on digital storage media—compact disks, tapes
- Networked video—video on ATM, Ethernet, and LANs
- Remote access video—video database, video on demand
- Professional video—editing, postproduction
- Special video—stereoscopic TV, video games, surgery.

The first step is to look at the requirements for these applications and then characterize the processing modules. Next, video processing algorithms need to be developed to completely define each module. The next step is to decide what kind of architecture makes sense. There is no reason to be purists. Mix-and-match often works best in most situations. For example, many video processing ICs have include all three types of engines: generic processors, custom data path engines, and application-specific processing engines. Consider AT&T's AVP-III video codec chip[8] illustrated in Fig. 16.7. The architecture contains multiple functional blocks connected with multiple buses. The main computing engines are at the top part of this figure. The control processor is a generic processor that not only controls the chip but also performs generic processing. The SIMD signal processor is a custom data path engine that can be programmed to perform DCT, IDCT, filtering, interpolation, and other tasks that have high processing regularity. The motion estimator engine, the variable length decoder, and the variable length encoder are application-specific processing engines. These application-specific engines are made relatively programmable so as to accommodate different video standards and processing requirements.

After the architecture has been decided the remaining steps are more mechanical. It is obvious that application-specific engines address processing needs for the tasks they are designed for. The remaining tasks are implemented in software on either generic processors or custom data path

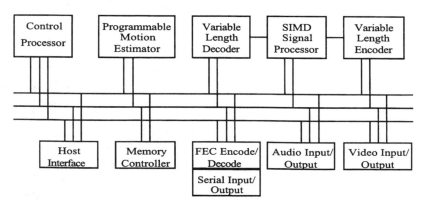

Fig. 16.7   Example architecture.

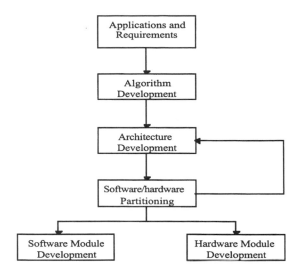

Fig. 16.8 Design flow.

engines. For example, audio processing tasks could be handled by the control processor or the SIMD signal processor in Fig. 16.7. After deciding the mapping between the tasks and hardware modules the designer could go ahead with software and hardware implementations. The rest of the design is nothing but a structured exercise for experienced software/hardware design team.

The design steps for implementing video processors are summarized in Fig. 16.8. Although the design steps appear to be quite straightforward, the design process itself involves many interdependent decisions. Design methodology plays an integral part of the design process; the end result is only as robust as the design methodology itself. A well-planned verification strategy is the only way to guarantee that the software modules and the hardware modules will work together as specified.

16.4 DIRECTIONS IN MPEG CHIP DEVELOPMENT

To date, many MPEG chips and video processors have been developed. Today most implementations are single-chip MPEG solutions as opposed to older chip-set solutions. Some of these video processor chips are described in survey papers.[9,10,15,16] The MPEG-2 encoder application

is probably one of the remaining few applications that still require multichip solutions,[11–13] although single-chip encoders are expected to be available soon. The programmable approaches demonstrated on a single chip include multimedia-enhanced microprocessors,[1] custom high-speed single-processor design,[2] crossbar-connected multiple DSPs,[3] crossbar-connected multiple programmable PEs,[4] and the processor–coprocessor pair design.[5,14] The performance achieved on a single chip is typically around a few Gops. Although the processing power is sufficient to deliver some multimedia services, the cost is still too high. Another more cost-effective approach is to use heterogeneous programmable engines mixed with application-specific processing engines, such as in Refs. 7 and 8. Data path engines with SIMD configuration are popular within this type of programmable solution.

Video processing system designs show a trend toward higher integration and flexibility. For consumer products such as digital video playback systems, a highly integrated IC helps reduce costs and sizes. Such an IC likely will subsume additional functions like audio processing, system-level processing, and other processing of digital media such as graphics or text if appropriate. On the other hand, given the diversity of MPEG applications, it is hard to imagine that a single video processing IC could address all these applications. With reasonable flexibility a video processing IC could be used in many applications. High integration and flexibility is common to many recent video IC developments, and the trend is likely to continue.

16.5 SUMMARY

In this chapter, we have reviewed various implementation choices for video processing. The discussion certainly applies to the MPEG-1/2 standards as well as other video standards. We started with a review of the MPEG processing requirements. Various architectural choices were then discussed. Next, we reviewed the procedure of putting together a complete design. Finally, we indicated directions in MPEG chip development in ptogress. The key points are listed as follows:

- MPEG processing could be subdivided into various processing modules.
- Data structures as well as functional processing need to be considered jointly.
- For a processing module there are three possible implementations: generic processor, custom data path engine, or application-specific processing engine.
- Generic processors are the most flexible but usually the least efficient.
- Application-specific engines are extremely efficient but not flexible.
- The design flow includes identifying the application requirements, developing algorithms, selecting architectures, and implementing software/hardware modules.
- MPEG chip development shows a trend toward single-chip solutions.

REFERENCES

1. A. CHAMAS et al., "A 64b Microprocessor with Multimedia Support," *ISSCC Digest of Technical Papers,* pp. 178–179, San Francisco (February 15–17, 1995).

2. T. INOUE et al., "A 300MHz 16b BiCMOS Video Signal Processor," *ISSCC Digest of Technical Papers,* pp. 36–37, San Francisco (February 24–26, 1993).

3. K. BALMER et al., "A Single Chip Multimedia Video Processor," *Proc. CICC,* San Diego, pp. 91–94 (May 1–4, 1994).

4. H. VEENDRICK et al., "A 1.5 GIPS Video Signal Processor (VSP)," *Proc. CICC,* San Diego, pp. 95–98 (May 1–4, 1994).

5. K. HERRMANN et al., "Architecture and VLSI Implementation of a RISC Core for a Monolithic Video Signal Processor," in *VLSI Signal Processing, VII,* pp. 368–377, IEEE, New York (1994).

6. S. C. PURCELL and D. GALBI, "C-Cube MPEG Video Processor," *SPIE, Image Processing and Interchange,* Vol. 1659 (1992).

7. INTEGRATED INFORMATION TECHNOLOGY, Video Communications Processor VCP, Preliminary Information, 1995.

8. D. BRINTHAUPT et al., "A Programmable Audio/Video Processor for H.320, H.324, and MPEG," *ISSCC Digest of Technical Papers,* pp. 244–245, San Francisco (February 8–10, 1996).

9. P. PIRSCH et al., "VLSI Architectures for Video Compression—A Survey," *Proc. IEEE,* 83 (2) (February 1995).

10. I. TAMITANI et al., "LSI for Audio and Video MPEG Standards," *NEC Res. Dev.* 35 (4) (October 1994).

11. T. MATSUMURA et al., "A Chip Set Architecture for Programmable Real-Time MPEG2 Video Encoder," *Proc. CICC,* pp. 393–396, Santa Clara (May 1–4, 1995).

12. J. ARMER et al., "A Chip Set for MPEG-2 Video Encoding," *Proc. CICC,* pp. 401–404, Santa Clara (May 1–4, 1995).

13. C-CUBE MICROSYSTEMS, VideoRISC Processor Product Information, 1995.

14. J. OTTERSTEDT et al., "A 16cm^2 Monolithic Multiprocessor System Integrating 9 Video Signal-Processing Elements," *ISSCC Digest of Technical Papers,* pp. 306–307, San Francisco (February 8–10, 1996).

15. B. ACKLAND, "The Role of VLSI in Multimedia," *IEEE J. Solid State Circuits* 29 (4) (April 1994).

16. H. D. LIN, "Multimedia Processing and Transmission in VLSI," in *Microsystems Technology for Multimedia Applications: An Introduction,* IEEE Press, New York (1995).

17

MPEG-4 and the Future

The MPEG-1 and the MPEG-2 standards are two remarkable milestones, the overwhelming impact of which, in defining the next generation of digital multimedia, products, equipment, and services is already being felt. MPEG-1 decoders/players are becoming commonplace for multimedia on computers. MPEG-1 decoder plug-in hardware boards have been around for a few years, and now, software MPEG decoders are already available with release of new operating systems or multimedia extensions for PC and Mac™ platforms. MPEG-2 is well on its way to making a significant impact in a range of applications such as digital VCRs, digital satellite TV, digital cable TV, HDTV, and others. So, you may wonder, what is next?

As mentioned earlier, MPEG-1 was optimized for typical applications using noninterlaced video of 30 (25, in European format) frames/s at bitrates in the range of 1.2 to 1.5 Mbits/s, although it can certainly be used at higher bitrates and resolutions. MPEG-2 is more generic in the sense of picture resolutions and bitrates although it is optimized mainly for interlaced video of TV quality in a bitrate range of 4 to 9 Mbits/s; it also includes MPEG-1-like noninterlaced video coding, and also supports scalable coding. Getting back to the question of what's next, a possible reply is in the form of yet another question: What about coding of video at lower (than 1 Mbit/s) bitrates?

Next, yet another question arises: Is it possible

for a standard to be even more generic than MPEG-2 and yet be efficient, flexible, and extendable in the future? If so, can it also be compatible in some way with the previous standards such as MPEG-1, MPEG-2, or other related standards. Further, can it also efficiently code or represent multiviewpoint scenes, graphics, and synthetic models, and allow functions such as interactivity and manipulation of scene contents? Indeed, we seem to have very high expectations from a new MPEG standard! To what extent this wish list may eventually be fullfilled is difficult to predict; however, it appears that ongoing work in MPEG-4 (the next MPEG standard) certainly seems to be addressing these issues. We will discuss this more a little later; for now we explore the relationship of MPEG-4 with other low-bitrate video standards.

During the early 1980s it was believed that for videoconferencing applications, reasonable quality was possible only at 384 kbits/s or higher and good quality was possible only at significantly higher bitrates, around 1 Mbit/s. Around 1984, the ITU-T (formerly CCITT) started an effort to standardize video coding primarily for videophone and video conferencing applications. This effort, although originally directed at video coding at 384 kbits/s and multiples of 384 kbits/s to about 2 Mbits/s, was refocused after a few years to address bitrates of 64 kbits/s and multiples of 64 kbits/s ($p \times 64$ kbits, where p can take values from 1 to

30). That standard was completed in late 1988 and is officially called the H.261 standard (the coding method is often referred to as $p \times 64$.). The H.261 standard, like the MPEG-1 and the MPEG-2 standards, is a decoder standard. In fact, the basic framework of the H.261 standard was used as a starting point in design of the MPEG-1 standard.

A few years ago, owing to a new generation of modems allowing bitrates of 28 kbits/s or so over PSTN, a new possibility of acceptable quality videophones at these bitrates arose. In early 1994, ITU-T launched a short-term effort to optimize and refine H.261 for videophone applications over PSTN; this effort is nearly complete and has resulted in a standard, called H.263.[5] Although, in this standard, video coding is optimized for the 10 to 24 kbits/s range, as is usual, the same coding methods have also been found to be quite efficient over a wider range of bitrates. Within ITU-T, effort is also in progress to come up with a version of this standard with more error resilience that would be robust for mobile applications as well. The ITU-T also has a plan for a long-term standard by 1998, capable of providing even higher coding efficiency; joint work is in progress with MPEG and it is expected that a subset of an ongoing MPEG-4 standard will satisfy this goal.

Besides videophone applications, at very low to low bitrates, several new applications have recently arisen. The requirements of such applications appear to be fairly diverse and not satisfactorily met by the H.263, the MPEG-1, the MPEG-2, or any other standards. To address such potential applications, in 1993, ISO started work toward the MPEG-4 standard, which is expected to reach a mature stage of Committee Draft (CD) by November 1997 and the final stage of International Standard (IS) by November 1998. Much of this chapter is an evaluation of the current status of MPEG-4 and what is foreseen beyond MPEG-4. However, before discussing MPEG-4, it is useful to have a good idea of the coding methods and the syntax of H.263, since H.263 has been found to offer a good starting basis for MPEG-4 (history repeats itself!). MPEG-4, as we shall see, is not only about increased coding efficiency but also about content-based functionalities, higher error resilience, and much much more.

17.1 H.263: THE ITU-T LOW-BITRATE STANDARD

In general, H.263, since it is derived from H.261 is based on the framework of block motion-compensated DCT coding. Both the H.261 and the H.263 standards, like the MPEG-1 and the MPEG-2 standards, specify bitstream syntax and decoding semantics. However, unlike the MPEG-1 and the MPEG-2 standards, these standards are video coding standards only and thus do not specify audio coding or systems multiplex, which can be chosen from a related family of ITU-T standards to develop applications requiring full systems for audio–visual coding. Also, unlike the MPEG standards, the ITU-T standards are primarily intended for conversational applications (low bitrates and low delay) and thus usually do not include coding tools needed for fully supporting applications requiring interactivity with stored data.

The H.263 standard[5] specifies decoding with the assumption of block motion-compensated DCT structure for encoding. This is similar to H.261, which also assumes a block motion-compensated DCT structure for encoding. There are, however, some significant differences in the H.263 decoding process as compared to the H.261 decoding process, which allow encoding to be performed with higher coding efficiency. For developing the H.263 standard, an encoding specification called the Test Model (TMN) was used for optimizations. TMNs progressed through various iterative refinements; the final test model was referred to as TMN5. Earlier, during the 1980s, when H.261 was developed, a similar encoder specification referred to as the Reference Model (RM) was used and went through iterative refinement leading to the final reference model, called RM8. The H.263 standard, although it is based on the H.261 standard, supports a structure that is significantly optimized for coding at lower bitrates such as several tens of kbit/s while maintaining good subjective picture quality at higher bitrates as well. We now discuss the syntax and semantics of the H.263 standard as well as how TMN5 encoding works.

17.1.1 Overview

The H.263 standard,[5] although intended for low bitrate video coding, does not specify a constraint on video bitrate; such constraints are given by the terminal or the network. The video coder generates a self-contained bitstream. The decoder performs the reverse operation. The codec can be used for either bidirectional or unidirectional visual communication.

The video coding algorithm is based on H.261 with refinements/modifications to enhance coding efficiency. Four negotiable options are supported to allow improved performance. The video coding algorithm, as in H.261, is designed to exploit both the spatial and temporal redundancies. The temporal redundancies are exploited by interpicture prediction, whereas the spatial redundancies are exploited by DCT coding. Again as in H.261, although the decoder supports motion compensation, the encoder may or may not use it. One difference with respect to H.261 is that instead of full-pixel motion compensation and loop filter, H.263 supports half-pixel motion compensation (as per MPEG-1/2), providing improved prediction. Another difference is in the Group-of-Block (GOB) structure, the header for which is now optional. Further, to allow improved performance, H.263 also supports four negotiable options which can be used together or separately, and are listed as follows:

- Unrestricted Motion Vector mode—This mode allows motion vectors to point outside a picture, with edge pixels used for prediction of nonexisting pixels.

- Syntax-based Arithmetic Coding mode—This mode allows use of arithmetic coding instead of variable length (huffman) coding.

- Advanced Prediction mode—This mode allows use of overlapped block motion compensation (OBMC) with four 8×8 motion vectors instead of 16×16 vectors per macroblock.

- PB-frames mode—In this mode, two pictures, one, a P-picture and the other a B-picture, are coded together as a single PB-picture unit.

17.1.1.1 Source Format

The video coder operates on noninterlaced pictures occurring 29.97 times per second (that is,

30000/1001, the same picture rate as NTSC). Pictures are represented by luminance, Y, and the two color difference signals (chrominance signals, Cb and Cr). Five picture formats are standardized and are referred to as sub-QCIF, QCIF, CIF, 4CIF, and 16CIF; the corresponding resolution of luminance signal for these formats is 128×96, 176×144, 352×288, 704×576, or 1408×1152. All H.263 decoders are mandated to operate using sub-QCIF and QCIF. Some decoders may also operate with CIF, 4CIF, or 16CIF. The information about which formats can be handled by decoders are signalled to encoder by external means. The two chrominance signals (Cb and Cr) have half the resolution of luminance in each direction. The chrominance samples are located at the center, equidistant from the four neighboring luminance samples.

17.1.1.2 Data Structures

Each picture is divided into GOBs. In H.263, a GOB consists of a row of $k \times 16$ lines with $k = 1$ for sub-QCIF, QCIF, and CIF, $k = 2$ for 4CIF and $k = 4$ for 16CIF; thus there are 6 GOBs for sub-QCIF, 9 for QCIF, and 18 for CIF, 4CIF, and 16CIF. Data for each GOB consists of a GOB header (which may be empty) followed by data for each of the macroblocks contained in a GOB. A macroblock in H.263 is the same as that in MPEG-1/2, that is, it consists of a 16×16 block of Y, and corresponding 8×8 blocks of each of the two chrominance components.

17.1.1.3 Coding and Decoding Basics

- Prediction—Motion-compensated interpicture prediction (as in H.261 and MPEG-1/2) is supported for high efficiency, although the use of motion compensation is optional at the encoder. If no prediction is applied, the coding mode is said to be Intra; otherwise it is Inter. There are two basic picture types, I-pictures and P-pictures. In I-pictures, macroblocks are coded in Intra mode only whereas in P-pictures, both Intra and Inter modes may be used as indicated by macroblock type. The optional PB-frames are always coded in the Inter mode. B-pictures are partly coded bidirectionally.

- Motion Compensation—The decoder normally accepts one motion vector per macroblock, or either one or four motion vectors per macroblock when Advanced Prediction is used. If the PB-frame mode is used, one additional delta motion vector is transmitted per macroblock of B-pictures and is used for refinement of motion vectors derived by scaling of motion vectors of the corresponding macroblock in P-pictures.

- Quantization—The number of quantizers is 1 for the first coefficient (DC coefficient) of Intra blocks and is 31 for all other coefficients of an Intra block as well as all coefficients in an Inter block. All coefficients except for the first coefficient use equally spaced reconstruction levels with a central dead zone around zero.

- Coding Control—A number of parameters can be varied to control the rate of generation of coded video data. These include preprocessing, quantizer selection, block coding mode selection, and temporal subsampling and are encoding issues and thus not standardized.

17.1.2 Syntax and Semantics

The video bitstream is arranged in a hierarchical structure composed of the following layers:

- Picture Layer
- Group of Blocks Layer
- Macroblock Layer
- Block Layer.

The syntax diagrams of each of these layers is shown in Fig. 17.1.

For each of the layers, the abbreviations used, their meaning, and semantics are discussed next.

17.1.2.1 Picture Layer

This represents the highest layer in bitstream; the structure of this layer is shown in Fig. 17.2. Each coded picture consists of a picture header followed by coded picture data, arranged as Group of Blocks (GOBs). After all pictures belonging to a sequence are sent, an End-of-Sequence (EOS)

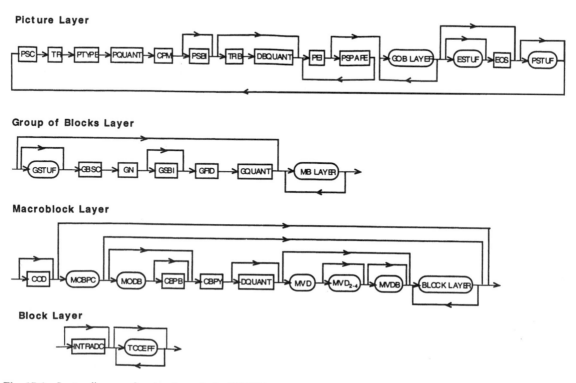

Fig. 17.1 Syntax diagram of various layers in the H.263 bitstream.

Fig. 17.2   Structure of Picture layer in the H.263 bitstream.

code and stuffing bits (ESTUF) is sent. The presence of some elements is conditional. For instance, if the Continuous Prescence Multipoint (CPM) indicates it, a Picture Sub-bitstream Indicator (PSBI) is sent. Likewise, if picture type (PTYPE) indicates a "PB-frame" Temporal Reference for B-pictures (TRB) and a B-picture Quantizer (DBQUANT) is sent. Combinations of Picture Extra Insertion Information (PEI) and Picture Spare Information (PSPARE) may not be present. EOS may not be present, while ESTUF may be present only if EOS is present. Further, if during the coding process some pictures are skipped, their picture headers are not sent. Some of the syntax elements in this layer are represented by variable length codes while others are of fixed length as shown in the figure.

The Picture Start Code (PSC) is 22 bits in length represented by 0000 0000 0000 0000 1 00000. All PSCs are byte aligned by using PSTUF bits before PSC.

Temporal Reference (TR) is 8 bits allowing 256 possible values. To generate TR for the current picture the TR of the previously transmitted picture is incremented by the number of nontransmitted pictures. When the optional PB-frame mode is employed, TR refers to P-pictures only, the TR of B-pictures is represented by TRB, a 3-bit code that represents the number of nontransmitted pictures since the last P- or I-picture and before the B-picture.

Picture type, PTYPE, is a 13-bit code, with bits 6 to 8 representing source format ("001" sub-QCIF, "010" QCIF, "011" CIF, "100" 4CIF, "101" 16CIF); bit 9 representing picture coding type ("0" I-picture, "1" P-picture); and bits 10, 11, 12, and 13 identifying the presence or absence of optional modes Unrestricted Motion Vector, Syn-

tax-Based Arithmetic Coding, Advanced Prediction Mode, and PB-frames. Other bits in PTYPE are as follows: bit 1 is always "1" to avoid start code emulation; bit 2 is always "0" for distinction with H.261; bits 3, 4, and 5 respectively correspond to split screen indication, document camera indication, and freeze picture release indication.

The Picture Quantizer (PQUANT) is a 5-bit code that represents quantizer values in the range of 1 to 31. DBQUANT is a 2-bit code representing BQUANT ("00" 5 × QUANT/4, "01" 6 × QUANT/4, "10" 7 × QUANT/4, "11" 8 × QUANT/4) such that QUANT is used in the P-block and BQUANT in the corresponding B-block.

17.1.2.2 Group of Blocks Layer

Data for the GOB layer consists of a GOB header followed by data for macroblocks. The structure of the GOB layer is shown in Fig. 17.3. For the first GOB (GOB number 0) in each picture, the GOB header is not transmitted, whereas for other GOBs, depending on the encoder decision, the GOB header may be empty. A decoder, if it so desires, can signal using an external means, the encoder, to send nonempty GOB headers. Group stuffing (GSTUF) may be present when Group of Blocks Start Code (GBSC) is present. Group Number (GN), GOB Frame ID (GFID), and GOB quantizer (GQUANT) are present when GSBSC is present. Further, the GOB Sub-bitstream indicator (GSBI) is present if the Continuous Presence mode is indicated by CPM.

GBSC is a 17-bit code with value 0000 0000 0000 0000 1 and may be byte aligned by using GSTUF. The group number is a 5-bit code identifying the GOB number, which can take values from 0 to 17, while values from 18 to 30 are

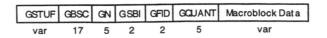

Fig. 17.3   Structure of Group-of-Blocks layer in H.263.

reserved for the future and a value of 31 indicates the end of sequence (EOS). The GOB number 0 is used in the PSC and for this value GOB header including GSTUF, GBSC, GN, GSBI, GFID, and GQUANT is empty. The length of GSBI and GFID codes is 2 bits each whereas for GQUANT it is 5 bits.

17.1.2.3 Macroblock Layer

Data for each macroblock consists of a macroblock header followed by data for blocks. The structure of the Macroblock layer is shown in Fig. 17.4. A coded macroblock indication (COD) is present for each macroblock in P-pictures. A macroblock type and Coded block pattern for chrominance (MCBPC) is present when indicated by COD or when PTYPE is an I-picture. A macroblock mode for B-pictures (MODB) is present for NonIntra MB types if PTYPE indicates a PB-frame. A coded block pattern for luminance (CBPY), codes for differential quantizer (DQUANT), motion vector data MVD, and MVD_{2-4} are present when indicated by MCBPC. The coded block pattern for B-blocks (CBPB) and motion vector data for the B-macroblock (MVDB) are present only if indicated by MODB. Block data are present when indicated by MCBPC and CBPY. MVD_{2-4} is present only in the Advanced Prediction mode. MODB, CBPB, and MVDB are present only in the PB-frame mode.

COD is a 1-bit codeword; its value when "0" indicates that a macroblock is coded and when "1" indicates that no further information is sent for that macroblock. A COD of "1" thus implies an Inter macroblock with zero motion vector and no DCT coefficient data. COD is present for each macroblock in pictures where PTYPE indicates a P-picture.

MCBPC is a variable length codeword that jointly represents macroblock type and coded block pattern for chrominance; it is always included for coded macroblocks. Macroblock type provides information about the coding mode and

which data elements are present. The coded block pattern for chrominance (CBPC) indicates whether at least one NonIntraDC DCT coefficient is transmitted (IntraDC is dc coefficient for Intra blocks). The first and second bits of CBPC indicate the status of each of the chrominance blocks in a macroblock; either of the bits when set to "1" indicates the presence of NonIntraDC coefficients for that block. Owing to different macroblock types in I- and P-pictures, separate VLC tables representing MCBPC for these pictures are used. For I-pictures, macroblock types allowed are Intra and Intra + Q (Intra with quantizer update). For P-pictures the macroblock types are Inter, Inter + Q, Inter4V, Intra, and Intra + Q. The Inter4V mode allows use of four motion vectors per macroblock when the Advanced prediction option is employed.

For PB-frames, the macroblock types allowed are the same as that for P-pictures; however, for these frames, MODB is present for each macroblock and indicates if CBPB or MVDB is also present. CBPB is a 6-bit code with each bit signalling the status of a B-block in the macroblock; if a corresponding bit is set to "1" it indicates the presence of DCT coefficients for that block.

CBPY is a variable length code that represents a pattern of Y blocks in the macroblock for which at least one NonIntraDC DCT coefficient is transmitted.

DQUANT is a 2-bit code indicating four possible changes in values of quantizers, -1, -2, 1, and 2, over the value of QUANT which ranges from 1 to 31. After addition of differential values the values are clipped to lie in the range of 1 to 31.

Motion vector data (MVD) are included for all Inter macroblocks and consist of a variable length code for the horizontal component followed by a variable length code for the vertical component. Three additional codewords, MVD_{2-4}, are also included if indicated by PTYPE and by MCBPC and each consists of a variable length code for the horizontal component followed by a variable length code for the vertical component. MVD_{2-4}

COD	MCBPC	MODB	CBPB	CBPY	DQUANT	MVD	MVD_2	MVD_3	MVD_4	MVDB	Block Data
1	var	var	6	var	2	var	var	var	var	var	var

Fig. 17.4 Structure of the macroblock layer in H.263.

are present when in the Advanced Prediction mode.

For B-macroblocks, MVDB is present if indicated by MODB and also consists of a variable length code for the horizontal component followed by a variable length code for the vertical component. MVDB is present in PB-frames only if indicated by MODB.

17.1.2.4 Block Layer

In normal mode (not PB-frame mode), a macroblock consists of four luminance blocks and two chrominance blocks. The structure of the Block layer is shown in Fig. 17.5. IntraDC is present for every block of the macroblock if MCBPC indicates the MB type of Intra or Intra + Q. TCOEF is present if indicated by MCBPC or CBPY.

In PB-frames, the macroblock can be thought of as being composed of 12 blocks. First, the data for six P-blocks is transmitted followed by data for the next six B blocks. An IntraDC is sent for every P-block of the macroblock if the MCBPC indicates an MB type of Intra or Intra + Q. IntraDC is not present for B-blocks. TCOEFF is present for P-blocks if indicated by MCBPC or CBPY; TCOEF is present for B-blocks as indicated by CBPB.

When the Syntax-Based Arithmetic Coding option is employed, the block layer is different and is explained later.

IntraDC is coded with 8 bits with values 0000 0000 and 1000 0000 not used. The reconstruction levels of 8, 16, 24, ... 1016 are represented by corresponding codewords 0000 0001, 0000 0010, 0000 0011, ... 0111 1111. The reconstruction levels of 1032, 1040, 1048, ... 2032 are represented by corresponding codewords 1000 0001, 1000 0010, 1000 0011, ... 1111 1110. The reconstruction level of 1024 is signalled by codeword 1111 1111.

The most frequently occurring EVENTs are variable length coded. An EVENT is a combination of a last nonzero coefficient indication (LAST; "0": there are more nonzero coefficients in this block, "1"; this is the last nonzero coefficient in this block), the number of successive zeros preceding the coded coefficient (RUN), and the nonzero value of the coded coefficient (LEVEL). In contrast, the EVENTs in VLC coding in MPEG-1 and MPEG-2 include the (RUN, LEVEL) combination and a separate code to signal end-of-block (EOB), instead of the (LAST, RUN, LEVEL) combination employed by the H.263. The remaining (less frequently occurring EVENTs) are coded by 22-bit codewords comprised of 7 bits of ESCAPE (0000 011), 1 bit of last (1 or a 0), 6 bits of RUN (0000 00 to 1111 11), and 8 bits of LEVEL. LEVELs of $-127, \ldots -1$ are represented by corresponding codewords 1000 0001, ... 1111 1111. LEVELs of $1, \ldots 127$ are represented by corresponding codewords 0000 0001, ... 0111 1111. Further, LEVELs of -128 and 0 are FORBIDDEN.

As a point of comparison, the dynamic range of LEVELs in MPEG-1 is -255 to $+255$ and in MPEG-2 it is $+2047$ to -2047 as compared to H.263 in which the range is only -127 to $+127$.

17.1.3 The Decoding Process

We now describe the decoding process which basically consists of motion compensation, coefficients decoding, and block reconstruction.

17.1.3.1 Motion Compensation

Motion compensation in H.263 may simply use the default prediction mode and in addition may use either one or both optional modes such as the Unrestricted Motion Vector mode or the Advanced Prediction mode. For now, we discuss the default prediction mode and will discuss the various optional modes a little later.

In default mode when coding P-pictures, each Inter macroblock has one motion vector. This vector is obtained by adding predictors to the vector differences indicated by MVD. Predictors for differential coding are taken from three surrounding macroblocks as shown in Fig. 17.6. The predictors are calculated separately for the horizontal and vertical components.

At the boundary of a GOB or picture, predictions are calculated according to a set of decision rules applied in the following sequence. The imme-

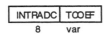

Fig. 17.5 Structure of the Block layer in H.263.

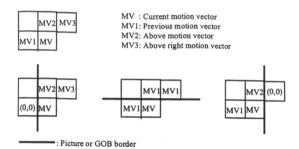

: Picture or GOB border

Fig. 17.6   Motion vector prediction in H.263.

diately previous candidate (left) predictor MV1 is set to zero if the corresponding macroblock is outside the picture. Then, the immediately above candidate (top) predictor MV2, and right of immediately above (top right) predictor MV3 are set to MV1 if the corresponding macroblocks are outside the picture (at the top) or outside the GOB (at the top) if the GOB header of the current GOB is nonempty. Next, the candidate predictor MV3 is set to zero if the corresponding macroblock is outside the picture (at the right side). Finally, if the corresponding macroblock is Intra coded or was not coded (COD = 1), the candidate predictor is set to zero.

Once the values of each of the three predictors MV1, MV2, and MV3 are determined, the actual predictor is obtained by taking median values of the three predictors for the horizontal and vertical components.

The value of each component of the motion vector is constrained to lie in the range of $[-16, +15.5]$. Each VLC word for MVD is used to represent a pair of difference values, only one of which is a valid value falling in the range $[-16, +15.5]$ depending on the value of the predictor.

A positive value of the horizontal or vertical component of the motion vector means that the prediction is being formed from pixels in the previous pictures which are spatially to the right or below the pixels being predicted. The same motion vector is used for all four luminance blocks in a macroblock. Motion vectors for chrominance blocks are obtained by dividing each of the motion vector components by 2, since chrominance resolution is half of that of the luminance resolution in each direction. Since, after division, the chromi-

nance vectors can now in fact be at quarter pixel resolution, they are truncated to the nearest half-pixel position.

Luminance or chrominance intensities at half-pixel position are calculated by bilinear interpolation as in MPEG-1/2.

17.1.3.2   Coefficient Decoding

If LEVEL = "0," the reconstructed value is given by REC = "0." The reconstructed values of IntraDC is simply 8 × LEVEL, except for decoded level of 255, which indicates REC = "1024." The reconstructed values of nonzero coefficients other than IntraDC are calculated as follows:

$$|REC| = (2 \times |LEVEL| + 1) \times QUANT$$
$$\text{when QUANT} = \text{"odd"}$$

$$|REC| = ((2 \times |LEVEL| + 1) \times QUANT) - 1$$
$$\text{when QUANT} = \text{"even"}$$

The coefficient reconstruction process disallows even values at the output to prevent IDCT mismatch errors. The value of REC is calculated from |REC| as follows:

$$REC = \text{sign(LEVEL)} \times |REC|;$$
$$\text{where sign(LEVEL) is given by}$$
$$\text{last bit of TCOEFF code.}$$

After inverse quantization, the reconstructed values of all coefficients other than IntraDC are clipped to lie in the range -2048 to 2047.

The reconstructed DCT coefficients are placed into an 8 × 8 block along the zigzag path, essentially undoing the zigzag scan performed at the encoder.

After inverse quantization and zigzag placement of coefficients, the resulting 8 × 8 block is inverse DCT transformed as in case of MPEG-2, discussed earlier in the book. The output of the inverse transform is 9 bits with a range of -255 to $+255$.

17.1.3.3   Block Reconstruction

After motion compensation and coefficients decoding, luminance and chrominance blocks are reconstructed. If the blocks belong to an Intra macroblock, the inverse transformation already provides the reconstruction. If the macroblock is Inter, the reconstructed blocks are generated by adding prediction on a pixel basis to the result of the inverse transformation. A clipping operation is then performed to ensure that the reconstructed pixels lie in the range of 0 to 255.

17.1.4 Optional Modes

Next, we discuss the optional modes in H.263. As mentioned earlier, there are four such modes and the capability of such modes is signalled by the decoder to the encoder by external means (such as ITU-T Recommendation H.245). The H.263 bitstream is self sufficient and carries information about the optional modes in use via the PTYPE information.

17.1.4.1 Unrestricted Motion Vector Mode

The Unrestricted Motion Vector mode removes the restriction of the default prediction mode which forces all referenced pixels in prediction to be from inside the coded picture area. Thus, in this mode, motion vectors are allowed to point outside of the picture. When a pixel referenced by a motion vector is outside of the coded picture area, an edge pixel is used instead and is found by restricting the motion vector to the last full pixel inside the coded picture area. The restriction of motion vector is performed on a pixel basis and separately for each component of the motion vector.

Thus, for example, if a picture size of QCIF (picture size 176 × 144) is used, and if prediction needs to access a pixel at an x-coordinate larger than 175, an x-coordinate of 175 is used instead; or if prediction needs to access a pixel at x-coordinate of less than 0, then an x-coordinate of 0 is used. For the same example, if prediction needs to access a pixel at a y-coordinate larger than 143, a y-coordinate of 143 is used, or if prediction needs to access a pixel at a y-coordinate of less than 0, then a y-coordinate of 0 is used.

Earlier, we noted that in the default prediction mode, the value of each component of motion vector is limited to $[-16, 15.5]$. This range is clearly insufficient in many cases and thus in Unrestricted Motion Vector mode, this range is extend-

ed to $[-31.5, 31.5]$ with the constraint that only values in the range of $[-16, 15.5]$ around the predictor for each motion vector component can be reached if the predictor is in the range $[-15.5, 16]$. If the predictor is outside this range, all values in the range of $[-31.5, 31.5]$ with the same sign as the predictor and the zero value can be reached.

With the unrestricted Motion Vector mode all motion vector data (MVD, MVD_{2-4}, MVDB) are interpreted as follows. If the predictor for the motion vector component is in the range $[-15.5, 16]$ only the first of the pair of vector differences applies. If the predictor is outside the motion vector component range $[-15.5, 16]$, the vector difference from pairs of vector differences will be used which results in a vector component in range $[-31.5, 31.5]$ with the same sign as the predictor, including zero.

17.1.4.2 Syntax Based Arithmetic Codes

In, MPEG-1, MPEG-2, and the default coding mode of H.263, each symbol is VLC encoded using a specific table based on the syntax of the coder. The table typically stores lengths and values of VLC codewords. The symbol is mapped to an entry by table look-up and the binary codeword specified by the entry is sent to the buffer for transmitting to the receiver. In VLD decoding, the received codewords, depending on the context, are compared against appropriate tables based on the syntax of the coder to determine matching entries and these actual symbols resulting from VLD decoding are an interim step in the video decoding process. Of course, the tables used at the encoder and the decoder must be exactly the same. The VLC/VLD process implies that each symbol must be coded in integer number of bits and removing this restriction can result in increase in coding efficiency; this can be accomplished by arithmetic coding.

In Syntax-Based Arithmetic coding (SAC), all the corresponding variable length coding/decoding operations of H.263 are replaced with arithmetic coding/decoding operations. The capability of this mode is signalled by external means while the use of this mode is indicated by PTYPE.

In the SAC encoder, a symbol is encoded by a specific array of integers (a model) based on the syntax of the coder and by calling a procedure called *encode_a_symbol (index, cumul_ freq)*. A FIFO (first-in–first out) called PSC_FIFO is used for buffering the output bits from the arithmetic encoder. The model is specified through cumul_freq[ ], and the symbol is specified using its index in the model.

In the SAC decoder, a symbol is decoded by using a specific model based on the syntax and by calling a procedure called *decode_a_symbol (cumul_ freq)*. As in the encoder, the model is specified via cumul_freq[ ]. The decoded symbol is returned through its index in the model. A FIFO similar to one at the encoder called PSC_FIFO is used for buffering the incoming bitstream. The decoder is initialized by calling a procedure called *decoder_reset ( )* to start decoding an arithmetic coded bitstream.

As discussed earlier, the H.263 syntax can be thought of as composed of four layers: Picture, Group-of-Blocks, Macroblock, and Block. The syntax of the top three layers, Picture, Group-of-Blocks, and Macroblock, remain exactly the same between variable length coding (default) or SAC mode. The syntax of the fourth layer, the Block layer, is slightly modified as shown in Fig. 17.7.

TCOEF1, TCOEF2, TCOEF3, and TCOEFr are symbols representing (LAST, RUN, LEVEL) EVENTs, and possibly can be the first, second, third, and the rest of the symbols respectively. Thus TCOEF1, TCOEF2, TCOEF3, and TCOEFr are present only when one, two, three, or more coefficients respectively are present in the block layer.

At the encoder or at the decoder, the PSC_FIFO, a FIFO of size >17 bits is used. In PSC_FIFO at the encoder, illegal emulations of

PSC and GBSC are located and are avoided by stuffing a "1" after every successive appearance of 14 "0"s, which are not part of PSC or GBSC). In PSC_FIFO of the decoder, the first "1" after every string of 14 "0"s is deleted; if instead of a string of 14 "0"s is followed by a "0," it indicates legal PSC or GBSC is detected with the exact location of PSC or GBSC determined by the next "1" following the string of zeros.

Fixed Length Symbols form three possible strings PSC–TR–PTYPE–PQUANT–CPM–(PSBI)–(TRB –DSQUANT)–PEI–(PSPARE–PEI–. . .), (GSTUF)–GBSC–GN–(GSBI)–GFID–GQUANT, and (ESTUF)–(EOS)–(PSTUF). These strings are directly sent to PSC_FIFO as in the normal VLC mode of the H.263 encoder, and are directly sent out from PSC_FIFO in the decoder after a legal PSC or GBSC or EOS is detected. If a fixed length string is not the first in a video session, the arithmetic encoder is reset before sending the fixed length string by calling a procedure called *encoder_ flush ( )*. This procedure is also called at the end of video session or before EOS. At the decoder side, after each fixed length symbol string, a procedure *decoder_reset ( )* is called.

Nonfixed Length Symbols employ precomputed models with a separate model for each symbol type. All models are stored as a one-dimensional array whose values are accessed by using the same indices as used in normal VLC coding. The models for COD and MCBPC in P-pictures are cumf_COD [ ] and cumf_MCBPC[ ]. The model for MCBPC in I-pictures is cumf_MCBPC_ intra[ ]. The model for MODB is cumf_MODB[ ]. For CBPB, two models are used, one for luminance and the other for chrominance and are respectively

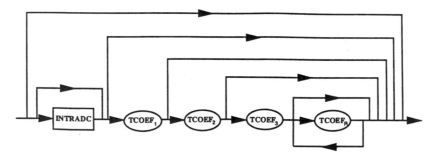

Fig. 17.7 Structure of SAC block layer in H.263.

called cumf_YCBPB [ ] and cumf_UVCBPB [ ]. The model for CBPY is cumf_CBPY [ ] in Inter macroblocks and cumf_CBPY_intra [ ] in Intra macroblocks. The model for DQUANT is cumf_DQUANT[ ]. For all MVDs, a single model, cumf_MVD[ ], is used. The model for INTRADC is cumf_INTRADC[ ]. The models for TCOEF1, TCOEF2, TCOEF3, and TCOEFr in Inter blocks are cumf_TCOEF1[ ], cumf_TCOEF2[ ], cumf_TCOEF3[ ], and cumf_TCOEFr[ ]; the corresponding models in Inter blocks are cumf_TCOEF1_intra[ ], cumf_TCOEF2_intra[ ], cumf_TCOEF3_intra[ ], and cumf_TCOEFr_intra[ ]. A separate model is used for coefficient sign and is cumf_SIGN[ ]. Finally, the models for LAST, RUN, LEVEL after ESCAPE in Inter blocks are cumf_LAST[ ], cumf_RUN [ ] and cumf_LEVEL [ ] and the corresponding models in Intra blocks are cumf_LAST_intra[ ], cumf_RUN_intra[ ], and cumf_LEVEL_intra[ ].

For details of procedures *encode_a_symbol (index, cumul_freq), decode_a_symbol (cumul_freq), encoder_flush ( ), decoder_reset ( )*, and the actual models used consult the H.263 standard.

17.1.4.3 Advanced Prediction Mode

H.263 allows yet another option to allow improved prediction; this option, like other options of H.263, is signalled through external means and when used is identified in the H.263 bitstream in the PTYPE information. This option is called the Advanced Prediction Mode and includes overlapped block motion compensation and the possibility of using four motion vectors per macroblock. In this mode, like in the Unrestricted Motion Vector Prediction mode, motion vectors can refer to an area outside of the picture for prediction. However, the extended motion vector range feature of the Unrestricted Motion Vector mode is active only if that mode is explicitly selected. If the Advanced Prediction mode is used in combination with the PB-frame mode, overlapped motion compensation is used only for prediction of P-pictures and not for B-pictures.

In the default mode of H.263, one motion vector is used per Inter macroblock in P-pictures. However, when the Advanced Prediction Mode is used, either one or four motion vectors can be used

as indicated by the MCBPC codeword. If information about only one motion vector MVD is transmitted, this is interpreted as four vectors each with the same value. If MCBPC indicates that four motion vectors are transmitted, the first motion vector is transmitted as codeword MVD and information about three additional vectors is transmitted as MVD_{2-4}. The vectors are obtained by adding predictors to vector differences MVD and MVD_{2-4} in a similar manner when only one motion vector is present with the caveat that candidate predictors are defined for each 8×8 block of a macroblock as shown in Fig 17.8.

The motion vector for chrominance is computed for both chrominance blocks by the sum of the four luminance vectors and dividing the sum by 8, separately for the horizontal and the vertical components. This division can result in chrominance vectors to $^1/_{16}$ pixel accuracy and have to be rounded to the nearest half-pixel position. The luminance or chrominance values at half-pixel position are calculated by bilinear interpolation.

We now discuss *Overlapped Motion Compensation*, a very important feature of the Advanced Prediction mode. In overlapped motion compensation, each pixel in an 8×8 luminance prediction block is computed as a weighted sum of three predictions. The three predictions are calculated by using three motion vectors: a vector of current luminance block and two out of four "remote" motion vectors. The "remote" vector candidates are the motion vector of the block to the left or the right side of the current luminance block, and the

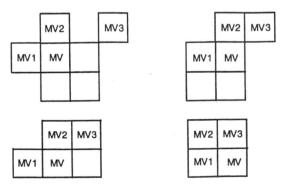

Fig. 17.8 Candidate predictors MV1, MV2, MV3 for each luminance block of a macroblock in H.263.

motion vector of the block above or below the current luminance block. Remote motion vectors from other GOBs are used in the same way as the motion vector within a GOB. For each pixel, remote motion vectors of the block corresponding to the two nearest block borders are used. Thus, for the upper half of the block, the motion vector corresponding to the block above the current block is used, while for the lower half of the block, the motion vector corresponding to the block below the current block is used. The weighting values used in conjunction with this type of prediction with motion vectors of luminance blocks are shown in Fig. 17.9b. Similarly, for the left half of the block the motion vector corresponding to the block on the left side of the current block is used, while for right half of the block the motion vector corresponding to the block on the right side of the current block is used. Again, the weighting values used in conjunction with this type of prediction with motion vectors of luminance blocks are shown in Fig. 17.9c.

Each pixel $p(i, j)$ in the 8×8 luminance prediction block is calculated as follows:

$$p(i, j) = (q(i, j) \times H_0(i, j) + r(i, j) \times H_1(i, j) + s(i, j) \times H_2(i, j) + 4)/8$$

where $q(i, j)$, $r(i, j)$, and $s(i, j)$ are pixels from referenced picture obtained using motion vectors MV^0, MV^1, and MV^2, such that MV^0 represents the motion vector of the current block, MV^1 represents the motion vector of the block either above or below, and MV^2 represents the motion vector either to the left or to the right of the current block. As always, each MV is in fact a pair (MV_x, MV_y) representing horizontal and vertical components.

If some motion vectors are not available such as when a surrounding block is not coded or it is coded as Intra, the corresponding remote motion vector is replaced by the motion vector of the current block (except in the PB-frame mode). If the scurrent block is at the border of a picture and a surrounding block is not present, the corresponding remote motion vector is replaced by the current motion vector.

The weighting values in terms of matrices $H_0(i, j)$, $H_1(i, j)$, and $H_2(i, j)$ are shown in Figure 17.9a, b, and c, respectively.

17.1.4.4 PB-Frames Mode

The last optional mode in H.263 is called the PB-frames mode. Like other optional modes discussed thus far, its capability is signalled by external means and when this mode is enabled it is indicated by PTYPE information as discussed earlier.

A PB-frame consists of a pair of pictures coded together as a single unit. A PB-frame consist of a normal P-picture coded with prediction from a previously decoded reference picture which may be an I- or a P-picture, and a B-picture coded with bidirectional prediction from a decoded past reference picture and a decoded future reference picture. In fact, the B-picture in the PB-frame mode of H.263 is a simplified and overhead reduced version of MPEG-1 (or MPEG-2) B-pictures. The prediction process of H.263 B-pictures is illustrated in Fig. 17.10.

When using PB-frames, the interpretation of Intra macroblocks is revised to mean P-blocks are Intra coded and B-blocks are Inter coded with prediction as for an Inter block. For Intra macroblocks in P-pictures, motion vectors are included for use by corresponding B-blocks in B-pictures of PB-

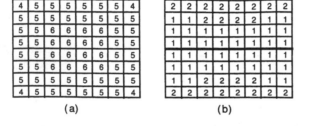

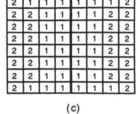

(a)   (b)   (c)

Fig. 17.9 Weighting values (a) H0, (b) H1, (c) H2, used for overlapped motion compensation in H.263.

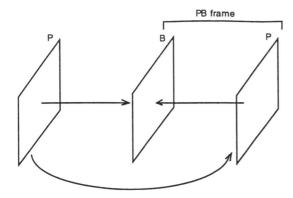

Fig. 17.10 Prediction in PB-frame mode of H.263.

frames. Further, when using the Advanced Prediction mode and the PB-frame mode, if one of the surrounding blocks is coded Intra, the corresponding remote motion vector is not replaced by the current block motion vector but uses the remote Intra motion vector instead.

As noted earlier, in the PB-frame mode a macroblock is considered to consist of 12 blocks. First the data for six P-blocks are transmitted, next followed by data for six B-blocks. If the macroblock is Intra in the P-picture, IntraDC is present for each P-block; however, IntraDC is not present for B-blocks. TCOEF is present for P-blocks if indicated by MCBPC or CBPY; TCOEF is present for B-blocks if indicated by CBPB.

The vectors for a B-picture are calculated using motion vectors available for the corresponding P-block in the P-picture. Assume the motion vector of a P-block is MV; then we calculate a forward motion vector MV_F and a backward motion vector MV_B from MV and enhance it by a delta motion vector given by MVDB. Further, let TR_D be the increment in temporal reference TR from the last picture header (if TR_D is negative, $TR_D = TR_D + 256$), and TR_B be the temporal reference for the B-picture and represents the number of non-transmitted pictures since the last I-or P-picture, just before the B-picture. Let MV_D be the delta motion vector for a B-block derived from MVDB and corresponds to the vector MV of a P-block. If MVDB is not present, MV_D is set to zero; otherwise, the same MV_D given by MVDB is used for each of the four luminance B-blocks in a macroblock.

The forward vector MV_F and the backward vector MV_B in half-pixel units are calculated as follows:

$$MV_F = (TR_B \times MV)/TR_D + MV_D$$

$$MV_B = ((TR_B - TR_D) \times MV)/TR_D$$
$$\text{if } MV_D \text{ is not equal to } 0$$

$$MV_B = MV_F - MV$$
$$\text{if } MV_D \text{ is not equal to } 0$$

Similar to that for MV, the range of values for MV_F is constrained. Also, a VLC word for MVDB represents a pair of difference values with only one value of MV_F falling in the default range of $[-16, 15.5]$ or $[-31.5, 31.5]$ in Unrestricted Motion Vector mode. For chrominance blocks, MV_F is calculated by sum of four luminance MV_F vectors, dividing the sum by 8, and truncating to nearest half-pixel; the MV is calculated similarly using four luminance MV_B vectors.

To calculate prediction for an 8×8 B-block in a PB-frame, first the forward and the backward vectors are calculated. It is assumed that the P-macroblock is first decoded and the reconstructed value P_{REC} is calculated. The prediction for B-block is calculated next and consists of two parts. First, for pixels where the backward vector, MV, points to inside P_{REC}, use bidirectional prediction which is computed as average of forward prediction obtained by MV_F and backward prediction obtained by MV_B. For all other pixels, forward prediction using MV_F relative to the previous decoded picture is used. Figure 17.11 illustrates the part of the block that is predicted bidirectionally and the remaining part that is predicted by forward prediction.

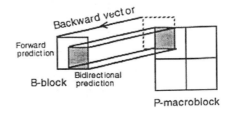

Fig. 17.11 Forward and bidirectional prediction for the B-block.

17.1.5 The Encoder

In Fig. 17.12 we show a high-level structure of a typical interframe encoder that can be used for the H.263 standard. This encoder is very similar to the encoder used for MPEG-1 encoding and uses motion estimation and compensation to generate an interframe prediction signal which is analyzed on a macroblock basis to decide if Intra or Inter coding should be employed. If Intra coding is employed, the prediction signal is ignored, and the block is transformed to a block of coefficients by DCT and quantized by Quant and Quant idices are sent along with the Inter/Intra decision and quantizer scale information to the video multiplex coder which generates the H.263 bistream. If Inter coding is employed, the motion-compensated prediction macroblock is differenced from the macroblock being coded and the prediction error signal is transformed by DCT, quantized by Quant, and results in Quant indices which along with motion vectors, Inter/Intra flag, transmit or not flag, and quantizer scale are sent to the video multiplex coder.

Details of motion estimation and compensation are not shown; depending on the coding options selected, motion estimation and compensation may involve computations specific to Unrestricted Motion Vectors, Advanced Prediction, or PB-frames. Also, the scaling of luminance motion vectors to generate chrominance motion vectors and calculation of bilinear interpolation for half-pixel motion compensation is not shown. Video multiplex uses various VLC tables and incoming flags to correctly format the bitstream to be H.263 compliant.

17.2 CODING ADVANCES

While various key standardization efforts related to image and video coding have been ongoing for the last ten years or so, outside of the standards arena, considerable progress has also been ongoing,[15] leading to a refinement of well-established coding schemes, gradual maturity of recent less established coding schemes, and invention of new promising coding schemes. Not only in terms of performance but also in terms of implementation complexity a number of coding schemes are beginning to be competitive with standards-based video coding. Thus, before proceeding with our discussion on MPEG-4, which is currently an ongoing standard, a brief review of the state of the art is necessary.

In reality, the MPEG standardization process has been fairly open such that whenever a new standard is initiated, a call for proposals is sent,

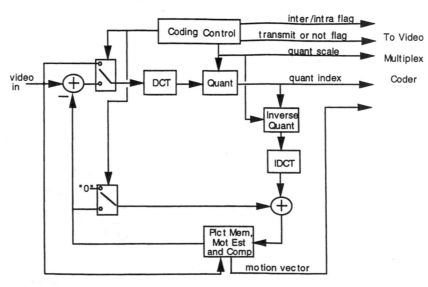

Fig. 17.12   Encoder structure.

inviting any candidate proposals to an open "contest" involving subjective testing of video quality, evaluation of functionalities it can achieve, and analysis of its implementation complexity. Further, even if a candidate proposal is not among the top rated proposals but consists of some promising components, as long as there are organizations willing to study them further, these components are evaluated and even improved via Core Experiments.

There are various ways of classifying image and video coding schemes. In one such technique for classification, terms such as first generation coding, second generation coding, and so forth have been used. The underlying implication of this technique for classification is that the second generation schemes analyze a scene and thus can code it more efficiently by using a contours (edges separating various regions) and texture (intensity variations in various regions) based representation, while the first generation schemes do not. Another technique for classification is based on clearly identifying the segmentation and motion models employed by a scheme. Various possibilities for segmentation models are: statistically dependent pixel blocks, 2D models based on irregular regions/deformable triangles/specific features, and 3D models such as wireframe representations. Likewise, possibilities for motion models are: 2D translation, sophisticated 2D models based on affine/bilinear/perspective transformations, and 3D models with global or local motion. Another technique for classification, besides coding efficiency, uses additional functionalities that a coding scheme can provide as a criterion for judging its value. Examples of functionalities are degree of scalability, error resilience, content based access, and others.

The performance, functionality, and complexity tradeoffs are different for different applications, and in general there are no easy answers. It is difficult for a single coding algorithm to be able to provide significantly higher compression than current standards while achieving desirable functionalities and somewhat reasonable complexity. Further, an important question is how general should a coding scheme be? Should it be very application specific, or should it be very generic? One way to answer these questions is in terms of coding efficiency/

functionalities achieved. If very significant coding/functionality improvements can be obtained by being application specific, it should be application specific; otherwise it should be generic. In fact one of the motivations in studying various techniques for classification is to determine trends and predict which methods are likely to lead to significant improvements in the future.

In presenting a comparative overview of various recent coding advancements, one of the main difficulties is that one is judging the performance of an overall coding system, and not that of individual components. To be more concrete, a typical image or video coding system consists of a number of components such as analyzer, quantizer, and entropy coder, with a number of possibilities for each component. Further, analysis may itself be done on an entire signal or on portions of the signal, on the original signal, or on the prediction error (including motion-compensated prediction error) signal, and moreover may use DCT transform, Sub-band, Discrete Wavelet Transform, Fractals, or others. The next component is quantization which may be perceptual visibility based or not, fixed or adaptive, linear or nonlinear, scalar or vector. The third major component is entropy coding which may again be fixed or adaptive, Huffman (VLC), or arithmetic. Recognizing the various pitfalls, we briefly review the state of progress of various promising coding schemes.

17.2.1 Transform Coding

Transform coding and in particular DCT coding has enjoyed significant popular owing to its ability to effectively exploit high spatial correlations in image data. It has even proven to be effective for coding of motion-compensated prediction error data which typically have fairly low correlations. All major image and video coding standards use DCT transforms of 8×8 block size. Besides DCT, other orthogonal block transforms have also been experimented with but from a practical standpoint significant benefits of other transforms over DCT have not yet been proven. From a coding efficiency standpoint, on the average, a block size of 8×8 has been proven to provide a suitable tradeoff in exploiting correlations and providing adaptations needed owing to local structures in images, and is thus commonly used in DCT coding.

Improvements in DCT coding have been reported by use of somewhat larger size transforms, which from the implementation standpoint are becoming feasible only now. Overall, to be sufficiently adaptive across images of different resolutions, variable block size (VBS) schemes[22] may be promising. Further, optimized VBS coding may involve a hybrid of DCT and pixel domain coding. By pixel domain coding, we mean schemes such as Vector Quantization (VQ) in which an approximation to the block being coded is obtained by lookup into a limited codebook, or by template coding (similar to VQ) but with a codebook satisfying specific properties. As an example of a VBS scheme, block sizes of 16×16, 8×8, and 4×4 may be used for DCT coding and block size of 4×4 may be used for pixel-domain template coding. The extent to which VBS coding can be successful depends on minimization of overhead, good codebook design, and use of optimized VLCs for different block sizes. Other directions for improvements have focused on adaptation of block DCT to coding of irregular shaped regions.

Some improvements have also been reported with VQ of DCT coefficients at the expense of loss of genericity over a wide range of bitrates. Before motion-compensated interframe coding became popular, 3D-DCT coding was also tried for coding of video but such attempts were not very successful. Recently, 3D-DCT of $8 \times 8 \times 4$ block size has been proposed for motion-compensated prediction error coding for B-pictures.[16]

17.2.2 Wavelet Coding

Sub-band coding decomposes an image signal into its frequency components similar to transform coding but also manages to retain spatial relationships (present in pixel domain representation) in its frequency components. Sub-band coding achieves critical sampling, unlike pyramid coding in the pixel domain, which requires oversampling. Sub-band encoding typically consists of applying a series of analysis filters over the entire image, downsampling of each band by a factor, quantization, and entropy coding. Sub-band decoding consists of complementary operations of entropy decoding, inverse quantization, upsampling of each band, and applying synthesis filters to generate the reconstructed image. The analysis and synthesis filters are generally chosen to allow perfect reconstruction. In general, sub-band coding can be viewed as transform coding with overlapped blocks so that it can exploit correlation over a larger set of pixels. While transform coding has the potential of producing blocking artifacts, sub-band coding may typically produce ringing or contouring artifacts. An advantage of sub-band coding is in terms of the flexibility it offers to allocate different rates to different sub-bands.

Typically, wavelet transform coding[15,18] has been treated as sub-band coding because sub-band analysis filtering and the transform operation are essentially equivalent and wavelet transform can be achieved with the sub-band filter bank. Without going into detail, the wavelet transform operation can be identified by properties such as shorter basis for high frequencies, higher resolution for high frequencies, orthogonality between basis, and multiresolution representation. Despite its relationship to sub-band decomposition, there is some benefit to treating wavelet coding as transform coding instead of subband coding owing to nonstationarity properties of images; these properties can be best exploited by locally adaptive quantization, scanning, and use of end of block code, such as often done in transform coding.

It has often been reported that wavelet coding can provide intraframe image coding with significantly higher performance[25], however, it is not completely clear as to how it can provide improvements in video coding. If simple intraframe wavelet techniques are applied for video coding, it is difficult to obtain sufficient performance. To improve performance, several ways of performing interframe coding with wavelets are being investigated by researchers and include coding of differences in wavelet transforms of current image and a previous image, wavelet coding over the entire image of block motion-compensated error signal generated by use of overlapped block motion compensation, and wavelet coding adapted to each region where motion compensation is performed on a region basis. Among the more promising techniques are those that employ overlapped block motion compensation and wavelet coding of the entire image, followed by arranging of coefficients into a number of blocks that can be quantized, scanned, and VLC (or arithmetic) coded much like transform coding. Some improvements have also been reported by vector quantization of wavelet coefficients instead of scalar quantization and scanning.

17.2.3 Fractal Coding

Fractal coding[15,17] exploits the property of self-similarities within an image; some of the self-similarities may exist at different resolution scales and may not be directly obvious. Fractal coding basically uses concepts such as affine transformations, iterated function systems, and the collage theorem. Affine transformations consist of mappings such as translation, rotation, and scaling that allows a part of source image to yield a target image. An iterated function system is a collection of contractive affine maps that when applied iteratively on a part of source image can generate a part of the target image using repeated parts (such as translations, rotations, and scales) of a source image. The collage theorem says that if an image can be described by a set of contractive affine transformations then these transformations specify an iterated function system that can be used to reconstruct an approximation after some number of iterations and that it will eventually converge.

Fractal coding and VQ seem to have distinct similarities owing to use of some codebooks and need for matching operations. Like VQ and transform coding practical fractal coding is also block based. In fractal transform coding, first an image is partitioned into nonoverlapping smaller blocks called domain blocks that cover the entire image. Next, range blocks are defined and are allowed to overlap and not necessarily cover the entire image. Affine transformations are then computed which allow mapping of range blocks to domain blocks. The compressed representation then involves affine transformation maps, domain bock geometry, and range block addresses. Fractal decoding then consists of generating two buffers, a domain buffer and a range buffer. Affine transforms are applied using the range buffer and filling the domain buffer. The range and domain block are identified in the opposite buffers and affine transforms are applied again; this process is repeated until differences in the two buffers are small and results in an image that is a close approximation to the original image.

Again, fractal coding is promising for coding of intraframes but has obvious difficulties in coding of video. Intraframe coding of video using fractals does not provide the desired compression. Fractal coding of difference images does not work well owing to the pulsive nature of the differential or motion-compensated differential signal. Fractal coding of video by using self-similarities in 3D blocks has also been investigated. Currently research is in progress to determine ways in which motion compensation and fractal coding can be combined to provide an effective solution for video coding.

17.2.4 Region-Object-Based Coding

Images and video of natural or artificial scenes are generally quite nonstationary. This is due to the presence of local structures (edges) that correspond to imaged objects in scenes. Earlier we discussed a classification of images in which second generation coding was meant to represent contours and textures in a scene. It has been believed for quite some time that to achieve very high compression, not only information theoretic (such as correlations) but also structural properties (such as organization of structures, edges, etc.) need to be exploited. Region-based coding can be derived as a generalization of contour/texture coding, where regions can be of arbitrary shapes. Consider a scene in which each region can be identified; then the image coding problem reduces to finding an efficient representation of region boundaries (shapes) and that for texture values (luminance and chrominance intensities) to fill in for each region. This can also be looked upon as generalization of block transform coding; however, in that case shape information is not needed since block sizes and locations are known a priori; the transform coding provides a means to specify texture information.

One of the main challenge in region-based coding[17,19] is to derive a few meaningful homogeneous regions in a scene; this is referred to as the segmentation problem. It is always possible to perform edge detection and try to define regions; however, this may result in too many regions which may cost considerable coding overhead. Ideally, one would like to segment a scene into a few principal objects that have real-world significance (semantic meaning). Each object could then be further subdivided into as many homogeneous regions as may provide overall good coding quality. However, automatic segmentation of scenes, except for the most trivial scenes, is a rather difficult problem for which considerable effort has been put forth by the computer vision community.

Next, for regions/objects that have been located, an efficient way of specifying shapes is necessary, since accurate shape representation causes a significant overhead, in effect reducing the number of bits available for texture coding. This is even more critical for low-bitrate video coding, in which the number of coding bitrates is limited. One of the simplest ways of representing shapes is via coding of chains of pixels, such that at each pixel a decision is made from among eight (or four) possible directions; with the use of differential coding and variable length coding chain coding can be made effective. Another method of shape coding is based on the use of polygon approximations[20] to actual shape via straight lines; this method is usually effective when some errors in shapes can be tolerated and does provide an overhead-efficient method for shape representation. Yet another method of shape representation is to use variable block size type segmentation using quadtree structures; this method also allows control of accuracy of shape representation to trade off shape coding overhead with texture coding overhead.

Finally, the texture for each region needs to be coded. For region-based texture coding, pixel domain coding (average values of region), transform coding, or other techniques such as wavelets can be employed. Recently, solutions for transform coding of arbitrary shaped regions have been proposed in the form of shape-adaptive DCT (SA-DCT), projection onto a convex set and DCT (POCS-DCT), and other schemes have been proposed. Attempts have also been made to jointly solve shape and texture coding problems by designing the texture coding scheme to implicitly specify shapes. Yet another direction has been to use block transform coding in conjunction with chroma keying to implicitly represent shapes, transferring some shape overhead into texture coding. Many of the aforementioned techniques can be applied for intraframe or motion-compensated interframe coding. In addition to texture coding improvements, attempts are also being made to reduce motion-compensated prediction error signal by use deformable triangles or quadrilaterals to perform advanced motion compensation. Such schemes can be used in conjunction with normal or region-based coding.

17.2.5 Model-Based Coding

By model-based coding, we primarily mean 3D model based coding.[21,26] Schemes that use 2D models were already discussed within the context of region-based coding. One of the main applications scenario in which research is ongoing in 3D model based coding is for representing the human head. Some advancements have also been made to extend this to represent a head and shoulders view as would be typical in a videophone scene. A generic wireframe model is used to represent the human head and provide the structure for key facial features.

A model-based coder then consists of three main components: a 3D head (facial) model, an encoder, and a decoder. The encoder segments an object of interest (a person) from the background, estimates motion of the object (translation, rotation, etc.), and then transmits these analysis parameters to the receiver. Further, as new previously unseen areas are uncovered at the encoder, it updates and corrects the model. This generates output images by using the 3D facial model and the analysis parameters. To model a person's face in sufficient detail, a deformable wireframe model consisting of several hundred triangles can be used. This model is adapted to a person's face by use of feature point positions. Then the front view of the person's face is mapped onto the wireframe model. Now this model can be manipulated and the resulting images appear natural. Research is ongoing to improve the accuracy of model by use of side views, modeling of muscles, and so forth. To regenerate natural looking images by animating this model, accurate synthesis of facial movements is necessary. Several methods to accomplish this exist, such as transmitting of fine details and pasting them, use of deformation rules to control vertices of the model by identifying the facial actions, and translating them into movement of a set of action units.

The problem of analysis is a lot more difficult as compared to the problem of synthesis and deserves separate mention. At present there is no system that can automatically work for modeling of objects and extraction of model parameters, even for head and shoulder images. Analysis actually includes a number of subproblems such as determining feature points, segmentation of objects, estimation of global and local motion, and so forth. A number of solutions are being investigated to allow better analysis by tracking marks plotted on the moving object and using that to compute global (head) motion as well as a number of local movements corresponding to important features. Researchers[27-32] are looking into ways by which additional information such as range data, two or more views, speech, facial muscle models etc can improve the overall performance and robustness of 3D model based video coding system.

17.3 INTRODUCTION TO MPEG-4

The MPEG-4 standard is the next generation MPEG standard which was started in 1993 and is currently in progress. It was originally intended for coding of audio–visual information with very high compression at very low bitrates of 64 kbits/s or under. When MPEG-4 video was started, it was anticipated that with continuing advances in advanced (non-block-based) coding schemes, for example, in region-based and model-based coding, a scheme capable of achieving very high compression, emerge. By mid 1994, two things became clear. First, video coding schemes that were likely to be mature within the timeframe of MPEG were likely to offer only moderate increase in compression (say, up to 2 at the most) over existing methods as compared to the original goal of MPEG-4. Second, a new class of multimedia applications was emerging that required greater levels of functionality than that provided by any other video standard at bitrates in the range of 10 kbits/s to 1024 kbits/s. This led to broadening of the original scope of MPEG-4 to larger range of bitrates and important new functionalities.[3]

17.3.1 Focus of MPEG-4

The mission and the focus statements of MPEG-4 explain the trends leading up to MPEG-4 and what can be expected in the future and are directly stated here from the MPEG-4 Proposal Package Description (PPD) document.[1]

The traditional boundaries between the telecommunications, computer, and TV/film industries are blurring. Video, sound, and communications are being added to computers; interactivity is being added to television; and video and interactivity are being added to telecommunications. What seems to be convergence in reality is not. Each of the industries approach audio–visual applications from different technological perspectives, with each industry providing its own, often incompatible solutions for similar applications.

Three important trends can be identified:

- The trend toward wireless communications
- The trend toward interactive computer applications
- The trend toward integration of audio–visual data into a number of applications.

At the intersection of the traditionally separate industries these trends must be considered in combination; new expectations and requirements arise that are not adequately addressed by current or emerging standards.

Therein lies the focus of MPEG-4, the convergence of common applications of the three industries. MPEG-4 will address these new expectations and requirements by providing audio–visual coding solutions to allow interactivity, high compression, or universal accessibility.

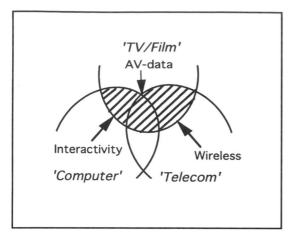

'TV/Film'
AV-data

Interactivity

Wireless

'Computer'

'Telecom'

Fig. 17.13 Applications area addressed by MPEG-4 (shaded region).

Figure 17.13 shows the applications of interest to MPEG-4 arising at the intersection of different industries.

17.3.2 MPEG-4 Applications

We now provide a few concrete examples of applications or application classes that the MPEG-4 video standard may be most suitable for:

- Video on LANs, Internet video
- Wireless video
- Video databases
- Interactive home shopping
- Video e-mail, home movies
- Virtual reality games, flight simulation, multi-viewpoint training.

17.3.3 Functionality Classes

Now that we have some idea of the type of applications MPEG-4 is aimed for we clarify the three basic functionality classes[1] that the MPEG-4 standard is addressing. They are as follows:

- *Content-Based Interactivity* allows the ability to interact with important objects in a scene. Currently such interaction is possible only for synthetic objects; extending such interaction to natural and hybrid synthetic/natural objects is important to enable new audio–visual applications.

- *High Compression* is needed to allow increase in efficiency in transmission or the amount of storage required. For low-bitrate applications, high compression is very important to enable new applications.

- *Universal Accessibility* means the ability to access audio–visual data over a diverse range of storage and transmission media. Owing to an increasing trend toward mobile communications, it is important that access be available to applications via wireless networks. This acceptable performance is needed over error-prone environments and at low bitrates.

17.3.4 MPEG-4 Work Organization

Currently the MPEG-4 work is subdivided into five parts:

- Video
- Audio
- Integration
  —Requirements
  —MPEG-4 Syntax Description Language (MSDL)
  —Synthetic and Natural Hybrid Coding (SNHC)
- Implementation studies
- Test.

Each part is handled by a separate subgroup, so in fact there are five parts. Further, some parts are divided into a number of subparts; for example, Integration work consists of Requirements, MSDL, and SNHC. Each subgroup, depending on the number of technical issues that need to be addressed, forms a number of ad hoc groups, which by definition are temporary in nature to ensure maximum flexibility.

17.4 MPEG-4 REQUIREMENTS

The process of collection of requirements for MPEG-4, although it was started in late 1993, is continuing at present,[8] in parallel with other work items. This is so because the process of development of an MPEG standard is intricate, tedious, thorough, and thus time intensive (about 3 to 5 years per standard with overlap between standards). To keep up with marketplace needs for practical standards at the right time and follow evolving trends, the requirements collection process is kept flexible. This ensures the ability to modify requirements as they evolve over time as well as to be able to add new ones when absolutely necessary. The work for various parts (or subparts) of MPEG-4, Video, Audio, MSDL, and SNHC is kept synchronized with the requirements collected at any given point in time, to the maximum extent possible. The major restructuring of the MPEG-4 effort in July 1994 to expand its scope to a number of useful functionalities and not to be limited to very low bitrates only was in part due to an exhaustive effort in requirements analysis which indicated changing marketplace and new trends.

Before we specify the new functionalities of MPEG-4, the standard functionalities it must sup port, and some general requirements, we need to clarify the terminology used in MPEG-4. Some of the terms used have evolved from MPEG-1 and MPEG-2 while other terms are new.

17.4.1 Terminology

Tool: A tool is a technique that is accessible or described by the MSDL. Tools may themselves consist of other tools. Motion compensation, transform filters, audio–visual synchronization, and so forth are some examples of tools.

Algorithm: An algorithm is an organized collection of tools that provides one or more functionalities. An algorithm may be composed of a number of tools tools, or even of other algorithms. DCT image coding, Code Excited Linear Prediction, Speech-Driven Image Coding, and so forth are some examples of algorithms.

Flexibility: Flexibility is the degree of programmability provided in an MPEG-4 implementation. Three degrees of flexibility can be defined: fixed set of tools and algorithms (Flex_0); configurable but fixed tools (Flex_1); and downloadable tools, algorithms, and configuration (Flex_2).

Profile: A profile addresses a cluster of functionalities that can enable one or more applications. At each flexibility, one or more profiles can be defined. Thus, at Flex_0, a profile is a set of tools

configured into an algorithm; at Flex_1, a profile is a set of tools; and at Flex_2, a profile enables the ability to download and configure tools and algorithms. MPEG-1 audio layer 3, MPEG-2 Video Profile, MPEG-2 System, and H.263 are all examples of profiles at Flex_0.

Level: A level is a specification of the constraints and performance criteria needed to satisfy one or more applications. Each profile is classified by a degree of flexibility and a value of level. At Flex_0, for a profile, a level simply indicates performance achieved. At Flex_1, for a profile profile, a level is the capability of a system to configure a set of tools and execute the resulting algorithms. At Flex_2, for a profile, a level specifies system capability relative to downloading, configuration of tools, and execution of resulting algorithms.

Conformance Points: Conformance points are specifications of particular Flexibility/Profile/Level combinations at which conformance may be tested. Like MPEG-2, MPEG-4 expects to standardize a set of conformance points.

17.4.2 MPEG-4 Functionalities

Earlier, we had discussed the various classes of functionalities being addressed by MPEG-4. In Table 17.1 we show a detailed list of eight key functionalities[1,2,8] and how they are clustered into three functionality classes. These functionalities are to be supported through definition of a set of coding tools and a mechanism (MSDL) to access, download, combine, and execute these coding tools.

17.4.3 Basic Functionalities

Besides the new functionalities that the ongoing work in MPEG-4 expects to support, a set of basic functionalities are also required and can be listed as follows.

Synchronization: The ability to synchronize audio, video, and other content for presentation.

Auxillary data capability: The ability to allocate channels for auxillary data streams.

Virtual channel allocation: The ability to dynamically partition video, audio, and system channels.

Table 17.1   Functionalities Supported by MPEG-4 and Their Explanation

Content Based Interactivity

Content-Based Multimedia Data Access Tools: The access of multimedia data with tools such as indexing, hyperlinking, browsing, uploading, downloading, etc.

Content-Based Manipulation and Bitstream Editing: The ability to provide manipulation of contents and editing of audio–visual bitstreams without the requirement for transcoding.

Hybrid Natural and Synthetic Data Coding: The ability to code and manipulate natural and synthetic objects in a scene including decoder controllable methods of compositing of synthetic data with ordinary video and audio, allowing for interactivity.

Improved Temporal Random Access: The ability to efficiently access randomly in a limited time and with fine resolution parts (frames or objects) within an audio–visual sequence. This also includes the requirement for conventional random access.

Compression

Improved Coding Efficiency: The ability to provide subjectively better audio–visual quality at bitrates compared to existing or emerging video coding standards.

Coding of Multiple Concurrent Data Streams: The ability to code multiple views/soundtracks of a scene efficiently and to provide sufficient synchronization between the elementary streams. To obtain high coding efficiency techniques for exploiting redundancies that exist between multiviewpoint scenes are required.

Universal Access

Robustness in Error-Prone Environments: The capability to allow robust access to applications over a variety of wireless and wired networks and storage media. Sufficient robustness is required, especially for low-bitrate applications under severe error conditions.

Content-Based Scalability: The ability to achieve scalability with fine granularity in spatial, temporal, or amplitude resolution; quality; or complexity. Content-based scaling of audio–visual information requires these scalabilities.

Low delay mode: The ability for system, audio, and video codecs to operate with low delay.

User controls: The ability to support user control for interactive operation.

Transmission media adaptibility: The ability to operate in a number of different media.

Interoperability with other systems: The ability to interoperate with other audio–visual systems.

Security: The ability to provide encryption, authentication, and key management.

Multipoint capability: The ability to have multiple sources or destinations.

Coding of audio types: The ability to deal with a range of audio and speech data such as wideband, narrowband, intelligible, synthetic speech, and synthetic audio.

Quality: The ability to provide audio–visual services at sufficient quality level.

Bitrates: The ability to efficiently operate over range of bitrates from 9.6 to 1024 kbits/s.

Low complexity mode: The ability to operate with low complexity (in hardware, software, or firmware).

17.4.4   Evolving Requirements

The various aforementioned advanced as well basic functionalities give rise to certain requirements that an MPEG-4 coding system must satisfy. Further, it is not surprising that several functionalities sometimes have common requirements and often, one or more such functionalities is simultaneously needed in an application. One way to derive specific requirements is to choose a number of representative applications and provide numerical bounds or characteristics for every single parameter; however, such an exercise is feasible for applications that are already in use and are well understood enough to allow full characterization. Another solution is to look for functional commonalities between applications to generate several clusters and specify parameter ranges for these clusters. Generating requirements for MPEG-4 is an ongoing exercise and is using a little bit of both approaches.

The requirements process is complicated further by the fact that requirements for MPEG-4 coding need to be translated into a set of requirements for Video, Audio, MSDL, and SNHC. The current approach is to specify requirements for Flexibility/Profile/Level combinations. In this manner, for each degree of flexibility, the requirements for one or more profiles are being described. Further, for each profile, requirements for one or more levels are also being specified.

17.5 MPEG-4 VIDEO

MPEG-4 video was originally intended for very high compression coding at bitrates of under 64 kbits/s. As mentioned earlier, in July 1994, the scope of MPEG-4 was modified to include a new set of functionalities, and as a consequence, it was no longer meaningful to restrict bitrates to 64 kbits/s. Consistent with the functionalities it intends to support, the MPEG-4 video effort is now aimed at very efficient coding optimized at (but not limited to) a bitrate range of 10 kbits/s to about 1 Mbit/s, while seeking a number of content-based and other functionalities.

Following the tradition of previous MPEG video standards, the MPEG-4 video effort is currently in its collaborative phase after undergoing a competitive phase. The competitive phase consisted of issuing an open call for proposals in November 1994 to invite candidate schemes for testing in October/November 1995. A proposal package description (PPD) was developed describing the focus of MPEG-4, the functionalities being addressed, general applications MPEG-4 was aimed at, the expected workplan, planned phases of testing, how Verification Models (VMs) would be used, and the time iternary of MPEG-4 development. The MPEG-4 PPD which was started in November 1994 underwent successive refinements until July 1995. In parallel to the PPD development effort, a document describing the MPEG-4 Test/Evaluation Procedures started in March 1995 underwent several iterations and was subsequently finalized by July 1995.

17.5.1 Test Conditions for First Evaluation

For the first evaluation tests,[2] proposers were required to submit algorithm proposals for formal subjective testing or tools proposals. Since not all functionalities were tested in the first evaluation, for the untested functionalities, tools submissions were invited. It turned out that some tools submissions were made by proposers who were unable to complete an entire coding algorithm because of limited time or resources. In a few cases, proposers also used tools submissions as an opportunity to identify and separately submit the most promising components of their coding proposals. Since tools were not formally tested they were evaluated by a panel of experts and this process was referred to as *evaluation by experts*.

The framework of the first evaluation involved standardizing test material to be used in the first evaluation. Toward that end, video scenes are classified from relatively simple to more complex by categorizing them into three classes: Class A, Class B, and Class C. Two other classes of scenes, Class D and Class E, were defined; Class D contained stereoscopic video scenes and Class E contained hybrids of natural and synthetic scenes.

Since MPEG-4 is addressing many types of functionalities and different classes of scenes, it was found necessary to devise three types of test methods. The first type of test method was called the Single Stimulus (SS) Method and involved rating the quality of coded scene on an 11-point scale from 0 to 10. The second type of test method was called the Double Stimulus Impairment Scale (DSIS) and involved presenting to assessors a reference scene (coded by a known standard) and after a 2-second gap, a scene coded by a candidate algorithm, with impairment of the candidate algorithm compared to reference using a five-level impairment scale. The third test method was called the Double Stimulus Continuous Quality Scale (DSCQS) and involved presenting two sequences with a gap of 2 seconds in between. One of the two sequences was coded by the reference and the other was coded by the candidate algorithm; however, the assessors were not informed of the order in which the reference scene

Table 17.2   List of MPEG-4 First Evaluation Formal Tests and Their Explanation

Compression

Class A Sequences at 10, 24, and 48 kbits/s: Coding to achieve the highest compression efficiency. Input video resolution is CCIR-601 and although any spatial and temporal resolution can be used for coding, the display format is CIF on a windowed display. The test method employed is SS.

Class B sequences at 24, 48, and 112 kbits/s: Coding to achieve the highest compression efficiency. The input video resolution is CCIR-601 and although any combination of spatial and temporal resolutions can be used for coding, the display format is CIF on a windowed display. The test method employed is SSM.

Class C Sequences at 320, 512, and 1024 kbits/s: Coding to achieve the highest compression efficiency. The input video resolution is CCIR-601 and although any combination of spatial and temporal resolution can be used for coding, the display format is CCIR-601 on a full display. The test method employed is DSCQS.

Scalability

Object Scalability at 48 kbit/s for Class A, 320 kbit/s for Class E, and 1024 kbits/s for Class B/C Sequences: Coding to permit dropping of specified objects resulting in remaining scene at lower then total bitrate; each object and the remaining scene is evaluated separately by experts. The display format for Class A is CIF on a windowed display and for Class B/C and Class E is CCIR-601 on a full display. The test method employed for Class A is SSM, for Class B/C DSCQS, and for Class E DSIS.

Spatial Scalability at 48 kbit/s for Class A, and 1024 kbits/s for Class B/C/E Sequences: Coding of a scene as two spatial layers with each layer using half of the total bitrate; however, full flexibility in choice of spatial resolution of objects in each layer is allowed. The display format for Class A is CIF on a windowed display and that for Class B/C/E is CCIR-601 on a full display. The test method employed for Class A is SSM, and that for Class B/C/E is DSCQS.

Temporal Scalability at 48 kbits/s for Class A, and 1024 kbits/s for Class B/C/E Sequences: Coding of a scene as two temporal layers with each layer using half of the total bitrate; however, full flexibility in choice of temporal resolution of objects in each layer is allowed. The display format for Class A is CIF on a windowed display and that for Class B/C/E is CCIR-601 on a full display.

Error Robustness

Error Resilience at 24 kbits/s for Class A, 48 kbits/s for Class B, and 512 kbits/s for Class C: Test with high random bit error rate (BER) of 10^{-3}, multiple burst errors with three bursts of errors with 50% BER within a burst, and a combination of high random bit errors and multiple burst errors. The display format for Class A and Class B sequences is CIF on a windowed display and for Class C sequences it is CCIR-601 on full display. The test method employed for Class A and Class B is SSM and that for Class C DSCQS.

Error Recovery at 24 kbits/s for Class A, 48 kbits/s for Class B, and 512 kbits/s for Class C: Test with long burst errors of 50% BER within a burst and a burst length of 1 to 2 seconds. Display format for Class A and Class B is CIF on a windowed display and for Class C it is CCIR-601 on full display. The test method employed for Class A and Class B is SSM and that for Class C DSCQS.

and the coded scene under test were presented to them. In the DSCQS method, a graphical continuous quality scale was used and was latter mapped to a discrete representation on a scale of 0 to 100.

17.5.2 First Evaluation and Results

Table 17.2 summarizes the list of informal tests and gives explanations of each test and the type of method employed for each test.

To ensure that the MPEG-4 video subjective testing process would provide a means of comparison of performance of new proposals to known standards it was decided to use existing standards

as anchors. In tests involving Class A and B sequences, it was decided to use the H.263 standard (with TMN5 based encoding results generated by a volunteer organization) as the anchor. The anchors for Class A and B used test sequences downsampled for coding and upsampled for display using filters that were standardized.

Likewise, in tests involving Class C sequences, it was decided to use the MPEG-1 standard (with encoding results generated by a volunteer organization) as the anchor. The anchor for Class C, however, used test sequences downsampled for coding and upsampled for display using proprietary nonstandardized filters. To facilitate scalability tests (object scalability, spatial scalability, and temporal scalability), standardized segmentation masks were generated so that all proposers could use the same segmentation, with the expectation of making the comparison of results easier.

The proposers were allowed to address one or more of the tests listed in the table. By the registration deadline of mid September 1995, more than 34 proposers had registered. By the end of the first week of October 1995, for the proposals being subjectively tested, proposers submitted D1 tapes with their results, the description of their proposals, coded bitstreams, and executable code of their software decoders. By the third week of October 1995, proposers of tools had also submitted demonstration D1 tapes and a description of their tool submissions. About 40 video tools submissions were received. The formal subjective tests of algorithm proposals as well as an informal evaluation by experts of tools were started by the end of October 1995. The results of the individual tests[23] and a thorough analysis of the trends were made available during the November MPEG meeting.

The results of compression tests in various categories revealed that the anchors performed quite well, in many cases among the top three or four proposals. In some cases, the differences between many top performing proposals were found to be statistically insignificant. In some categories there were, however, innovative proposals that beat the anchors or provided as good a picture quality as the anchors while providing additional (untested) functionalities. It was also found that since spatial and temporal resolutions were not prefixed, there was some difficulty in comparing subjective and objective (SNR) results, owing to differences in the choices made by each proposer. It seemed that subjective viewers had preferred very low temporal resolutions as long as the spatial quality looked good, over a more balanced tradeoff of spatial quality and temporal resolution. Besides compression, the proposers had also managed to successfully achieve other functionalities tested.

Besides the results from subjective tests, the tools evaluation experts presented their results,[24] about 16 tools were judged as promising for further study.

17.5.3 Core Experiments Formulation

In the period from November 1995 through January 1996, the process of definition of core experiments was initiated. A total of 36 core experiments were defined prior to January 96 MPEG meeting;[7] another five experiments were added at the meeting itself, bringing the total number of experiments to about 41. Since the total number of experiments is rather large, these experiments are classified into a number of topics:

1. Prediction
2. Frame Texture Coding
3. Quantization and Rate Control
4. Shape and alpha Channel Coding
5. Object/Region Texture Coding
6. Error Resilience and Error Correction
7. Bandwidth and Complexity Scalability
8. Multiview and Model Manipulation
9. Pre-, Mid-, and Postprocessing

A set of common experimental conditions was also agreed to. A number of ad hoc groups were initiated each focusing on one or more topics as follows:

- Coding Efficiency—topics, 1, 2, and 3
- Content-Based Coding—topics 4 and 5
- Robust Coding—topic 6
- Multifunctional Coding—topics 7, 8, and 9.

Between January 1996 and March 1996 MPEG meetings these ad hoc groups undertook the responsibility of producing an improved description of each core experiment consistent with the VM 1.0, seeking volunteer organizations for performing core experiments and finalizing experiment conditions. In the meantime, a few new experiments were also proposed. At the MPEG meeting in March 1996, the ad hoc group on content-based coding was subdivided into two groups. At the same meeting, there were also minor changes in rearrangement of experiments and an addition of three or four experiments.

17.5.4 Verification Model (VM)

It was agreed that MPEG-4 only have a single verification model for development of the standard to address the various functionalities.[6] The first MPEG-4 Video VM (VM 1.0) was released on January 24, 1996. It supports the following features:

- VOP structure
- Motion/Texture coding derived from H.263
- Separate Motion/Texture syntax as an alternative for error resilience
- Binary and Grayscale Shape Coding
- Padding.

The second MPEG-4 Video VM (VM 2.0)[9] was released on March 29, 1996. It refines some existing features of VM 1.0 and adds a few new features as follows:

- B-VOPs derived from H.263 B-pictures and MPEG-1/2 B-pictures
- DC coefficients prediction for Intra-Macroblocks as per MPEG-1/2
- Extended Motion Vector Range
- Quantization Visibility Matrices as per MPEG-1/2

17.5.4.1 VM Description

We now describe the important elements of the latest MPEG-4 Video VM.[9] An input video sequence consists of a sequence of related pictures separated in time. Further, each picture can be considered as consisting of a set of flexible objects, that from one picture to the next picture undergo a variety of changes such as translations, rotations, scaling, brightness and color variations, and so forth. Moreover, new objects enter a scene and existing objects depart, leading to the presence of certain objects only in certain pictures. Sometimes, scene changes occur, and thus the entire scene may either get reorganized or initialized.

Many of MPEG-4 functionalities require access not only to individual pictures but also to regions or objects within a picture. On a very coarse level, access to individual objects can be thought of as generalization of the slice structure of MPEG-1/2 or that of the GOB structure of H.263, although such structures do not have a semantic meaning in these standards and although potentially there is a way of accessing them, they are not intended for individual access or display. MPEG-4, to allow access to contents of a picture, has introduced the concept of Video Object planes (VOPs).

A VOP can be a semantic object that is represented by texture variations (a set of luminance and chrominance values) and (explicit or implicit) shape information. In natural scenes, VOPs are obtained by semiautomatic or automatic segmentation, and the resulting shape information can be represented as a binary mask. On the other hand, for hybrid (of natural and synthetic) scenes generated by blue screen composition, shape information is represented by an 8-bit component.

In Fig. 17.14 we show the decomposition of a natural scene into a number of VOPs. The scene consists of two objects (head and shoulders view of a human, and a logo) and the background. The objects are segmented by semiautomatic or automatic means and are referred to as VOP1 and VOP2, while the background without these objects is referred to as VOP0. Each picture of interest is segmented into VOPs in this manner. The same object in a different picture is refered to by the VOP number assigned to it when it first appeared in the scene. The number of VOPs selected in a scene is dependent on the total available bitrate for coding, the degree of interactivity desired, and of course, the number and size of unique objects in the scene.

The VOPs are encoded separately and multiplexed to form a bitstream that users can access and manipulate (cut, paste, . .). The encoder sends, together with VOPs, information about scene composition to indicate where and when VOPs are to be displayed. This information is however optional and may be ignored at the decoder which may use user-specified information about composition.

In Fig. 17.15 we show a high-level structure of a VOP-based Codec. Its main components are VOP Formatter, VOP Encoders, Mux, Demux, and VOP Decoders and VOP Compositor. The VOP Formatter segments the input scene into VOPs and outputs formatted VOPs for encoding. Although separate encoders (and decoders) are shown for

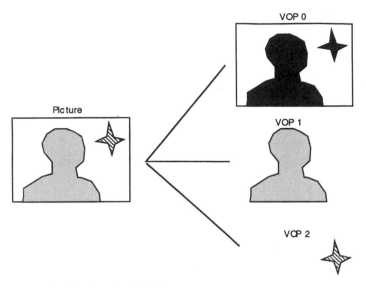

Fig. 17.14 Semantic segmentation of a picture into VOPs.

each VOP, this separation is mainly a logical one rather than a physical one. In principle even though different VOPs can use different methods of coding, it is anticipated that this will mainly be the case when VOPs contain different data types such as natural data, synthetic data, etc. Thus, although there may be many VOPs, there may be only a few (typically, one or two) coders. The multiplexer, Mux, multiplexes coded data from different VOPs into a single bitstream for transmission or storage; the demultiplexer, DeMux, performs the inverse operation. After decoding, individual VOPs are fed to a VOP compositor which either utilizes the composition information in the bitstream or composes VOPs under user control.

In Fig. 17.16 we show a block diagram structure of a VOP coder; this coder is meant for coding of natural video data. Its main components are: Texture Coder, Motion Coder, and Shape Coder.

The Texture Coder codes the luminance and chrominance variations of the region bounded by the VOP. This region may either be subdivided into macroblocks and blocks and coded by DCT coding, or other non-block-based techniques such as region-oriented DCT or wavelet coding may be used. Further, the coded signal may in fact be either an Intra- or motion-compensated Inter-frame signal. The Texture Coder is currently assumed to be a block-based DCT coder involving block DCT, quantizer, scan (and complementary operations of inverse

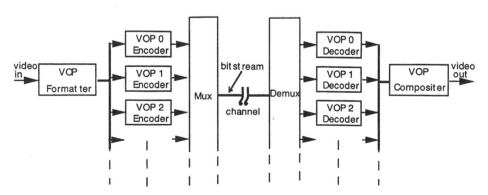

Fig. 17.15 Structure of VOP-based Codec for MPEG-4 Video.

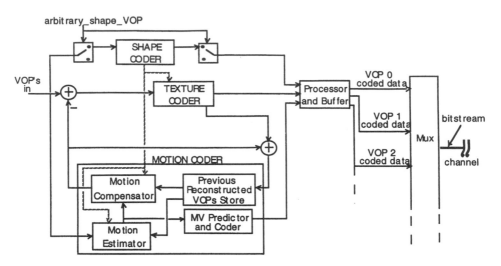

Fig. 17.16 Detailed structure of VOP Encoder.

DCT, inverse quantizer, and inverse scan) as per H.263, MPEG-1, and MPEG-2 standards. Since VOPs can be of irregular shape, padding of VOPs (see discussion of Motion Compensation) is needed.

The Motion Coder consists of a Motion Estimator, Motion Compensator, Previous VOPs Store, Motion Vector (MV) Predictor, and Coder. The Motion Estimator computes motion vectors using the current VOP and temporally previous reconstructed version of the same VOP available from the Previous Reconstructed VOPs Store. The Motion Compensator uses these motion vectors to compute motion-compensated prediction signal using the temporally previous reconstructed version of the same VOP (reference VOP). The MV Predictor and Coder generates prediction for the MV to be coded using H.263-like median prediction of MVs. Currently, the motion estimation and compensation in the VM is block based, basically quite similar to that in H.263. The current MPEG-4 VM employs not only default motion mode but also Unrestricted Motion Vector mode and Advanced Prediction modes of H.263; these modes are however not options. In the current VM, at the VOP borders, owing to irregular shape of VOPs, polygon matching instead of block matching is employed. This requires use of repetitive padding on the reference VOP for performing motion compensation and for texture coding when using block DCT.

In repetitive padding each pixel outside the object boundary is considered as zero pixel, each line is scanned horizontally for zero pixel segments, and if the zero pixels occur between nonzero pixel segments they are filled by the average of pixels at endpoints; otherwise they are filled by pixels at endpoints of the nonzero segment. Each line is then scanned vertically, repeating the filling operation as for horizontal scanning. If zero pixels can be filled by both horizontal and vertical scanning an average of the two values is used, and finally the remaining zero pixels are filled by the average of horizontally and vertically closest nonzero pixels.

While we are still on the subject of motion estimation and compensation, it is worth clarifying that the MPEG-4 VM now supports B-pictures. B-pictures currently in VM are derived by combining B-picture modes from H.263 and MPEG-1/2. Thus motion estimation and compensation also have to be performed for B-pictures. In the current MPEG-4 B-pictures, a choice of one out of four modes that differ mainly in the type of motion compensation is allowed on a macroblock basis. The motion compensation modes allowed are Direct, Forward, Backward, and Interpolated. The Direct mode is derived from H.263 and uses scaled motion vectors and a delta update vector. The Forward, Backward, and Interpolated are modes from MPEG-1/2. with one vector each in Foreward and Backward modes and two vectors in Interpolated mode.

The Shape Coder codes the shape information of a VOP and may involve a quadtree representation, a chain representation, a polygonal representation, or some other representation. If a VOP is of the size of a full picture, no shape information is sent. Thus, inclusion of the shape information is optional depending on the value of the binary signal, arbitary_shape_VOP, which controls the associated switches. The Shape Coder in the current version of the VM is assumed to be Quadtree based. Efforts are in progress to replace it by a Chain or a Polygon Approximation based technique.

The complete VOP Coder just described allows either combined coding of motion and texture data as in H.263, MPEG-1/2, or separate coding of motion and texture data for increased error resilience. The main difference is whether different types of data can be combined for entropy coding; this is not possible for separate motion and texture coding which assumes separate entropy coding to belong to individual Motion, Texture, and Shape coders rather than to a single entropy coder. Further, since in our discussion we have assumed VOPs derived from natural image or video data, all VOPs are coded by the same VOP Coder and the coded data of each VOP is output (by Processor and Buffer) for multiplexing to Mux which generates the bitstream for storage or transmission.

Earlier, although we mentioned the quadtree based shape coding method currently in the MPEG-4 VM, a few words of explanation are in order. The shape information may be either binary or grey scale and is correspondingly referred to as binary alpha plane or grey scale alpha plane. Currently in VM, binary alpha planes are coded with quadtree (without vector quantization) and grey scale alpha planes are coded by quadtree with vector quantization. An alpha plane is bounded by the tightest rectangle that includes the shape of a VOP. The bounding rectangle is extended on the right-bottom side to multiples of 16×16 samples, extended alpha samples set to zero, and the extended alpha plane is partitioned to blocks of 16×16 samples and encoding/decoding is done per 16×16 block. We now briefly explain binary alpha plane coding.

Figure 17.17 shows the quadtree structure employed for binary alpha plane coding. At the bottom level (level 3) of quadtree, a 16×16 alpha block is partitioned into 64 sub-blocks of 2×2 samples. Each higher level of quadtree is then formed by grouping 4 sub-blocks at the lower level as shown. The following sequence of steps is then employed.

1. Rounding process for bitrate control

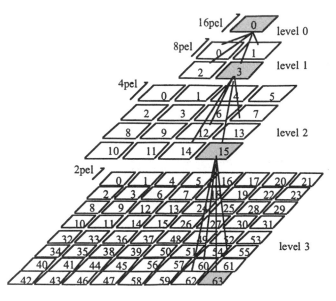

Fig. 17.17 Quadtree structure for binary shape (alpha plane) coding.

2. Indexing of sub-blocks at Level 3
3. Grouping process for higher levels
   • grouping of sub-blocks from Level 3 to Level 2
   • grouping of sub-blocks from Level 2 to Level 1
   • grouping of sub-blocks from Level 1 to Level 0
4. Encoding process.

The rounding process is needed only for lossy coding of binary alpha planes and can be ignored for loss-less coding.

The indexing of sub-blocks at level 3 consists of assigning an index to each 2×2 sub-block by first performing a swapping operation on alpha pixels of each sub-block, comparing their values against values of upper-block alpha pixels and left-block alpha pixels, except for sub-blocks belonging to the uppermost row of 2×2 sub-blocks or left-most column of 2×2 sub-blocks in Level 3 block comprised of a 8×8 array of sub-blocks (each of size 2×2). After swapping, a numerical index is calculated for the 2×2 sub-block as follows:

$$index = 27 \times b[0] + 9 \times b[1] + 3 \times b[2] + b[3]$$

where $b[i] = 2$ if the sample value $= 255$; $b[i] = 0$ if the sample value $= 0$.

The current sub-block is then reconstructed and used as a reference when processing subsequent sub-blocks.

The grouping process of blocks at the higher level first starts at Level 2 where four sub-blocks from Level 3 are grouped to form a new sub-block. The grouping process involves swapping and indexing similar to that discussed for Level 3. The current sub-block is then reconstructed and used as a reference when processing subsequent sub-blocks. At the decoder swapping is done following a reverse sequence of steps as at the encoder. The grouping process is also performed similarly for level 1 where four sub-blocks from Level 2 are grouped to form a new sub-block. The swapping, indexing, and reconstruction of sub-block follows grouping, the same as that for other levels.

The encoding process involves use of results from the grouping process which produces a total of 85 ($= 1 = 4 + 16 + 64$) indices for a 16×16 alpha block. Each index is encoded from topmost level (Level 0). At each level, the order for encoding

and transmission of indices is shown by numbers in Fig. 17.17. Indices are encoded by variable length codes.

As a final note on the current Video VM, it now supports a larger range of motion vectors (H.263 range was too limited even when using the Unrestricted Motion Vector mode), MPEG-1/2 style quantization visibility matrices for Intra- and Inter-macroblocks, MPEG-1/2 style DC prediction for Intra-macroblocks, and of course, B-VOPs. The MPEG-4 Video VM is currently undergoing an iterative process which is eventually expected to lead to convergence as core experiments get resolved and the successful techniques get adopted.

17.5.5 Video Directions

From our discussion of the MPEG-4 Video VM development process, it may appear that the goal of VM is solely to address all functionalities via a single coding algorithm, the various parts of which are selected based on best results of core experiments. In reality, the goal of MPEG-4 is to select a set of coding tools (and thus the resulting algorithms) to address the specified functionalities, in the best possible way. Tools are expected to be selected to minimize redundancy in problems they solve; however, at times a tool may be selected if it significantly outperforms another tool, which was originally selected for solving a different problem.

Although it is a bit difficult to completely predict which tools will eventually be selected in the MPEG-4 Video standard, the trends are quite clear. Block motion-compensation and DCT-based schemes still remain fairly competitive and in many cases outperform any other alternative schemes. The continuing core experiments will most likely be able to achieve up to about 40% to 50% improvement over existing standards in specific bitrate ranges of interest. A number of useful tools on content-based scalability, error-resilient coding, synthetic model based coding, multiviewpoint coding, and others will offer functionalities not offered by any other standard in an integrated manner.

The MPEG-4 Video effort is expected to contribute to tools and algorithms for part 2 of the MPEG-4 Working Draft-1 scheduled for November 1996. After the first revision, Working Draft-2 is scheduled for release in March 1997, and after the second revision, Working Draft Version-3 is scheduled for release in July 1997; both these revisions will involve refinements in tools and algorithms contributed to part 2. Also, in July 1997, the final MPEG-4 Video VM will undergo Verification Tests for its validation; any new candidate proposals at that time are also allowed to participate in those tests. The MPEG-4 effort (MSDL, Video, and Audio) is expected to lead to a stable stage of Committee Draft by November 1997 and be finally approved as the International Standard in November 1998.

17.6   MPEG-4 AUDIO

The MPEG-4 Audio coding effort is also in progress in parallel with the MPEG-2 NBC Audio coding effort which is now reaching a mature stage. The MPEG-4 Audio effort is targeting bitrates of 64 kbits/s or under. Besides coding efficiency, content-based coding of audio objects, and scalability are also being investigated.

The MPEG-4 Audio effort also underwent subjective testing recently. Three classes of test sequences—Class A, B, and C—were identified. Class A sequences were single-source sequences consisting of a clean recording of a solo instrument. Class B sequences were single-source with background sequences consisting of a person speaking with background noise. Class C sequences were complex sequences consisting of an orchestral recording. All sequences were originally sampled at 48 kHz with 16 bits/sample and were monophonic in nature. For generating reference formats filters were specified to downsample them to 24, 16, and 8 kHz. A number of bitrates such as 2, 6, 16, 24, 40, and 64 kbits/s were selected for testing of audio/speech. The first three bitrates are obviously suitable only for speech material. The audio test procedures used were as defined in ITU-R Recommendation 814.

The submissions to tests included variants of MPEG-2 NBC Audio coding, improvements on MPEG-1 coding, and new coding schemes. For specific bitrates, improvements over existing coding

solutions were found. The collaborative work has been started and an initial MPEG-4 Audio VM has been developed. MPEG-4 Audio development is expected to undergo a core experiment process similar to the MPEG-4 Video development process. The MPEG-4 Audio effort is expected to contribute to tools and algorithms for part 3 of the MPEG-4 standard. The schedule of various versions of the Working Draft, the Committee Draft, and the International Standard is the same as that mentioned while discussing MPEG-4 Video.

17.7   MPEG-4 SYNTAX DESCRIPTION LANGUAGE (MSDL)

Owing to the diverse nature of the various functionalities to be supported by MPEG-4 and the flexibilities offered by software based processing, it was envisaged that the MPEG-4 standard would be basically different from traditional audio–visual coding standards such as MPEG-1, MPEG-2, or others. While the traditional standards are fairly fixed in the sense of the bitstream syntax, and the decoding process they support, the MPEG-4 standard expects to relax these constaints and offer more flexibility and extensibility.

The MPEG-4 Syntax Description Language, MSDL, was originally conceived to deliver flexibility and extensibility by providing a flexible means to describe algorithms (including new ones in the future) and related syntax. With time, however, the MSDL[10] is evolving into the role of systems layer, addressing system capabilities needed to support MPEG-4 functionality classes such as content-based manipulation, efficient compression, universal accessibility, and basic functionalities, in addition to its original role.

17.7.1   MSDL Scope

To support flexibility and extensibility, MSDL[10] defines three types of decoder programmability as follows:

- Level 0 (nonprogrammable) decoder incorporates a prespecified set of standardized algorithms which must be agreed upon by the decoder during the negotiation phase.

- Level 1 (flexible) decoder incorporates a prespecified set of standardized tools which can be flexibly configured into an algorithm by the encoder during the scripting phase.
- Level 2 (extensible) decoder provides a mechanism for the encoder to download new tools as well as algorithms.

The aforementioned definitions of various levels are nested, that is, Level 2 programmability assumes Level 1 capabilities and Level 1 programmability assumes Level 0 capabilities. Currently MPEG-4 is planning to address Level 0 and Level 1 capabilities only and Level 2 may be addressed later.

To support MPEG-4 functionalities to the fullest, MSDL deals with presentable *audiovisual objects* (audio frames, video frames, sprites, 3D objects, natural or synthetic objects, . .) and their *representation methods* (waveform, splines, models, . .). The decoder decodes each object and either based on prespecified format or under user control *composes* and *renders* the scene. All audio–visual objects are assumed to have the necessary interfaces.

In Fig. 17.18 we show the position of MSDL in a vertical stack of protocols as was envisaged earlier. Then, the MSDL was considered as in between the channel coding protocols and the presentation protocols, with some overlap with these protocols.

Owing to MSDL recently taking over the role of systems layer also, it is expected that the overlap with presentation protocols will be even more significant.

Earlier, while discussing MPEG-4 requirements, we mentioned one role of MSDL as the glue between Tools, Algorithms, and Profiles. Since MSDL uses object-oriented methodology, we can represent Tools, Algorithms, and Profiles as objects, each with a clearly defined interface for input/output. In the context of MPEG-4, a number of Tools when connected together in a meaningful manner result in an Algorithm, and a Profile contains a number of related Algorithms. Figure 17.19 more clearly illustrates the role of MSDL as a glue between Tools, Algorithms, and Profiles.

17.7.2   MSDL Requirements

Our discussion of the scope of MSDL, although quite helpful in clarifying the range of functions MSDL is addressing, is still not specific enough. We now list various categories of requirements[10] that MSDL is expected to satisfy and list detailed requirements in each category:

- *General Requirements*
- *Structural Requirements*
- *Interface Requirements*
- *Construction Requirements*
- *Downloading Requirements.*

17.7.2.1   *General Requirements*

1. The MSDL should allow for genericity up to Level 2

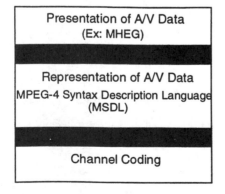

Fig. 17.18   Position of MSDL in a vertical stack of protocols.

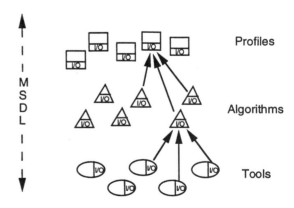

Fig. 17.19   Role of MSDL in binding Tools, Algorithms, and Profiles.

2. The MSDL shall provide rules for parsing the bitstream. This will define the procedure for understanding the ensuing data.

3. The MSDL should define a Structure Module, the Interface Description Module, an Object Construction Module, a Downloading Description Module, and a Configuration Module.

4. The MSDL should provide solutions for a configuration phase, a learning phase, and a transmission phase.

5. The MSDL should provide means for synchronization.

6. The MSDL should provide a means for error-free transmission of its data files (which contain set-up information between encoder and decoder, downloading of new tools/logarithms, etc.)

17.7.2.2   Structural Requirements

1. The MSDL should provide a structural library for the objects (tools, algorithms, and profiles).

2. The MSDL should provide different ranges of programmability for different applications.

3. The MSDL objects should be extensible with a mechanism to add in the future, ISO standardized data.

4. The MSDL structure shall use constructs, notations, and formats in a consistent manner.

5. The MSDL should provide for efficient coding of object identifiers.

17.7.2.3   Interface Requirements

1. The MSDL should provide ways to describe objects, this means describing their interfaces, what information fields exist, and which fields have to be set up each time the object is used and which ones do not.

2. The MSDL shall use constructs, notation, and formats in a consistent manner.

3. The MSDL shall code data with maximum efficiency. This means that dynamic and static parameters shall be coded with maximum efficiency. Further, there shall be an efficient mechanism to default such parameters when their coding is necessary.

17.7.2.4   Construction Requirements

1. The MSDL should provide a language to link elementary objects or already built complex objects to create new complex objects.

2. The MSDL via OCM should allow for easy mapping from the description of the objects used and connections among these objects to a real coding/decoding processor.

17.7.2.5   Downloading Requirements

1. The MSDL should be a language that has a flexibility and expressiveness similar to that of other general-purpose programming languages.

2. The MSDL should have as a goal to construct a machine-independent downloading language.

3. The MSDL via DDM should support dynamic binding of objects during the run-time to allow adaptation and adjustment during communication.

4. The MSDL should provide for mechanisms to ensure security and preservation of intellectual property rights.

17.7.3   MSDL Parts

Having reviewed the requirements that the MSDL must satisfy, it is clear that the MSDL effort needs to be clearly partitioned into a number of areas.[10] Currently, the ongoing MSDL work expects to address the following areas:

Architecture (MSDL-A): Specifies the global architecture of the MPEG-4 system. This includes the role of MPEG-4 system to support complete audio–visual applications as well as conceptual objects and their data content exchanged between encoder and decoder.

Class Hierarchy Definition (MSDL-O): Specifies particular classes of objects that are useful for specific audio–visual applications. It includes the definition of class libraries of MSDL.

Readable Language Specification (MSDL-R): Describes a readable format for transmission of decoder scripts.

Binary Language Specification (MSDL-B): This is a specification of a binary executable format for scripts or descriptions. This is the executable binary language understood by the decoder.

Syntactic Description Language (MSDL-S): This is a specification of a language that can be used to describe the bitstream syntax. It is expected to be an extension of the MPEG-2 syntax description method into a formal well-defined language that can be interpreted by machines.

Multiplex Specification (MSDL-M): Describes a procedure for multiplexing of encoded information.

Work on all the aforementioned parts of MSDL is currently in progress and thus it is best to discuss only the basic concepts behind these methods rather than any details.

17.7.4 MSDL-A

The MSDL-A specifies the global architecture of an MPEG-4 system; an MPEG-4 system is a system for communicating audio–visual objects. An audiovisual (AV) object may be aural or visual, 2D or 3D, static or time varying, natural or synthetic, or combinations of these types. The architecture for communicating these objects is as follows. The coded AV objects and their spatiotemporal relationships if any are transmitted by an MPEG-4 encoder to an MPEG-4 decoder. At the encoder, the AV objects are compressed, error protected, multiplexed, and transmitted (downstream) to the decoder. Actual transmission may take place over multiple channels each with

different service qualities. At the decoder, AV objects are demultiplexed, error corrected, decompressed, composited, and presented to the user. The user may interact with the presentation; the user interaction information may be used locally at the decoder or transmitted back (upstream) to the encoder. Figure 17.20 shows the architecture of the MPEG-4 system from the MSDL aspect.

The encoder and the decoder exchange configuration information prior to the encoder transmitting AV objects. The encoder determines the classes of algorithms, tools, and other objects that the decoder must have to process coded AV objects it intends to send to the decoder. Each class of objects can be defined by a data structure and the executable code. The definitions of missing classes are downloaded to the decoder, where they supplement or override existing or predefined classes at the decoder. If the decoder needs new class definitions in response to the user interaction it can request these from the encoder to download additional class definitions, in parallel with transmitted data. Figure 17.20 clarifies these various aspects just discussed.

For the new classes that the encoder communicates to the decoder, class definition can be considered as the header and the data component as the

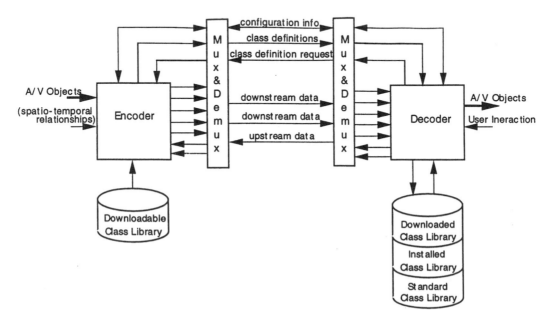

Fig. 17.20 Overall System architecture from the MSDL aspect.

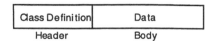

Class Definition	Data
Header	Body

Fig. 17.21 Audio–visual objects specification.

body of a transmitted A/V objects. The class definition specifies the data structure and the methods that are used to process the data component. The specification of an AV object is illustrated in Fig. 17.21.

Not all AV objects need to transmit their class definition along with the data, for example, in the case of an MPEG-1 movie, its class definition can be found in a standard library. Likewise, if an AV object uses the same class definition as one already sent, it does not need to be sent again. Data are used to control the behavior of objects. For some AV objects, there may be no data component, as the behavior of the object may be predetermined. It is expected that all class definitions may be transmitted lumped separate from the data components.

17.7.5 MSDL-O

The MSDL-O specifies classes of MPEG-4 objects needed in specific applications. In particular it addresses classes of objects that would need to be constructed during decoding to render and manipulate the audio–visual scene.

Audio–visual objects correspond to data structures that can actually be rendered by the compositor. In a coded bitstream audio–visual objects are represented as a sequence of content objects of various types.

Process Objects correspond to a process that will be applied on content or presentable objects and will return a new presentable object. Hence, each process object contains an *Apply* method.

A partial class hierarchy for MSDL-O is as follows:

- Audio–visual and Content Objects
  —Audio–visual Scene Model
  —Hierarchical Model for Spatiotemporal Scene
  —Modeling 2D Objects
  —Texture Information
  —Shape Information

  —Motion Information
  —Depth and Opacity Information
  —Location Information
- Process Objects
  —Quantizer Object
  —Transformer Object
  —Compensator Object.

17.7.6 MSDL-R

The MSDL-R specifies a readable format or a textual representation of the algorithm. It uses a subset of (C and) C++ as the starting point; however, it is envisaged that new capabilities not present in C++ such as multithreading and message passing will need to be added at some point.

The MSDL-R features include the following.

- C variable names, expressions, and function calls
- C constructs: if–else, for, while, for which conditions are limited to constants, variables, and function calls.
- Return statements.

The MSDL-R excludes the following:

- Definition of functions, procedures, and global variables
- Preprocessor directives
- Statements such as goto, break, and continue
- Virtual functions
- Structures or Classes and access to them
- Type casts.

Overall, for security reasons, the MSDL-R is strongly typed, that is, no casts, function parameter type checking, no implicit type change.

An MSDL-R program is a procedure. It has local parameters, local variables, global variables, and instructions. No new data structure can be defined in MSDL-R; only the standard elementary types and decoder known types are allowed.

17.7.7 MSDL-B

The MSDL-B is the specification of the binary executable format and can be generated from MSDL-R. Two approaches have been identified. In the first approach, MSDL-B is a coded version of the high-level language MSDL-R; for example, using Huffman coding of MSDL-R keywords. The second approach involves compilation of MSDL-R to a platform-independent low-level representation. Currently both approaches are being evaluated and there are concrete proposals for each.

In the first approach, MSDL-B is obtained by direct compression of the MSDL-R code. MSDL-R keywords, arithmetic operators, and variables are mapped to predefined codewords. The decoder can reconstruct the original MSDL-R code, but without the symbolic names used for class instances, variables, and methods. MSDL-B can be executed directly or translated to native machine code.

In the second approach, MSDL-B is obtained by compilation at the encoder. Compilation can be done based on "lex" or "yacc" which are textual syntax analysis tools. Initially, the resulting MSDL-B code may be suboptimal in terms of efficiency but its efficiency can be improved. Instructions are 32 bits long and are interpreted by Forth (programming language) like stack-based interpreters. The interpreter and tools are written in C++ and linked together. If required, a direct compilation to the native machine language is also possible.

17.7.8   MSDL-S

The goal of MSDL-S is to provide a language that can be used to describe the bitstream syntax. It is intended to somewhat separate the definition of bitstream syntax from the decoding/rendering process. Since MPEG-4 intends to provide flexibility and programmability it becomes necessary to separate syntax from processing to allow content developers to disclose only the bitstream structures they may use but not the details of any processing methods. Another advantage of this separation is that the task of the bitstream architect is greatly simplified, allowing focus on decoding and preparing it for display rather than the mundane task of extracting bits from the bitstream. Further, this separation also eases the process of compliance with the overall bitstream architecture.

Some explanation is in order regarding the relationship of MSDL-S and MSDL-R. The MSDL-R, as discussed earlier, addresses the general pro-gramming facilities of MSDL, and is the language in which decoding and generic processing operations are described. On the other hand, MSDL-S is an orthogonal subset of MSDL-R; orthogonality implies that the two are independent, that is, the specification of MSDL-S does not affect the specification of MSDL-R. There is, however, an assumption of commonality at the level of their capability to define object hierarchies. Currently, MSDL-S assumes a C++ or Java like approach to be employed for MSDL-R. For example, MSDL-S can be thought of as generalizing the concept of declaring constants with hard coded values to declaring constants that derive values from the bitstream. An MSDL-R programmer can assume that a constant is parsed from the bitstream before it is accessed, similar to the way a programmer is not concerned about how the initialization of a constant occurs, as long as it does.

We now briefly introduce the main elements of Syntax Description Language, the work on which is currently in progress.

- Elementary Data Types
  —Constant-length direct representation bit-fields
  —Variable-length direct representation bit-fields
  —Constant length indirect representation bit-fields
  —Variable length indirect representation bit-fields
- Composite Data Types
- Arrays
- Arithmetic and Logical Expressions
- Temporary Variables
- Control Flow Structures
- Parsing Modes.

17.7.9   MSDL-M

The goal of MSDL-M is to specify multiplexing and demultiplexing procedures for coded audio–visual information. In particular this goal can be broken down into a number of functions, such as interleaving of multiple compressed streams into a single stream, recovery of a system time base, managing of a decoder buffer, synchronization of multiple compressed data streams on decoding, and others. Owing to a large variety of anticipated MPEG-4 applications, such as broadcast, real-time bidirectional communication, database retrieval, and others, substantial flexibility needs to be provided.

The starting point for MSDL-M is the MPEG-2 systems specification (13818–1), the ITU-T multiplex specification H.223, and the evolving ITU-T multiplex specification H.22M. This approach is intended to maintain commonality with various current and evolving standards. To proceed further with the design of MSDL-M, a number of categories of requirements, each with its own list of requirements, are taken into account. The main categories of requirements are as follows:

- General Requirements
- Application-Dependent Requirements
- Timing/Synchronization Requirements
- Error Resilience Requirements
- Network Adaptation Requirements
- Compatibility Requirements.

The specification of MSDL-M is currently being developed taking into consideration the detailed requirements in each of these categories and the system functionalities already in existence in other standards. A partial list of specific issues that are being addressed is as follows:

- Clock Synchronization
- Stream Synchronization
- Error Resilience
- Configuration
- Periodic Retransmission
- Buffer Management Models.

17.7.10 MSDL Directions

The MSDL effort is expected to result in part 1 of the MPEG-4 Working Draft-1 by November 1996, coincident with the schedule for parts 2 and 3. The schedule for following iterations of the Working Draft (Working Draft-2 and Working Draft-3), the Committee Draft, and the International Standard was discussed earlier while discussing the schedule for Video.

17.8 SYNTHETIC AND NATURAL HYBRID CODING (SNHC)

In recent years, with advances in digital video technology and fueled by ever increasing demand for sophisticated multimedia, traditionally separate fields of natural images/video and synthetic images/animations have been merging at a breathtaking speed. For example, several recent movies have included composites of natural and quite realistic looking synthetically generated scenes, with composition performed using chroma keying (Chapter 5). The key to even more sophisticated multimedia is not only to produce more sophisticated composites but also to provide the ability to interact with audio–visual objects in these scenes; this ability is severely limited at the present time. Traditional audio–visual presentation has simply consisted of displaying a frame of video and playing with its accompanying audio soundtrack/s. The increasing popularity of the World Wide Web, which offers the opportunity to interact with and thus control the presentation of data in a highly nonlinear fashion, implies that increasingly in the future, multimedia data will also be interacted with and presented in the same way. Thus, coding of natural and synthetic image data, ability to interact with and manipulate objects in the coded domain, and the ability to control the order of presentation are related important functionalities that MPEG-4 is addressing via the Synthetic and Natural Hybrid Coding (SNHC)[12] effort.

The most obvious way of providing access to individual objects in the coded domain is by coding each object individually. In addition, it is anticipated that by coding objects separately some gain in overall compression efficiency may also be obtained as different coding strategies can be more easily used for different objects, as appropriate. A normal way to proceed would then simply appear to be to separately encode segmented or prerendered objects. However, there are certain applications such as video games, CAD, medical, and geographical where many different types of image data are employed and no universal standards exist, the volume of data generated is usually too high, and user interactivity places additional synchronization requirements. In such cases, a different coding paradigm is better suited and involves sending coded representation of parameters to the decoder followed by rendering at the decoder prior to display. In general, when dealing with both natural and synthetic data, a combination of the two paradigms appears to offer a practical solution.

The MPEG-4 SNHC effort is aimed at addressing the needs of sophisticated multimedia applications involving synthetic and natural, audio and video data in an efficient, yet practical way. The SNHC effort is expected to undergo two major phases, a competitive phase followed by a collaborative phase, much like the ongoing effort in MPEG-4 Video and Audio. MPEG-4 is currently seeking candidate proposals for consideration for standardization and is currently in its competitive phase.

17.8.1 SNHC Goal and Motivation

A more precise goal of MPEG-4 SNHC[12] work is to establish a standard for efficient coding, representation, and integration of 2D/3D, synthetic–natural hybrid data of an audio–visual nature.

This standard is intended to support a wide variety of multimedia experiences on a range of current and future platforms. SNHC is expected to

be employed in a variety of applications such as video games, multimedia entertainment, educational, medical surgery, industrial design and operations, geographical visualizations, and so forth.

17.8.2 SNHC Architecture and Requirements

The functional architecture for the SNHC Decoding System[12] is shown in Fig. 17.22. It consists of System Layer Decoder, a Video Decoder, an Image Synthesizer, an Audio Decoder, an Audio Synthesizer, a Cache, and a Display Processor. The Display Processor is controlled by the user to control compositing of audio, video, and graphics objects for presentation. A Cache is used to store objects in repeated use such as 3D geometry, texture, audio clips, and segmentation masks.

The SNHC effort, by defining what is referred to as a media model, is expected to focus on providing building blocks for efficient coding and rendering in interactive environments, which synchronize 3D embedding of various objects such as images, real-time video, audio, and static or animated shapes. These objects may be either natural or synthetic in nature. We now list the various issues that need to be resolved in completely defining a media model:

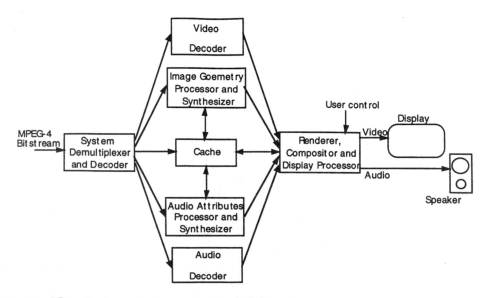

Fig. 17.22 Overall Decoding System Architectures from the SNHC aspect.

- Basic visual and audio primitives (polygons, sprites, textures, wavetables)
- 2D and 3D spatial models including geometric and structural properties
- Surface and material properties relative to scene illuminants
- Texture mapping attributes for still and moving images
- Combining natural and synthetic, static and moving, images and audio
- Combining synthetic spatial structures with traditional audio and video
- Structures for compositing scenes from objects
- Conventions for coordinate systems
- Parameterized models and free-form deformations
- Compressed and uncompressed data storage formats
- Static and dynamic behavior of objects
- Representation of viewers, cameras, and sound sources for rendering
- Naming and hyperlinking conventions for servers and libraries
- Modeling for different class of terminal resources and scene degradation.

17.8.3   SNHC Test Conditions

SNHC currently is in the process of establishing a standardized digital 2D/3D data set referred to as the Virtual Playground (VP). The VP will include critical data sets needed to discern performance differences between various competing approaches expected to be submitted to MPEG-4 for SNHC evaluation.

The VP is a collection of data sets representing agreed upon objects. These objects will be stored in popular formats such as VRML 1.0 for geometric objects, JPEG for image textures, and AIFF for audio clips. These objects, when selected, will be downloadable from MPEG World Wide Web home page. Various compositions of these VP objects will be used to demonstrate applications of MPEG-4 as well as uncover areas for study and standardization.

The VP is expected to include objects with features such as animations, complex geometry, texture mapping, and various sound sources to test complexity and synchronization issues. CAD models of various items such as furniture and vehicles may be used

to represent complex geometry; talking synthetic characters may exercise the use of parametrized models, audio synchronization, and 3D localization.

17.8.4   SNHC Evaluation

SNHC focuses on coding for storage and communication of 2D and 3D scenes involving synthetic–natural images, sounds, and animated geometry. Further, the coded representation should facilitate various forms of interactions. The SNHC group is seeking algorithm and tool proposals for efficient coding and interactivity in the coded domain; the following is a partial list[11] of topics of potential interest:

- Compression and simplification of synthetic data representations—synthetic and natural texture, panoramic views, mapping geometry, mapping photometry, animation, and deformation
- Parametrized animated models—encoding of parameterized models and encoding of parameter streams
- New primitive operations for compositing of natural and hybrid objects
- Scalability—extraction of subsets of data for time-critical use and time-critical rendering
- Real-time interactivity with hybrid environments
- Modeling of timing and synchronization
- Synthetic Audio

The MPEG-4 SNHC effort, in the competitive phase, similar to the MPEG-4 video effort, has made an official call for proposals to undergo evaluation. The evaluation of proposals is likely to take place later in 1996. These proposals are expected to be evaluated by a group of experts. The evaluation criteria are likely to be based on functionality addressed, coding efficiency, quality of decoded model, real-time interactivity, anticipated performance in the future, and implementation cost. Each proposer is required to submit the following items:

- Technical Description detailing scope, advantages, description, and statistics
- Coded Bitstream of test data set and an executable decoder capable of decoding submitted streams

- A D1 tape showing results of compression/ decompression, and simplification or any other process that modifies the data.

17.8.5 SNHC Directions

The SNHC proposers are expected to provide algorithms and tools to address certain functionalities, and provide associated media models and bitstream format. After the evaluation of SNHC proposals, the selected proposals will undergo further testing by core experiments, similar to that being used for MPEG-4 video. These experiments are expected to cover many categories such as compression, scalability, new media primitives, timing and synchronization, and real-time interactivity. A collection of algorithms, tools, and associated media models and bitstream formats is expected to result from the SNHC standardization effort. It is also expected that during the convergence phase there will be a need to harmonize SNHC, MSDL, MPEG-4 Video, and MPEG-4 Audio. In response to the recent SNHC Call for Proposals,[11] the evaluation of proposals is expected to take place in September 1996.

The MPEG-4 SNHC effort is expected to contribute to tools and algorithms for part 2 and part 3 of the MPEG-4 Working Draft-1. Part 2 of MPEG-4 deals with synthetic and natural visual representation and coding tools and algorithms, while part 3 of MPEG-4 deals with synthetic and natural aural representation and coding tools and algorithms. The MSDL is expected to provide the framework and glue for uniformly addressing natural and synthetic representations. The SNHC effort is thus expected to contribute to tools and algorithms in parts 2 and 3 of the MPEG-4 Working Draft-1 by November 1996, coincident with the tools and algorithm contributions from natural video to part 2 and natural audio to part 3. The schedule for following iterations of the Working Draft (Working Draft-2 and Working Draft-3), the Committee Draft, and the International Standard was discussed earlier while discussing the schedule for Video.

17.9 THE FUTURE

The MPEG-4 standard, owing to its flexibility and extensibility, is expected to remain a very useful standard for many years to come. Although MPEG-4, when completed, is expected to support many functionalities, only the very basic functionalities may be applied initially for specific applications. This may occur on next generation video signal processors or on custom VLSI chips. The more advanced functionalities are likely to be deployed only at a much later stage when very powerful and flexible video processors that can take full benefit of the MSDL capabilities become possible.

In the very near term, a number of applications, products, and services are expected to be based on MPEG-1, MPEG-2, and the MPEG-4 standards. In the United States, it appear that even before the introduction of HDTV, digital TV may be introduced; currently the FCC Advisory Committee on Advanced Television Standards (ACATS) is finishing up specification of digital TV standard, the video and systems part of which are based on the MPEG-2 standard. A number of types of devices (e.g., set-top boxes) allowing access and interactivity to a new generation of services are expected to be deployed. Relating to these devices, besides MPEG standards, the work done by DAVIC (Digital Audio and Video Council) and MHEG (Multimedia and Hypermedia Experts Group) is quite relevant.

17.9.1 Future Standards

The MPEG-4 standard is promising unparalleled flexibility and extensibility. This would seem to imply that if it succeeds in achieving its goals, there would not be a need for a future audio–visual standard. In reality, there will still be a need for audio–visual standardization activity, although it may not be mainly related to standardization of coding methods. We briefly discuss one such topic on which there may be a need for future MPEG standards activity.

17.9.1.1 Man–Multimedia Service Interface

Man–Multimedia Service Interface (MMSI)[14] can be defined as a set of high-level functions that enable user to access, utilize, and manage multimedia services. Examples of specific operations may be selection, browsing, navigation, customization, and so forth. This activity may use existing and emerging multimedia standards and may provide solutions for a wide range of multimedia access devices and mechanisms including minimal resource devices.

Multimedia services are typically employed by a number of types of industries such as content providers, service operators, network operators, platform manufacturers, and end users. Each industry has its own set of requirements; however, taking into account a common set of requirements across various industries the following trends seem clear.

- The need for standardized navigation and browsing facilities
- The need for clear separation between contents and applications on one hand, and actual platform and configurations on the other hand
- The need for standardization of multimedia information, balanced by the need for differentiation between competing systems.

It is envisaged that multimedia services under MMSI may be categorized as follows:

- *Composition-Oriented Services:* These services include providing access to decoded representations of objects or actual decoding.
- *Distribution-Oriented Services:* These services may provide distribution of objects, scripts, content data, and so forth in a distributed environment.
- *Presentation-Oriented Services:* These services support the development of a dedicated environment where published and distributed multimedia information encoded according to a given representation will be perceived as an application-specific man–machine communication.
- *Utility Services:* This implies the need for user service groups to regulate access to multimedia information. This includes accounting, billing, security, and so forth.

This future work item[14] has recently been brought to MPEG for consideration and will undergo discussion of its relevance to MPEG and possible reinterpretation. There appears to be some commonality between this topic and topics addressed by MPEG-4, which can provide a smooth transition to the next MPEG standard.

17.9.2 Example Applications

Now it is time for some predictions about the future of audio–visual applications. This is where we the authors get to look into our crystal ball and with a fairly straight face try to make sense of where the technology is heading (a generally impossible task!). And yes, even knowing well in advance that we are going to get a good-natured ribbing from our colleagues and friends, we still dare to stick our necks out!! So, here goes. . .

17.9.2.1 Short Term (Within 3 to 5 years)

- Digital TV to home via terrestrial broadcast and cable. Interactive features allowing viewer participation
- Multimedia of high quality via Internet and other means. Variety of multimedia services such as multimedia e-mail, shared real-time multimedia work spaces, VOD (video on demand), MOD (movies on demand), variety of databases, multimedia information access, and retrieval tools
- Wireless multimedia on a number of devices such as personal phones that can be used anywhere, cellular phones, Personal Digital Assistants (PDAs), wireless Internet access terminals, and so forth
- Portable multimedia terminals that can be used for advanced networked games, video phones, or for database access

17.9.2.2 Medium Term (Within 5 to 12 years)

- Digital HDTV via satellite, terrestrial and cable, for reception by home TV, PC-TVs etc. Advanced interactive features allowing viewer participation
- Specialized services offering customized multimedia newspapers, customized magazines, customized movies, and special events
- Advanced, personal wireless multimedia communicators providing seamless access to a variety of services and systems locally, regionally, nationally, and globally.
- High-quality wireless multimedia headsets bringing capabilities of networked video game playing, video database access, Internet access, telepresence and others via functionalities such as

stereoscopic viewing, high-end graphics, 3D sound, and others

- PCs and high end game/Internet systems offering high-quality stereoscopic/multiviewpoint access, graphics capabilities, and interactivity.

17.9.2.3   Long Term (Beyond 15 years)

- User configurable personalized information multimedia services, devices, and networks
- Virtual entertainment gear (suits) combining advanced multimedia and variety of sensors to provide advanced communication, networking, games, and experiences
- Holographic TV/display systems offering surreal interactivity and immersion experiences.

17.10   SUMMARY

In this chapter we have taken a step beyond MPEG-2 and introduced the next MPEG (MPEG-4) standard currently in progress. The necessary background information leading up to the MPEG-4 standard and the multiple facets of this standard are introduced showing relationships to the MPEG-1 and the MPEG-2 standards. In particular we have discussed the following points:

- The bitstream syntax, decoding process, optional modes, and encoder configuration of the low-bitrate ITU-T H.263 standard which is the starting basis for the MPEG-4 standard
- Recent advances in Transform Coding, Wavelet Coding, Fractal Coding, Region-/Object-Based Coding, and Model-Based Coding that may offer promising solutions for MPEG-4 Video coding
- An introduction to MPEG-4 including a discussion of its focus, applications, functionality classes, and how it is organized
- The requirements for MPEG-4 including terminology, advanced functionalities, basic functionalities, and analysis of evolving requirements
- The MPEG4 Video development process including test conditions, first evaluation, results of evaluation, formulation of core experiments, definition of the Verification Model, and directions of ongoing work
- A brief introduction to the MPEG-4 Audio development process

- Discussion about the MPEG-4 Synax Description Language including its scope, requirements, its various parts, architecture, details of its parts, and directions of ongoing work.
- The MPEG-4 Synthetic and Natural Hybrid Coding effort, its goals, architecture and requirements, test conditions, evaluation process, and directions of ongoing work
- The Future—what are the implications of MPEG-4 and what lies beyond? Example applications involving audio–visual coding that seem promising for the short term, the medium term, and the long term.

REFERENCES

1. AOE GROUP, "MPEG-4 Proposal Package Description (PPD)—Rev. 3," ISO/IEC JTC1/SC29/WG11 N0998, Tokyo (July 1995).

2. AOE GROUP, "MPEG-4 Testing and Evaluation Procedures Document," ISO/IEC JTC1/SC29/WG11 N0999, Tokyo (July 1995).

3. L. CHIARIGLIONE, "MPEG-4 Call for Proposals," ISO/IEC JTC1/SC29/WG11 N0997, Tokyo (July 1995).

4. A. PURI, "Status and Direction of the MPEG-4 Standard," *International Symposium on Multimedia and Video Coding*, New York (October 1995).

5. ITU-T, "Draft ITU-T Recommendation H.263: Video Coding for Low Bitrate Communication" (December 1995).

6. T. EBRAHIMI, "Report of Ad Hoc Group on Definition of VMs for Content Based Video Representation," ISO/IEC JTC1/SC29/WG11 MPEG 96/0642, Munich (January 1996).

7. A. PURI, "Report of Ad Hoc Group on Coordination of Future Core Experiments in MPEG-4 Video," ISO/IEC JTC1/SC29/WG11 MPEG 96/0669, Munich (January 1996).

8. MPEG-4 REQUIREMENTS AD HOC GROUP, "Draft of MPEG-4 Requirements," ISO/IEC JTC1/SC29/WG11 N1238, Florence (March 1996).

9. MPEG-4 VIDEO GROUP, "MPEG-4 Video Verification Model Version 2.0," ISO/IEC JTC1/SC29/WG11 N1260, Florence (March 1996).

10. MSDL AD HOC GROUP, "MSDL Specification V1.1," ISO/IEC JTC1/SC29/WG11 N1246, Florence (March 1996).

11. MPEG-4 INTEGRATION GROUP, "MPEG-4 SNHC Call for Proposals," ISO/IEC JTC1/SC29/WG11 N1195, Florence (March 1996).

12. MPEG-4 INTEGRATION GROUP, "MPEG-4 SNHC Proposal Package Description," ISO/IEC JTC1/SC29/WG11 N1199, Florence (March 1996).

13. L. CHIARIGLIONE, "Resolutions of 34th WG11 Meeting," ISO/IEC JTC1/SC29/WG11 N1186, Florence (March 1996).

14. ISO/IEC JTC1/SC29, "Proposal for a New Work Item: Information Technology—Man–Multimedia Service Interface," ISO/IEC JTC1 N3333 (December 1994).

15. DELTA INFORMATION SYSTEMS, "Technical Work in the Area of Video Teleconferencing: Advanced Coding Techniques," Final Report Contract DCA100-91-C-0031 (December 1994).

16. A. PURI, R. L. SCHMIDT, and B. G. HASKELL, "SBASIC Video Coding and Its 3D-DCT Extension for MPEG-4 Multimedia," *Proc. SPIE Visual Commun. Image Proc.,* pp. 1331–1341 (March 1996).

17. T. EBRAHIMI, E. REUSENS, and W. LI, "New Trends in Very Low Bitrate Video Coding," *Proc.* IEEE 83 (June 1995).

18. J. KATTO et al., "A Wavelet Codec with Overlapped Motion Compensation for Very Low Bit-Rate Environment," *IEEE Trans. Circuits Syst. Video Technol.,* pp. 328–338 (June 1994).

19. H. CHEN, C. HORNE, and B. G. HASKELL, "A Region Based Approach to Very Low Bitrate Video Coding," AT&T Internal Technical Memorandum (December 1993).

20. J. OSTERMAN, "Object-Based Analysis–Synthesis Coding Based on the Source Model of Moving Rigid 3D Objects," *Signal Proc. Image Commun.,* pp. 143–161 (May 1994).

21. K. AIZAWA, "Model-Based Image Coding," *Proc. SPIE Conf.* 2308: 1035–1049 (1994).

22. A. PURI, "Efficient Motion-Compensated Coding for Low Bit-rate Video Applications," Ph.D. Thesis, City Univ. of New York, 1988.

23. H. PETERSON, "Report of the Ad Hoc group on MPEG-4 Video Testing Logistics," ISO/IEC JTC1/SC29/WG11 Doc. MPEG95/0532 (November 1995).

24. J. OSTERMAN, "Report of the Ad Hoc group on Evaluation of Tools for non tested Functionalities of Video submissions," ISO/IEC JTC1/SC29/WG11 Doc. MPEG95/0488 (November 1995).

25. W. LI, Y.-Q. ZHANG, "Vector-Based Signal Processing and Quantization for Image and Video Compression," Proc. of the IEEE, vol. 83, No. 2, pp. 317–335 (February 1995).

26. D. E. PEARSON, "Developments in Model-Based Video Coding," Proc. of the IEEE, vol. 83, No. 6, pp. 892–906 (June 1995).

27. M. LI and R. FORCHHEIMER, "Two-View Facial Movement Estimation," IEEE Transac. on Circuits and Systems for Video Technology, vol. 4, No. 3 (June 1994).

28. F. LAVAGETTO and S. CURINGA, "Object-oriented Scene Modelling for Interpersonal Video Communication at Very Low Bit-rate," Signal Processing: Image Commun., pp. 380–395 (Oct. 1994).

29. T. CHEN et al., "Speech Assisted Lip Synchronization in Audio-Visual Communications," Int. Conf. on Image Proc. (Oct. 1995).

30. G. BOZDAGI, A. M. TEKALP, and L. ONURAL, "3-D Motion Estimation and Wireframe Adaptation Including Photometric Effects for Model-Based Coding of Facial Image Sequences," IEEE Transac. on Circuits and Systems for Video Technology, vol. 4, No. 3, pp. 246–256 (June 1994).

31. O. LEE, and Y. WANG, "A Region-Based Video Coder Using an Active Mesh Representation," Proc. IEEE Workshop on Vis. Signal Proc & Commun., pp. 168–172, Sept. 1994.

32. T. CHEN and R. RAO, "Audio-Visual Interaction in Multimedia: From Lip Synchronization to Joint Audio-Video Coding," to appear IEEE Circuits and Devices magazine.

Appendix A

Resource Guide

We now provide a list of useful resources for those interested in further learning, exploring, or (yes) simply having fun with the MPEG 1 and the MPEG-2 standards. As can be expected, any such list of resources is continually changing and as such can never really be fully up-to-date, accurate, or complete. Any omissions or mistakes in these lists we attribute to our ignorance, pure coincidence, or unfortunate editorial errors.

A. GENERAL

1. News groups on Internet—comp.compression, comp.graphics, comp.multimedia, alt.comp. compression, alt.binaries.multimedia, comp. os.ms-windows.misc, comp.os.ms-windows. programmer.graphics, comp.os.ms-windows. programmer.multimedia, comp.os.ms-windows. video, alt.binaries.multimedia, alt.binaries. pictures.utilities

2. MPEG-related web resources at http://www. cs.tu-berlin.de/~phade/mpegwww.html

3. MPEG Archive at http://www.cs.tu-berlin. de/~phade/mpeg.html

4. MPEG FAQ at BTU-Berlin at http://www. cs.tu-berlin.de/~phade/mpegfaq/index.html

5. The Internet MPEG CD-Rom—Version 1.0 at http://www.cs.tu-berlin.de/~phade/ mpegcd/mpegcd.html

6. MPEG-1 CD-Rom at http://www.bbtt.de/ harti/mpegcd.html

7. MPEG FAQ at CRS4 at http://www. crs4.it/~luigi/MPEG/mpegfaq.html and ftp://ftp.crs4.it; Addresses: 156.148.1.4, 156.148.3.4, 156.148.4.4, 156.148.5.4 Contains misc. informative MPEG articles, code, FAQs.

8. Video on the Internet: The MPEG Technology, at http://www.netvideo.com/ technology/technology.html

9. MPEG resources, trivia and tricks by Tristan Savatier, at http://sugar.la.tce.com/~ tristan/MPEG.html (http://198.211.69. 127/~tristan/MPEG.html) and http:// www.bok.net/~tristan (http://www.creative. net/~tristan) http://www.bok.net/~tristan/ MPEG.html (http://www.creative.net/~ tristan/MPEG.html) and http://www.bok. net/~tristan/leonardo/mp2/

10. Hans Bakhuizen's Broadcasting Corner at http://www.wp.com/hansbakhuizen/

11. Compression Times at http://www. webpress.net/doceo/

413

12. Workshop on Advanced Digital Video in the National Information Infrastructure, at http://www.eeel.nist.gov/advnii/

13. Advanced Television Systems Committee (ATSC) at http://atsc.org

14. International Standards Organization (ISO) for purchase of MPEG Standard at http://www.iso.ch/welcome.html

15. Win32 Software patch executables and drivers at ftp://ftp.microsoft.com/developers/DEVTOOLS/WIN32SDK; address: 198.105.231.1

Win32®'s patch executables and drivers which allows 32-bit programs to be executed on 386, 486, and Pentiums from within the traditional 16-bit Windows® (e.g., version 3.1) environment. This patch is needed to execute the Windows® display versions of the MPEG Software Simulation Group's programs if you do not already have Windows 4.0® (Chicago) or Windows NT®.

B.   HARDWARE

1. MPEG Mania, PCs, MPEG cards list at http://www.ziff.com:8009/~cshopper/features/9504/hw2_0495/

2. MPEG Related companies and products at http://www.crs4.it/~luigi/MPEG/mpegcompanies.html

C.   SOFTWARE, BITSTREAMS, AND OTHER RESOURCES

1.   System and Tools

1. MPEG2 Video/Audio/Systems Software at ftp://mm-ftp.cs.berkeley.edu/pub/multimedia/mpeg2

2. Home of mplex (MPEG-1 public domain Systems multiplexer) at ftp://ftp.informatik.tu-muenchen.de/pub/comp/graphics/mpeg/mplex/

3. Public-domain real-time MPEG-1 system software player (MCL Software Release) at http://spiderman.bu.edu/pubs/software.html or via anonymous ftp from spiderman.bu.edu in /pub/code.

MPEG-1 Multistream System Layer Player

Developed by J. Boucher (jboucher@spiderman.bu.edu), Z. Yaar, J. Palmer, and E. Rubin.

The player decompresses the MPEG system layer audio and video and attempts playout in real time. Because system layer streams can have up to 16 video streams interleaved with up to 32 stereo audio streams, the system can be easily overloaded. The player currently supports playback of the system layer stream on the following machines: SGI Indigo/Indigo-2®, Sun Sparc® (SunOS 4.1.1) with 8KHz μlaw AMD audio device, and Linux with a 16 bit soundcard (/dev/dsp). (NOTE: You can compile the software with the -DNONE option if you don't have an audio device. You will need MOTIF and SYSTEM V extensions for shared memory and semaphores. A user's guide is provided in postscript format in the docs. Test streams are also provided.

To achieve satisfactory audio and video playout, a machine with sufficient performance is required. The SGI Indy had suitable performance, but the 486/33 Linux and IPX platforms were marginal. The newer Suns or faster PCs will likely yield satisfactory playout.

As a by-product of this development, a modified mpeg_play v2.0 that provides for the real-time playback of MPEG-1 video streams was produced. Here, real-time means that the player adapts dynamically to drop frames to match the number of frames passed through the player to the frame rate given in the sequence header of the video stream. (These are in the range 24 fps to 60 fps). To create this executable, recompile the video player without the -DSYSTEM_PLAY switch. Any platform that plays mpeg_play v2.0 can use the real-time video player.

The player was designed with a Mosaic interface in mind so you can easily add it to the set of your Mosaic file players.

For more information, contact Jim Boucher (jboucher@spiderman.bu.edu) or mcl@spiderman.bu.edu.

The decoder distribution is in mpeg_system_play_v1.0.tar.gz.

The encoder/decoder distribution is in mpeg_system_source_v1.0.tar.gz.

The following binary player versions are available:

SGI_IRIX5.2: mpeg_system_irix5.2_v1.0.tar.gz.
SUNOS 4.1.1: mpeg_system_sun411_v1.0.tar.gz.
LINUX: mpeg_system_play_linux_v1.0.tar.gz.

4. MPEG1 Systems Multiplexer V1.1 by Munich University of Technology (Chris Moar, moar@informatik.tu-muenchen.de)

Announcing a newer, bug-fixed version of MPEG1/Systems multiplexer to code one or two MPEG A/V streams into one MPEG1/Systems stream is ready and put on our FTP site. It generates fixed bitrate MPEG1 streams for playback on MPEG cards or software players and keeps track of all buffer requirements by the ISO spec.

The software is available as C source code and/or DOS executable, for those that do not have a C development platform.

Check out:

ftp.informatik.tu-muenchen.de
in/pub/comp/graphics/mpeg/mplex/*.idx

The URL is:

ftp://ftp.informatik.tu-muenchen.de/pub/comp/graphics/mpeg/mplex/00-index.html

mplex-1.1.tar.gz is the C source with small README files.

mplex11.zip is the DOS® executable only (no documentation)

You might want to get both if you want to use the DOS executable.

The 46-page Postscript Document mpeg_systems_paper_0.99.ps.gz tries to explain all issues on MPEG/Systems, but *careful*, this is in *GERMAN* only.

5. MPEG Video and Audio Player / MCI driver, VMPEG 1.7 Lite, Windows® at Chromatic Research (Stefan Eckart, stefan@chromatic.com)

VMPEG 1.7 Lite is available via FTP from ftp://ftp.netcom.com:/pub/cf/cfogg/vmpeg/vmpeg17.exe

VMPEG is a fast video and audio MPEG player for Windows 3.1® and 95®. The attached README file describes some of its features.

Major changes to the previous version (1.6a, May '95) are:

- Support of arbitrary image scaling
- Video CD support (even in the Lite version)
- Stereo audio support (even in the Lite version)
- WAVE audio support (xyz.mpg + xyz.wav combinations)
- Integer arithmetic audio decoder (for 486 CPUs)

Retrieval Instructions:

site: ftp.netcom.com

type: anonymous (login: anonymous, password: [your E-mail address])

IP address: 192.100.81.1

directory: pub/cf/cfogg/vmpeg/

file: vmpeg17.exe (self-extracting zip, about 500 kbyte)

e-mail Retrieval:

To send a message to: ftp-request@netcom.com in the body of the message say:

SEND cf/cfogg/vmpeg/vmpeg17.exe

To split the 500 kbyte file into smaller, friendlier pieces:

SEND cf/cfogg/vmpeg/vmpeg17.exe [split size]

Example:

SEND cf/cfogg/vmpeg/vmpeg17.exe 50

.. will split the vmpeg 17.exe file into several pieces, each 50 Kbytes in size.

General Information:

mail a message to ftp-request@netcom.com with only the word "HELP" in the body of the message.

VMPEG V1.7 Lite
Windows 3.1® MPEG player
by Stefan Eckart
August 1995

Features:

- Full MPEG-1 video standard (ISO 11172–2): I, P, B frames of arbitrary size
- Plays system layer (ISO 11172–1) and video compression layer files
- Decodes layer II MPEG-1 audio in high quality stereo
- Plays MPEG video/WAVE audio file pairs
- High speed, e.g., up to 33 frames/s on a Pentium 90 for a 352 × 240 video only sequence
- Display options:
  4 × 4 ordered dither normal size (8 bit)
  4 × 4 ordered dither double size (8 bit)
  Grayscale (8 bit)
  True color (24 bit)
- Arbitrary scaling of the video output
- Supports DCI enabled graphics cards to achieve full-screen 24-bit real-time display
- Plays Video CDs
- MCI (Media Control Interface) provided
- Generates stream information and decode speed statistics
- Frame rate control:
  Full speed (no rate control)
  Synchronous (real-time)
  Manual (rate set by user)

Requirements:

- '386 (or better) processor
- 4 MB RAM (8 MB recommended)
- VGA or Super VGA
- Windows® 3.1 or Windows® 95

For video and audio playback of full resolution MPEG files a Pentium-based PC and a PCI graphics card is required. True-color full-screen display additionally requires a graphics card with color–space conversion and scaling hardware accessible via DCI (Display Control Interface).

The Win32s extensions are not required.

This release of VMPEG is a demonstration version. For this reason audio and Video CD playback are restricted to 60 seconds.

VMPEG is copyrighted software, © Stefan Eckart, 1995. You may use, copy, and distribute this program solely for demonstration purposes, only in unmodified form, and without charging money for it. Commercial use of this demonstration version is strictly prohibited.

Disclaimer: This program comes without any warranty. You are using it at your own risk

6. MPEG Software Decoder by Applied Vision at ftp://ftp.netcom.com/pub/apvision

MPEG software decoder running on Sun, about three or four times faster than mpeg_play on a Sparc 10, with synchronized audio and bitstream editing capabilities.

SGI (Silicon Graphics) workstation: Web-FORCE®, MPEG encoding tools for IRIX 5.3®, MpegExpert software, documentation, and demo version are available online via anonymous ftp from ftp.portal.com or ftp.netcom.com at the /pub/apvision directory.

For licensing information contact Kilicaslan Mertan (apvision@netcom.com

7. Systems multiplexer and demux for MPEG-1 at ftp://flash.bu.edu in directory pub/code

8. MPEG Tools by U. Pennsylvania at ftp://atum.ee.upenn.edu/dist/pkg/MPEGTool/

10. MPEG Utilities at ftp://ftp.best.com:/pub/bryanw/wwwfaq.zip and http://www.best.com/~bryanw/

Bryan Woodworth's (bryanw@best.com) Word Wide Web (WWW) Graphics Archive with hypertext and one-stop-shopping and retrieval of MPEG utilities. Also includes a hypertext MPEG FAQ written in HTML.

11. MPEG-2 Systems Transport Layer Source Code and Sample Bitstreams at ftp://wuarchive.wustl.edu/graphics/x313/pub/bitstreams/systems/

Site: wuarchive.wustl.edu {128.252.135.4}

login id: anonymous
password: {your E-mail address}
directory: graphics/x313/pub/bitstreams/systems/

12. Audio/Video MPEG Encoding/Decoding and Tools from North Valley Research at ftp://nvr.com/pub/NVR-software

Audio/Video MPEG encoding/decoding, audio/image conversion software. Requires IRIX

5.2. Any SGI platform will support encoding/ playback; Audio playback requires AL-compatible audio-equipped system. 100MHz or faster processor recommendcd for playback.

Status: Version 2.0 available via FTP to nvr.com. License required; demo licences available: ftp://nvr.com/pub/NVR-software

For Price list: ftp://nvr.com/pub/NVR-datasheets/price.list.txt

Contact: NVR (support@nvr.com)

13. CDDA tool to read CD-ROM files to hard disk at ftp://ftp.cdrom.com/pub/cdrom/ incoming/da2wav11.zip

CDDA: reads CD-ROM files (Red Book audio streams or WhiteBook Video CD streams) onto hard disk.

Site: ftp.cdrom.com
Directory: pub/cdrom/incoming/da2wav11.zip
Contact: Jim McLaughlin (jmclaugh@bnr.ca)

14. Integrated Audio/Video MPEG Player for workstations at http://cse.ogi.edu/DISC/ projects/synthetix/Player/ and ftp://ftp.cse. ogi.edu/pub/dsrg/Player/

Integrated audio/video MPEG player for workstations
Site: ftp.cse.ogi.edu (129.29.20.2)
Directory: /pub/dsrg/Player/.
WWW: http://cse.ogi.edu/DISC/projects/ synthetix/Player/
Contact: Shanwei Cen (scen@cse.ogi.edu)

15. Video File Analysis and Edit tool at ftp:// pkl.lancs.ac.uk/pub/mpegUtil/

MPEG-1 video file analysis and edit program
Site: pkl.lancs.ac.uk (194.80.36.104),
Directory: /pub/mpegUtil.
Contact: Phillip Lougher, (phillip@comp.lancs. ac.uk)

16. MPEG-1 Video Editor for X windows at ftp://ftp.demon.co.uk/pub/unix/tools/

Announcing the release of interactive Mpeg-1 video stream editor for X windows.

It has been uploaded to ftp.demon.co.uk in the directory /pub/unix/tools

The file is called mpegedit_v2.2.tar.gz
Contact: Alex Ashley (milamber@ripley. demon.co.uk)

17. MPEG on Xing Distributed Media Architecture (XDMA) at http://www.xingtech.com. products/xdma/xdma_general_info.html

XDMA was developed for client–server media distribution architecture which can operate independently or complement existing WWW HTTP/HTML architectures on local area networks, private data wide area networks, and public data networks (internet).

XDMA delivers "streaming" multimedia— pictures video and sound—based on MPEG international standards for video and audio compression from Unix and Windows NT® servers.

2. Video

1. Home of MPEG Research at U.C. Berkeley (mpeg_play, mpeg_encode ...) at http:// www-plateau.cs.Berkeley.EDU:80/mpeg/ and ftp://s2k-ftp.cs.berkeley.edu/multimedia/ mpeg; Address: 128.32.149.157

Fast MPEG-1 encoder and decoder/viewer with user-selectable encoding strategies. Good code. Network video server programs too. MPEG-1 video bitstream verification and analysis tools.

2. Home of Chad Fogg (mpeg2play, mpeg2decode, vmpeg etc.) at ftp://ftp. netcom.com/pub/cf/cfogg

3. Stanford University at ftp://havefun.stanford. edu/pub/mpeg, address: 36.2.0.35

Very flexible implementation. Well documented. JPEG and H.261 too. Good code; a tutorial on MPEG included. YUV to X-windows viewing utility also available at this site.

4. Source Image sequences at ftp://ipl.rpi.edu/; address 128.113.14.50

Raw image sequences (table tennis, flower garden, football, etc) in Sun Raster RGB format can be retrieved from: ipl.rpi.edu. The PBM plus package (ftp.uu.net pub/graphics/) contains a utility for converting .ras files to other formats usable by MPEG encoders.

5. Fast MPEG-1 player by Michael Simmon at ftp://ftp.netcom.com/pub/ms/msimmons/ and ftp.ecel.uwa.edu.au (130.95.4.2) /users/michael/

New site: ftp.netcom.com in /pub/ms/msimmons/

Michael Simmon's fast MPEG-1 player for Windows® based on the Berkeley player. Advanced VCR-like user interface and features. Directory also contains CD-I video player source code. This program comes recommended for Windows® World Wide Web browsers.

6. MPEG-1 video player for OS/2 by Mike Brown in ftp://ftp.cdrom.com/pub/os2/32bit/mmedia/PMMPEG21.ZIP.

7. The MPEG-1 Video Player and MPEG-2 Decoder for Macs by Maynard J. Handley

Version 2.3.1

Sparkle plays MPEGs, PICTs, and QT movies and converts between them. It is multifinder friendly and, with enough memory, will open multiple documents at once. It is free. Also in this package are a VERY rough quick and dirty port of some Unix MPEG-2 decoding code. MPEG demuxer included. This will split a muxed MPEG file (one containing both based on code by Rainer Menes. REQUIRES:—System 7 or greater. QuickTime 1.6 or greater. Sound Manager 3.0 or greater. An 020 based mac or greater. Works much better with SCSI Manager 4.3 installed. Thread Manager.2.0

Contact: maynard@elwing.otago.ac.nz (Maynard Handley)

3.   Audio

1. Playing MPEG Compressed sound files, at http://www.io.org/~cme/MPEG_Utils/mpeg.html

2. MPEG Audio Software Solutions by Kauai Media, at http://www.kauai.com/~bbal

3. MPEG-1 and MPEG-2 Audio software and data at ftp://ftp.tnt.uni-hannover.de/pub/MPEG/audio/; address: 130.75.31.73

The MPEG conformance audio test bitstreams are available by ftp from ftp.tnt.uni-hannover.de in the directory: /pub/MPEG/audio as compressed tar file: mpeg-testbitstreams.tar.Z (Hendrik Fuchs, fuchs@tnt.uni-hannover.de)

4. MPEG-1 fast Layer-II audio decoder source by Tobias Bading (MAPLAY) at ftp://ftp.iuma.edu/audio_utils/converters/source/mpegaudio.tar.Z; Address: 152.2.22.81 and ftp://ftp.cs.tu-berlin.de/; address: 130.149.17.7

Code in C++. The IUMA also contains several megabytes of Layer II bitstreams for audio. The ISO MPEG Committee audio codec source code is in pub/electronic-publications/IUMA/audio_utils/converters/

5. MPEG-1 Audio:Layer III shareware executables for MS-DOS at ftp://fhginfo.fhg.de; address: 153.96.1.4

Site for the Layer III MPEG-1 audio shareware executables for MS-DOS. Layer III was designed for efficient coding at lower bit rates than Layer II and Layer I applications. A copy of the audio conformance bitstreams can also be found here.

6. Information about MPEG-1 Audio Layer-3 (Mini FAQ), at ftp://fhginfo.fhg.de/pub/layer3/MPEG_Audio_L3_FAQ.html

7. MPEG Facts and Info (MPEG-1 Audio), at http://www.dreamscape.com/putz/mpegfact.html

4.   MPEG-1 Video Bitstreams

http://artemide.cselt.stet.it/MPEGS/

ftp://ftp.crs4.it/mpeg/meteo

gopher://dokudami.phys.tohoku.ac.jp:70/11/pub/mpeg/misc

ftp://artemis.earth.monash.edu.au/pub/Graph-ics/Mpeg-Anims

http://info.arl.mil/ACIS/ACD/SciVis/MPEG/

http://www.cs.dal.ca/movies

gopher://seds.lpl.arizona.edu:70/11/.ftp/pub/astro/SL9/animations

http://www.crs4.it/HTML/stat_top_13.html

http://www.mcs.csuhayward.edu/tebo/Anims

http://homer.nta.no/movies/movie_clips/index.html

http://jefferson.village.virginia.edu/pmc/

http://www.dsg.cs.tcd.ie:/dsg_people/afcondon/windsurf/video/video.html

gopher://gasnet.med.nyu.edu:70/11/Video

http://lal.cs.byu.edu/mpegs/mpegs.html

ftp://seds.lpl.arizona.edu/pub/images/clementine/mpeg

http://www.atmos.uiuc.edu/wxworld/mpegs/MPEGirl2/

5.   MPEG-2 Bitstreams

1.   MPEG2 Video/Audio/Systems Bitstreams at ftp://mm-ftp.cs.berkeley.edu/pub/multime-dia/mpeg2

2.   MPEG-2 Video Bitstreams at http://www.cstv.to.cnr.it/UK/cstvhp.html and ftp://ftp.cstv.to.cnr.it/outgoing/bitsadhoc

3.   MPEG-2 Video Conformance bitstreams at anonymous ftp site ftp://ftp.cstv.to.cnr.it/pub/ MPEG2/conformance-bitstreams/

Normative MPEG-2 video conformance bit-stream test suites are now available by anonymous on the server of CSTV (Centro di Studio per la Televisione) in Italy.

The bitstream test suites currently available include nonscalable 4:2:0 profiles up to Main Level (SP@ML, MP@ML and MPEG-1 compatibility tests).

Only bitstreams and trace files are made avail-able. Reconstructed frames are not provided and can be obtained with the software verifier of the technical report.

A relatively recent version of the source code of the technical report verifier (mpeg2decode) has been placed in verifier/mpeg2decode_950608.tar.gz. This version has been verified and works well for non-scalable profiles. However this verifier does not support the frame buffer intercept method that should be used to verify decoder com-pliance.

Future plans are:

*   To provide an updated (and verified) version of the verifier that works properly with scalable pro-files and that implements the frame buffer inter-cept method.

*   To provide a set of shellscripts that could be used to perform the static check of part 4, assuming that the frames reconstructed by the decoder under test are available.

*   To provide the test suites for 4:2:0 and nonscal-able profiles. However those bitstreams are much larger and ftp may not be the appropriate means to distribute them.

*   To provide additional bitstreams that are not part of the normative test suites but that would be useful for better testing decoders.

5.   MPEG-2 Systems Bitstreams at anonymous ftp site, retrieve bitstreams as follows:

ftp mpeg.cablelabs.com

ftp>login: anonymous

ftp>passwd: {you need to specify your email address here}

ftp>binary

ftp>cd/pub/bitstreams/working

ftp> (now you can download bitstreams by using 'get' command}

6.   MPEG-1 Audio:Layer III Conformance bit-streams at ftp://fhginfo.fhg.de; address: 153.96.1.4

7.   MPEG-2 Video verification bitstreams at ftp://wuarchive.wustl.edu/graphics/x313/

pub/video_verification/paris; address: 128.252.135.4

Copy of the 109 MBytes worth of verification bitstreams and picture reconstructions exchanged by MPEG committee participants. You do not want to download these files unless you are very serious about verifying your decoder implementation (not a recreational activity).

8. Sample MPEG-1 and MPEG-2 streams in graphics/x313/pub/video_verification/samples.

13 Mbytes of sample MPEG-2 and MPEG-1 streams.

9. MPEG-1 bitstream, "Burn Cycle," (brncycad.mpg) with audio and video. (5 MByte file ftp.cdrom.com, pub/cdrom/cdi directory) or (temporary location: pub/cdrom/incoming)

D. INTELLECTUAL PROPERTY

1. MPEG Intellectual Property Rights Licensing related information, at: http://www.cablelabs.com/PR/950327mpeg_ipr.html and http://www.cablelabs.com/PR/lausanne.html

E. MPEG-RELATED COMPANIES AND ORGANIZATIONS

1. MPEG Companies list at http://www.crs4.it/~luigi/MPEG/mpegcompanies.html

2. Bell Atlantic (Center for Networked Multimedia) at http://www.cnm.bell-atl.com/

3. Chromatic Research at http://www.mpact.com

4. DAVIC (Digital Audio Visual Council at http://www.cnm.bell-atl.com/davic/davic.html

5. FutureTel at http://www.ftelinc.com/

6. General Instrument at http://www.gi.com/

7. Hewlett-Packard at http://www.hp.com/

8. IBM at http://www.ibm.com/

9. Imedia Corporation at http://image.mit.edu/~pshen/imedia.html

10. Kauai Media at http://www.kauai.com/~bbal

11. Lucent Technologies at http://www.lucent.com/micro

12. MPR Teltech at http://www.mpr.ca/

13. Netvideo—Internet Video Services, Inc at http://www.netvideo.com

14. North Valley Research at http://nvr.com/htdocs/NVRHome.html

15. Optibase at http://www.optibase.com/

16. Optivision at http://www.optivision.com/

17. Sigma Designs at 1-800-845-8086

18. Siemens at http://www.eng.monash.edu.au/Commercial/Siemens/siemens.html

19. Sony Electronics at http://www.sel.sony.com/SEL/index.html

20. Telenor Research (Digital Video Coding group) at http://www.nta.no/brukere/DVC/

21. TNC at http://tnc.www.com/

22. VideoLogic at 1-800-494-4938 US Government (Digital Video in NII) at http://www.eeel.nist.gov/advnii/

23. Xing Technology at http://www.xingtech.com

24. AT&T at http://www.att.com

ARTICLES OF GENERAL INTEREST

1. R. JURGEN, "Putting the Standards to Work: Digital Video," *IEEE Spectrum mag.,* pp. 24–30 (March 1992).

2. R. A. QUINNELL, "Image Compression: Part 1," *EDN Mag.,* pp. 62–71 (January 21, 1993).

3. R. A. QUINNELL, "Image Compression: Part 2," *EDN Mag.,* pp. 120–126 (March 4, 1993).

4. R. A. QUINNELL, "Image Compression: Part 3," *EDN Mag.,* pp. 114–120 (May 13, 1993).

5. G. Legg, "New Chips Give PCs TV-Quality Video," *EDN Mag.,* pp. 38–42 (March 31, 1994).

6. D. Kim et al., "A Real-time MPEG Encoder Using a Programmable Processor," *IEEE Trans. Consum. Electron. J.,* pp. 161–170 (May 1994).

7. L. J. Nelson, "MPEG-1, MPEG-2 & You: Light in the Middle of the Tunnel," *Adv. Imag. Mag.,* pp. 28–31 (November 1994).

8. L. Yencharis, "Harder to PEG: Video Compression and Processing Chips Now," *Adv. Imag. Mag.,* pp. 20–26 (November 1994).

9. L. Yencharis, "Len's Top Ten Turned-On Chips: Video Compression/Processing Silicon Now," Adv. Imag. mag., pp. 14–18 (August 1995).

10. L. J. Nelson, "The Latest Introductions for Image and Video Compression," *Adv. Imag. Mag.,* pp. 72–79 (August 1995).

11. "MPEG Special Report," *Digital Video Mag.,* (August 1995); digital video website at http://dv.com/dv/Mag.html

12. K. Mills, "Compression It's All In The 'Rithmetic," Multimedia Producer magazine, pp. 34–36 (March 1995).

13. A. W. Davis, "The Latest Introductions for Still and Video Image Conferencing," Advanced Imaging magazine, pp. 72–79 (October 1995).

14. M. Magel, "The Box That Will Open Up Interactive TV," Multimedia Producer magazine, pp. 30–31, 34 (April 1995).

15. M. Berger, "Video Capture Boards: A Resource Directory," Computer Pictures magazine, pp. s-16–s-17, (November/December 1994).

16. L. Yencharis, "Meeting Video I/O Demands: One Size Doesn't Fit All," Advanced Imaging magazine, pp. 16–20 (April 1995).

17. L. Bielski, "Coming Soon to a Theater Near You: HDTV Digital Movie Transmission" Advanced Imaging magazine, pp. 30–32 (April 1995).

18. R. Doherty, "Here Comes MPEG 2. But which MPEG2?," Advanced Imaging magazine, pp. 85–86 (April 1995).

19. R. Doherty, "Digital Video Disk: An Inside Look at where it Really Stands," Advanced Imaging magazine, pp. 97–98 (March 1995).

20. "Special Issue on Digital Television," IEEE Spectrum magazine, April 1995.

21. H. Green, "Digital Video Disk For Multiple Image Forms," Advanced Imaging magazine, pp. 10–14 (Nov. 1995).

22. L. Yencharis, "Networked Digital Video," Advanced Imaging magazine, pp. 25–28 (Nov. 1995).

23. P. Anderson, "Into '96: Imaging Concerns Dominate Computer & Telecom Developments" pp. 90–91 (Nov. 1995).

24. P. Worthington, "Catching Codecs," Multimedia Producer magazine, pp. 19–20, July 1995

25. P. Casey, "DVD Gets Rolling," AV Video magazine, pp. 51–54 (December 1995)

26. M. Drabick, "Power of The Press: Video Compression's Expanding Influence," AV Video magazine, pp. 84–89 (April 1996)

27. M. Ely, "Digital Video Disc Pre-Mastering," Advanced Imaging, pp. 26–29 (April 1996)

28. P. Barnett, "Implementing Digital Compression: Picture Quality Issues for Television," Advanced Imaging, pp., 33–33 (April 1996)

29. F. Hamit, "Moving Images from Cable TV to the Web," Advanced Imaging, pp. 34–36 (April 1996)

30. A. W. Davis, "Cable Modems: Trojan Horse for Internet Imaging & Video," Advanced Imaging magazine, pp. 48–54 (April 1996)

Appendix B

MPEG Committee

Directly participating in MPEG standards activity (MPEG-1, MPEG-2 or MPEG-4) has been a challenging experience for us and hopefully for hundreds of others as well.

The scope of MPEG activity has grown significantly since the early MPEG-1 days of few tens of people per meeting to a few hundreds of people, nowadays in MPEG-4. This increase in scope, the desire to produce the highest quality standard, the desire for timeliness, the desire for thorough testing, the desire to keep the process open but fair and so forth have led to MPEG standards being held in high regard and MPEG as a standards committee that can deliver. The leadership of our esteemed convenor Dr. Leonardo Chiariglione as well as individual contributions from the many of us who have participated in this process have been phenomenal and often beyond the call of normal job duties for most of us. In the end, and many times much earlier than that, the standard becomes a commodity (software on the web/internet?) and there may be little to recognize the many individual efforts, some of which may have succeeded in the sense of getting their technical proposals accepted into the standard, but far more of you for which the many equally good ideas did not make it due to one reason or the other. If all of this sounds a bit philosophical, it is, although one can ask how this situation is any different from anything else in this fast-paced society where to make technological progress some sacrifices have to be made. We believe that those who have participated in this process or in similar endeavors know the answer to that or to similar questions.

As mentioned earlier in the preface, MPEG has been a collective activity and the credit belongs to not only those who directly participated but also to those colleagues who supported us through deadlines for each meeting. While it really is difficult to say how many people were actually working on MPEG-2 worldwide during its development, our guess is probably around a thousand if not more. The number of people involved in MPEG related activities directly or indirectly must now be much much more. While it would be literally infeasible for us to know or list all of you, we did want to list people that we have directly or indirectly known due to participation in any of the MPEG standards. Although we are taking a risk in missing out a few due to difficulties in compiling an accurate list, we could not complete this book without thanking each of you for your contributions to MPEG. Since we had difficulty in obtaining accurate information on affiliations due to job changes etc, and in some cases even the first names (we remembered it once or just had used last names in other cases), we have chosen to use your first initials and last names to list you. Further the list enclosed here is alphabetically sorted for last names. We would be happy to correct any omissions or typos in the next edition of our book, so please feel free to contact us to have the situation rectified.

MPEG MEMBERSHIP LIST

C. Adams
T. Ahn
K. Akagiri
D. Anastassiou
C. Anderson
M. Anderson
W. Andriessen
R. Aravind
J. Arnold
K. Asai
P. Au
C. Auyeung
O. Avaro
D. Bailey
M. Balakrishnan
G. Bar-On
V. Baroncini
L. Bartelletti
S. Battista
G. Beakley
R. Beaton
K. Bechard
R. Bedford
M. Bellanger
G. Benbassat
L. Bengtsson
C. Bertin
M. Biggar
C. Birch
G. Bjoentegaard
F. Bompard
C. Boon
P. Borgwardt
M. Bosi
J. Boyce
R. Bramley
K. Brandenburg
H. Brusewitz
B. Burman
B. Burns
B. Canfield
W. Carter

P. Cassereau
M. Chan
S. Chang
T. Chao
K. Chau
C. Chen
H. Chen
G. Cherry
P. Chiang
T. Chiang
L. Chiariglione
T. Chinen
C. Chiu
T. Cho
D. Choi
S. Chong
C. Chu
S. Cismas
M. Civanlar
L. Claudy
H. Clayton
L. Conte
A. Cook
I. Corset
P. Crosby
D. Curet
M. Danielsen
C. Declerc
P. Decotignie
Y. Dehery
M. Deiss
M. Delahoy
D. Delcoigne
J. Devine
G. Dimino
S. Dixit
L. Dobson
J. Dolvik
M. Draves
S. Dunstan
B. Dyas
S. Eckart

H. Ecker
R. Egawa
R. Eifrig
T. Erdem
G. Eude
M. Faramarz
S. Farkash
T. Fautier
F. Fechter
F. Feige
R. Ferbrache
G. Fernando
M. Field
J. Fiocca
C. Fogg
J. Fontaine
S. Forshay
G. Franceshini
C. Frezal
J. Fritsch
H. Fuchs
H. Fukuchi
T. Fukuhara
B. Futa
D. Gardner
A. Gaspar
T. Geary
A. Giachetti
A. Gill
R. Gluecksmann
M. Gold
M. Goldman
M. Goldstein
C. Gonzales
J. Gooding
B. Grill
V. Guaglianone
J. Guichard
P. Haavisto
G. Haber
S. Haigh
J. Hamilton
B. Hammer
T. Hanamura
M. Haque
B. Haskell
P. Haskell
A. Hekstra

B. Helms
D. Hepper
C. Herpel
K. Hibi
T. Hidaka
M. Hion
Y. Ho
R. Hodges
C. Holborow
H. Holtzman
C. Homer
T. Homma
C. Horne
Y. Hozumi
H. Hsieh
W. Hsu
S. Hu
Y. Hu
S. Huang
V. Huang
L. Hui
K. Illgner
M. Ishikawa
Y. Itoh
R. Ivy
M. Iwadare
D. Jadeja
B. Jeon
H. Jeon
J. Jeon
J. Jeong
R. Johnson
J. Jolivet
J. Juhola
W. Kameyama
I. Kaneko
Y. Katayama
M. Kato
Y. Kato
A. Katsumata
N. Katsura
G. Keesman
N. Kenyon
L. Kerkhof
R. Kermode
S. Kian
J. Kim
K. Kim

R. Kim
Y. Kim
J. Kimura
K. Kitamura
D. Klenke
K. Kneib
A. Knoll
T. Kogure
A. Komly
A. Koster
S. Koto
H. Koyama
H. Koyama
H. Koyanagi
M. Koz
F. Krause
M. Kuehn
O. Kwon
F. Laczko
W. Lam
R. Lampach
F. Lane
F. Lavagetto
P. Leach
D. Lecomte
M. Leditschke
C. Lee
D. Lee
K. Lee
N. Lee
S. Lee
T. Lee
Y. Lee
D. LeGall
L. Leske
J. Lhuillier
C. Li
H. Li
K. Lillevold
B. Lindsley
A. Lippman
P. List
S. Liu
G. Logston
T. Lookabaugh
B. Loret
V. Lunden
A. Luther

K. Ma
Y. Machida
A. MacInnis
G. Madec
A. Maekivirta
B. Maison
F. Markhauser
P. Marlier
I. Martins
M. Maslaney
H. Matsumoto
S. Matsumura
J. Mau
K. McCann
C. McGrath
D. Mead
P. Meron
I. Messing
M. Mcthiwalla
K. Metz
R. Mickos
W. Middleton
T. Mimar
C. Min
Y. Miyamoto
M. Modena
R. Montagna
W. Moog
W. Mooij
H. Mook
J. Moon
T. Morika
P. Moroney
J. Morris
G. Morrison
F. Mueller
K. Mundkur
T. Murakami
J. Nachman
T. Nagata
S. Naimpally
Y. Nakajima
E. Nakasu
J. Nam
S. Narasimhan
D. Nasse
T. Naveen
J. Nelson

S. Ng
A. Nicoulin
M. Nilsson
M. Nishida
Y. Nishida
H. Nishikawa
G. Nocture
S. Nogaki
Y. Noguchi
S. Noh
P. Noll
K. O'Connell
T. Odaka
N. Oishi
S. Okubo
K. Oosa
J. Ostermann
S. Othman
K. Ozawa
M. Ozkan
W. Paik
D. Pan
K. Pang
G. Park
H. Park
S. Park
G. Parladori
L. Pearlstein
T. Peng
B. Penney
F. Pereira
M. Perkins
H. Peterson
N. Pham
D. Pian
M. Piech
J. Pineda
O. Poncin
P. Pont
M. Porter
B. Powell
F. Pranpolini
R. Prodan
A. Puri
B. Quandt
I. Rabowsky
J. Raina
N. Randall

P. Rao
J. Rault
C. Reader
A. Reibman
U. Riemann
K. Rijkse
P. Roberts
P. Rodi
J. Rosengren
J. Rowlands
G. Russo
T. Ryden
R. Saeijs
R. Saint-Girons
T. Saitoh
T. Sakaguchi
K. Sakai
H. Sallic-Dupont
P. Sarginson
A. Satt
T. Savatier
A. Sawe
G. Schamel
R. Schaphorst
D. Schinkel
P. Schirling
P. Schreiner
S. Searing
I. Sebestyen
T. Selinger
T. Senoo
Y. Seo
L. Sestrienka
M. Sezan
I. Shah
P. Shen
K. Shimizu
J. Shin
T. Sikora
H. Silbiger
K. Sjaunja
E. Sklaeveland
G. Smith
J. Smolinske
K. Sohn
J. Spille
J. Stampleman
H. Stephansen

J. Steurer
P. Stevens
G. Stoll
M. Stroppiana
K. Sugaya
M. Sugihara
A. Sugiyama
K. Sugiyama
G. Sullivan
M. Sun
H. Suzuki
P. Szychowiak
K. Tahara
K. Takahashi
T. Takahashi
T. Takeuchi
Y. Takishima
R. Talluri
A. Tan
T. Tanaka
T. Tanigawa
N. Tanton
M. Tayer
D. Teichner
R. Horst
L. Thibault
G. Thom
J. Tiernan
C. Todd
A. Tom
H. Tominaga
Y. Tong-Tse
T. Tong
L. Tranchard
Y. Tse
Y. Tsuboi
H. Ueno
J. Urano
R. Urano
K. Uz
M. Uz
M. Vaananen
M. Vakalopoulou
S. Valli
L. van-den-Berghe
L. van-der-Kerhhoff

J. van-der-Meer
R. van-der-Waal
P. Vasseur
M. Veltman
V. Venkateswar
C. Verreth
J. Vial
J. Villasenor
A. Vincent
E. Viscito
G. Vlasov
J. Vollbrecht
H. Waerk
G. Wallace
F. Wang
I. Wang
T. Wasilewski
H. Watanabe
T. Watanabe
A. Wechselberg
N. Wells
W. Welsh
O. Werner
T. Werner
F. Whittington
J. Williams
R. Wilson
A. Wise
A. Wong
D. Wood
Y. Yagasaki
Y. Yamada
A. Yamakata
H. Yamauchi
N. Yanagihara
H. Yasuda
M. Yim
J. Yonemitsu
T. Yoshimura
Y. Yu
T. Yukitake
H. Yun
B. Yung
J. Zdepski
S. Zindal

Index